AF148056

Integrable Systems, Quantum Groups, and Quantum Field Theories

NATO ASI Series

Advanced Science Institutes Series

A Series presenting the results of activities sponsored by the NATO Science Committee, which aims at the dissemination of advanced scientific and technological knowledge, with a view to strengthening links between scientific communities.

The Series is published by an international board of publishers in conjunction with the NATO Scientific Affairs Division

A Life Sciences **B Physics**	Plenum Publishing Corporation London and New York
C Mathematical and Physical Sciences **D Behavioural and Social Sciences** **E Applied Sciences**	Kluwer Academic Publishers Dordrecht, Boston and London
F Computer and Systems Sciences **G Ecological Sciences** **H Cell Biology** **I Global Environmental Change**	Springer-Verlag Berlin, Heidelberg, New York, London, Paris and Tokyo

NATO-PCO-DATA BASE

The electronic index to the NATO ASI Series provides full bibliographical references (with keywords and/or abstracts) to more than 30000 contributions from international scientists published in all sections of the NATO ASI Series.
Access to the NATO-PCO-DATA BASE is possible in two ways:

– via online FILE 128 (NATO-PCO-DATA BASE) hosted by ESRIN, Via Galileo Galilei, I-00044 Frascati, Italy.

– via CD-ROM "NATO-PCO-DATA BASE" with user-friendly retrieval software in English, French and German (© WTV GmbH and DATAWARE Technologies Inc. 1989).

The CD-ROM can be ordered through any member of the Board of Publishers or through NATO-PCO, Overijse, Belgium.

Series C: Mathematical and Physical Sciences - Vol. 409

Integrable Systems, Quantum Groups, and Quantum Field Theories

edited by

L. A. Ibort

and

M. A. Rodríguez

Departamento de Física Teórica,
Universidad Complutense de Madrid,
Madrid, Spain

Springer Science+Business Media, B.V.

Proceedings of the NATO Advanced Study Institute and
XXIII GIFT International Seminar on
Recent Problems in Mathematical Physics
Salamanca, Spain
June 15–27, 1992

A C.I.P. Catalogue record for this book is available from the Library of Congress.

ISBN 978-0-7923-2396-9 ISBN 978-94-011-1980-1 (eBook)
DOI 10.1007/978-94-011-1980-1

Printed on acid-free paper

PREFACE

This volume contains the proceedings of the NATO Advanced Institute and the XXIII edition of the GIFT International Seminar on Theoretical Physics, *Recent Problems in Mathematical Physics*, that was held at Salamanca, Spain, from 15 to 27 June 1992. The Advanced Institute was organized by the editors (L.A. Ibort and M.A. Rodriguez) and Professors A. Bohm and P. Winternitz.

The Institute was conceived as a preparatory School for the *XIX International Colloquium on Group Theoretical Methods in Physics*, which was held in Salamanca the week after the Advanced Institute. That imposed that the School covered a wide range of problems in Mathematical Physics, from modern aspects of Integrability on Quantum Field Theories to Geometrical Phases in Quantum Mechanics. Several courses were devoted to Conformal Field Theory, integrability, Quantum Groups and Topological Field Theory. Gravitation in low dimensions and Quantum Gravity were discussed in another series of lectures and finally, some other topics like, symmetries in partial differential equations, geometric phases, etc. were addressed along the School.

The level of the School was intended for advanced graduate students and recent Ph. D's and the invited lecturers are leading experts in their fields. The organizers would like to express their gratitude to the lecturers for their beautiful and stimulating talks as well as for the careful preparation of their contributions to these proceedings.

Apart from the organizers of the Advanced Institute, the editors want to thank the invaluable help and support obtained from the rest of the Organizing Committee of the *XIX ICGTMP*; Professors Aldaya, C.S.I.C. and Universidad de Granada; J. A. de Azcárraga, Universidad de Valencia and IFIC (C.S.I.C.); L. J. Boya, Universidad de Zaragoza; J. F. Cariñena, Universidad de Zaragoza; M. Lorente, Universidad de Oviedo; J. Mateos Guilarte, Universidad de Salamanca; M. A. del Olmo and M. Santander, Universidad de Valladolid.

The directors of the Advance Institute wish to thank the NATO Science Committee for their support of the School and to the following Institutions for their partial financial contribution:

Dirección General Interministerial de Ciencia y Tecnología (DGICYT); Real Sociedad Española de Física, G.I.F.T; Universidad Complutense de Madrid, Universidad de Salamanca and Apple España.

The help of Mrs. Ascensión Iglesias during the School has been invaluable. The Dean and the staff of the Facultad de Ciencias, where most of the sessions of the School were held, kindly put many facilities at our disposal. We should like to thank Prof. F. Fernández and to the Professors of the Department of Theoretical Physics, specially J. M. Cerveró and J. Martín.

L.A. Ibort
M.A Rodriguez

CONTENTS

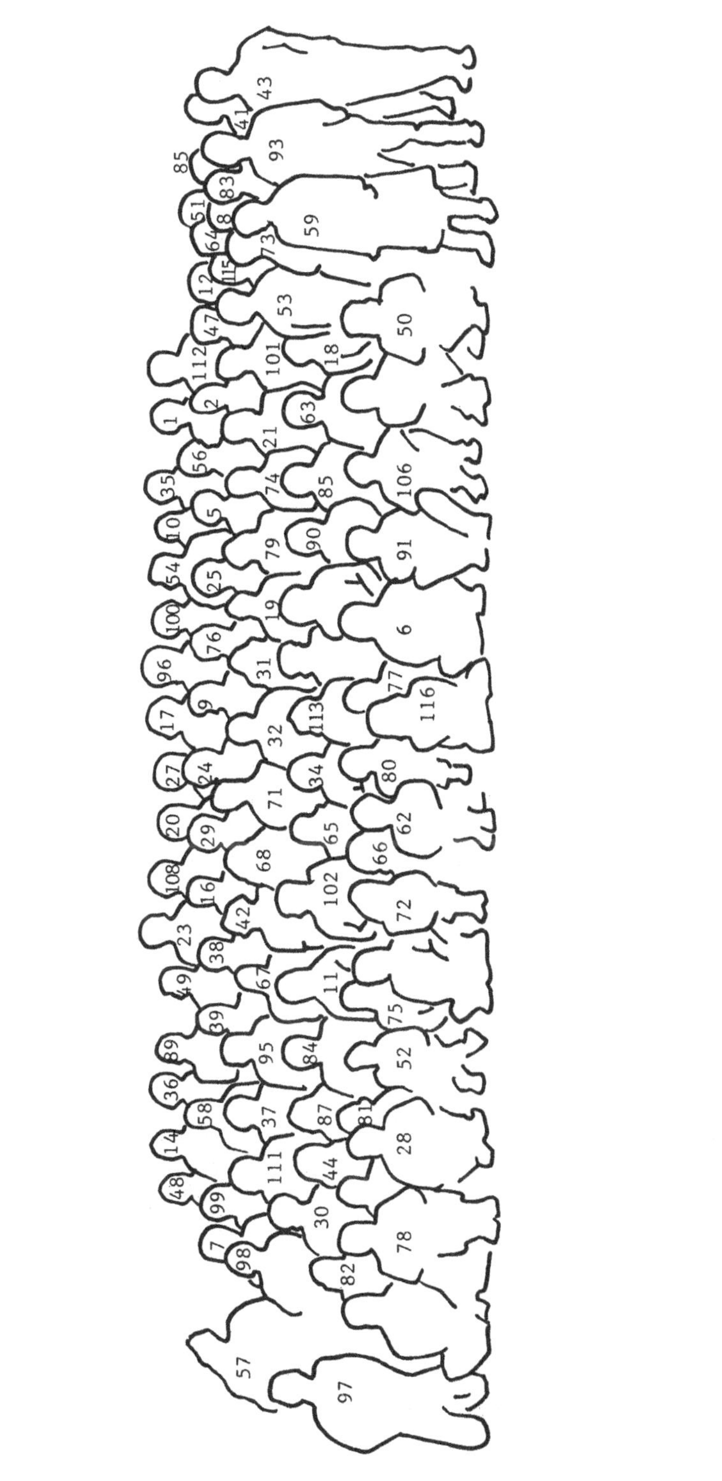

LIST OF PARTICIPANTS

1. Ahluwalia, K.S. (Cambridge Univ.., UK)
2. Alber, M. S. (Univ. of Notre Dame, USA)
3. Aldaya, V. (Univ. de Granada, Spain)
4. Álvarez Gimenez, M. (Univ. de Santiago, Spain)
5. Anderson, A. (McGill Univ., Canada)
6. Azcárraga, A. de. (Univ. de Valencia, Spain)
7. Barnich, G. (Univ. Libre de Bruxelles, Bélgique)
8. Batlle Arnau, C. (Univ. Politécnica de Catalunya, Spain)
9. Benoit, L. (Univ. de Montreal, Canada)
10. Bergqvist, G. (Univ. Southampton, UK)
11. Berkovich, A. (CSIC, Spain)
12. Bohm, A. (Univ. of Texas at Austin , USA)
13. Boya, L.J. (Univ. de Zaragoza, Spain)
14. Cangemi, D. (MIT, USA)
15. Cariñena, J.F. (Univ. de Zaragoza, Spain)
16. Cuerno Rejado, R. (CSIC, Spain)
17. Chruscinski, D. (Nicholas Copernicus Univ., Poland)
18. Debergh, N. (Univ. de Liège, Bélgique)
19. Deckmyn, A. (K. U. Leuven, Bélgique)
20. Demichev, A. P. (Moscow State Univ., Russia)
21. El Gradechi, A. (Univ. de Montréal, Canada)
22. Ellinas, D. (Univ. of Helsinki, Finland)
23. Emparán, R. (Univ. del País Vasco, Spain)
24. Ezquerra Larrode, F. (Univ. de Zaragoza, Spain)
25. Faddeev, L.D. (Steklov Mathematical Institute, Russia)
26. Faucher, M. (Univ. of Montréal, Canada)
27. Fecko, M. (Comenius Univ. Bratislava, Czechoslovakia)
28. Fernández Cabrera, D. (Univ. de Valladolid, Spain)
29. Fernández Jambrina, L. (Univ. Complutense de Madrid, Spain)
30. Fernández Nuñez, J. (Univ. de Valladolid, Spain)
31. Figueroa González, H. (Univ. de Zaragoza, Spain)
32. Freed, D. (Univ. of Texas at Austin , USA)
33. Gadella, M. (Univ. de Valladolid, Spain)
34. García Fuertes, W. (Univ. de Oviedo, Spain)
35. Gingras, F (Univ. of Montréal, Canada)
36. Gómez Nicola, A. (Univ. Complutense de Madrid, Spain)
37. Gómez, C. (CSIC, Spain)
38. González Ruiz, A. (Univ. Complutense de Madrid, Spain)
39. Güngör, F. (Istanbul Tech. Univ., Turkey)
40. Hall, R. A. (Durham Univ., UK)
41. Harrington, R. J. (Univ. of Texas at Austin, USA)
42. Hernández Heredero, R. (Univ. Complutense de Madrid, Spain)
43. Ibort, L.A. (Univ. Complutense de Madrid, Spain)
44. Illieva-Litova, N. (Bulgarian Acad. of Sci., Bulgaria)
45. Isham, C. (Imperial College, UK)
46. Izquierdo Rodriguez, J.M. (Univ. de Valencia. Spain)
47. Jackiw, R. (Columbia Univ., USA)
48. Jain, S. (Derbyshire College, UK)
49. Jiménez Lorenzo, F. (Univ. Politécnica de Madrid, Spain)
50. Jiracek, P. (Fac.of Nucl. Sci. and Phys. Eng., Czechoslovakia)
51. Kabat, D. (MIT, USA)
52. Kausch, H. (Imperial College, UK)
53. Kimura, T. (Univ. of Texas at Austin, USA)
54. Kopecky, Z. (Masaryk Univ,Czechoslovakia)
55. Kucera, J. (Masaryk Univ., Czechoslovakia)
56. Laartz, J. (Harvard Univ., USA)
57. Larson, J. (College of William and Mary, USA)

58. Leblanc, M. (MIT, USA)
59. Leng, X. (Univ. of Montréal, Canada)
60. LeTourneux, J. (Univ. de Montreal. Canada)
61. London, L. A. J. (DAMTP, UK)
62. López García, J. L. (Univ. de Zaragoza, Spain)
63. López Manzanares, E. (CSIC, Spain)
64. Lorente, M. (Univ. de Oviedo, Spain)
65. Lozano Gómez, Y. (Univ. Autónoma de Madrid, Spain)
66. Luzón Marco, G. (Univ. de Zaragoza, Spain)
67. Llanos Vázquez, R. (Univ. Complutense de Madrid, Spain)
68. Manuel Hidalgo, C. (Univ. de Barcelona, Spain)
69. Mañas Baena, M. (Univ. Complutense de Madrid, Spain)
70. Marín Solano, J. (Univ. Complutense de Madrid, Spain)
71. Marshall, I. (Loyola Campus. Canada)
72. Martinez Ontalba, C. (Univ. Complutense de Madrid, Spain)
73. Mateos Guilarte, J. (Univ. de Salamanca, Spain)
74. Matías Espona, J. (Univ. de Barcelona, Spain)
75. Mc Isaac, N. (Univ. de Montreal, Canada)
76. Méndez Llatas, P. (Univ. Autónoma de Madrid)
77. Molina Paris, C. (Univ. of Texas at Austin, USA)
78. Montigny, M. de. (Univ. de Montreal, Canada)
79. Mostafazadeh, A. (Univ. of Texas at Austin, USA)
80. Navarro Navarro, M. (Univ. de Valencia, Spain)
81. Negro Vadillo, J. (Univ. de Valladolid, Spain)
82. Nieto Calzada, L. M. (Univ. de Valladolid, Spain)
83. Olmo, M. del. (Univ. de Valladolid, Spain)
84. Paranjape, M. (Univ. de Montreal, Canada)
85. París Mollo, J. (Univ. de Barcelona, Spain)
86. Perelomov, A. (Phys. Inst. Univ. Bonn, Germany)
87. Ptukha, A. R. (Russian Friendship Univ., Russia)
88. Rau, J. (Duke Univ., USA)
89. Rivero García, A. (Univ. de Zaragoza, Spain)
90. Roca, J. (Univ. de Barcelona, Spain)
91. Ródenas, F. (Univ. de Valencia, Spain)
92. Rodriguez Plaza, M. J. (Univ. Complutense, Spain)
93. Rodríguez, M.A. (Univ. Complutense de Madrid, Spain)
94. Rozenberg, M. (Univ. of Texas at Austin, USA)
95. Ruegg, H. (Univ. de Genève, Switzerland)
96. Sa, N. (Portugal)
97. Samuel, S. (City College of New York, USA)
98. Santander, M. (Univ. de Valladolid, Spain)
99. Schulze, J. (Univ. Karlsruhe, Germany)
100. Shabanov, S. V. (Joint INR, Russia)
101. Siebelink, R. (K. U. Leuven, Bélgique)
102. Sierra, G. (CSIC, Spain)
103. Strachan, I.A.B. (Math. Inst. Oxford, UK)
104. Strasburger, A. (Univ. of Warsaw, Poland)
105. Sutcliffe, P. M. (Univ. of Durham. UK)
106. Talavera Usano, C. (Univ. de Valencia, Spain)
107. Teitelboim, C. (Centro de Estudios Científicos, Chile)
108. Toldrá Sabater, R. (Univ. de Barcelona, Spain)
109. Townsend, P.K. (DAMPT, UK)
110. Varilly, J. (Univ. de Costa Rica, Costa Rica)
111. Velazquez Campoy, L. (Univ. de Zaragoza, Spain)
112. Vinet, L. (Univ. de Montreal. Canada)
113. Viña Escalar, A. (Univ. de Oviedo, Spain)
114. Voropaev, S.A. (Moscow Phys. and Tech. Inst., Russia)
115. Winternitz, P. (Univ. de Montréal, Canada)
116. Iglesias, A. (Secretary)

FROM INTEGRABLE MODELS TO CONFORMAL FIELD THEORY VIA QUANTUM GROUPS

L.D. FADDEEV

St. Petersburg Branch of Steklov Mathematical Institute
St. Petersburg, Russia

and

Research Institute for Theoretical Physics
University of Helsinki, Finland

Abstract

In these lectures, which are a variant of ones given during the last year (see references), I present a historical development in the 1+1 dimensional integrable models, leading to the notion of quantum groups. I also give a modern exposition of this notion in the R-matrix language and explain a new application to Conformal Field Theory.

The mathematical theory of solitons is about 25 years old. It started with the invention of the so-called inverse scattering method. The inverse scattering method was based on the introduction of the Lax equation, a very important additional idea was the Hamiltonian interpretation of these concepts. The Hamiltonian interpretation of the Korteweg-de Vries equation was first given by Gardner, Zakharov and Faddeev, in 1971. In our approach we were mainly led by the idea of a future quantization of this subject, which was completely classical at that time. To quantize something one has to know first of all the Hamiltonian structure of the corresponding classical problem. Going step by step deeper into quantum mechanics an algebraic structure evolved, which was on one side very simple and on the other side quite universal.

Our framework will be a $1+1$ dimensional quantum field theory. We will consider mainly systems with discrete space variable and continuous time variable.

Let $x = n \cdot \Delta$ be the space, t the time variable, with $n = 1 \ldots N$ and $N + 1 \equiv 1$, (one dimensional chain with periodic boundary conditions). With each site n we connect a Hilbert space $\mathcal{H}_n$. The total space of physical states is thereby given by

L. A. Ibort and M. A. Rodríguez (eds.), Integrable Systems, Quantum Groups, and Quantum Field Therapy 1–24.

$$\mathcal{H} = \otimes_{n=1}^{N} \mathcal{H}_n.$$

We are given some dynamical algebra, generated by the operators

$$X_n^a = 1 \otimes \ldots \otimes \overset{(n\ site)}{X^a} \otimes \ldots \otimes 1,$$

where a is some additional index, e.g. with respect to a Lie algebra basis, and all dynamical variables are required to be functions of the X_n^a.

Next impose the condition of ultralocality condition:

$$[X_n, X_m] = 0 \quad \text{for} \quad n \neq m.$$

The dynamical equations are given as the usual Heisenberg equations:

$$\dot{X}_n^a = [H, X_n^a]$$

(a dot denoting derivation with respect to the time variable t). The idea of soliton theory is to associate to the Hamiltonian H a large series of commuting integrals of motion. To achieve this we introduce a new object, the so-called *Lax operator*

$$L_n(\lambda) = ((L_{n,ij}(\lambda)))_{m\times m},$$

which is an $m \times m$ matrix in the auxiliary space $V = \mathbb{C}^m$ and with matrix entries which are operators on the Hilbert space $\mathcal{H}_n$. It depends also on the additional parameter λ, called the spectral parameter.

We look for a nice instrument to display the various commutation relations between the matrix elements of the Lax operator in a compact way. By ultralocality these commutation relations do not depend on the index n. Skipping therefore the n-dependence we may express everything by terms of the form

$$L_{pq}(\lambda)L_{ij}(\mu) = (L(\lambda) \otimes L(\mu))_{pi|qj}$$

and the products with reversed order of factors. Therefore it seems convenient to define the following operators on $V \otimes V$ constructed in terms of L:

$$L^1 = L \otimes 1 \quad \text{and} \quad L^2 = 1 \otimes L$$

The commutation relations among the matrix elements may now be written as:

$$R(\lambda - \mu)L^1(\lambda)L^2(\mu) = L^2(\mu)L^1(\lambda)R(\lambda - \mu) \tag{1}$$

with a matrix $R(\lambda) : V \otimes V \to V \otimes V$. We will call this relation the *fundamental commutation relations* (FCR). Considering L_n as a local transport matrix, we put now:

$$\psi_{n+1} = L_n \psi_n, \quad \text{where} \quad \psi_n \in V.$$

This relation is called *the auxiliary problem.* It should be understood as a system of matrix equations with operator coordinates, i.e. as a system of equations in a noncommutative space.

If L_n for small Δ is of the form:

$$L_n = 1 + \Delta \cdot L(x) + O(\Delta^2),$$

the auxiliary problem becomes in the continuous limit $\Delta \to 0$

$$\frac{d\psi}{dx} = L(x)\psi.$$

Furthermore we have

$$\psi_{k+1} = L_k L_{k-1} \dots L_1 \psi_1.$$

As *monodromy* it is natural to define the operator

$$M \stackrel{\text{def}}{=} L_N \dots L_1.$$

The matrix entries of M are global operator on $\mathcal{H}$. The special feature of the FCR is, that a pure local relation Eq. (1) gives a global relation (global on $\mathcal{H}$), for the monodromy M:

$$R(\lambda-\mu)M^1(\lambda)M^2(\mu) = M^2(\mu)M^1(\lambda)R(\lambda-\mu) \tag{2}$$

The proof is as follows: (In the following, we suppress the arguments λ and μ). By the FCR, Eq. (1) and ultralocality we have for the products $L_k L_{k-1}$:

$$\begin{aligned} RL_k^1L_{k-1}^1L_k^2L_{k-1}^2 &= RL_k^1L_k^2L_{k-1}^1L_{k-1}^2 \\ &= L_k^2L_k^1RL_{k-1}^1L_{k-1}^2 \\ &= L_k^2L_k^1L_{k-1}^2L_{k-1}^1R \\ &= L_k^2L_{k-1}^2L_k^1L_{k-1}^1R \end{aligned}$$

and the rest follows by induction.

The matrix entries of the monodromy are global operators. In order to get scalar commuting operator on the full Hilbert space $\mathcal{H}$ we take the trace of the monodromy:

$$F(\lambda) \stackrel{\text{def}}{=} \text{tr}M(\lambda).$$

Assuming $R(\lambda)$ to be invertible and using Eq. (2) we get

$$F(\lambda)F(\mu) = F(\mu)F(\lambda).$$

The $F(\lambda)$ are therefore generators of an infinite dimensional algebra of commuting operators. Now we will illustrate the above framework by examples.

EXAMPLE 1.

Let $\mathcal{H}_n = \mathbb{C}^2$ be the spin $\frac{1}{2}$ quantum Hilbert space and $\vec{S} = \frac{1}{2}\vec{\sigma}$ the spin operator where $\sigma^a, a = 1,2,3$, are the Pauli matrices. The S_n^a satisfy the commutation relations of $Sl(2,\mathbb{C})$: $[S_n^a, S_m^b] = i\epsilon^{abc}S_n^c\delta_{nm}$. The Lax operator in our example is now:

$$L_n = \begin{pmatrix} \lambda + iS_n^3 & iS_n^+ \\ iS_n^- & \lambda - iS_n^3 \end{pmatrix} = \lambda \mathbf{1} \otimes \mathbf{1} + i\vec{S}_n \otimes \vec{\sigma} = \lambda \mathbf{1} \otimes \mathbf{1} + i\sum_{a=1}^{3} S_n^a \otimes \sigma^a. \tag{3}$$

So in this example $V = \mathcal{H}_n$, which is not true in general. With the R-matrix given by:

$$R(\lambda) = \frac{1}{\lambda + i}(\lambda \mathbf{1} + iP) = \begin{pmatrix} 1 & 0 & 0 & 0 \\ 0 & b(\lambda) & c(\lambda) & 0 \\ 0 & c(\lambda) & b(\lambda) & 0 \\ 0 & 0 & 0 & 1 \end{pmatrix} \tag{4}$$

where $b = \frac{\lambda}{\lambda+i}$ and $c = \frac{i}{\lambda+i}$, the Lax operator L_n satisfies the FCR as may be verified by direct computation. As a Hamiltonian we choose the following element of the abelian algebra generated by the $F(\lambda)$:

$$H = \left.\frac{d}{d\lambda}\right|_{\lambda=\frac{i}{2}} \ln F(\lambda) = \text{const.}(\sum_n S_n^a S_{n+1}^a + \text{const.}),$$

which turns out to be the Hamiltonian of the isotropic Heisenberg magnet. In this particular case the FCR holds for any spin representation, so that the example might be easily extended to higher spin j, with the corresponding Hilbert space $\mathcal{H}_n = \mathbb{C}^{2j+1}$.

EXAMPLE 2.

The previous example will now be used to find in an analogous way a Lax operator and Hamiltonian for the nonlinear Schrödinger equation. For that purpouse we construct an "infinite dimensional representation" for compact Lie groups in terms of oscillator variables. Starting from the usual commutation relations for annihilation and creation operators: $[\psi_n, \psi_m^\dagger] = \delta_{nm}$, we build up now spin operators, which are satisfying the $Sl(2,\mathbb{C})$ commutation relations,

$$\begin{aligned} S_n^+ &= \psi_n^\dagger(2S - \psi_n^\dagger\psi_n)^{\frac{1}{2}} \\ S_n^- &= (2S - \psi_n^\dagger\psi_n)^{\frac{1}{2}}\psi_n \\ S_n^3 &= \psi_n^\dagger\psi_n - S \end{aligned}$$

where S is any complex number. In the case where $2S$ is a positive integer one gets subrepresentations (just the Verma modules of $Sl(2,\mathbb{C})$), which are finite dimensional.

If S tends to infinity ("the quasiclassical limit"), we have the following behaviour of the spin operators in the vicinity of the lowest weight state (playing here the role of the ground state).

S_n^+ behaves like $\sqrt{S}\psi^\dagger$,

S_n^- behaves like $\sqrt{S}\psi$,

S_n^3 behaves like S.

With this, one obtains the asymptotic behavior of the Lax operator Eq. (3).

$$\frac{1}{S}L_n = \begin{pmatrix} 1+\frac{\lambda}{S} & \frac{\psi_n^\dagger}{\sqrt{S}} \\ -\frac{\psi_n}{\sqrt{S}} & 1-\frac{\lambda}{S} \end{pmatrix}\sigma^3 = \left(\mathbf{1} + \Delta\begin{pmatrix} \lambda & \psi_n^\dagger(x) \\ -\psi_n(x) & -\lambda \end{pmatrix}\right)\sigma^3 + O(\Delta^2)$$

if we put $S = \frac{1}{\Delta}$ and take into account that

$$\psi_n = \Delta^{\frac{1}{2}}\psi(x).$$

The continuous quasiclassical limit of the Lax operator of the isotropic Heisenberg magnet is therefore nothing else than the well known Lax operator for the nonlinear Schrödinger equation.

EXAMPLE 3.

Another model might be obtained by substituting λ by $\sinh\lambda$, so we can define the Lax operator

$$L_n = \begin{pmatrix} \sinh\lambda\cot\gamma + i\cosh(\lambda)S_n^3 & iS_n^+ \\ iS_n^- & \sinh\lambda\cot\gamma - i\cosh(\lambda)S_n^3 \end{pmatrix}$$

for the spin $\frac{1}{2}$ operator S_n^a. One realizes, that after the substitution $\frac{\lambda}{\gamma} \to \lambda$ one gets the same form of the Lax operator as in example 1 in the limit $\gamma = 0$. The R-matrix has the same form as in example 1, Eq. (4) but the coefficients b, c are now substituted by:

$$b = \frac{\sinh\lambda}{\sinh(\lambda + i\gamma)} \quad \text{and} \quad c = \frac{i\sin\gamma}{\sinh(\lambda+i\gamma)}.$$

The Hamiltonian may now be calculated to be

$$H = \text{const.}\sum(S_n^1 S_{n+1}^1 + S_n^2 S_{n+1}^2 + \cos\gamma S_n^3 S_{n+1}^3),$$

which is the Hamiltonian of the anisotropic Heisenberg model (XXZ-model). The Lax operator may alternatively be written as

$$L_n(\lambda) = \frac{1}{\sin\gamma}\begin{pmatrix} \sinh(\lambda + i\gamma S_n^3) & i\sin\gamma S_n^+ \\ \sin\gamma S_n^- & \sinh(\lambda - i\gamma S_n^3) \end{pmatrix} \tag{5}$$

Now the fundamental commutation relations do not hold in the case of higher spin. But a slight modification leads again to fundamental commutation relations. Instead of the $Sl(2,\mathbb{C})$ commutation relations, we impose the following deformed $Sl(2,\mathbb{C})$ relation:

$$[S_n^3, S_n^\pm] = \pm S_n^\pm \quad \text{and} \quad [S_n^+, S_n^-] = \frac{\sin(2\gamma S_n^3)}{\sin\gamma}. \tag{6}$$

If these relations are satisfied, then the $L_n(\lambda)$ of the form (5) satisfy the fundamental commutation relations but the operators $S_n^\pm$, S_n^3 no longer form a Lie algebra. Instead they "generate" a new algebraic structure which gives in the limit $\gamma \to 0$ a Lie algebra. A realization of the relations (6) by usual spin operators:

$$[\pi, \phi] = i\lambda \mathbf{1}$$

is given by:

$$S^\pm = \frac{1}{2a\sin\gamma} e^{\pm i\frac{\pi}{2}}(1 + 2e^{\pm 2i\phi})e^{\pm i\frac{\pi}{2}}, \quad S^3 = \frac{\phi}{\gamma}.$$

Here now one obtains the quantum Sine-Gordon model as a special case of the Heisenberg XXZ-chain. Indeed, after multiplication of Eq. (5) by $2a\sin\gamma\begin{pmatrix} 0 & 1 \\ 1 & 0 \end{pmatrix}$, which do not change the FCR Eq. (1), we get the L-operator:

$$\begin{aligned} \mathrm{L}_n^{SG}(\lambda) &= \begin{pmatrix} e^{i\pi_n/2}(1+2a^2e^{2i\phi_n})e^{i\pi_n/2} & 2a\sinh(\lambda + i\phi_n) \\ 2a\sinh(\lambda - i\phi_n) & e^{-i\pi_n/2}(1+2a^2e^{-2i\phi_n})e^{-i\pi_n/2} \end{pmatrix} \\ &\sim \mathbf{1} + i\Delta\begin{pmatrix} \pi(x) & -2i\sinh(\lambda + i\phi(x)) \\ -2i\sinh(\lambda - i\phi(x)) & -\pi(x) \end{pmatrix} + O(\Delta^2) \end{aligned} \tag{7}$$

If one puts $2a = \Delta$ and

$$\pi(x) = \Delta\pi_n; \quad \phi(x) = \phi_n$$

this leads to continuous auxiliary problem for the Sine-Gordon equation.

These examples cover the case of "rank 1", namely group $Sl(2,\mathbb{C})$. One can use now groups of higher rank, there are generalizations of Lax operators for them and corresponding integrable models, the parameters in this list are: group, representation, anisotropy parameter (i.e. γ). It will be interesting to see if that is a classification.

EXAMPLE 4.

Let us return to the S-G L-operator (7) and consider the "Liouville limit":

$$\begin{aligned}\phi_n &\to i(\phi_n + c),\ c \to \infty, \pi_n \to -i\pi_n \\ ae^c &= 1\end{aligned}$$

We get a new L-operator

$$L_n^\infty(\lambda) = \begin{pmatrix} e^{\pi_n/2}(1+e^{2\phi_n}) & e^\lambda e^{\phi_n} \\ -2e^{-\lambda}e^{\phi_n} & e^{\pi_n} \end{pmatrix}$$

with a trivial dependence on λ. The λ independent L-operator:

$$\begin{aligned} L_n^{\text{Liou}} &= \begin{pmatrix} e^{\pi_n/2}(1+e^{2\phi_n})e^{\pi_n/2} & e^{\phi_n} \\ e^{\phi_n} & e^{-\pi_n} \end{pmatrix} \\ &= Q(\lambda)L_n^\infty(\lambda)Q(\lambda)^{-1} \end{aligned}$$

where

$$Q(\lambda) = \begin{pmatrix} e^{-\lambda/2} & 0 \\ 0 & e^{\lambda/2} \end{pmatrix}$$

satisfy the modified FCR

$$\tilde{R}(\lambda-\mu)L_n^1L_n^2 = L_n^2L_n^1\tilde{R}(\lambda-\mu)$$

where (after changing the normalization)

$$\tilde{R}(\lambda) = Q^1(\lambda)Q^2(\mu)R(\lambda-\mu)Q^1(\lambda)^{-1}Q^2(\mu)^{-1}$$

The λ-dependence in $\tilde{R}(\lambda)$ is also separated. We have

$$\begin{aligned} \tilde{R}(\lambda) &= \begin{pmatrix} \sinh(\lambda+i\gamma) & 0 & 0 & 0 \\ 0 & \sinh\lambda & i\sin\gamma e^{-\lambda} & 0 \\ 0 & i\sin\gamma e^\lambda & \sinh\lambda & 0 \\ 0 & 0 & 0 & \sinh(\lambda+i\gamma) \end{pmatrix} \\ &= \frac{1}{2}(e^\lambda R_q - e^{-\lambda}PR_q^{-1}P) \end{aligned}$$

where R_q is a simple triangular matrix

$$R_q = \begin{pmatrix} q & 0 & 0 & 0 \\ 0 & 1 & 0 & 0 \\ 0 & q-\frac{1}{q} & 1 & 0 \\ 0 & 0 & 0 & q \end{pmatrix}$$

P-permutation matrix in $V \otimes V$

$$P = \begin{pmatrix} 1 & 0 & 0 & 0 \\ 0 & 0 & 1 & 0 \\ 0 & 1 & 0 & 0 \\ 0 & 0 & 0 & 1 \end{pmatrix}$$

and we denoted

$$q = e^{i\gamma}$$

It is clear, that L_n^{Liou} satisfy the λ-independent FCR

$$R_q L_n^1 L_n^2 = L_n^2 L_n^1 R_q. \tag{8}$$

Thus the development of Lax operators presents us with the generalization of Lie-algebra relations (5), the relation Eq. (8) and their "stability" under the multiplication of matrices in the auxiliary space with independent matrix elements. All this leads to the general definition of Quantum Groups and Quantum Lie Algebras, to which we turn now.

Let us begin with the first one, and consider the relation.

$$RT^1T^2 = T^2T^1R \tag{9}$$

where the matrix T in the auxiliary space V has as entries the generators T_{ij} of the associative algebra $\mathcal{A}$. Define a comultiplication in $\mathcal{A}$

$$\Delta: \mathcal{A} \to \mathcal{A} \otimes \mathcal{A}$$

on the generators T_{ij} by:

$$\Delta(T_{ij}) = \sum_k T_{ik} \otimes T_{kj}.$$

The FCR Eq. (9) guaranties that this indeed defines an algebra homomorphism as already was shown on the example of Lax operators L_{n+1} and L_n.

We have to require certain conditions on R for the algebra $\mathcal{A}$ to be rich enough. Consider the operator $T^1T^2T^3$ acting as a matrix on $V \otimes V \otimes V$, where T^i acts as T on the i-th component of $V \otimes V \otimes V$ and as the identity on the others. There are two possibilities to interchange these three operators according to the paths:

$$\begin{array}{ccccc} & & T^2T^1T^3 \to T^2T^3T^1 & & \\ & \nearrow & & \searrow & \\ T^1T^2T^3 & & & & T^3T^2T^1 \\ & \searrow & & \nearrow & \\ & & T^1T^3T^2 \to T^3T^1T^2 & & \end{array}$$

Define

$$R^{12}R^{13}R^{23} \equiv R^{123}$$

and:

$$R^{23}R^{13}R^{12} \equiv R^{321}$$

The realization of (9) gives:

$$\begin{aligned} T^1T^2T^3 &= (R^{123})^{-1}T^3T^2T^1R^{123} \\ &= (R^{321})^{-1}T^3T^2T^1R^{321}. \end{aligned}$$

In this way one obtains higher and higher relations on the matrix entries of T. But it turns out, that it is enough to require $R^{123} = R^{321}$ to get rid of all higher order relations simultaneously.

Thus we impose the main condition

$$R^{12}R^{13}R^{23} = R^{23}R^{13}R^{12} \tag{10}$$

on the structure matrix R to define a deformed (or "quantized") matrix algebra $\mathcal{A}$. Here the term "deformed" or "quantized" is used in the spirit of noncommutative geometry. The relation (10) is called the Yang-Baxter equation.

One may look at the Yang-Baxter equation as a kind of Jacobi equation for the "structure constants" of the quantized matrix algebra. It appeared previously in statistical mechanics as well as in the theory of factorizable S-matrices.

The Liouville example has given us a solution R_q of the Yang-Baxter equation for the two-dimensional space V. It will be convenient in the following to use R_q with a different normalization

$$R_+ \;=\; q^{-1/2}PR_qP = \begin{pmatrix} q^{1/2} & 0 & 0 & 0 \\ 0 & q^{-1/2} & q^{1/2} - q^{-3/2} & 0 \\ 0 & 0 & q^{-1/2} & 0 \\ 0 & 0 & 0 & q^{1/2} \end{pmatrix}$$

and

$$R_- = q^{1/2}R_q^{-1}.$$

The defining relation (9) is now:

$$T^1T^2R_\pm = R_\pm T^2T^1$$

(where both signs can be used).

Let us look at them more closely and introduce the matrix elements of T:

$$T = \begin{pmatrix} a & b \\ c & d \end{pmatrix}$$

The six nontrivial conditions resulting out of the FCR are:

$$\begin{aligned} ab &= qba, \\ dc &= q^{-1}cd, \\ ad - da &= (q - q^{-1})bc, \\ ac &= qca, \\ db &= qI{-}1bd, \\ bc &= cb. \end{aligned}$$

The matrix algebra generated by the a, b, c, d together with these relations will be called $Gl_q(2, \mathbb{C})$. Here we think of the entries of a matrix as the generators of the polynomial algebra over the matrices, i.e. we rather quantize the algebra of functions over the Lie group. One may look at q as a deformation parameter, as one gets for $q = 1$ a commutative algebra, corresponding to usual matrices T.

As we have seen these relations were found by looking at the quantized Liouville model on the lattice.

A natural algebraic question which arises, is whether there exist central elements of the algebra $\mathcal{A}$. We find

$$\det_q T \equiv ad - qbc$$

which we may fix to be 1 to get a subalgebra, which we call $Sl_q(2, \mathbb{C})$. Another formal central element is $\frac{b}{c}$, which may be singular as c is not required to be invertible. T has an inverse with respect to matrix multiplication:

$$T^{-1} = \begin{pmatrix} d & -q^{-1}b \\ -qc & a \end{pmatrix} = S.$$

$T \to S = T^{-1}$ is therefore an antiautomorphism from $Sl_q(2, \mathbb{C})$ to $Sl_{\frac{1}{q}}(2, \mathbb{C})$ i.e. S satisfies the following fundamental commutation relations:

$$R_{\pm} S^1 S^2 = S^2 S^1 R_{\pm}.$$

We expect now, that there will exist such quantum deformations for all classical groups. Corresponding R-matrices were found by Jimbo and Bazhanov.

Let us turn now to another algebra $\mathcal{B}$ associated with the relations (5) that appeared in the XXZ model. Once more we will consider first a two-dimensional example, but this time we take two matrices $L_{\pm}$ as generator matrices. In order to get not too many generators we restrict ourselves to triangular matrices, which corresponds to the choice of Borel subalgebras in a matrix Lie algebra. The matrices

$$L_+ = \begin{pmatrix} q^{\frac{H}{2}} & (q-q^{-1})X_+ \\ 0 & q^{-\frac{H}{2}} \end{pmatrix}, \quad L_- = \begin{pmatrix} q^{-\frac{H}{2}} & 0 \\ -(q-q^{-1})X_- & q^{\frac{H}{2}} \end{pmatrix},$$

contain three generators $X_\pm, H$ and satisfy the following FCR's:

$$R_\pm L_+^1 L_+^2 = L_+^2 L_+^1 R_\pm \tag{11}$$
$$R_\pm L_-^1 L_-^2 = L_-^2 L_-^1 R_\pm \tag{12}$$
$$R_+ L_+^1 L_-^2 = L_-^2 L_+^1 R_+ \tag{13}$$
$$R_- L_-^1 L_+^2 = L_+^2 L_-^1 R_- \tag{14}$$

only three of which are independent.

By direct computation of the FCR we get

$$q^{\frac{H}{2}} X_+ = q X_+ q^{\frac{H}{2}}, \tag{15}$$
$$X_+ X_- - X_- X_+ = \frac{q^H - q^{-H}}{q - q^{-1}} \tag{16}$$

as defining relations on a new algebra $\mathcal{B}$, which is generated by $X_\pm$ and H. These relations coincide with (5).

This example also naturally generalizes to higher rank classical Lie algebras. The components of the corresponding matrices $L_\pm$ constitute the analog of the Cartan-Weyl basis.

In a different abstraction from relation (5) Drinfeld and Jimbo use generators, corresponding to the Chevalley basis. So they have additional Serre relations, which are absent in our formulation.

The comultiplication is given by

$$\Delta L_\pm = L_\pm \otimes L_\pm$$

in accordance with Eqs. (11)(12). The determinant of $L_\pm$ is equal to 1 by construction, so that $L_\pm^{-1}$ are defined analogously to T^{-1}. For example

$$L_-^1 = \begin{pmatrix} q^{\frac{H}{2}} & 0 \\ q^{-1}(q-q^{-1})X_- & q^{-\frac{H}{2}} \end{pmatrix}.$$

An alternative way of description the quantum Lie algebra uses only one matrix

$$L = L_+ L_-^{-1}$$

The commutation relations for L follow from (11)-(14) and look as follows

$$L^1 R_-^{-1} L^2 R_- = R_+^{-1} L^2 R_+ L^1. \tag{17}$$

The comultiplication can not be written explicitely in terms of L only. However this formulation is useful in many respects.

Let us show how the undeformed Lie-algebraic relations follow from (17) in the limit $q \to 1$ on the rank one example.

Posing as usual $q = e^{i\gamma}$ we have for $\gamma \to 0$

$$R_{\pm} = I + i\gamma r_{\pm} + O(\gamma^2)$$

where the matrix $r_{\pm}$ is given by:

$$r_{+} = \begin{pmatrix} \frac{1}{2} & 0 & 0 & 0 \\ 0 & -\frac{1}{2} & 2 & 0 \\ 0 & 0 & -\frac{1}{2} & 0 \\ 0 & 0 & 0 & \frac{1}{2} \end{pmatrix}$$

and

$$r_{-} = -Pr_{+}P.$$

Now we look for the expansion of L in the form:

$$L = I + \gamma l + \dots \tag{18}$$

and then Eq. (17) leads to the relation

$$[l^1, l^2] + i[C, l^2] = 0 \tag{19}$$

in order γ^2. Here

$$C = r_{+} - r_{-} = \begin{pmatrix} 1 & 0 & 0 & 0 \\ 0 & -1 & 2 & 0 \\ 0 & 2 & -1 & 0 \\ 0 & 0 & 0 & 1 \end{pmatrix} \tag{20}$$

is a realization of the Casimir operator:

$$C = \sum_a \sigma^a \otimes \sigma^a$$

in terms of Pauli matrices σ^a.

One recognizes in Eq. (19) the ordinary Lie-algebraic relations for the generators l^a, combined into the matrix:

$$l = \sum_a l^a \sigma^a.$$

The formula (18) shows that after deformation the matrix L is more "group-type" than a "Lie-algebra type" matrix l. So the difference between the Lie group and Lie-algebra is much less in the deformed case than in the classical undeformed situation.

Now we shall introduce the important notion of q-trace:

$$\mathrm{tr}_q A = \mathrm{tr}\mathcal{D}A$$

where $\mathcal{D}$ is a diagonal matrix, which is attributed to any classical group. In the $Sl(2,\mathbb{C})$ case, $\mathcal{D}$ can be chosen as:

$$\mathcal{D} = \begin{pmatrix} q & 0 \\ 0 & q^{-1} \end{pmatrix}$$

One can show now, that

$$\mathrm{tr}_q^1 R_\pm A^1 R_\pm = \mathrm{tr}_q A \cdot I^2$$

where tr_q^1 means trace over the first auxiliary space and I^2 is a unit matrix in the second auxiliary space. Using this relation and Eq. (17) we see that $\mathrm{tr}_q L$ commutes with all the elements of the matrix L and thus is a central element. It plays the role of the lowest Casimir operator.

In $Sl(2,\mathbb{C})$ example we have:

$$\begin{aligned} \mathrm{tr}_q L &= q^H + \frac{1}{q}q^{-H} + \left(q - \frac{1}{q}\right)^2 X_+ X_- \\ &= \frac{1}{2}\left[\left(q + \frac{1}{q}\right)\left(q^H + q^{-H}\right) + \left(q - \frac{1}{q}\right)^q (X_+X_- + X_-X_+)\right] \end{aligned}$$

Another way to find elements in the center of the quantum Lie-algebra consists in diagonalizing the matrix L. One can show (see i.e., my Cargèse lectures) that L can be represented in the form

$$L = u \begin{pmatrix} q^\theta & 0 \\ 0 & q^{-\theta} \end{pmatrix} u^{-1}$$

with the matrix u satisfying the relations:

$$R_+ u^1 u^2 = u^2 u^1 R(\theta)$$

where

$$R(\theta) = \begin{pmatrix} q^{1/2} & 0 & 0 & \\ 0 & q^{-1/2}\left(1 - \frac{q-q^{-1}}{q^\theta - q^{-\theta}}\right)^{1/2} & -q^{-\theta-1/2}\frac{q-q^{-1}}{q^\theta-q^{-\theta}} & 0 \\ 0 & q^{\theta-1/2}\frac{q-q^{-1}}{q^\theta-q^{-\theta}} & q^{-1/2}\left(1 - \frac{q-q^{-1}}{q^\theta-q^{-\theta}}\right)^{1/2} & 0 \\ 0 & 0 & 0 & q^{1/2} \end{pmatrix}.$$

The operator matrix u has an interpretation as a combination of deformed Clebsch-Gordan coefficients and $R(\theta)$ is the corresponding $6-j$ symbol.

The central elements θ and $\mathrm{tr}_q L$ are related as follows

$$\mathrm{tr}_q L = q^{\theta} + q^{-\theta}$$

which formally coincides with the usual trace of the diagonal form of L.

In the case when the deformation parameter γ is real, so that $|q| = 1$, we have:

$$R_+^* = R_-$$

Hence we can introduce the real unitary form of $Sl(2, \mathbb{C})$ by the requirement:

$$L_+ = L_-^* \tag{21}$$

or:

$$X_+^* = X, \quad H^* = H.$$

The Casimir $\mathrm{tr}_q L$ is then selfadjoint. Its eigenvalues are real for θ real or θ imaginary. The first case can be called elliptic, and the second, hyperbolic. For the elliptic case θ acquires quantized values:

$$\theta = 2j + 1, \quad j = 0, \frac{1}{2}, \ldots$$

in the interval $0 < \theta < \frac{\pi}{\gamma}$. In particular, for:

$$\gamma = \frac{\pi}{k}, \tag{22}$$

where k is an integer, j varies between 0 and $\frac{k}{2} - 1$. The corresponding representations V_j have dimensions $2j + 1$.

In the hyperbolic case the representations are infinite dimensional; however for γ having the form (22) they are cyclic and have dimension k.

That finishes the description of the quantum groups and the quantum Lie algebras. We proceed to describe the action of the latter on the former.

In the classical nondeformed case the generators l can be realized as vector fields acting on functions on the group variables T_{ij}. In particular the action ∇_l of l on the coordinate functions T can be described as

$$\nabla_{l^1} T^2 = T^2 C$$

where C is a matrix Casimir (20). Alternatively, we can say, that vector fields l and coordinates T have the commutation relation:

$$[l^1, T^2] = T^2 C \tag{23}$$

characteristic for the regular representation of the group.

The corresponding relation in the deformed case can be written as:

$$L^1_\pm T^2 = T^2 R_\pm L^1_\pm$$

or in terms of L,

$$L^1 T^2 = T^2 R_+ L^1 R_-^{-1}. \tag{24}$$

It is clear that Eq. (24) reduces to Eq. (23) in the classical limit $\gamma \to 0$ if we take into account (13), (18) and (20).

The relations:

$$\begin{aligned} T^1 T^2 R_\pm &= R_\pm T^2 T^1 \\ L^1 R_-^{-1} L^2 R_- &= R_+ L^2 R_+^2 L^1 \\ L^1 T^2 &= T^2 R_+ L^1 L_-^{-2} \end{aligned}$$

give us an algebra which can be called $(T^*G)_q$. It represents the quantization and deformations of the classical contangent group. Putting:

$$q = e^{i\hbar\gamma}$$

when $\hbar$ is the Planck constant of quantum mechanics we can draw a diagram:

$$\begin{array}{ccc} \text{classical top} & \xrightarrow{\hbar} & \text{regular representation of } G \\ \gamma\downarrow & & \downarrow\gamma \\ \text{deformed classical top} & \xrightarrow{\hbar} & (T^*G)_q \end{array}$$

so that the regular representation of the group, classical deformed and nondeformed top, could be obtained in the limit $\gamma \to 0$ or $\hbar \to 0$, or both.

One can extend the involution (21) onto $(T^*G)_q$. It would be nice to be able to restrict this noncommutative manifold to the case of elliptic L. One could hope, that the corresponding Hilbert space would be finite dimensional and have the structure:

$$\mathcal{H} = \sum_i V_i \otimes V_i \tag{25}$$

with the finite sum over j's corresponding to real θ . The condition (22) on γ must be essential for that. The work in this direction is in progress.

Now we turn to the generalization of our object on the loops over the chain with sites n, $n = 1, \ldots N$, $n+N \equiv n$. The ultralocal FCR for L_n independently prescribed to any site n reads

$$\begin{aligned} L_n^1 R_-^{-1} L_n^2 R_- &= R_1^{-1} L_n^2 R_+ L_n^1 \\ L_n^1 L_m^2 &= L_m^2 L_n^1, \quad n \neq m \end{aligned}$$

can be written in one line as

$$L_m^1 R_-^{-1}(m-n) L_n^2 R_-(m-n) = R_+^{-1}(m-n) L_n^2 R_+(m-n) L_m^1$$

with:

$$R(m-n) = \begin{cases} R & m = n \\ I & m \neq n \end{cases}.$$

We shall show that a simple twist changes these relations into:

$$L_m^1 R_-^{-1}(m-n-1) L_n^2 R_-(m-n) = R_+^{-1}(m-n) L_n^2 R_+(m-n+1) L_m^1 \tag{26}$$

which accounts for a central extension of the loop algebra. Indeed in the continuous classical limit $\Delta \to 0$, $\hbar \to 0$ if we put:

$$\begin{aligned} L_n &= I + \Delta l(x) + \ldots, \quad x = \hbar\Delta \\ R_\pm(m-n) &= I + i\hbar\gamma r_\pm \delta_{m,n} + \ldots \end{aligned}$$

and we take into account that

$$\begin{aligned} \delta_{m,n} &= \Delta\delta(x-y) \\ \delta_{m+1,n} - \delta_{m,n} &= \Delta^2\delta'(x-y) \end{aligned}$$

then we get in order Δ^2

$$\frac{1}{i\hbar}[l^1(x), l^2(y)] + \gamma[C, l^2(y)]\delta(x-y) + \gamma C\delta'(x-y) = 0$$

which give us the classical realization of the Kac-Moody algebra with level

$$l_{\mathrm{cl}} = \frac{\pi}{\gamma}$$

if we identify $\frac{1}{i\hbar}[\ ,\ ]$ with the Poisson bracket.

Thus the relation (26) could be interpreted as giving a lattice deformation of a Kac-Moody algebra. This opens to us the way in the direction of Conformal Field Theory.

Unfortunately in this purely algebraic setting there is no simple way to describe a continuous limit $\Delta \to 0$ without recourse to the classical limit $\hbar \to 0$. However some indirect considerations, based on the operator expansion, lead to the identification of the level l and deformation parameter γ as follows:

$$\gamma = \frac{\hbar}{l + c}$$

where c is the Coxeter number of the group G, $c = 2$ for $Sl(2,\mathbb{C})$.

The nontrivial relations in (26) are:

$$\begin{aligned} R_+ L_n^1 L_n^2 &= L_n^2 L_n^1 R_-^{-1} && (27) \\ L_n^1 L_{n+1}^2 &= L_{n+1}^2 R_+ L_n^1 && (28) \\ L_n^1 R_-^{-1} L_{n-1}^2 &= L_{n-1}^2 L_n^1 && (29) \end{aligned}$$

It follows from these relations that L_n commutes with the product of its three neighbors.

$$L_n^1 L_{n+1}^2 L_n^2 L_{n-1}^2 = L_{n+1}^2 L_n^2 L_{n-1}^2 L_n^1 \tag{30}$$

This allows to compute the commutation relations between products of L_n. In particular for the monodromy:

$$M = L_N \cdots L_1$$

we obtain the relation

$$M^1 R_-^{-1} M^2 R_- = R_+^{-1} M^2 R_+ M^1 \tag{31}$$

which is nothing but the defining relation of the quantum lie algebra. So we see that the lattice Kac-Moody algebra (Latt K-M) contains a quantized lie algebra.

$$\mathcal{B} \subset \text{Latt K-M}$$

One can calculate the relations between M and the generators L_n. The only nontrivial ones are for L_1 and L_N. We have:

$$\begin{aligned} R_- M^1 R_-^{-1} L_N^2 &= L_N^2 M^1 \\ M^1 L_2^1 &= L_1^2 R_+ M^1 R_+^{-1} \end{aligned}$$

This means that the center of $\mathcal{B}$, namely $\mathrm{tr}_q M$, (or more generally, invariants of M) is also a center of Latt K-M. Thus any representation V of the quantum Lie algebra $\mathcal{B}$ labels a representation $\mathcal{H}_j$ of Latt K-M with the same values for the center:

$$\mathcal{H}_j = V_j \otimes \mathcal{H}_0, \quad j = 0, \frac{1}{2}, \ldots \frac{l}{2}. \tag{32}$$

We are ready now to turn to Conformal Field Theory. We shall consider the typical model of CFT, namely the WZNW model.

First, I will describe the formulation of the WZNW model as a classical field theory. The space time M is a cylinder $S^1 \times \mathbb{R}^1$ with coordinates x, t, $0 \leq x \leq 2\pi$,

$-\infty < t < \infty$ with Minkowskian metric. The field $g(x,t)$ is a unitary unimodular 2×2 matrix

$$g : M \rightarrow SU(2).$$

The action functional looks as follow

$$A = -\frac{1}{8\gamma}\int \mathrm{tr}(\partial g g^{-1})^2 dx dt + \frac{1}{12\gamma}\int (d^{-1}\mathrm{tr}(dgg^{-1})^3) \tag{33}$$

where the pullback of the integrand of the closed, but not exact, 3-form $\mathrm{tr}(dgg^{-1})^3$, is used to define the WZ term in the action. This nonunivalent action (in the terminology of Novikov) can be used in quantum theory only if the coupling constant γ is quantized, namely

$$l = \frac{\pi}{\gamma}$$

must be an integer.

The phase space of the model is spanned by the Cauchy data $g(x), J_0(x)$ for fixed time, where we use the notation J_μ for right invariant currents

$$J_\mu = \partial_\mu g g^{-1}.$$

To specify their Poisson brackets it is convenient to use the first order formalism when g and J_0 are considered as independent variables. The action, equivalent to (33) can be taken in the form.

$$A = -\frac{1}{4\gamma}\int \mathrm{tr}\left[\partial g g^{-1} J_0 - \frac{1}{2}(J_0^2 + J_1^2)\right] dx dt + \mathrm{WZ}$$

where WZ is a WZ-term, i.e. the second term in the RHS of (33). It is considered as a functional of g alone as it contains $\partial_0 g$ only linearly. Furthermore,

$$J_1 = \partial_x g g^{-1}$$

The variation of the canonical 1-form in A leads to the symplectic form

$$\Omega = \frac{1}{4\gamma}\int \mathrm{tr}(dgg^{-1} \wedge dJ_0 + (J_0 - J_1) dgg^{-1} \wedge dgg^{-1}) dx dt. \tag{34}$$

In the derivation, one has to use the property of WZ

$$\delta \mathrm{WZ} = \frac{1}{4\gamma}\int \mathrm{tr}(J_1 \delta g g^{-1} \partial_0 g g^{-1} dx dt$$

which follows from

$$\delta \mathrm{tr}(dgg^{-1})^3 = 3d\mathrm{tr}(\delta g g^{-1}(dgg^{-1})^2).$$

The symplectic form Eq. (34) leads to the Poisson brackets:

$$\{g^1(x), g^2(x)\} = 0$$

$$\{\mathcal{J}_0^1(x), g^2(y)\} = -2\gamma C g^2(y)\delta(x-y)$$

and

$$\{\mathcal{J}_0^1(x), \mathcal{J}_0^2(x)\} = \gamma[\mathcal{J}_0^1(x) - \mathcal{J}_0^2(y) + \mathcal{J}_1^2(y), C]\delta(x-y).$$

Furthermore we have

$$\begin{aligned}\{\mathcal{J}_0^1(x), \partial g^2(y)(g^2(y))^{-1}\} &= \\ &= -2\gamma C(\partial g^2(y)\delta(x-y)(g^2(y))^{-1} \\ &\quad +2\gamma C g^2(\delta'(x-y)(g^2(y))^{-1} \\ &\quad +2\gamma\partial g^2(y)(g^2(y))^{-1}C\delta(x-y)\end{aligned}$$

or

$$\{\mathcal{J}_0^1(x), \mathcal{J}_1^2(x)\} = 2\gamma[C, \mathcal{J}_1^2(y)]\delta(x-y) + 2\gamma\delta'(x-y).$$

If we define $l = \frac{1}{2}(\mathcal{J}_0 + \mathcal{J}_1)$ to be the current with respect to light cone variables, it follows

$$\{l^1(x), l^2(y)\} = \frac{1}{2}\gamma[C, l^1(x) - l^2(y)]\delta(x-y) + \gamma C\delta'(x-y)$$

and in coordinates with respect to the Pauli matrices, $l = l^a\sigma^a$

$$\{l^a(x), l^b(y)\} = \gamma\epsilon^{abc}l^c\delta(x-y) + \gamma\delta^{ab}\delta'(x-y)$$

These are the commutation relations for a Kac-Moody algebra. A part of the phase space constitutes therefore a Kac-Moody algebra. Let us define a right current by

$$r = \frac{1}{2}g^{-1}(\mathcal{J}_0 - \mathcal{J}_1)g,$$

for which we have

$$\{r^1(x), r^2(y)\} = -\frac{1}{2}\gamma[C, r^1(x) - r^2(y)]\delta(x-y) - \gamma C\delta'(x-y)$$

and

$$\{r^1(x), r^2(y)\} = 0.$$

So we have a chiral decomposition of the phase space in left and right movers, l and r respectively. However, it will be shown later on, that these variables are not enough to characterize the phase space completely. We are to introduce another scalar cyclic coordinate. With this we have the coordinate transformation:

$$(g, \mathcal{J}) \rightarrow (l(x), r(x), q)$$

and the Hamiltonian H decouples into two currents:

$$H = H_0^2 + \mathcal{J}_1^2 = l^2 + r^2$$

The equations of motion are:

$$\begin{aligned} \partial_0 l + \partial_1 l &= 0 \\ \partial_0 r - \partial_1 r &= 0 \end{aligned}$$

Substituting the expressions for $\mathcal{J}_0$ and $\mathcal{J}_1$, into l and r and subtracting after multiplication by g from the right and left, respectively, we get for the old field $g(x,t)$

$$\partial_x g = lg - gr$$

which is solved by:

$$g = u(x) K v(x),$$

where u and v satisfy $\partial_x u = lu$ and $\partial_x v = -vr$ with initial conditions $u(0) = \vec{1}$, $v(0) = \vec{1}$, where we have chosen a fixed point $x_0 \in S^1$. We define the monodromy for u and v by:

$$u(x + 2\pi) = u(x) M_L, \quad \text{and} \quad v(x + 2\pi) = M_R v(x)$$

From the periodicity of $g(x)$ it follows that $K = M_L K M_R$. Therefore M_L and M_R^{-1} are conjugated by K and have the same spectrum. We set

$$M_L = Z_L D Z_L^{-1}, \quad M_R = Z_R^{-1} D^{-1} Z_R.$$

Finally we have:

$$\{l^1(x), g^2(y)\} = \gamma g^2(y) C \delta(x - y) \tag{35}$$

and

$$\{r^1(x), g^2(y)\} = -\gamma C g^2(y) \delta(x - y) \tag{36}$$

where D is a diagonal matrix:

$$D = \begin{pmatrix} e^{i\beta} & 0 \\ 0 & e^{-i\beta} \end{pmatrix}$$

with $0 \leq \beta \leq \pi$.

This implies that D and $Z_L^{-1} K Z_R^{-1}$ commute and the only freedom we have for K is:

$$K = Z_L Q Z_R$$

where

$$Q = \begin{pmatrix} e^{i\alpha} & 0 \\ 0 & e^{-i\alpha} \end{pmatrix}$$

and α is the looked cyclic variable.

The above equations bear a strong resemblance to the equations for the quantized top, so that we may suspect, that the quantum top is hidden inside this model of conformal field theory.

Now we turn to the quantization. We introduce the lattice $x \rightarrow h\Delta$ and use the algebra (26) as quantization of currents $l(x)$ and $r(x)$. Thus we come to lattice analogues of the chiral fields $u(x)$ and $v(x)$

$$\begin{aligned} u_n &= L_n \ldots L_1 \\ v_n &= R_1 \ldots R_n \end{aligned}$$

Using the rules (27) and property (30) we can easily get the commutation relations:

$$\overset{1}{u}_n \overset{2}{u}_m R_\pm = \overset{2}{u}_m \overset{1}{u}_n \begin{cases} + & \text{for } n > m \\ - & \text{for } n < m \end{cases}$$

and

$$R_+ \overset{1}{u}_n \overset{2}{u}_n = \overset{2}{u}_n \overset{1}{u}_n R_-^{-1}.$$

Similarly we obtain for v:

$$\overset{1}{v}_n \overset{2}{v}_m = R_\pm \overset{2}{v}_m \overset{1}{v}_n \begin{cases} + & \text{for } n > m \\ - & \text{for } n < m \end{cases}$$

and

$$R^{-1} \overset{1}{v}_n \overset{2}{v}_n = \overset{2}{v}_n \overset{1}{v}_n R_\pm,$$

while u_n and v_m commute for all m, n.

Now let us turn to the local field

$$g_n = u_n K v_n,$$

where K plays the role of the initial condition.

$$K = g_0 = g_N.$$

In the classical case we have constructed K classically through the matrices Z_L, Z_R, diagonalizing M_L and M_R and an additional variable α. This process could be repeated here. However we shall propose instead some characterization of K in terms of simple commutation relations and check their consistency.

We require that the following relations take place:

$$\begin{aligned} K^1 L_2^1 &= L_1^2 R_+ K^1 \\ R_- K^1 L_N^2 &= L_N^2 K^1 \end{aligned}$$

and K commutes with L_n, $n = 2, \ldots, N-1$. This is the quantum counterpart of the Poisson commutation relation between $J_0(x)$ and $g(y)$ (at $y = 0$). Moreover, K itself constitues a quantum group

$$RK^1K^2 = K^2K^1R, \tag{37}$$

where R stands for R_+ or R_-. It follows that K has similar relations with u_n,

$$K^1 u_n^2 = u_n^2 R_+ K^1, \quad n = 1, \ldots, N-1$$

and monodromy $M_L = u_N$:

$$R_- K^1 M_L^2 = M_L^2 R_+ K^2. \tag{38}$$

The corresponding relations with v_n look as follows:

$$K^1 R_+ v_n^2 = v_n^2 K^1$$

(one reads the relations for u from right to left). Now by simple algebra one can check that the local fields g_n satisfy the relations:

$$\begin{aligned} g_m^1 g_n^2 &= g_n^2 g_m^1; \quad n \neq m \\ R g_n^1 g_n^2 &= g_n^2 g_n^1 R \end{aligned}$$

among itself and

$$\begin{aligned} g_n^1 L_{n+1}^2 &= L_{n+1}^2 R^+ g_n^1, \\ R_- g_n^1 L_n^2 &= L_n^2 g_n^1, \\ g_n^1 L_m^2 &= L_m^2 g_n^1, \quad m \neq n, n+1 \end{aligned}$$

between g_n and L_m.

These relations turn into Eqs. (35),(36) in the classical continuous limit.

The consistency with the periodicity condition is now checked in the following manner, $M_L K M_R$ has exactly the same relations with K, L_n and M_L and itself (and their right counterparts) as K. It follows that

$$K = \alpha M_L K M_R$$

where α commutes with all dynamical variables, so it is a constant and must be equal to 1 by comparison of determinants.

Now we observe that the relations (37), (38) together with the relation of type (32) for M_L define us an algebra $(T^*G)_q$ with the identification:

$$M_L \to L; \qquad K^{-1} \to T$$

Thus the local field at a fixed point and the monodromy of a chiral component around the circle from this point give us a representation of $(T^*A)_t$. These data comprise zero modes of the WZNW model and essentially define its full structure. This establishes the intimate connection of Quantum Groups and Kac-Moody algebras.

In particular the Hilbert space of the top (25) generates the Hilbert space of the WZNW model:

$$\mathcal{H}_{WZNW} = \sum_j \mathcal{H}_j \otimes \mathcal{H}_j$$

in accordance with (26). The oscillators, generating $\mathcal{H}_0$ do not influence the nontrivial part of the dynamics.

We believe, that the unravelled structure will be useful in the calculation of the correlation functions of the local field. However this truly dynamical problem is out of the scope of these lectures.

References:

I add now some comments on the literature.

Lattice Lax operators were introduced in the course of development of the method of quantum inverse problem, for a survey see:

[1] L.D. Faddeev: Sov. Sci. Rev. Sec. C: Math. Phys.**1**,107 (1980),

[2] A.G. Izergin, V.E. Korepin: Fisika Echaya, (JINR, Dubna) **13**, 501 (1982).

The deformed Lie algebra $Sl_q(2)$ appeared first in

[3] P.P.Kulish, N.Yu. Reshetikhin: Zap. nauch. seminarov LOMI **101**, 101 (1981) (in Russian); J.Sov.Math. **23**, 2435 (1983),

and the Lie groups $SL_q(2)$ in

[4] L.D.Faddeev, L.A. Takhtajan: Lectures Notes in Physics **246**, 166 (1986),

in connection with XXZ and Liouville models. The general notion of quantum groups was developed by Drinfeld and Jimbo:

[5] V.G. Drinfeld: *Dokl. AN SSSR* **273**, 531 (1883),

[6] V.G. Drinfeld: *Dokl. AN SSSR* **283**, 1060 (1985),

[7] V.G. Drinfeld: in *Proc. ICM Berkeley*, 786 (1986),

[8] M.Jimbo: *Lett. Math. Phys.* **11**, 247 (1986).

The description of the lecture follows

[9] L.D.Faddeev, N. Yu. Reshetikhin, L.A. Takhtajan: *Algebra and Analysis* **1**, 178 (1989).

The notion of $(T^*G)_q$ is introduced in a paper with A. Yu. Alekseev:

[10] A.Yu. Alekseev, L. D. Faddeev: *Commun. Math. Phys.* **141**, 413-422 (1991).

Its role in WZNW was indicated in

[11] A. Alekeseev, L. Faddeev, M. Semenov-Tjan-Shansky, A. Volkov: preprint CERN-TH-5981/91 (1991).

[12] A.Alekeseev, L.Faddeev, M. Semenov-Tjan-Shansky: LOMI preprint E-5-91 (1991)

and elucidated by me in Cargese and Berlin lectures in 1991.

[13] L.Faddeev: preprint HU-ITF 92-5 and

[14] L.Faddeev, preprint SFB 922 N1 Berlin.

A BRIEF HISTORY OF HIDDEN QUANTUM SYMMETRIES IN CONFORMAL FIELD THEORIES[1]

CÉSAR GÓMEZ and GERMÁN SIERRA

Instituto de Matemáticas y Física Fundamental, CSIC
Serrano 123. Madrid, Spain

Abstract

We review briefly a stream of ideas concerning the role of quantum groups as hidden symmetries in Conformal Field Theories, paying particular attention to the field theoretical representation of quantum groups based on Coulomb gas methods. An extensive bibliography is also included.

[1]Lecture presented in the XXI DGMTP-Conference (Tianjin ,1992) and in the NATO-Seminar Salamanca,1992).

A. Ibort and M. A. Rodríguez (eds.), Integrable Systems, Quantum Groups, and Quantum Field Therapy 25–43.
1993 *Kluwer Academic Publishers.*

Integrability, Yang-Baxter equation and Quantum Groups

In the last few years it has become clear the close relationship between conformal field theories (CFT's), specially the rational theories , and the theory of quantum groups.From a more general point of view this is just one aspect of the connection between integrable systems and quantum groups. In fact if one recalls the history, quantum groups originated from the study of integrable models and more precisely from the quantum inverse scattering approach to integrability. The basic relation:

$$R(u-v)\,T_1(u)\,T_2(v) = T_2(v)\,T_1(u)\,R(u-v) \tag{1}$$

which is at the core of this approach may also be taken as a starting point in the definition of quantum groups [FRT].

In statistical mechanics the operator $T(u)$ is nothing but the monodromy matrix which depends on the spectral parameter u and whose trace gives the transfer matrix of a vertex model. Meanwhile the matrix $R(u)$ stands for the Boltzmann weights of this vertex model. The RTT equation guarantees the commutativity of a one parameter family of transfer matrices:

$$[\,tr\,T(u),\ tr\,T(v)\,] = 0 \tag{2}$$

This equation is equivalent to the existence of an infinite number of conserved quantities when taking the thermodynamic limit. In these setting the Yang-Baxter equation:

$$R_{12}(u)R_{13}(u+v)R_{23}(v) = R_{23}(v)R_{13}(u+v)R_{12}(u) \tag{3}$$

which underlies the integrability of the model appears as the associativity condition of the "quantum algebra" (1).

RCFT's, exchange algebras and the braid group : the polynomial equations

Let us turn now to conformal field theories. There, it was soon realized that the braiding matrices of the conformal blocks of a RCFT also satisfy a Yang-Baxter like equation. In fact these matrices provide a representation of the braid group in the space of conformal blocks. The braid group and the exchange algebras appeared suddenly to play an important role in CFT's and more generally in 2D quantum field theories [TK,Ko,Fr,FRS,Rh,MS]. However there are some structural differences with regard to the vertex models in statistical mechanics. In CFT the braiding matrices do not depend on any spectral parameter while in statistical mechanics the Boltzmann weights certainly do depend on it and only when taking the limits $u \to \pm\infty$ one gets non trivial solutions of the Yang-Baxter equation without spectral parameter. In addition ,the Boltzmann weights of a vertex model $R_{ij}^{kl}(u)$ depend on four labels while the braiding matrices $B_{pp'}\begin{bmatrix} k & l \\ \vdots & \vdots \\ i & j \end{bmatrix}$ depend on six labels which are the primary fields of a RCFT and consequently are subject to certain constraints given by the fusion rules. These later conditions are equivalent to the restrictions on the heights of a restricted solid on solid model (RSOS) where the Coxeter diagram is playing the role of the fusion rules in RCFT.

Having in mind the RTT equation of integrable models as well as the braiding properties of the conformal blocks it was proposed in reference [AGS1] that quantum groups should be present in RCFT's as "hidden symmetries" underlying and explaining the origin of the polynomial equations of Moore and Seiberg. The motivation of these authors in writing the polynomial equations was to give an axiomatic definition of RCFT's which would, among another things, give a rigorous proof of the Verlinde conjecture concerning the diagonalization of the fusion rules N_{ij}^{k} by means of the modular matrix S_{ij} which acts on the Virasoro characters [V].

Reference [AGS1] was in this way a first attempt to lay down a "quantum group" interpretation of the duality properties of RCFT's, which are encoded in the polynomial equations. Indeed one of these equations is nothing but the Yang-Baxter equation for the braiding matrices of the conformal blocks mentioned above. A second attempt in the same direction was taken by Moore and Reshetikhin in reference [MR].

A first clue of the problem is suggested by the classical limit of a WZW model based on a group G which is obtained when taking the level k going to infinity. This model has an infinite number of primary fields labelled by their spin $j = 0, 1/2, 1, \ldots$ whose braiding and fusion properties can be computed entirely from the representation theory of the classical group G [MS]. Thus for example the fusion matrices F of this classical WZW model become the 6-j symbols of the group G , while the fusion rules are nothing but the usual Clebsch-Gordan decomposition of the tensor product of irreps of G. More generally, what one calls the duality data of a RCFT, namely: fusion rules, braiding matrices, fusion matrices, modular matrices, etc. are computable in the classical limit using ordinary group theory. Of course not all the RCFT's known admit such a classical limit where all the conformal weights of the primary fields vanish and consequently the braiding properties become almost trivial. One example of this is provided by the $c < 1$ theories. However what this limit suggested was that some "deformation" of the "classical" group theory could in principle explain the whole duality data of genuine quantum RCFT's.

Quantum group phenomenology

At that time quantum groups have already taken off from the land of integrable systems. The general notion of quantum groups were developed by Drinfeld and Jimbo in references [Dr1,Ji]. The first non classical example of this definition was a q-deformation of the Lie algebra $Sl(2)$ which was found earlier by Kulish and

Reshetikhin [KuR]. Drinfeld and Jimbo generalized this quantum deformation to include all the classical Lie groups affine or not. Later on in a series of papers Kirillov and Reshetikhin [KR,R] constructed the representation theory of $U_q(Sl(2))$ and others groups for a generic value of the deformation parameter q.

After these works it was rather clear that quantum groups were the correct "deformation" of classical groups needed to explain the duality properties of RCFT's. In reference [AGS2] it was considered the case of the WZW model based on the Kac-Moody algebra $\widehat{SU}(2)$ at level k, and it was shown that the braiding matrices , computed by Tsuchiya and Kanie [TK] solving the Knizhnik-Zamolodchikov equation for the conformal blocks [KZ], were in fact given by the representation theory of $U_q(Sl(2))$ with $q = e^{\pi i/k+2}$. This work was later generalized to other groups, where the previous relation between the quantum deformation parameter and the data of the WZW model reads:

$$q = e^{i\pi/k+g} \tag{4}$$

with k the level and g the dual Coxeter number of G. It was also possible to understand the modular properties and in particular the Verlinde theorem using exclusively quantum-group tools [AGS3]. In what concerns modular properties, these lead to the concept of ribbon Hopf algebras [RT].

Similar results were obtained for the minimal models of type (p,p'), where the braiding matrices essentially factorize into the product of two quantum 6-j symbols with deformation parameters: $q = e^{i\pi p/p'}$ and $q' = e^{i\pi p'/p}$ [AGS1, FFK]. Also some orbifold models where shown to have a quantum group structure [DPR] given by quasi-Hopf algebras [Dr2].

The previous examples showed that the deformation parameters coming from RCFT's were given by roots of unity. The representation theory of quantum groups for generic values of q is essentially the same as the classical one [Ro].However quantum groups at roots of unity were shown to display unexpected and interesting properties [L,RA]. The trouble with roots of unity is that many of the formulas in the representation theory for q generic break down or become ambiguous. A solution to this problem, which may look a bit ad hoc, is to restrict the representations to the good ones, so that one has a well defined highest weight theory, and throwing away the bad representations [L,PS,AGS2-3]. For example for $U_q(Sl(2))$ the good representations ,also called regular or of type II, have spin j between zero and k/2, and are in one to one correspondence with the integrable irreps of the WZW model $\widehat{SU}(2)_k$. This truncation was also seen to occur in the representation theory of the Hecke algebra $H_N(q)$ which is the centralizer of the tensor product of the spin 1/2 representation of $U_q(Sl(2))$ [Wz,AGS2]. The relation between fusion rules of RCFT's and quantum groups have been further studied in [GP,FGP,FD,T].

These efforts of trying to match bits of data from different pieces of mathematics and physics were in a way "phenomenological", because it was missing a neat understanding of the close relationship between RCFT's and quantum groups. Further research has deepen and broaden the knowledge of the topic, which cannot yet considered as being closed.

One astonishing aspect of the connection between RCFT's and quantum groups is that on one side one is dealing with an infinite dimensional symmetry i.e. the Virasoro algebra , Kac-Moody algebra ,etc. and on the other the symmetries, although a bit peculiar, are finite dimensional. How could it be possible that some finite dimensional structure would know so much about an infinite dimensional one?. Making a very rough analogy we could say that given a function of the form: $f(z) = z^{\alpha} \sum a_n z^n$ the information that quantum groups are telling us is the integer part of α, which is what determines the monodromy of the function ,i.e. $f(e^{2\pi i}z) = e^{2\pi i\alpha} f(z)$. In RCFT's the role of the function $f(z)$ is played by the conformal blocks while the phase factors $e^{2\pi i\alpha}$ or $e^{\pi i\alpha}$ become the monodromy or braiding matrices respectively. To complete this analogy we could say that the chiral algebra (i.e. Virasoro, Kac-Moody ,etc) would imply a differential equation for the function $f(z)$, equation that would contain in a disguised way the monodromy of their solutions.

In this context the connection between RCFT's and quantum groups is the relation between a differential equation and the monodromy or braiding properties of their solutions. The later problem is deeply connected to the classic and well known Riemann-Hilbert problem , which consist in the characterization of a differential equation and the set of the solutions from the knowledge of their monodromy properties. What the polynomial equations establish are the consistency conditions to be satisfied by the braiding and fusion matrices associated to a set of differential equations which are in principle unknown. The "reconstruction program", advocated by Moore and Seiberg , which tries to determine from every solution to the polynomial equations a set of differential equations whose solutions give the conformal blocks of the theory, is nothing else but a Riemann-Hilbert problem. This program has only been sketched but, as one can imagine, the task of finding a solution to the polynomial equations seems already quite formidable without the help of some guiding principle.

An important thing that we must not forget is the meaning of the differential equations satisfied by the conformal blocks. They guarantee the decoupling of null vectors of the degenerate primary fields. The very existence of these fields is what makes the theory rational (finite number of primary fields) and solvable. We may expect that quantum groups would also have a lot to say about the decoupling of null vectors of the corresponding primary fields.

Vertex formulation of CFT's

From the previous discussion it seemed natural that the ordinary formulation of RCFT's as the representation of a chiral algebra (or rather the product of a holomorphic $\mathcal{A}_L$ times an antiholomorphic chiral algebra $\mathcal{A}_R$) should be enlarged in order to accommodate the quantum group symmetry, which otherwise remains hidden or invisible. A signal of this hidden nature was that objects like 3j-symbols or the R-

matrix itself did not show up in RCFT's while for example 6j-symbols did. In general the IRF content of quantum groups was clearly present in RCFT's. This suggested that what was needed, to uncover the quantum group structure of RCFT's, was a "vertex formulation" of conformal field theories. In statistical mechanics it is well known that some models admit both kinds of formulations, either as vertex or as IRF models, in which case there exist a vertex-IRF map which establishes the correspondences [Ba].It happens also, when going from the vertex to the IRF formulation, that one misses along the way the degrees of freedom on which the quantum group is acting.These kind of arguments were also used in [Wi] in an attempt to derive a quantum group structure from the 3D Chern-Simon theory.

In group theoretical language the vertex formulation is like working with tensor products of representations of a group, while the IRF formulation is like performing the tensor product of these irreps and keeping only the spaces on which the centralizer of the group is acting:

$$V_{1/2} \otimes \cdots^{N} \cdots \otimes V_{1/2} = \bigoplus_{j} (W_j^N \otimes V_j) \tag{5}$$

In the vertex formulation the group or q-group is acting on each individual vector space, $V_{1/2}$ in the example above, by means of the comultiplication. The vector space W_j^N is formed by the invariant tensors of the q-group and serves as the representation space of the centralizer $C_{1/2}^N$ which is defined as the set of endomorphisms of $V_{1/2}^{\otimes N}$ that commute with the action of the group. The dimension of W_j^N equals the multiplicity of the spin j irrep into the tensor product $V_{1/2}^{\otimes N}$. In RCFT's W_j^N can be identified with the space of conformal blocks of N legs of spin 1/2 and one leg of spin j, and its dimension can be computed using the fusion rules of the theory.From the previous identification we see that the centralizer of the quantum group is precisely realized on the space of conformal blocks , while they remain invariant under the action of the quantum group. This explains in a neat way the meaning of hidden quantum symmetries when talking about the role of quantum groups in RCFT's.

In more physical terms the centralizer has the structure of a braid group when acting on the space of conformal blocks. In the case of a WZW model based on the group $SU(2)$ it was shown by Kanie and Tsuchiya that the braid group representation that one obtains from the conformal blocks with spin 1/2 primary fields at the external legs gives a representation of the Hecke algebra $H_N(q)$ or more precisely of the Temperley-Lieb-Jones algebra. On the other hand Kirillov and Reshetikhin showed that the centralizer of the quantum group $U_q(Sl(2))$ in the tensor product of spin 1/2 representations was also a Hecke algebra $H_N(q)$ [KR]. This connection was again part of the phenomenology that we mention before, however this time one could see that the missing objects to complete the link with quantum groups were the q-group spaces $V_{1/2}$ and V_j of equation(5). These spaces should not be confused with the Verma modules $\mathcal{H}_j$ of the Kac-Moody algebra. If this were the case then the q-group $U_q(Sl(2))$ would already be contained in the Kac-Moody algebra $\widehat{SU(2)}$.

This suggest that the full symmetry algebra should be something like a tensor product $\widehat{SU}(2) \otimes U_q(Sl(2))$. For a general conformal field theory the symmetry would be $\mathcal{A}_L \otimes Q_L$ for the holomorphic degrees of freedom and $\mathcal{A}_R \otimes Q_R$ for the antiholomorphic ones. Q_L and Q_R denote the quantum groups associated to the chiral algebras $\mathcal{A}_L$ and $\mathcal{A}_R$ respectively. Later on we shall discuss in more detail the structure of the tensor products $\mathcal{A}_L \otimes Q_L$ which turns out to be a bit more subtle than just a direct tensor product. The key idea in this respect is that the q-group spaces V_j contain fields or states that cannot be obtained from the primary fields by the action of an operator in the chiral algebra but by that of a genuine quantum group operator. We should then distinguish between two kind of descendants, those of the chiral algebra $\mathcal{A}_L$ and those of the quantum group Q_L.

Having settled these conceptual matters the problem was then to give an "explicit" construction of these quantum group fields which could not be conformal (either primary or descendants)in the usual sense. This technical problem was overcome by the use of the Feigin-Fuchs or Coulomb gas formalism which has proved to be extremely important in the study of conformal field theories [FF].

The Coulomb Gas formalism of CFT's

A Coulomb gas is nothing but a free field realization of a conformal field theory. The first known CFT's admiting a Coulomb gas version are the BPZ $c < 1$ theories [BPZ], which were studied by Dotsenko and Fateev using only one free scalar field [DF]. The CFT's based on the W_n algebras [Z] require the introduction of n-1 scalar fields [FZ,FL], while the WZW models need both scalar fields and $\beta - \gamma$ systems [Wa,FFr]. An open problem is to know wether any conformal field theory admits a free field realization. The fact that a highly interacting field theory allows a free field description is on the other hand a mistery with very deep consequences. Some of them are the ability of a systematic computation of the conformal blocks, the characterization of the null vectors, a straightforward derivation of the Kac's formula, the construction of a BRST formulation *à la* Felder [Fe], and , as we shall show next, the construction of quantum groups.

The basic feature of a Coulomb gas is that the chiral algebra generators can be constructed entirely in terms of a set of free fields which have very simple operator product expansions. Moreover the primary fields, which are essentially vertex operators as in string theory, also satisfy simple OPE's. All this reduces enormously the task of computating correlators which are given by integrals of products of monomials. If we are studying the $c < 1$ or a WZW model we know that these integrals come from the solution of hypergeometric type equations. However in the Coulomb gas these integrals come from the screening charge that one has to introduce in order to balance the background charge. A screening charge is an integral $\int dt J_a(t)$ which has the property of commuting with all the operators of the chiral algebra O_n up to a total derivative:

$$[\, O_n \,, \int dt J_a(t) \,] = \int dt \frac{\partial}{\partial t}(X_n) \tag{6}$$

If there are not boundary contributions to (6) then the screening charges fully commute with the chiral algebra.

Another ingredient of this construction is the use of Fock spaces of free fields instead of Verma modules . The later are recovered as cohomological classes associated to a BRST charge constructed using screening operators [Fe]. The philosophy adopted in references [GS1-2] was to use these Fock spaces to enlarge, rather than to restrict, the number of fields in the theory in order to give room to q-groups. We could say in this sense that quantum groups are hidden in a RCFT in much the same manner that the corresponding free field version is hidden, which lead us in the long run to ask why the solution to the null vector decoupling equations of a RCFT admit an integral representation. This is a highly non trivial fact and a mistery which certainly needs more study.

Field theory formulation of quantum groups

The starting point of the work of reference [GS1] is the definition of a special kind of screened vertex operators which form the basis for the representation spaces of quantum groups. They are defined integrating a collection of screening vertex operators J_a around a primary field ϕ_α:

$$e^{\alpha}_{a_1,\ldots,a_n}(z_P, z_\infty) = \int_{C_1} dt_1 \; J_{a_1}(t_1) \cdots \int_{C_n} dt_n \; J_{a_n}(t_n) \;\; \phi_\alpha(z_P) \tag{7}$$

In the choice of the integration contours $C_1, \cdots, C_n$ we have taken into account the fact that there are branch cuts in the integrand coming from the OPE's between the screening operators with the primary field as well as among the screening themselves.They are choosen as a nested set of Hankel's contours, similar to the ones that appear in the definition of the Γ function, i.e. they go from infinity to infinity encercling the point where the primary field is inserted. Strictely speaking the operators $e^{\alpha}_{a_1,\cdots,a_n}$ depend on two points, one is the point z_P where the primary field is located and the other is the point at infinity z_∞, which plays the role of a base point.

In this construction the primary field itself ϕ_α , is identified with the highest weight vector of a q-group space V_α. The remaining vertices obtained integrating screening operators are identified with q-group descendants. This leads naturally to the interpretation of the screening charges as quantum lowering operators : $F_a = \int_C dt J_a(t)$. This interpretation was already anticipated in references [Sa,BMP]. In the $c < 1$ theory there are two screening operators J_+ and J_- hence we have two lowering operators F_+ and F_- [GS2]. In the Wakimoto construction of $\widehat{SU(n)}$ there are n-1 screenings operators which then lead to n-1 lowering operators each one associated to a positive simple root [RRR]. Under this interpretation each screening charge J_a give rise to a quantum group $U_{q_a}(Sl(2))$, where the deformation parameter q_a is given by the braiding factor among two screenings of the same type. If there are various screening operators they will define in general larger q-groups whose defining relations would follow from their braiding properties. For example for a WZW theory the braiding between screening operators: satisfies:

$$J_a(z)\, J_b(w) = q^{C_{ab}}\, J_b(w)\, J_a(z) \tag{8}$$

where C_{ab} is the symmetrized Cartan matrix and q is given by eq.(4).

Equation (8) reflects the non-local nature of the screening operators and in that sense the quantum deformation parameter q acquires a conformal field theory meaning.

From this equation one can also derive the quantum deformation of the Serre relations satisfied by the lowering operators F_a [BMP,RRR].

After applying a sufficient number of screening operators to a primary field one eventually gets zero or a null vector. To understand why this happens we have to recall that the q-vertex operators do depend in general on the point at infinity z_∞ and the point z_P where the primary field is inserted, however under certain circunstances the branch cut effectively disappears and the contour integrals shrink to the neighbour of the point z_P. This property holds for primary fields within the Kac's formula and it is extensively used in the BRST construction. In our q-group interpretation this means that the primary field give rise to a finite dimensional representation. In other words, a null vector in the CFT is at the same time a null vector for the q-group. This is the core of the connection between quantum groups and RCFT's.

It would be of some interest to find a more intrinsic definition of the q-vertex operators (7) , i.e. a definition independent of the use of the Coulomb gas formalism. A possibility is to consider these vertices as the solutions of the null vector decoupling equations.The primary fields themselves would of course be a solution, the one with conformal properties, while the others will span the rest of a representation space. The order of the equation ,which is nm for the primary field ϕ_{nm} for a $c < 1$ theory, would then give the dimension of the associated representation space of the quantum group.

We conclude that screening charges are lowering operators, similarly the Cartan operators are identified with the Coulomb charges. These two sets of operators form the Borel subalgebra which is traditionally associated with the lower triangular matrices. To complete the RCFT picture of q-groups one needs a definition of the raising operators E_a. These cannot be represented as integrals of screening operators with opposite charge because they do not have the correct conformal properties. What one really needs is an operator that would destroy integrals of screening operators and this can be achieved with the action of the chiral algebra. Take for example the Virasoro operator L_{-1} , acting on a q-vertex operator it integrates each of the screening charges leaving a boundary term as can be seen in eq.(6). Explicit computations show that eq.(6) implies the q-group relation:

$$[O_n, \int dt J_a(t)] = \int dt \frac{\partial}{\partial t}(X_n) \Rightarrow [E_a\ , F_b] = \delta_{a,b} \frac{K_a - K_a^{-1}}{q_a - q_a^{-1}} \tag{9}$$

In this sense the q-group generators E_a are contained in the chiral algebra generators O_n, which shows that the relation between $\mathcal{A}_L$ and Q_L is more intricated than expected. From equation (9) we also see that the shrinking condition for the contours which define a q-vertex operator is equivalent to the statement that this q-vertex operator is a highest weight vector of the quantum group:

$$\frac{\partial}{\partial z_\infty} e^\alpha_{a_1,\cdots,a_n} = 0 \Rightarrow E_a\ e^\alpha_{a_1,\cdots,a_n} = 0\ \forall a \tag{10}$$

It is also curious to observe some parallelism between this construction and the construction of the quantum double by Drinfeld [GS1,BL1].

Having defined the q-group spaces and the q-group generators, it is easy to obtain the comultiplication, the R-matrix, the 3j-symbols, the 6j-symbols , that is all the ingredients of quantum groups in a field theoretical framework. These ideas and techniques, which were first applied to the $c < 1$ theories, have been extended successfully to WZW models [RRR], W_n algebras [C] , N=1 and 2 superconformal field theories [Jz] and a WZW model based on the supergroup $Gl(1,1)$ [RSa] . Moreover the whole construction has a geometrical and topological flavor [GS3]. Everything follows from contour manipulations involving elementary complex analysis and q-combinatorics. Group theory when quantum deformed becomes a topology of contours (for a more elaborated version of this idea see [FeW,SV]).

Towers of algebras

Until now we have proceed to unravel the quantum group structure of RCFT's from the integral representation of the decoupling equations, we can refer to this line of thought as the analytic approach. A complementary point of view partially inspirated in Pasquier's ethiology of IRF-models [P] consist in starting the construction with the fusion rules and from that piece of information to obtain the duality (braiding and fusion) matrices in the very same way one associates in the Jone's fundamental construction a Temperley-Lieb-Jones algebra to a given graph [GHJ]. In this algebraic approach the RCFT defines a model of the Bratelli diagram in terms of the fusion rules[GS4]. Recent progress by Ocneanu [Oc] indicates ,as it should be expected, that associated with the RCFT tower of algebras it is possible to define in a unique way a quantum group structure. For a different version of the algebraic approach based on operator algebras see [FrK]. We believe that it would be fruitful both from a mathematical and a physical point of view to get a deeper understanding of the interplay between these two apparently different mathematical approaches, namely the monodromy of differential equations with their related Riemann-Hilbert problem and the theory of subfactors.

Cannonical quantization approach to CFT's

The discussions above about the role of quantum groups in CFT's have been done in the general framework of the bootstrap program of BPZ, where actions for the local fields are not strictely needed. In doing so we have omitted another important approach to the role of quantum groups in quantum field theories, namely the one based on the use of local actions or hamiltonians which are quantized in a cannonical way. The traditional methods were applied to Liouville theory which was shown to exhibit quantum group attributes in a time when quantum groups were not known [GN,FT].In fact in reference [FT] first appeared the defining relations of the quantum Lie group SL_q, not to be confused with the quantum Lie algebra sl_q which is the same as the quantized universal enveloping algebra $U_q(Sl(2))$.

After quantum groups came to fashion this approach was renewed again in connection with Liouville theory [Bb,Ge,ST] , Toda field theory [BG,HM] and the Wess-Zumino-Witten model [AS,Fa,Ga,AF,AFS,AFSV] . The use of a different language has made a bit difficult to stablish the connections with the, let's say, bootstrap approach although they certainly do exist. Particularly interesting are the lattice formulations of the Liouville and the WZW theories which may perhaps allow a Bethe ansantz analysis analogous to the one applied in the study of magnetic chains.We refer to the lectures of prof.L.D.Fadeev for a discussion of these matters.

Integrable quantum field theories

So far we have considered applications of the quantum groups to conformal field theories but their range of applicability includes also massive integrable 2D field theories. An archetype of the later is the well known sine-Gordon theory. In references [RS,BL2] it was shown that there exist a hidden quantum symmetry which governs the scattering of the solitons, antisolitons and bound states of the model, which explains in particular the S-matrix obtained in [ZZ]. This symmetry is affine and can also be given a field theory realization in terms of non-local conserved charges [BL3,FL,BFL].

q-RCFT's: New hidden symmetries ?.Elliptic quantum groups

We would like now to discuss briefly some new directions of research. As we have seen, a particular interesting class of RCFT's are the WZW models based on Kac-Moody algebras. A centrally extended affine Lie algebras can also be deformed in a very non trivial way. These q-RCFT's preserve much of the structure of the ordinary RCFT's and can be interpreted as some kind of massive field theories. In reference [FR] Frenkel and Reshetikhin have found a q-deformation of the Kniniznik-Zamolodchikov equation which governs the correlators of these q-WZW model. This is a finite difference equation whose solutions are q-hypergeometric functions. The connection matrices, or in CFT language, the braiding matrices of this "q-conformal blocks" turns out to be given by the elliptic solution to the Yang-Baxter equation of the RSOS model of Andrews, Baxter and Forrester [ABF]. These authors also pose the question of the possible existence of a hidden quantum symmetry which would underlie the elliptic Boltzmann weights, much in the same way that quantum groups underlie the trigonometric solutions. It has also been suggested as a candidate for this "elliptic quantum symmetry" the Sklyanin algebra which is intimately related to the Baxter's eight vertex model [Sk].

The solution to these questions is not known to us. We would like however to propose a way to attack the problem which consist in following the same steps that were pursued in the unravelling of quantum groups in RCFT's, namely : find a free field realization of the q-Kac-Moody algebra, then look for the q-screening operators and finally define in terms of q-integrals the "elliptic version" of quantum groups. The first two steps of this program have already been achieved in references [FJ,M,Sh,KQS,ABG] where various free field versions of the q deformed Kac-Moody algebra have been given. In [Sh] it is also constructed a q-screening operator whose q-integral *à la* Jackson commutes with the generators of q-Kac-Moody . Therefore one has in principle all the ingredients to unravel a new hidden quantum symmetry.

Spin chains at roots of unity

To finish this brief history we would like to make some comments in connection to recent progress by De Concini and Kac [DK] in the study of the representation theory of quantum groups at roots of unity (see also [DKP]. As we have already mentioned the quantum deformation parameter of the quantum group associated to a RCFT is a root of unity , for example $q = e^{\pi i/k+2}$ for a SU(2) WZW model. This implies that strictely speaking the quantum group associated to this RCFT should be $U_q(SU(2))$ moded out by the central Hopf which appears when q is a root of unity [McSo].

Another phenomena that appears for quantum groups at roots of unity is the existence of non restricted representations, i.e. representations which transform non trivially under the central Hopf subalgebra. In a series of papers [GRS,GS5,BGS,CGS] we have started the study of a family of these representations called semicyclic and nilpotents, which have the property that their tensor product is decomposable [GS5,A] so that one can define 3j and 6j symbols. An important question is wether this new class of braiding and fusion matrices may become the duality data of a new class of decoupling equations defining a new hierarchy of conformal field theories.

Acknowledgements

We would like to thanks Prof.M.L.Ge for his kind invitation to participate in the XXI DGMTP Conference at Tianjin.

References

[ABG] A.Abada,A.H.Bougourzi and M.A.El Gradechi: Deformation of the Wakimoto construction. Preprint (1992).

[AF] A.Yu.Alekseev and L.D.Fadeev: $(T^*G)_t$: A toy model for conformal field theory. Commun.Math.Phys. 141 (1991) 413.

[AFSV] A.Yu.Alekseev,L.D.Fadeev,M.Semenov-Tian-Shansky: Hidden quantum groups inside Kac-Moody algebra.Commun.Math.Phys. 149 (1992) 335.

[AFSV] A.Yu.Alekseev,L.D.Fadeev,M.Semenov-Tian-Shansky and A.Volkov: The unravelling of the quantum group structure in the WZW theory. CERN preprint Th 5981/91 (1991).

[AS] A.Alekseev and S.Shatashvili: From geometric quantization to conformal field theory. Commun.Math.Phys. 128 (1990) 197; Quantum groups and WZNW models. Commun.Math.Phys. 133 (1990) 353 .

[AGS1] L.Alvarez-Gaume,C.Gomez and G.Sierra:Hidden quantum symmetries in rational conformal field theories. Nucl.Phys. B319 (1989) 155.

[AGS2] L.Alvarez-Gaume,G.Gomez and G.Sierra:Quantum group interpretation of some conformal field theories. Phys.Lett. 220B (1989) 142.

[AGS3] L.Alvarez-Gaume,G.Gomez and G.Sierra:Duality and quantum groups.Nucl.Phys. B330 (1990) 347.

[ABF] G.Andrews,R.Baxter and P.Forrester: Eight-vertex SOS model and generalized Rogers-Ramanujan-type identities. J.Stat.Phys. 35 (1984) 193.

[A] D.Arnaudon: Fusion rules and R-matrix for the composition of regular spins with semi-periodic representations of $SL(2)_q$. Phys.Lett. 268B (1991) 217.

[Bb] O.Babelon:Extended conformal algebra and the Yang-Baxter equation. .Phys.Lett. 215B (1988) 523.

[Ba] R.Baxter: Exactly solved models in statistical mechanics.Academic Press (1982).

[BPZ] A.A.Belavin,A.N.Polyakov and A.N.Zamolodchikov: Infinite conformal symmetries in two-dimensional quantum field theory.Nucl.Phys. B241 (1984) 333.

[BGS] A.Berkovich,C.Gomez and G.Sierra: On a new class of integrable models. Inter.J.Mod.Phys. B6 (1992) 1939; q-Magnetism at roots of unity. to appear in J. of Phys.A.

[BFL] D.Bernard,G.Felder and A.LeClair: Non-local conserved currents: the lattice approach.

[BL1] D.Bernard and A.LeClair: The quantum double in integrable quantum field theory. Preprint CLNS 92/1147 (1992).

[BL2] D.Bernard and A.LeClair: Residual quantum symmetries of the restricted sine-Gordon theories. Nucl.Phys. B340 (1990) 721.

[BL3] D.Bernard and A.LeClair: The fractional supersymmetric sine-Gordon models.Phys.Lett. 247B (1990) 309; Quantum group symmetries and non-local conserved currents in 2D QFT .Commun.Math.Phys. 142 (1991) 99.

[BG] A.Bilal and J.L.Gervais: Systematic approach to conformal systems with extended Virasoro symmetries.Phys.Lett. 206B (1988) 412.

[BMP] P.Bouwknegt.J.McCarthy and K.Pilch: Free field realization of WZNW-models; BRST complex and its quantum group structure. Phys.Lett. 234B (1990) 297; Quantum group structure in the Fock space resolutions of $\widehat{sl(n)}$ representations. Commun.Math.Phys. 131 (1990) 125.

[C] R.Cuerno: Quantum symmetries in the free field realization of W_n algebras. Phys.Lett. 271B (1991) 314.

[CGS] R.Cuerno,C.Gomez and G.Sierra:On integrable quantum group invariant antiferromagnets.To appear in Journal of Geometry and Physics.

[DK] C.De Concini and V.G.Kac:Representations of quantum groups at roots of 1. Progress in Math. 92 (1990) 471.

[DKP] C.De Concini,V.G.Kac and C.Procesi: Quantum coadjoint action. Pisa preprint (1991).

[DPR] R.Dijkgraaf,V.Pasquier and P.Roche: Quasi-quantum groups related to orbifold models.Tata Inst. preprint 1990.

[DF] V.S.Dotsenko and V.A.Fateev: Conformal algebra and multipoint correlation functions in 2D statistical models. Nucl.Phys. B240[FS12](1984) 312, B251 (1985) 691.

[Dr1] V.G. Drinfeld: Hopf algebras and the quantum Yang-Baxter equation. Soviet.Math.Dokl.32 (1985) 254; A new realization of Yangians and quantized affine algebras.Soviet. Math.Dokl.36 (1988) 212; Quantum groups. Proc. ICM-86 (Berkeley),Vol.1 AMS (1986) 798.

[Dr2] V.G.Drinfeld: Quasi-Hopf algebras and Knizhnik-Zamolodchikov equation.Preprint ITP-89-43E.

[Fa] L.D.Fadeev: On the exchange matrix for WZNW models. Commun.Math.Phys.132 (1990) 131.

[FRT] L.D.Fadeev, N.Yu.Reshetikhin and L.A.Takhtajan. Quantization of Lie Groups and Lie Algebras. Leningrad LOMI preprint E-14-87. Algebra and Analysis 1 (1989) 178 (in russian).

[FT] L.D.Fadeev and L.Takhtajan: Liouville model on the lattice. Lecture Notes in Physics,vol.246,p.166, Springer (1986).

[FL] V.A.Fateev and S.L.Lykyanov: The models of two-dimensional conformal quantum field theory with Z_n symmetry. Inter.J.Mod.Phys. A3 (1988) 507.

[FZ] V.A.Fateev and A.B.Zamolodchikov: Conformal quantum field theory models in two dimensions having Z_3 symmetry. Nucl.Phys B280[FS18] (1987) 644.

[FFr] B.L.Feigin and E.V.Frenkel: Affine Kac-Moody algebras and semi-infinite flag manifolds. Commun.Math.Phys. 128 (1990) 161.

[FF] B.L.Feigin and D.B.Fuchs: Invariant skew-symmetric differential operators on the line and Verma modules over the Virasoro algebra.Func.Anal.Appl. 16 (1982) 114; Verma modules over the Virasoro algebra. Func.Anal.Appl. 17 (1983) 241; Representations of the Virasoro algebra. In: Topology,Proc. Leningrad 1982, L.D. Fadeev ,A.Malçev (eds.).Lecture Notes in Mathematics,vol 1060. Springer 1984.

[Fe] G.Felder: BRST approach to minimal models. Nucl.Phys. B317 (1989) 215.

[FFK] G.Felder, J.Frohlich and G.Keller: Braid matrices and structure constants for minimal conformal models.Commun.Math.Phys. 124 (1989) 647; On the structure of unitary conformal theory,II: Representation theoretic approach. Commun.Math.Phys. 130 (1990) 1.

[FL] G.Felder and A.LeClair: Restricted quantum affine symmetry of perturbed minimal conformal models. Intern.J.Mod.Phys. A7,Suppl.1A (1992) 239.

[FeW] G.Felder and C.Wieczerkowski: Topological representations of the quantum group $U_q(sl_2)$. Commun.Math.Phys.138 (1991) 583.

[FRS] K.Fredenhagen,K.-H.Rerhen and B.Schroer: Superselection sectors with braid group statistics and exchange algebras, I: General theory. Commun.Math.Phys. 125 (1989) 201.

[FJ] I.B.Frenkel and N.Jing: Vertex representations of quantum affine algebras. Proc.Natl.Acad.Sci.USA 85 (1988) 9373.

[FR] I.B.Frenkel and N.Yu.Reshetikhin: Quantum affine algebras and holonomic difference equations. Commun.Math.Phys.146 (1992) 1.

[Fr] J.Frohlich: Statistics of fields, the Yang-Baxter equation and the theory of knots and links. Cargese lectures, Plenum Press (1987).

[FrK] J.Frohlich and T. Kerler: Preprint ETH-92.

[FD] J.Fuchs and P. van Driel: Some symmetries of quantum dimensions.J.Math. Phys. 31 (1990) 1770 ; WZW fusion rules , quantum groups and the modular matrix S. Nucl.Phys. B346 (1990) 632.

[FGP] P.Furlan,A.G.Ganchev and V.B.Petkova: Quantum groups and fusion rule multiplicities. Nucl.Phys. B343 (1990) 205.

[GP] A.C.Ganchev and V.B.Petkova: $U_q(Sl(2))$ invariant operators and minimal theory fusion rules. Phys. Lett 233B (1989) 374.

[Ga] K.Gawedzki: Classical origin of quantum group symmetries in Wess-Zumino-Witten conformal field theory. Commun.Math.Phys. 139 (1991) 201.

[Ge] J.-L.Gervais: The quantum group structure of 2D gravity and minimal models. Commun.Math.Phys.130 (1990) 257.

[GN] J.-L.Gervais and A.Neveu: New quantum treatment of Liouville field theory.Nucl.Phys. B224 (1983) 329 ; Novel triangle relation and absence of tachyons in Liouville string field theory. Nucl.Phys. B238 (1984) 125.

[GRS] C.Gomez,M.Ruiz-Altaba and G.Sierra: New R-matrices associated with finite dimensional representations of $U_q(sl(2))$ at roots of unity. Phys.Lett. 265B (1991) 95.

[GS1] C.Gomez and G.Sierra: Quantum group meaning of the Coulomb gas.Phys.Lett. 240B (1990) 149.

[GS2] C.Gomez and G.Sierra: The quantum symmetry of rational conformal field theories. Nucl.Phys. B352 (1991) 791.

[GS3] C.Gomez and G.Sierra: Contour representation of quantum groups in RCFT's. Proc.Triestre 1990 Summer School. World Scientific (1991).

[GS4] C.Gomez and G.Sierra: Towers of algebras in rational conformal field theories. Inter.J.Mod.Phys. A6 (1991) 2045.

[GS5] C.Gomez and G.Sierra: A new solution to the star-triangle equation based on $U_q(sl(2))$ at roots of unity. Nucl.Phys. B373 (1992) 761; New integrable deformations of higher spin Heisenberg-Ising chains. Phys.Lett. 285B (1992) 126.

[GHJ] F.M.Goodman,P.de la Harpe and V.F.R.Jones: Coxeter-Dynkin diagrams and towers of algebras.MSRI Publications , Springer-Verlag,1989.

[HM] T.J.Hollowood and P.Mansfield: Quantum group structure of quantum Toda conformal field theories (I). Nucl. Phys. B330 (1990) 720.

[Ji] M.Jimbo:Quantum R matrix for generalized Toda system. Commun.Math.Phys. 102 (1986) 537; A q-difference analogue of $U(g)$ and the Yang-Baxter equation. Lett.Math.Phys. 10 (1985) 63; A q-analogue of $U(gl(N+1)$, Hecke algebra and the Yang-Baxter equation.Lett.Math.Phys. 11 (1986) 247.

[Jz] F.Jimenez: Quantum group structure of N=1 superconformal field theories. Phys.Lett. 252B (1990) 577; The quantized symmetry of N=2 superconformal field theories. Phys.Lett. 273B (1991) 399.

[KQS] A.Kato,Y.H.Quano and J.Shiraishi: Free boson representation of q-vertex operators and their correlation functions. Univ. Tokio preprint UT-618 (1992).

[KR] A.N.Kirillov and N.Yu.Reshetikhin: Representations of the algebra $U_q(sl(2))$, q-orthogonal polynomials and invariants of links. In infinite-dimensional Lie algebras and groups.V.G.Kac (ed.) World Scientific (1989).

[KZ] V.G.Knizhnik and A.B.Zamolodchikov: Current algebra and Wess-Zumino model in two dimensions. Nucl.Phys. B247 (1984) 83.

[Ko] T.Khono: Monodromy representations of braid groups and Yang-Baxter equations. Ann.Inst.Fourier 37 (1987) 139.

[KuR] P.P.Kulish and N.Yu.Reshetikhin: Quantum linear problem for the Sine-Gordon equation and higher representations.Zap.Nauch. Semi. LOMI 101 (1981); J.Sov.Math. 23 (1983) 2435.

[L] G.Lusztig: Modular representations and quantum groups. Contemp.Math. 82 (1989) 58 ;Quantum groups at roots of unity.Geom.Dedicata(1990).

[McSo] G.Mac and V.Schomerus: Quasi-Hopf quantum symmetry in quantum theory. Nucl.Phys. B370 (1992) 185; Quasi-quantum group symmetry and local braid relations in the conformal Ising model.Phys.Lett. 267B (1991) 207.

[M] A.Matsuo: Free field representation of quantum affine algebra $U_q(\widehat{sl(2)})$. Univ.Nagoya preprint (1992).

[MR] G.Moore and N.Yu.Reshetikhin: A comment on quantum group symmetry in conformal field theory. Nucl.Phys. B328 (1989) 557.

[MS] G.Moore and N.Seiberg: Polynomial equations for rational conformal field theories. Phys.Lett. 212B (1988) 451; Naturality in conformal field theory. Nucl.Phys. B313 (1989) 16; Classical and quantum conformal field theory. Commun.Math:Phys. 123 (1989) 177,

[Oc] A. Ocneanu. Lectures at College de France 1992.

[P] V.Pasquier: Etiology of IRF models. Commun.Math.Phys. 118 (1988) 335.

[PS] V.Pasquier and H.Saleur: Common structures between finite systems and conformal field theories though quantum groups. Nucl.Phys. B330 (1990) 523.

[RRR] C.Ramirez,H.Ruegg and M.Ruiz-Altaba:Explicit quantum symmetries of WZNW theories. Phys.Lett. 247B (1990) 499; The contour picture of quantum groups: conformal field theories. Nucl.Phys. B364 (1991) 195.

[Rh] K.-H.Rehren: Locality of conformal fields in two dimensions:Exchange algebra on the light-cone.Commun.Math. Phys.116 (1988) 675.

[R] N.Yu.Reshetikhin: Quantized universal enveloping algebras,the Yang-Baxter equation and invariants of links,I. LOMI preprint E-4-87, 1988; II. LOMI preprint E-17-87,1988.

[RS] N.Yu.Reshetikhin and F.Smirnov: Hidden quantum group symmetry and integrable perturbations of conformal field theories. Commun.Math.Phys. 131 (1990) 157.

[RT] N.Yu.Reshetikhin and V.G.Turaev: Ribbon graphs and their invariants derived from quantum groups. Commun.Math.Phys. 127 (1990) 1.

[RA] P.Roche and D.Arnaudon:Irreducible representations of the quantum analogue of $SU(2)$. Lett.Math.Phys. 17 (1989) 295.

[Ro] M.Rosso: Finite dimensional representations of the quantum analogue of the enveloping algebra of complex simple Lie algebras. Commun.Math.Phys.117 (1988) 581.

[RSa] L.Rozansky and H.Saleur: Quantum field theory for the multi-variable Alexander-Conway polynomial. NuclPhys. B376 (1992) 461.

[Sa] H.Saleur: Lattice models and conformal field theories. Phys.Rep. 184 (1989) 177.

[SV] V.V.Schechtman and A.N.Varchenko: Arrangement of hyperplanes and Lie algebra homology.Invent.Math. 106 (1991) 139; Quantum groups and homology of local systems. IAS preprint (1990).

[Sh] J.Shiraishi: Free boson representation of $U_q(\widehat{sl(2)})$. Univ. Tokio preprint UT-617 (1992).

[ST] F.Smirnov and L.Takhtajan: Towards a quantum Liouville theory with $c < 1$. Colorado preprint 1990.

[Sk] E.K.Sklyanin: Some algebraic structures connected with the Yang-Baxter equation. Func.Anal.Appl. 16 (1982) 27; Some algebraic structures connected with the Yang-Baxter equation. Representations of quantum algebras. Func.Anal.Appl. 17 (1983) 273.

[T] I.T.Todorov: Quantum groups as symmetries of chiral conformal algebras.Lecture Notes in Physics.Vol. 370 p.231,Springer-Verlag 1990.

[TK] A.Tsuchiya and Y.Kanie: Vertex operators in conformal field theory on P^1 and monodromy representations of braid group. Adv.Stud.Pure Math. 16 (1988) 297, Lett.Math.Phys. 13 (1987) 303.

[V] E.Verlinde: Fusion rules and modular transformations in 2D conformal field theories.Nucl.Phys.B300 (1988) 360.

[Wa] M.Wakimoto: Fock representations of the affine Lie algebra $A_1^{(1)}$. Commun.Math.Phys. 104 (1986) 605.

[Wz] H.Wenzl: Hecke algebras of type A_n and subfactors. Invent.Math. 92 (1988) 349.

[Wi] E.Witten: Gauge theories,vertex models and quantum groups. Nucl.Phys. B330 (1990) 285.

[Z] A.B.Zamolodchikov: Infinite additional symmetries in 2-dimensional conformal quantum field theory. Theor.Math.Phys.65 (1986) 1205.

[ZZ] A.B.Zamolodchikov and Al.Zamolodchikov: Factorized S-matrices in two dimensions and the exact solutions to certain relativistic quanttum field theory models.Ann.Phys. 120 (1979) 253.

q–DEFORMATION OF SEMISIMPLE AND NON–SEMISIMPLE LIE ALGEBRAS

HENRI RUEGG
Département de Physique Théorique
Université de Genève
CH-1211 GENEVE 4, Switzerland

ABSTRACT: We give an elementary introduction to the Drinfeld-Jimbo procedure of the quantum deformation $U_q(g)$ of semisimple Lie algebras g. The q–Serre relations are discussed in some detail, in the Chevalley and the Cartan–Weyl basis. We review the real forms of g and $U_q(g)$. The general procedure is illustrated by the examples $U_q(s\ell(2))$ and $U_q(so(5,\mathbf{C}))$. After some general considerations on non-semisimple Lie algebras, we discuss in detail the quantum deformation of the Poincaré algebra, involving the contraction of $U_q(so(3,2))$. Some physical consequences are mentioned.

1. Introduction

The aim of these lectures is to give an elementary introduction to some problems in "quantum groups" and "quantum algebras". These terms do not denote groups and Lie algebras, but particular deformations of Lie groups and Lie algebras, and belong to the more general category of Hopf algebras. We shall speak of "q–deformation" G_q of a Lie group G and <u>q–deformation $U_q(g)$</u> of a Lie algebra g. The emphasis will be on the latter.

We shall first describe the well established theory of q–deformation of <u>semisimple</u> Lie algebras, particularly $s\ell(2)$. Then we shall discuss real forms of q–algebras. Finally, examples of the less well known quantum deformation of non-semisimple L ie algebras will be given, especially the Poincaré algebra.

Quantum groups were first introduced in physics in two–dimensional integrable systems and soon applied to conformal field theories (see the lectures in these proceedings by L. Faddeev, C. Gomez, G. Sierra, and references [1, 2, 3, 4]).

Another application of q deformed algebras is to provide a new possibility of <u>symmetry breaking</u>. The first example is the Heisenberg model of a one–dimensional metal, or XXX quantum spin chain, solved by Bethe. It is invariant under $s\ell(2)$, whereas the XXZ chain is only invariant under the classical $U(1)$ but actually also under the quantum $U_q(s\ell(2))$ [5, 6, 7]. In these lectures we shall discuss the possible breaking of Lorentz invariance at high energy by the quantum deformation of the

L. A. Ibort and M. A. Rodríguez (eds.), Integrable Systems, Quantum Groups, and Quantum Field Therapy 45–82.

Poincaré algebra in 4 dimensions [8,9,10]. Although at present there is no experimental evidence for such violation, it is interesting to get a "figure of merit" of a theory by comparing it to another theory with similar power of prediction, the latter being obtained by continuous deformation of the former. In addition, the corresponding field theory could provide a new way of regularizing the quantum field theories of particle physics. Finally, it is likely that space-time is drastically changed at the Planck mass [11]. For other examples see the Firenze group [12].

As mentioned above the q-deformation theory of semisimple Lie groups G (Faddeev *et al.*[13], Woronowicz [14]) and semisimple Li e algebras g (Drinfeld [15], Jimbo [16]) is well established. They are special examples of Hopf algebras. We shall adopt the procedure given by the two last authors. However, there is no such general scheme for non-semisimple Lie algebras. Examples are known [17,18,19] and hopefully they will be helpful in finding their general features. In these lectures we shall discuss in detail the non-trivial example of the quantum deformation of the Poincaré algebra [8,9].

1.1. THREE METHODS TO DEFORM INHOMOGENEOUS LIE ALGEBRAS

Most of the above examples are semidirect sums of a semisimple and an abelian algebra, such as rotations and translations in 3 dimensions = $E(3)$ [18], the inhomogeneous Lorentz algebra = Poincaré algebra [8,9], etc. There exist several methods to quantum deform these algebras :

a) Contraction. One starts with a simple or semisimple Lie algebra, deforms it according to Drinfeld–Jimbo and rescales some of the generators, which in the limit correspond to an abelian invariant translation algebra. It is then shown that the deformed $E(3)$ [18] or Poincaré [8,9] are Hopf algebras.

b) Subalgebra. The simple algebra $so(4,2)$ contains the Weyl algebra, that is Poincaré plus dilatations. A deformation of $so(4,2)$ yields a q–Weyl Hopf algebra [20,21]. Unfortunately, the deformation thus obtained of the Poincaré algebra is not a Hopf algebra by itself, *i.e.* the subalgebra hierarchy is not preserved by the deformation.

c) Non-commutative space [22]. As an example consider the q–deformed Lorentz algebra acting on a four–dimensional Manin space. The differential operators in this q-geometry correspond to q–translations. In this way one gets q–Weyl as a Hopf algebra [23].

We shall discuss in detail only method (a), the contraction.

1.1.1. Classical contraction procedure. Let us sketch the contraction procedure for getting Poincaré in the classical case, that is for $q = 1$.

Start with the complex simple algebra $so(5, \mathbf{C})$. It has the three real forms $so(5)$, $so(4,1)$ and $so(3,2)$ corresponding respectively to the diagonal metrics $(++++ +), (-++++)$ and $(-+++-)$. The generators

$$M_{AB} = -M_{BA} \quad , \quad A, B = 0, \cdots, 4 \tag{1.1}$$

satisfy the commutation relations

$$[M_{AB}, M_{CD}] = g_{AD}M_{BC} + g_{BC}M_{AD} - g_{AC}M_{BD} - g_{BD}M_{AC}. \tag{1.2}$$

involving the metric tensor

$$g_{AB} = g_{BA} \tag{1.3}$$

with ± 1 on the diagonal as above. The de Sitter and anti–de Sitter algebras $so(4,1)$ and $so(3,2)$ contain a Lorentz subalgebra with generators

$$M_{\mu\nu} \quad , \quad \mu, \nu = 0, \cdots, 3 \; . \tag{1.4}$$

Putting

$$M_{\mu 4} = RP_{\mu} \quad , \tag{1.5}$$

and letting the "radius of the fifth dimension" R tend to infinity, $R \to \infty$, it follows that

$$\begin{aligned} [P_{\mu}, P_{\nu}] &= \frac{1}{R^2}[M_{\mu 4}, M_{\nu 4}] \to 0 \\ [M_{\mu\nu}, P_{\lambda}] &= g_{\mu\lambda}P_{\nu} - g_{\nu\lambda}P_{\mu} \; . \end{aligned} \tag{1.6}$$

which is the customary Poincaré algebra.

1.1.2. Quantum contraction. The actual practical program for obtaining the q–deformation of Poincaré can be summarized in the following six steps :

1) For $q = 1$, start with the simple Lie algebra $so(5, \mathbf{C})$ and its 10 generators M_{AB}. It is convenient to introduce right away some nomenclature. We shall see below the exact relationship between the usual M_{AB} and the generators h_i, e_i which we shall refer to as follows:

$$\begin{array}{rcc} h_1, h_2 & : & \text{Cartan basis} \\ h_1, h_2, e_{\pm 1}, e_{\pm 2} & : & \text{Chevalley basis} \\ h_1, h_2, e_{\pm 1}, e_{\pm 2}, e_{\pm 3}, e_{\pm 4} & : & \text{Cartan} - \text{Weyl basis} \end{array} \tag{1.7}$$

2) To introduce the quantum deformation, q–deform $so(5, \mathbf{C})$ à la Drinfeld–Jimbo in the Chevalley basis to get $U_q(so(5, \mathbf{C}))$. Introduce the more cumbersome but necessary Cartan–Weyl basis.

3) Find the real form of $U_q(so(5, \mathbf{C}))$ corresponding to $U_q(so(3,2))$ or $U_q(so(4,1))$. (The "real form" is a fancy name for some specific linear combinations of the Cartan–Weyl generators with special properties under complex conjugation.) Calculate the (anti) hermitean "physical" generators M_{AB}, which provide the unitary representation.

4) Compute the q–deformation of $so(3,2)$ in the "physical" basis.

5) Contract to get the quantum Poincaré algebra.

6) Check that quantum Poincaré is indeed a Hopf algebra.

1.2. PLAN OF THESE LECTURES

The above introduction motivates the course. My main goal is to describe the above six steps. In order to do so, I will first introduce some notation and basic formalism about quantum groups. Then I will develop the construction program applied to the de Sitter algebra and obtain the quantum Poincaré Hopf algebra. The reader is invited to check the assertions pencil in hand to acquire the necessary technical skill in algebraic manipulations. At the end, our reward will be to explore some physical consequences of the quantum-deformed Poincaré algebra. Accordingly, the lectures are organized as follows:

- Drinfeld–Jimbo deformation in the Chevalley and Cartan–Weyl basis (chapter 2).
- Real forms of (q–deformed) semisimple Lie algebras (chapter 3).
- quantum deformation of the Poincaré algebra using contraction (chapter 4).
- Physical elucubrations (chapter 5).

2. q–Deformation of semisimple Lie algebras

2.1. THE PARADIGMATIC QUANTUM GROUP $U_q(s\ell(2))$

We start with the q–deformation of $s\ell(2)$, which has already many of the essential features of the general case.

The simple Lie algebra $s\ell(2)$, the symmetry algebra of spin $\frac{1}{2}$ systems, has three generators with commutation relations

$$\begin{aligned} [h, e_{\pm}] &= \pm 2 e_{\pm} \\ [e_+, e_-] &= h \end{aligned} \tag{2.1}$$

We shall see further ahead (chapter 3) that by taking linear combinations of e_+ and e_- one gets the two real forms $su(2)$ and $su(1,1)$, but for the time being let us remain general and talk about $s\ell(2)$.

The q–deformation of $s\ell(2)$ was proposed long ago [24]; only the third commutation relation is changed :

$$\begin{aligned} [h, e_{\pm}] &= \pm 2 e_{\pm} \\ [e_+, e_-] &= [h]_q \end{aligned} \tag{2.2}$$

where the square bracket (with a subindex q, often omitted) stands for the following q–deformation of any quantity (number or operator)

$$[x]_q \equiv \frac{q^x - q^{-x}}{q - q^{-1}} \tag{2.3}$$

We understand the formal parameter q as a complex number. When $q \to 1$, $[x]_q \to x$: the limit $q \to 1$ will often be referred to as classical, in the sense that there is no deformation whatsoever.

Remarkably, in equation (2.2) all positive and negative powers of h appear. Hence (2.2) does not define a Lie algebra in the purest sense, but rather the q–deformation of the universal enveloping algebra of $s\ell(2)$. This q-deformation of $U(s\ell(2))$ is noted, not suprisingly, $U_q(s\ell(2))$. We shall see that it is necessary to include also powers of e_+ and e_-, so that $U_q(s\ell(2))$ contains all the formal power series in h, e_+ and e_- modulo (2.2).

It is useful to introduce

$$k = q^{h/2} \tag{2.4}$$

and replace the first equation (2.2) by the equivalent "commutation" relation

$$k e_\pm k^{-1} = q^{\pm 1} e_\pm \,. \tag{2.5}$$

For most purposes it is enough to consider only polynomials in k, k^{-1}, e_+, e_-.

The fundamental representation of $s\ell(2)$, corresponding to spin $\frac{1}{2}$, is identical to the fundamental representation of $U_q(s\ell(2))$. One obtains higher spin states (where q–dependent terms will appear explicitly) through tensoring, so we must specify how to introduce the tensor product of representations. For $s\ell(2)$, the generators act trivially on one factor (like a derivative) :

$$\Delta(h) = h \otimes I + I \otimes h \tag{2.6a}$$
$$\Delta(e_\pm) = e_\pm \otimes I + I \otimes e_\pm \tag{2.6b}$$

This operation Δ, which tells us how to act with the generators of $s\ell(2)$ on the tensor product of two representations, is called the co-product or co-multiplication, and I is the identity. The reader is familiar with this co-product from quantum mechanics, where the action of an angular momentum operator on a many–particle system is the sum of its actions on the one–particle states.

For k defined by (2.4), the operation (2.6a) implies

$$\Delta(k) = k \otimes k \tag{2.6c}$$

The co-product Δ must be a homomorphism of Lie algebras, meaning that it must preserve the commutation relations (2.1). Since the commutation relations of $U_q(s\ell(2))$ are different from those of $s\ell(2)$, we should expect that the co-product for $U_q(s\ell(2))$ is also different from that for $s\ell(2)$. This is indeed the case. For $U_q(s\ell(2))$, (2.6a), or equivalently (2.6c), is retained, but (2.6b) is replaced by

$$\Delta(e_\pm) = e_\pm \otimes k + k^{-1} \otimes e_\pm . \tag{2.7}$$

This co-product respects the commutation relations (2.2), i.e. it is a homomorphism of $U_q(s\ell(2))$:

$$\Delta(ab) = \Delta(a)\Delta(b) \qquad \forall a, b \in U_q(s\ell(2)). \tag{2.8}$$

Perhaps the most noteworthy feature of the co-product Δ for a q–deformed algebra $U_q(s\ell(2))$ is that it is not symmetric, i.e. that $\Delta(e_\pm)$ acts non-symmetrically on the two factors. Formally, the co-product is a map

$$\Delta : U_q(s\ell(2)) \to U_q(s\ell(2)) \otimes U_q(s\ell(2)). \tag{2.9}$$

Motivated by the concept of Hopf algebra (see next paragraph) we define the <u>co-unit</u> map ε :

$$\begin{aligned} \varepsilon :\ \ & U_q \to \mathbf{C} \\ & \varepsilon(ab) = \varepsilon(a)\varepsilon(b) \qquad \forall a, b \in U_q(s\ell(2)) \\ & \varepsilon(I) = 1 \\ & \varepsilon(k) = 1 \\ & \varepsilon(h) = 0 \\ & \varepsilon(e_\pm) = 0 \end{aligned} \tag{2.10}$$

and the <u>antipode</u> map S (which at the group level G_q is the q–deformation of the inverse) :

$$\begin{aligned} S :\ \ & U_q(s\ell(2)) \to U_q(s\ell(2)) \\ & S(ab) = S(b)S(a) \quad \forall a, b \in U_q(s\ell(2)) \\ & S(k) = k^{-1} \\ & S(h) = -h \\ & S(e_\pm) = -q^{\pm 1} e_\pm \ . \end{aligned} \tag{2.11}$$

Because the co-product Δ is not symmetric, one can define a second co-product Δ' as its transpose

$$\begin{aligned} \Delta' &= P \circ \Delta \\ P(a \otimes b) &= b \otimes a \quad a, b \in U_q \end{aligned} \tag{2.12}$$

Explicitly, Δ' of the generators of $U_q(s\ell(2))$ reads as follows:

$$\begin{aligned} \Delta'(k) &= k \otimes k \\ \Delta'(e_\pm) &= e_\pm \otimes k^{-1} + k \otimes e_\pm \end{aligned} \tag{2.13}$$

The two co-products Δ and Δ' are related through a "similarity transformation" by the so-called <u>R-matrix</u>, $R \in U_q(s\ell(2)) \otimes U_q(s\ell(2))$

$$R \Delta R^{-1} = \Delta' \tag{2.14}$$

It is often useful to use the "braid–group $\hat{R}$–matrix" which differs from the above by a permutation:

$$\begin{aligned} \hat{R} &= P \circ R \\ \hat{R}\Delta\hat{R}^{-1} &= \Delta\,. \end{aligned} \tag{2.15}$$

An explicit expression for R can be given [25] in terms of $e_\pm$ and h (but not k) by

$$R = q^{\frac{1}{2}h\otimes h} \sum_{n\geq 0} \frac{(1-q^{-2})^n}{[n]_q!} q^{\frac{(n-n^2)}{2}} q^{\frac{n}{2}h}(e_+)^n \otimes q^{-\frac{n}{2}h}(e_-)^n \tag{2.16}$$

This universal R–matrix can be shown [25] to satisfy the famous <u>Yang–Baxter</u> equation, a key property of two–dimensional integrable models [1]

$$R_{12}R_{13}R_{23} = R_{23}R_{13}R_{12} \tag{2.17}$$

or equivalently

$$\hat{R}_{12}\hat{R}_{23}\hat{R}_{12} = \hat{R}_{23}\hat{R}_{12}\hat{R}_{23} \tag{2.18}$$

where, if the general form of R is

$$R = \sum_i x_i \otimes y^i \quad , \quad x_i, y^i \in U_q(s\ell(2)) \tag{2.19}$$

then the subindexed notation means

$$\begin{aligned} R_{12} &= \sum_i x_i \otimes y^i \otimes I \\ R_{13} &= \sum_i x_i \otimes I \otimes y^i \\ R_{23} &= \sum_i I \otimes x_i \otimes y^i \end{aligned} \tag{2.20}$$

Note that the indices i go up or down without any implication about a dual.

2.2. HOPF ALGEBRA

We now have all the ingredients necessary for defining the general concept of a Hopf algebra, of which $U_q(s\ell(2))$ is a particular example. We shall do it in several steps [26, 27].

Let A be an associative algebra with unity I over $\mathbf{C}$. This means that the product

$$\begin{aligned} m:\quad & U_q(s\ell(2)) \otimes U_q(s\ell(2)) \to U_q(s\ell(2)) \\ & (x,y) \mapsto m(x,y) = xy \\ & (x,I) \mapsto x \\ & (I,y) \mapsto y \end{aligned} \tag{2.21}$$

is associative,

$$(xy)z = x(yz) \tag{2.22}$$

A bi-algebra is a set of two algebras sharing the same base set: the first algebra proper (A, m, i) with product m and unit i, the second algebra or co-algebra (A, Δ, ε) with co-product Δ and co-unit ε, defined by :

$$\begin{aligned} m :& A \otimes A \to A \\ & x \otimes y \mapsto xy \\ i :& \mathbf{C} \to A \\ & \lambda \mapsto \lambda I \end{aligned} \tag{2.23}$$

and

$$\begin{aligned} \Delta :& A \to A \otimes A \\ & x \mapsto \sum_i x_i \otimes x^i \\ \varepsilon :& A \to \mathbf{C} \\ & I \mapsto 1 \end{aligned} \tag{2.24}$$

Again, the short–hand above does not mean that x^i belongs to the dual of A.

Just like the product m is taken to be associative, the co-product Δ is assumed to be co-associative, namely

$$(\Delta \otimes I)\Delta = (I \otimes \Delta)\Delta \tag{2.25}$$

or more explicitly

$$\sum_i \Delta(x_i) \otimes x^i = \sum_i x_i \otimes \Delta(x^i) \tag{2.26}$$

The co-product Δ, and by extension the co-algebra (A, Δ, ε), is called co-commutative if $\Delta = \Delta'$.

The two algebras are related if Δ and ε are homomorphisms of the first algebra, that is of (A, m, i):

$$\begin{aligned} &\Delta(xy) = \Delta(x)\Delta(y) \\ &\Delta(I) = I \otimes I \\ &\varepsilon(xy) = \varepsilon(x)\varepsilon(y) \\ &(I \otimes \varepsilon) \circ \Delta = I \\ &(\varepsilon \otimes I) \circ \Delta = I \ . \end{aligned} \tag{2.27}$$

In this case the algebra and co-algebra assemble into a bi-algebra $(A, m, i, \Delta, \varepsilon)$.

A Hopf algebra is a bi-algebra with an antipode S, *i.e.* it is given by the set $(A, m, i, \Delta, \varepsilon, S)$ with the bi-algebra conditions above and furthermore :

$$m \circ (S \otimes I) \circ \Delta(x) = m \circ (I \otimes S) \circ \Delta(x) = i \circ \varepsilon(x) \tag{2.28}$$

which relate all elements of the set. The antipode can be thought of as a co-inverse.

We can verify (2.28) on the example of $U_q(s\ell(2))$. Using (2.7) for Δ and (2.11) for S we first get

$$\begin{aligned} m \circ (S \otimes I) \circ \Delta(e_\pm) &= m \circ (S \otimes I) \circ (e_\pm \otimes k + k^{-1} \otimes e_\pm) \\ &= m \circ (S(e_\pm) \otimes k + S(k^{-1}) \otimes e_\pm) \\ &= m \circ (-q^{\pm 1} e_\pm \otimes k + k \otimes e_\pm) \\ &= -q^{\pm 1} e_\pm k + k e_\pm \; . \end{aligned} \tag{2.29}$$

Similarly, we obtain

$$m \circ (I \otimes S) \circ \Delta(e_\pm) = e_\pm k^{-1} - k^{-1} q^{\pm 1} e_\pm \tag{2.30}$$

Using now Eq. (2.10) and (2.23) for ε and i, we get

$$i \circ \varepsilon(e_\pm) = 0 \; . \tag{2.31}$$

With the help of the commutation relations (2.5) we find the equality of (2.29), (2.30) and (2.31).

Notice that we could have used (2.29) and (2.31) to calculate S from Δ, m, i and ε, and use (2.30) to check the consistency of the Hopf algebra structure.

Finally, one defines the concept of quasi-triangular Hopf algebra to be a Hopf algebra $(A, m, i, \Delta, \varepsilon, S, R)$ with an R–matrix satisfying (2.14), *i.e.*

$$R\Delta = \Delta' R \tag{2.32}$$

and [*cf.* (2.20)]

$$\begin{aligned} (\Delta \otimes I)R &= R_{13} R_{23} \\ (I \otimes \Delta)R &= R_{13} R_{12} \\ (S \otimes I)R &= R^{-1} \; . \end{aligned} \tag{2.33}$$

One can verify [25] that $U_q(s\ell(2))$ is indeed a quasi-triangular Hopf algebra.

Crucially, the Yang–Baxter equation (2.17) for R follows from the requirements of quasi-triangularity, namely equations (2.32) and (2.33).

Quite often, a quasi-triangular Hopf algebra is referred to simply as a quantum group.

2.3. $U_q(g)$ IN THE CHEVALLEY AND CARTAN–WEYL BASIS

Let us sketch the general procedure due to Drinfeld [15] and Jimbo [16] for q–deforming a semisimple Lie algebra g. It also applies to Kac-Moody algebras [28] and, with some minor and obvious modifications, to superalgebras [28, 29].

2.3.1. Lie algebra g. In the Chevalley basis, one uses only the Cartan subalgebra $h_1, \cdots, h_\ell$, $\ell =$ rank of g, and the raising and lowering operators $e_{\pm 1}, \cdots, e_{\pm \ell}$ corresponding to the simple roots $\alpha_1, \cdots, \alpha_\ell$. Nevertheless, the commutation relations in the set $(h_i, e_{\pm i})$ must be supplemented by the so-called <u>Serre relations</u> which ensure that g is semisimple [30]: this is the price to pay for using a small basis.

The main new ingredient is the <u>Cartan matrix</u> a_{ij} which gives the scalar products of the simple roots α_i. For the simply–laced algebras A_ℓ, D_ℓ and E_ℓ, the Cartan matrix a_{ij} is symmetric. This is not the case for the non-simply laced algebras B_ℓ, C_ℓ, F_4 and G_2. In the latter case it is convenient to introduce the symmetrized Cartan matrix a^S_{ij}.

The commutation relations for any semisimple Lie algebra or Kac–Moody algebra with symmetrizable Cartan matrix [31] are, in the Chevalley basis,

$$\begin{aligned} [h_i, h_j] &= 0 \\ [h_i, e_{\pm j}] &= \pm a^S_{ij} e_{\pm j} \end{aligned} \tag{2.34}$$

$$[e_{+i}, e_{-j}] = \delta_{ij} h_i \tag{2.35}$$

where

$$i, j = 1, \cdots, \ell = \text{ rank of } g \qquad e_{\pm i} \equiv e_{\pm \alpha_i} \tag{2.36}$$

<u>Remark</u> : One can write the commutation relations in terms of the non-symmetrized Cartan by a simple rescaling of the generators.

Note that equations (2.34) and (2.35) define a semisimple Lie algebra iff the following <u>Serre relations</u> are satisfied :

$$(\operatorname{ad} e_{\pm i})^{1-a_{ij}} e_{\pm j} = 0 \quad i \neq j \tag{2.37}$$

where

$$(\operatorname{ad} x) y = xy - yx = [x, y] \tag{2.38}$$

and a_{ij} is the <u>non</u>-symmetrized Cartan matrix.

2.3.2. An example: C_2. We shall illustrate the general procedure on the example of the simple algebra $B_2 = so(5) = sp(4) = C_2$, which will be used for the quantum deformation of the Poincaré algebra via contraction (see the Introduction and chapter 4).

The Lie algebra C_2 has two simple roots α_1, α_2 with lengths squared one and two:

$$\alpha_1^2 = 1 \quad , \quad \alpha_2^2 = 2 \; . \tag{2.39}$$

The Cartan matrix is

$$a_{ij} = \frac{2(\alpha_i, \alpha_j)}{\alpha_i^2} = \begin{pmatrix} 2 & -2 \\ -1 & 2 \end{pmatrix} \tag{2.40}$$

The notation (α_i, α_j) stands, as usual, for the scalar product of the simple roots. The symmetrized Cartan matrix is obtained from a_{ij} by left multiplication with the diagonal matrix D:

$$a_{ij}^S = D a_{ij} \quad ; \quad D = \begin{pmatrix} \frac{1}{2} & 0 \\ 0 & 1 \end{pmatrix} \tag{2.41}$$

$$a_{ij}^S = (\alpha_i, \alpha_j) \equiv \alpha_{ij} = \begin{pmatrix} 1 & -1 \\ -1 & 2 \end{pmatrix} \tag{2.42}$$

The Cartan subalgebra is h_1, h_2 and the "simple" generators are $e_{\pm 1}, e_{\pm 2}$.

The Serre relations for C_2 are the following:

$$\begin{aligned} (\operatorname{ad} e_1)^{1-a_{12}} e_2 &= (\operatorname{ad} e_1)^3 e_2 = [e_1, [e_1 [e_1, e_2]]] \\ &= e_1^3 e_2 - 3 e_1^2 e_2 e_1 + 3 e_1 e_2 e_1^2 - e_2 e_1^3 = 0 \end{aligned} \tag{2.43}$$

$$\begin{aligned} (\operatorname{ad} e_2)^{1-a_{21}} e_1 &= (\operatorname{ad} e_2)^2 e_1 = [e_2 [e_2, e_1]] \\ &= e_2^2 e_1 - 2 e_2 e_1 e_2 + e_1 e_2^2 = 0 \; . \end{aligned} \tag{2.44}$$

The Serre relations have a simple interpretation in the <u>Cartan–Weyl basis</u> where one considers all the generators corresponding to the root diagram, and not only those associated with simple roots. For C_2 we have, in addition to h_i and $e_{\pm i}$ $(i = 1, 2)$, the four generators $e_{\pm 3}$ and $e_{\pm 4}$ corresponding to the remaining non–simple roots $\pm\alpha_3, \pm\alpha_4$. From the root relations

$$\begin{aligned} \alpha_3 &= \alpha_1 + \alpha_2 \\ \alpha_4 &= \alpha_1 + \alpha_3 \end{aligned} \tag{2.45}$$

one defines the associated generators

$$\begin{aligned} e_3 &= [e_1, e_2] \\ e_4 &= [e_1, e_3] \; . \end{aligned} \tag{2.46}$$

The important observation is that $\alpha_1 + \alpha_4$ an d $\alpha_2 + \alpha_3$ are <u>not</u> roots, hence

$$\begin{aligned} [e_1, e_4] &= 0 \\ [e_3, e_2] &= 0 \end{aligned} \tag{2.47}$$

which are precisely the Serre relations (2.43) and (2.44), as the reader will check with the help of (2.46).

2.3.3. The quantum group $U_q(g)$. We now come to the q-deformation of g à la Drinfeld–Jimbo. The scheme is a simple generalization of $U_q(s\ell(2))$.

The commutation relations (2.34) are unchanged :

$$\begin{aligned} &[h_i, h_j] = 0 \\ &[h_i, e_{\pm j}] = \pm a^S_{ij} e_{\pm j} \equiv \pm \alpha_{ij} e_{\pm j} \ . \end{aligned} \tag{2.48}$$

Defining $k_i = q^{\frac{1}{2}h_i}$ these relation can be rewritten as

$$\begin{aligned} &k_i k_j = k_j k_i \\ &k_i e_{\pm j} k_i^{-1} = q^{\pm \frac{1}{2}\alpha_{ij}} e_{\pm j} \end{aligned} \tag{2.49}$$

On the other hand, the commutators (2.35) between raising and lowering operators do change, and they become

$$[e_i, e_{-j}] = \delta_{ij} [h_i]_q \ . \tag{2.50}$$

The technical problem is now to q–deform the Serre relations. Recall that in the limit $q = 1$, that is for classical g, we used the adjoint representation in equations (2.37) and (2.38). So we should try to q–deform the adjoint operator ad. The main property we want to maintain is associativity. The following definitions, rather involved, do the trick [17, 32, 33]:

$$\begin{aligned} &\text{ad}^+(x) \stackrel{\text{def}}{=} (m_L \otimes m_R) \circ (I \otimes S) \circ \Delta(x) \\ &\text{ad}^-(x) \stackrel{\text{def}}{=} (m_L \otimes m_R) \circ (S \otimes I) \circ \Delta(x) \ . \end{aligned} \tag{2.51}$$

Here, the left and right multiplications m_L resp. m_R are given by

$$(m_L \otimes m_R) \circ (x \otimes y)(z) \stackrel{\text{def}}{=} xzy \ . \tag{2.52}$$

In general, the co-product Δ is a sum of terms, as in equations (2.6) and (2.7) for the simplest case of $U_q(s\ell(2))$, which can be formally written as

$$\Delta(x) = \sum_\mu x_\mu \otimes x^\mu \equiv x_\mu \otimes x^\mu \ . \tag{2.53}$$

Using (2.52) and (2.53), $\text{ad}^+ x$ applied on y gives

$$((m_L \otimes m_R)(I \otimes S)\Delta(x))(y) = (m_L \otimes m_R)(x_\mu \otimes S(x^\mu))(y) = x_\mu y S(x^\mu) \tag{2.54}$$

Hence

$$(\mathrm{ad}^+x)(y) = x_\mu y S(x^\mu)$$
$$(\mathrm{ad}^-x)(y) = S(x_\mu) y x^\mu \; . \tag{2.55}$$

Associativity of $\mathrm{ad}^\pm$ is defined by

$$(\mathrm{ad}^+xy)(z) = (\mathrm{ad}^+x)((\mathrm{ad}^+y)(z))$$
$$(\mathrm{ad}^-xy)(z) = (\mathrm{ad}^-y)((\mathrm{ad}^-x)(z)) \tag{2.56}$$

from which follows the representation property :

$$(\mathrm{ad}^\pm[x,y])(z) = \pm\{(\mathrm{ad}^\pm x)((\mathrm{ad}^\pm y)(z)) - (\mathrm{ad}^\pm y)((\mathrm{ad}^\pm x)(z))\} \; . \tag{2.57}$$

The proof of the associativity of the adjoint goes as follows :

$$\begin{aligned}
(\mathrm{ad}^+xy)(z) &= (m_L \otimes m_R) \circ (I \otimes S) \circ \Delta(xy)(z) \\
\Delta(xy) &= \Delta(x) \circ \Delta(y) = (x_\mu \otimes x^\mu) \circ (y_\nu \otimes y^\nu) = x_\mu y_\nu \otimes x^\mu y^\nu \\
(I \otimes S)\Delta(xy) &= x_\mu y_\nu \otimes S(x^\mu y^\nu) = x_\mu y_\nu \otimes S(y^\nu) S(x^\mu) \\
(\mathrm{ad}^+xy)(z) &= x_\mu y_\nu z S(y^\nu) S(x^\mu) = x_\mu (\mathrm{ad}^+y)(z) S(x^\mu) = \\
&= (\mathrm{ad}^+x)((\mathrm{ad}^+y)(z)) \; .
\end{aligned} \tag{2.58}$$

Clearly, in order to compute the adjoint action we need the explicit forms of the co-product and the antipode. For g semisimple, in the Chevalley basis, the co-product Δ of $U_q(g)$ is very similar to the Δ of $U_q(s\ell(2))$:

$$\begin{aligned}
\Delta(h_i) &= h_i \otimes I + I \otimes h_i \\
\Delta(k_i) &= k_i \otimes k_i \\
\Delta(e_{\pm i}) &= e_{\pm i} \otimes k_i + k_i^{-1} \otimes e_{\pm i} \; .
\end{aligned} \tag{2.59}$$

The co-unit is again

$$\begin{aligned}
\varepsilon(h_i) &= \varepsilon(e_{\pm i}) = 0 \\
\varepsilon(k_i) &= \varepsilon(I) = 1 \; .
\end{aligned} \tag{2.60}$$

The relation $m \circ (S \otimes I) \circ \Delta(x) = m \circ (I \otimes S) \circ \Delta(x) = i \circ \varepsilon(x)$ gives for the antipode S :

$$\begin{aligned}
S(h_i) &= -h_i \\
S(k_i) &= k_i^{-1} \\
S(e_{\pm i}) &= -k_i e_{\pm i} k_i^{-1} = -q^{\pm\frac{1}{2}a_{ii}} e_{\pm i} \; .
\end{aligned} \tag{2.61}$$

We now have all the ingredients to calculate $\mathrm{ad}^{\pm}$. Using (2.55), (2.59) and (2.61) we get

$$\begin{aligned}(\mathrm{ad}^{\pm}h_i)(y) &= [h_i, y]\\ (\mathrm{ad}^{+}e_{\pm i})(y) &= (e_{\pm i}y - k_i^{-1}yk_ie_{\pm i})k_i^{-1}\\ (\mathrm{ad}^{-}e_{\pm i})(y) &= k_i(ye_{\pm i} - e_{\pm i}k_i^{-1}yk_i)\ .\end{aligned} \tag{2.62}$$

Let us specialize (2.62) to $y = e_{\pm j}$:

$$\begin{aligned}(\mathrm{ad}^{+}e_{\pm i})(e_{\pm j}) &= (e_{\pm i}e_{\pm j} - q^{\mp\frac{1}{2}\alpha_{ij}}e_{\pm j}e_{\pm i})k_i^{-1}\\ (\mathrm{ad}^{+}e_{\pm i})(e_{\mp j}) &= (e_{\pm i}e_{\mp j} - q^{\pm\frac{1}{2}\alpha_{ij}}e_{\mp j}e_{\pm i})k_i^{-1}\\ (\mathrm{ad}^{-}e_{\pm i})(e_{\pm j}) &= k_i(e_{\pm j}e_{\pm i} - q^{\mp\frac{1}{2}\alpha_{ij}}e_{\pm i}e_{\pm j})\ .\end{aligned} \tag{2.63}$$

In order to get the q–Serre relations, we could try to replace in Eq. (2.37) ad by $\mathrm{ad}^{\pm}$. It turns out that the factors k_i and k_i^{-1} in (2.62) complicate matters. We can eliminate them by a change of variables which leaves the algebra untouched but modifies the co-product and the antipode :

$$\begin{aligned}E_i &\stackrel{\mathrm{def}}{=} e_ik_i^{-1}\\ E_{-i} &\stackrel{\mathrm{def}}{=} k_ie_{-i}\\ H_i &\stackrel{\mathrm{def}}{=} h_i\ .\end{aligned} \tag{2.64}$$

Hence, from (2.49) and (2.50) we get

$$\begin{aligned}k_iE_{\pm j}k_i^{-1} &= q^{\pm\frac{1}{2}\alpha_{ij}}E_{\pm j}\\ [E_i, E_{-j}] &= \delta_{ij}[H_i]_q\end{aligned} \tag{2.65}$$

which have the same structure as the commutation relations (2.49) and (2.50) for $e_{\pm i}$. The generators $E_{\pm i}$ are the natural ones in the contour representation of quantum groups [3, 4].

From (2.59) and (2.61) we obtain

$$\begin{aligned}\Delta(E_i) &= E_i \otimes I + k_i^{-2} \otimes E_i\\ \Delta(E_{-i}) &= E_{-i} \otimes k_i^2 + I \otimes E_{-i}\\ S(E_i) &= -k_i^2E_i\\ S(E_{-i}) &= -E_{-i}k_i^{-2}\ .\end{aligned} \tag{2.66}$$

From (2.53), (2.55) and (2.66) we calculate

$$\begin{aligned}(\mathrm{ad}^{+}E_i)(E_{\pm j}) &= E_iE_{\pm j} - q^{\mp\alpha_{ij}}E_{\pm j}E_i \equiv [E_i, E_{\pm j}]_{q'}\\ (\mathrm{ad}^{-}E_{-i})(E_{\pm j}) &= E_{\pm j}E_{-i} - q^{\mp\alpha_{ij}}E_{-i}E_{\pm j} \equiv [E_{\pm j}, E_{-i}]_{q'}\end{aligned} \tag{2.67}$$

where $q' = q^{\pm 1}$. The right–hand side is called a q-commutator.

Thanks to the above change of basis, one can write the q–Serre relations in the simple form

$$(\mathrm{ad}^+ E_{\pm i})^{1-a_{ij}}(E_{\pm j}) = 0 \quad i \neq j \tag{2.68}$$

For conciseness, from now on we display the explicit expressions only for ad^+; it is left as an exerciser for the reader to carry on with ad^-. For actual computations, the following iterative formula is useful:

$$(\mathrm{ad}^+ E_i)^{P+1}(E_j) = E_i(\mathrm{ad}^+(E_i))^P(E_j) - q^{-P\alpha_{ii}-\alpha_{ij}}(\mathrm{ad}^+(E_i))^P(E_j)E_i \tag{2.69}$$

so that, for instance,

$$\begin{aligned}(\mathrm{ad}^+ E_i)^2(E_j) &= [E_i[E_i, E_j]_q]_q = \\ &= E_i[E_i, E_j]_q - q^{-\alpha_{ii}}q^{-\alpha_{ij}}[E_i, E_j]_q E_i \ .\end{aligned} \tag{2.70}$$

To establish these formulae, we use (2.55),(2.66) and (2.67) to find

$$(\mathrm{ad}^+ E_i)(E_j E_k) = E_i(E_j E_k) - q^{-\alpha_{ij}}q^{-\alpha_{ik}}(E_j E_k)E_i \ . \tag{2.71}$$

Notice that this can also be written :

$$(\mathrm{ad}^+ E_i)(E_j E_k) = (\mathrm{ad}^+ E_i)(E_j)E_k + q^{-\alpha_{ij}}E_j(\mathrm{ad}^+ E_i)(E_k) \ . \tag{2.72}$$

Hence, ad^+ can be considered as a q–derivation obeying a q–Leibniz rule [34, 35].

For the simple Lie algebra C_2 we have seen [recall equations (2.40) and (2.42)]

$$a_{ij} = \begin{pmatrix} 2 & -2 \\ -1 & 2 \end{pmatrix} \qquad \alpha_{ij} = a_{ij}^S = \begin{pmatrix} 1 & -1 \\ -1 & 2 \end{pmatrix} . \tag{2.73}$$

Hence, the q–Serre relations for $U_q(C_2)$ are, using (2.67) to (2.70) :

$$\begin{aligned}&E_2^2 E_1 - (q + q^{-1})E_2 E_1 E_2 + E_1 E_1^2 = 0 \\ &E_1^3 E_2 - (q + 1 + q^{-1})(E_1^2 E_2 E_1 - E_1 E_2 E_1^2) - E_2 E_1^3 = 0\end{aligned} \tag{2.74}$$

which reduce to (2.43) and (2.44) for $q = 1$.

Notice that the q–Serre relations are invariant under $q \leftrightarrow q^{-1}$. The same is true for the whole algebra (see Eq. (2.65) and also (2.48) to (2.50)), but not for the co-algebra.

2.3.4. Cartan–Weyl basis. As was the case for $q = 1$, the Serre relations have a much simpler interpretation in the Cartan-Weyl basis. This is defined in a similar way for $q = 1$ and $q \neq 1$, replacing the operator ad by $\mathrm{ad}^{\pm}$. But since q enters in a non-symmetric way into the definition of $\mathrm{ad}^{\pm}$, it is necessary to order the positive (negative) roots in a specific way [36, 28]. Namely :

- Definition 1 : The system Δ_+ of positive roots is in normal-order if each non-simple root $\gamma = \alpha + \beta \in \Delta_+$, where $\alpha \neq \lambda\beta$, $\alpha, \beta \in \Delta_+$, is written between α and β.

The q-analog of the Cartan-Weyl basis is constructed using the following inductive algorithm [37].

- Definition 2 : Fix some normal ordering in Δ_+. Let $\alpha, \beta, \gamma \in \Delta_+$ be pairwise non-collinear roots, such that $\gamma = \alpha + \beta$. Suppose, moreover, that between α and β (in the normal ordering at hand) there are no other roots α' and β' such that $\alpha' + \beta' = \gamma$.

  Then, if $E_{\pm\alpha}$ and $E_{\pm\beta}$ have already been constructed, we set (cf. (2.82))

$$E_\gamma = [E_\alpha, E_\beta]_q \quad , \quad E_{-\gamma} = [E_{-\beta}, E_{-\alpha}]_{q^{-1}} \tag{2.75}$$

and we get the commutation relations

$$\begin{aligned} [h_i, E_\gamma] &= (\alpha_i, \gamma) E_\gamma \\ [E_\gamma, E_{-\gamma}] &= a_\gamma [H_\gamma]_q \\ H_\gamma &= H_\alpha + H_\beta \end{aligned} \tag{2.76}$$

where a_γ is a function of q. We say that $\alpha < \beta$ if α is located to the left of β in the normal-ordered Δ_+.

Now the <u>q-Serre relations</u> are equivalent to the statement that, if $\alpha + \beta$ is not a root, the q-commutator (2.82) is zero :

$$[E_\alpha, E_\beta]_q = 0 \quad , \quad \alpha + \beta \notin \Delta_+ \quad , \quad \alpha < \beta \, . \tag{2.77}$$

To complete the Cartan-Weyl basis in the co-algebra sector, one uses the homomorphism and antihomomorphism properties applied to (2.75) :

$$\begin{aligned} \Delta(xy) &= \Delta(x)\Delta(y) \\ \varepsilon(xy) &= \varepsilon(x)\varepsilon(y) \\ S(xy) &= S(y)S(x) \, . \end{aligned} \tag{2.78}$$

The diligent reader will check that

$$m \circ (S \otimes I) \circ \Delta = m \circ (I \otimes S) \circ \Delta = i \circ \varepsilon \tag{2.79}$$

2.3.5. The example of C_2. As an example, consider again $U_q(C_2)$. Choose the normal order : $\alpha_1, \alpha_4, \alpha_3, \alpha_2$ where $\alpha_4 = \alpha_1 + \alpha_3$, $\alpha_3 = \alpha_1 + \alpha_2$.

First define, according to (2.67) and (2.75),

$$\begin{aligned} E_3 &\overset{\text{def}}{=} [E_1, E_2]_q = E_1E_2 - q^{-\alpha_{12}}E_2E_1 = E_1E_2 - qE_2E_1 \\ E_4 &\overset{\text{def}}{=} [E_1, E_3]_q = E_1E_3 - q^{-\alpha_{13}}E_3E_1 = [E_1, E_3] \end{aligned} \tag{2.80}$$

recalling that $\alpha_{ij} = (\alpha_i, \alpha_j)$.

Then, since $\alpha_1 + \alpha_4$, $\alpha_4 + \alpha_3$, $\alpha_3 + \alpha_2$ are <u>not</u> roots :

$$\begin{aligned} &[E_1, E_4]_q = [E_1[E_1, [E_1, E_2]_q]_q]_q = 0 \\ &[E_4, E_3]_q = 0 \\ &[E_3, E_2]_q = [[E_1, E_2]_q, E_2]_q = [E_2, [E_2, E_1]]_q]_q = 0 \ . \end{aligned} \tag{2.81}$$

these are the same equations as (2.74), (the second equation is a consequence), where the q-commutator is defined by

$$[E_\alpha, E_\beta]_q = E_\alpha E_\beta - q^{-(\alpha,\beta)}E_\beta E_\alpha \ . \tag{2.82}$$

Similarly,

$$\begin{aligned} E_{-3} &= [E_{-2}, E_{-1}]_{q^{-1}} \\ E_{-4} &= [E_{-3}, E_{-1}]_{q^{-1}} \end{aligned} \tag{2.83}$$

and

$$[E_{-2}, E_{-3}]_{q^{-1}} = [E_{-3}, E_{-4}]_{q^{-1}} = [E_{-4}, E_{-1}]_{q^{-1}} = 0 \ . \tag{2.84}$$

It turns out that it is possible to obtain the q–Serre relations and the Cartan–Weyl basis in terms of the generators e_i substituting E_i by e_i in the equations (2.74) to (2.84). The reader is invited to verify this.

3. Real forms

3.1. DEFINITION OF REAL FORMS

The simple Lie algebra $s\ell(2)$ is generated by h and $e_\pm$. It is a vector space over complex numbers. In physics one is often interested in the spin operators $S_i (i = 1,2,3)$ which are hermitean ($S^+ = S$) and generate the unitary representations (finite rotations) $U = \exp(iS)$. In order to get S_i in the framework of $s\ell(2)$, one should first define the hermitean conjugation + acting on the generators of $s\ell(2)$ and then find the complex linear combinations of $h, e_\pm$ which are hermitean. The result is called a <u>real form</u> of $s\ell(2)$ [38]. A real form is a vector space over real numbers. (The representation matrices may be, however, complex). There are different choices of conjugations, and therefore different real forms. One of them is <u>compact</u> ($su(2)$ or $so(3)$ in our example) the others are <u>non-compact</u> ($su(1,1)$ or $so(2,1)$ or $s\ell(2,\mathbb{R})$). The numbers inside the parentheses (,) refer to the signature of the <u>metric</u>.

The situation is similar for the q-deformation $U_q(g)$, with the difference that some real forms which are equivalent for g, are not so for $U_q(g)$. For example $U_q(su(1,1)) \neq U_q(s\ell(2,\mathbb{R}))$, as shown below. This is due to the fact that the co-product Δ now contains the complex number q. Two main choices are possible : $q \in \mathbb{R}$ or $|q| = 1$. (The choice $q \in i\mathbb{R}$ is not essentially new [39]). We shall show that $U_q(su(1,1))$ corresponds to $q \in \mathbb{R}$, whereas $U_q(s\ell(2,\mathbb{R}))$ corresponds to $|q| = 1$.

We define the conjugation $+$ as a morphism of the algebra, by its action on the generators x, y, etc., which satisfies

(i) It is an involution:

$$(x^+)^+ = I\ . \tag{3.1}$$

(ii) It leaves the commutation relations invariant.

(iii) It is an (anti) involution:

$$(xy)^+ = y^+x^+ \tag{3.2}$$

like the hermitean conjugation and is called <u>standard</u>.

One can also define a <u>non-standard</u> conjugation $*$ by

(iii')

$$(xy)^* = x^*y^* \tag{3.3}$$

This is like the ordinary complex conjugation.

(iv) It acts like a complex conjugation on the number field of the Lie algebra, resp. enveloping algebra vector space.

For $U_q(g)$ one requires in addition for the co-product the following

(v) <u>Standard</u> conjugation

$$\Delta(x)^+ = \Delta(x^+)\ . \tag{3.4}$$

(v') <u>Non-standard</u> conjugation

$$\Delta(x)^\oplus = \Delta'(x^\oplus)\ . \tag{3.5}$$

The choices (iii), (iii') and (v), (v') allow four possibilities [40, 41, 42]. The action on the antipode S is then fixed, namely :

$$(xy)^+ = y^+x^+ \quad ; \quad \Delta(x)^+ = \Delta(x^+) \quad ; \quad (S \circ +)^2 = I \tag{3.6}$$

$$\begin{array}{lll} (xy)^* = x^* y^* & \Delta(x)^* = \Delta'(x^*) & (S \circ *)^2 = I \\ (xy)^{\oplus} = y^{\oplus} x^{\oplus} & \Delta(x)^{\oplus} = \Delta'(x^{\oplus}) & S \circ \oplus = \oplus \circ S \\ (xy)^{\circledast} = x^{\circledast} y^{\circledast} & \Delta(x)^{\circledast} = \Delta(x^{\circledast}) & S \circ \circledast = \circledast \circ S \,. \end{array} \tag{3.7}$$

The conjugation (3.6) will be called standard, the others, non-standard.

The standard involution + is used by [13, 14, 39] and in the second version of quantum Poincaré [9]. Non-standard conjugations $\oplus$ are natural in conformal field theory when q is a root of unity [43, 44, 45], and they were also used in the first version of quantum Poincaré [8, 46].

3.2. STANDARD REAL FORMS OF $s\ell(2)$ AND $U_q(s\ell(2))$

Recall the commutation relations (2.2) of $U_q(s\ell(2))$:

$$\begin{aligned} [h, e_\pm] &= \pm 2 e_\pm \\ [e_+, e_-] &= [h]_q = \frac{q^h - q^{-h}}{q - q^{-1}} \,. \end{aligned} \tag{3.8}$$

For the standard involution +,

$$[x, y]^+ = -[x^+, y^+] \,. \tag{3.9}$$

Therefore, the following conjugations leave (3.8) invariant, for $q = 1$ and $q \neq 1$

I

$$h^+ = h \Rightarrow (e_\pm)^+ = \lambda e_\mp \quad , \ \lambda = \pm 1 \tag{3.10}$$

II

$$h^+ = -h \Rightarrow (e_\pm)^+ = \varepsilon e_\pm \ , \quad \varepsilon = -1 \tag{3.11}$$

whereas $\varepsilon = +1$ gives nothing new. For $|q| = 1$, we get $q^* = q^{-1}$, which does not affect $[h]_q$.

It is elementary to find the linear combinations y which are antihermitean, such that $U = \exp(y)$ is unitary.

Next, we shall give the action of the conjugation on the generators $e_{\pm i}$ and h_i, and we shall display the antihermitean generators y_i. We shall also compute the quadratic Casimir C_2 in the limit $q = 1$, from which we shall derive the signature of the metric, and thereby deduce whether the real form is compact or non-compact. In this way we get [47] the three real forms of $U_q(s\ell(2))$:

I$_1$

$$h^+ = h \ ; \ (e_\pm)^+ = e_\mp \tag{3.12}$$

$$y_1 = -\frac{i}{2}(e_+ + e_-) \; ; \; y_2 = \frac{1}{2}(e_+ - e_-) \; ; \; y_3 = \frac{i}{2}h \tag{3.13}$$

$$\begin{aligned} [y_1, y_2] &= \frac{i}{2}[-2iy_3]_q \\ [y_2, y_3] &= y_1 \\ [y_3, y_1] &= y_2 \end{aligned} \tag{3.14}$$

Since we define the signature of the metric according to the form the quadratic Casimir takes in the limit $q \to 1$, and in this case, $C_2 = y_1^2 + y_2^2 + y_3^2$ is invariant for $q = 1$ (remember $[x]_q = x$ for $q = 1$), we say that the metric in case I_1 is $(+++)$, and it follows that all the generators are compact. For $q \in \mathbb{R}$ the algebra is $U_q(su(2)) = U_q(so(3))$ (see equ. (3.22)).

I$_2$

$$h^+ = h \quad ; \quad (e_\pm)^+ = -e_\mp \tag{3.15}$$

$$y_1 = \frac{1}{2}(e_+ + e_-) \; ; \; y_2 = \frac{-i}{2}(e_+ - e_-) \; ; \; y_3 = \frac{i}{2}h \tag{3.16}$$

$$\begin{aligned} [y_1, y_2] &= \frac{i}{2}[-2iy_3]_q \\ [y_2, y_3] &= -y_1 \\ [y_3, y_1] &= -y_2 \; . \end{aligned} \tag{3.17}$$

The invariant for $q = 1$ is $-y_1^2 - y_2^2 + y_3^2$, with metric $(--+)$, the algebra is non-compact. It is called, for $q \in \mathbb{R}$, $U_q(su(1,1))$.

II

$$h^+ = -h \quad ; \quad (e_\pm)^+ = -e_\pm \tag{3.18}$$

$$y_1 = \frac{1}{2}(e_+ + e_-) \; , \; y_2 = \frac{1}{2}(e_+ - e_-) \; ; \; y_3 = \frac{1}{2}h \; . \tag{3.19}$$

All the coefficients are real. The algebra is called $U_q(s\ell(2,\mathbb{R}))$ and $|q| = 1$ (see equ. (3.23))

$$\begin{aligned} [y_1, y_2] &= -\frac{1}{2}[2y_3]_q \\ [y_2, y_3] &= -y_1 \\ [y_3, y_1] &= y_2 \; . \end{aligned} \tag{3.20}$$

The invariant for $q = 1$ is $-y_1^2 + y_2^2 - y_3^2$, the metric is thus $(-+-)$, and the algebra is non-compact.

In the three situations above, the numbering of the generators is arbitrary, therefore $s\ell(2,\mathbb{R})$ is equivalent to $su(1,1)$. But for $q \neq 1$, the co-product will distinguish between the two real forms. Actually, even for $q = 1$, there is a slight difference between cases I_2 and II, namely in the former y_3 is compact, and y_1 and y_2 non-compact whereas in the latter y_2 is compact [20,48]. Consider the standard action on the co-product, where we require $\Delta(x)^+ = \Delta(x^+)$. Since

$$\Delta(e_\pm) = e_\pm \otimes q^{h/2} + q^{-h/2} \otimes e_\pm \tag{3.21}$$

there are, again, two cases

I

$$h^+ = h \Rightarrow q \in \mathbb{R} \tag{3.22}$$

II

$$h^+ = -h \Rightarrow |q| = 1 \tag{3.23}$$

This shows that $U_q(su(1,1)) \neq U_q(s\ell(2,\mathbb{R}))$ for $q \neq 1$.

The choices (3.22) and (3.23) describe standard real Hopf alge bras. The Hopf algebras with reality conditions using non standard involutions (see (3.7)) define non standard real Hopf algebras.

3.3. STANDARD REAL FORMS FOR $U_q(g)$

For g semisimple, the general discussion in the Chevalley basis has been given by Twietmeyer [39]. Dobrev [20] gives a canonical procedure in the Cartan–Weyl basis, with emphasis on the (non)-compactness of the real forms.It is different from the one followed by [39]. One may check if he gets standard real forms in all cases.

The features are very similar to $U_q(s\ell(2))$, with the additional freedom of choosing an involutive automorphism η of the Dynkin diagram D. For example for $A_2 = su(3)$, one can interchange the two simple roots, *i.e.* $\eta(1) = 2$, $\eta(2) = 1$. For $A_3 = su(4)$, $\eta(1) = 3$, $\eta(2) = 2$, $\eta(3) = 1$. For $D_4 = so(8)$, three roots can be interchanged (triality). For B_n and C_n there is no such symmetry.

The result is the following, in the Chevalley basis [39]. There are two main categories of real forms

I

$$q \in \mathbb{R} \; ; \; h_i^+ = h_{\eta(i)} \; ; \; e_{\pm i}^+ = \lambda_i e_{\mp\eta(i)} \tag{3.24}$$

where $\eta \in Aut(D)$, $\eta^2 = I$, $\lambda_i = 1$ for $\eta(i) \neq i$, $\lambda_i = \pm 1$ for $\eta(i) = i$.

II

$$|q| = 1 \; ; \; h_i^+ = -h_{\eta(i)} \; ; \; e_{\pm i}^+ = -e_{\pm\eta(i)} \tag{3.25}$$

where $\eta \in Aut(D)$, $\eta^2 = I$ for q not a root of unity, whereas if q is an ℓ-th root of unity, then η is allowed to be a permutation of the Dynkin diagram (provided ℓ divides $\alpha_{ij} - \alpha_{\eta(i)\eta(j)}$ for all i and j).

A third possible case, with $q \in \sqrt{-1}\,\mathbb{R}$, is essentially equivalent to I [39].

For more details on the special role played by the non-equivalent Cartan subalgebras see [39, 48] and [20]. The latter classifies the (non-) compact Cartan generators.

3.4. REAL FORMS OF $U_q(so(5,\mathbf{C})) = U_q(C_2)$

We shall need these real forms for the discussion of quantum Poincaré. We shall also consider non-standard conjugations and state which metrics one obtains. For details see [40, 9, 49].

There are again two standard forms, for which

$$\begin{aligned} (xy)^+ &= y^+ x^+ \\ \Delta(x)^+ &= \Delta(x^+) \\ (S \circ +)^2 &= I \end{aligned} \tag{3.26}$$

In the Chevalley basis :

I

$$q \in \mathbb{R} \; ; \; h_i^+ = h_i \; ; \; e_{\pm i}^+ = \lambda_i e_{\mp i} \; ; \; \lambda_i = \pm 1 \; . \tag{3.27}$$

For different values of λ_i one gets the metrics related to $U_q(so(5))$, $U_q(so(4,1))$ and $U_q(so(3,2))$. The last one will be contracted to quantum Poincaré (see chapter 4 where the hermitean linear combinations are explicitly given). This conjugation leads outside the Cartan-Weyl basis [40, 9, 49] but o f course inside $U_q(g)$.

II

$$|q| = 1 \; ; \; h_i^+ = -h_i \; ; \; e_{\pm i}^+ = \lambda_i e_{\pm i} \; . \tag{3.28}$$

Whatever values $\lambda_i = \pm 1$ are chosen, one always gets $U_q(so(3,2))$. This conjugation can be extended to the Cartan-Weyl basis.

We now list three non-standard conjugations which can all be extended to the Cartan-Weyl basis. For notations see paragraph (3.1)

a)

$$q \in \mathbb{R} \; ; \; h_i^* = -h_i \; ; \; e_{\pm i}^* = \lambda_i e_{\mp i} \tag{3.29}$$

b)

$$|q| = 1 \; ; \; h_i^\oplus = h_i \; ; e_{\pm i}^\oplus = \lambda_i e_{\mp i} \tag{3.30}$$

c)

$$q \in \mathbb{R} \; ; \; h_i^\circledast = h_i \; ; \; e_{\pm i}^\circledast = \lambda_i e_{\pm i} \; . \tag{3.31}$$

It turns out that if one plays with the different values $\lambda_i = \pm 1$ one gets the same three metrics in cases (a) and (b) as in I. The common features of these conjugations is to interchange raising and lowering operators. In cases (c) and II one only gets $U_q(so(3,2))$. Case (b) was used in [8](see also [45, 46]).

4. Deformation of the Poincaré algebra

We now want to give a physically interesting non-trivial example of a deformation of a non-semisimple Lie algebra, the Poincaré (inhomogeneous Lorentz) algebra. The procedure was announced in chapter 1. We recall the main steps :

1) q–deformation of $so(5, \mathbf{C})$, following Drinfeld-Jimbo.
2) Cartan Weyl-basis.
3) Choice of the real form $U_q(so(3, 2))$.
4) Choice of physical generators M_{AB}.
5) Contraction.
6) Proof that we obtained a Hopf algebra.
7) Casimir operators and representations.

The contraction procedure will force us to replace the dimension-less deformation parameter q by the parameter κ, with the dimension of an energy. Physically this will imply that Lorentz invariance is broken. Rotation and translation invariance will be maintained at the algebra level, however the co-product for space translations will be non-trivial.

For a fixed value of κ, the amount of Poincaré invariance breaking will increase with energy. A new κ–Poincaré invariance will emerge. For $\kappa \to \infty$, the usual relativistic (classical) invariance is restored.

A realization of κ–Poincaré with derivatives acting on a space-time with commutative co-ordinates will be given. The energy–dependent breaking of Lorentz invariance manifests itself in the appearance of finite–difference time operators. Finite κ–Lorentz boosts are given by elliptic, instead of hyperbolic, functions. The κ–deformed Klein-Gordon and Dirac equations will be displayed.

4.1. q–DEFORMATION OF $so(5, \mathbf{C})$

We start from the commutation relations (2.49),(2.50) for $U_q(so(5, \mathbf{C}))$ in the Chevalley basis, with $(i, j) = (1, 2)$:

$$\begin{aligned} k_i e_{\pm j} k_i^{-1} &= q^{\pm \frac{1}{2}\alpha_{ij}} e_{\pm j} \\ [e_i, e_{-j}] &= \delta_{ij} [h_i]_q \ . \end{aligned} \tag{4.1}$$

where $k_i = q^{\frac{1}{2}h_i}$. The Cartan matrix a_{ij} and its symmetrized form $\alpha_{ij} = (\alpha_i, \alpha_j)$, *i.e.* the scalar product of simple roots, are given by

$$a_{ij} = \begin{pmatrix} 2 & -2 \\ -1 & 2 \end{pmatrix} \qquad \alpha_{ij} = \begin{pmatrix} 1 & -1 \\ -1 & 2 \end{pmatrix} . \tag{4.2}$$

The non-trivial co-product Δ and the antipode S are (see Eqs. (2.59) and (2.61)

$$\begin{aligned} \Delta(e_{\pm i}) &= e_{\pm i} \otimes k_i + k_i^{-1} \otimes e_{\pm i} \\ S(e_{\pm i}) &= -q^{\frac{\pm\alpha_{ii}}{2}} e_{\pm i} \end{aligned} \tag{4.3}$$

It turns out that the Cartan–Weyl basis can be defined as in (2.80) by

$$\begin{aligned} e_3 &\stackrel{\text{def}}{=} [e_1, e_2]_q = e_1 e_2 - q^{-\alpha_{12}} e_2 e_1 = e_1 e_2 - q e_2 e_1 \\ e_4 &\stackrel{\text{def}}{=} [e_1, e_3]_q = e_1 e_3 - q^{-\alpha_{13}} e_3 e_1 = [e_1, e_3] \end{aligned} \tag{4.4}$$

and

$$\begin{aligned} e_{-3} &\stackrel{\text{def}}{=} [e_{-2}, e_{-1}]_{q^{-1}} = e_{-2} e_{-1} - q^{-1} e_{-1} e_{-2} \\ e_{-4} &\stackrel{\text{def}}{=} [e_{-3}, e_{-1}]_{q^{-1}} = [e_{-3}, e_{-1}] \end{aligned} \tag{4.5}$$

Notice the opposite normal order for the lowering operators.

The q–Serre relations are given by the analogue of (2.81) by

$$[e_1, e_4]_q = [e_3, e_2]_q = 0 \tag{4.6}$$

which agrees with (2.74) using the definitions (2.64).

Similarly.

$$[e_{-2}, e_{-3}]_{q^{-1}} = [e_{-3}, e_{-4}]_{q^{-1}} = 0 \tag{4.7}$$

4.2. REAL FORMS OF $U_q(so(5, \mathbf{C}))$

There are different real forms which can, a priori, be contracted to Poincaré. One needs at least one Minkowski metric, which still leaves the choice of $so(3,2)$ or $so(4,1)$. Furthermore, one can permute the $+$ and $-$ signs in the metric, for example $(+ - - - +)$ or $(- - + - +)$. For $q = 1$, these permutations give essentially the same result, except for some subtleties [48].

For $q \neq 1$ this is not the case. Indeed, the Cartan subalgebra generators h_i play a special role since they have undeformed commutation relations and furthermore enter in the definition of the co-product. The relation between Cartan subalgebra and physical generators will depend on a given permutation. In [8, 9] we imposed the requirement that the rotation subalgebra $so(3)$ and its co–product remain undeformed after contraction. This is satisfied by the choice of the

real form $so(3,2)$ with signature $(+ - - - +)$, and the selection of h_1 and $e_{\pm 1}$ for the generators of space rotations $so(3)$, as will be shown b elow.

There are still several possibilities for obtaining this real form [40, 49]. In [8] and [9] we chose the hermitean conjugation +, which is standard in the algebra sector. In [8] we chose $|q| = 1$, which is non-standard in the co-algebra sector, while the conjugate generators remained inside the Cartan–Weyl basis. More explicitly, we chose

$$\begin{aligned} |q| = 1 \quad &; \quad h_i^{\oplus} = h_i \, , \; i = 1,2 \; ; \\ e_1^{\oplus} = e_{-1} \quad &; \quad e_2^{\oplus} = -e_{-2} \\ e_3^{\oplus} = -e_{-3} \quad &; \quad e_4^{\oplus} = -e_{-4} \end{aligned} \tag{4.8}$$

In [9] we chose $q \in \mathbb{R}$, which is standard in both sectors, *i.e.*

$$\begin{aligned} q \in \mathbb{R} \quad &; \quad h_i^{+} = h_i \, , \; i = 1,2 \\ e_1^{+} = e_{-1} \quad &; \quad e_2^{+} = -e_{-2} \\ e_3^{+} = q\tilde{e}_{-3} \quad &; \quad e_4^{+} = -q\tilde{e}_{-4} \end{aligned} \tag{4.9}$$

the generators $\tilde{e}_i$ being outside the Cartan-Weyl basis for $q \neq 1$

$$\begin{aligned} \tilde{e}_3 = e_2 e_1 - q e_1 e_2 \; ; \; \tilde{e}_4 = [\tilde{e}_3, e_1] \\ \tilde{e}_{-3} = e_{-1} e_{-2} - q^{-1} e_{-2} e_{-1} \; ; \; \tilde{e}_{-4} = [e_{-1}, \tilde{e}_{-3}] \end{aligned} \tag{4.10}$$

Notice that (4.8) and (4.9) agree in the Chevalley basis, but not in the Cartan-Weyl basis, because q^* distinguishes the two conjugations.

In these lectures we shall pursue only the choice of a standard real Hopf algebra (4.9), which satisfies all the stringent physical requirements.

4.3. PHYSICAL GENERATORS AND CONTRACTION

We want to express the physical generators M_{AB} of $U_q(so(3,2))$ in terms of the Cartan–Weyl generators h, e of $U_q(so(5, \mathbb{C}))$, where

$$M_{AB} = -M_{BA} = M_{AB}^{+} \; ; \; A, B = 0, \ldots, 4 \tag{4.11}$$

and the metric tensor

$$g_{AB} = \text{ diag } (+ - - - +) \, . \tag{4.12}$$

For $q = 1$, there are two Lorentz subalgebras (see Eq. (1.2)).

The Cartan generators h_1 and h_2 commute, which allows for a certain freedom of choice. The choice $h_1 = M_{12}$ and $h_2 = M_{34}$ would be a good candidate for a q-deformation of the Lorentz algebra (without contraction) [50, 3, 51, 52, 20]. For the deformation of the Poincaré algebra (with contraction), the following choice [9] (q real) satisfies the Jacobi identities and the standard reality condition (4.9) :

space rotations :

$$\begin{aligned} M_3 &= M_{12} = h_1 \\ M_1 &= M_{23} = \frac{1}{\sqrt{2}}(e_1 + e_{-1}) \\ M_2 &= M_{31} = \frac{-i}{\sqrt{2}}(e_1 - e_{-1}) \end{aligned} \tag{4.13}$$

boosts :

$$\begin{aligned} L_1 &= M_{14} = \frac{1}{2}(e_4 - q\tilde{e}_{-4} - e_2 + e_{-2}) \\ L_2 &= M_{24} = \frac{-i}{2}(e_4 + q\tilde{e}_{-4} + e_2 + e_{-2}) \\ L_3 &= M_{34} = -\frac{1}{\sqrt{2}}(q^{-\frac{i}{2}} e_3 + q^{\frac{i}{2}} q\tilde{e}_{-3}) \end{aligned} \tag{4.14}$$

translations :

$$\begin{aligned} RP_0 &= M_{04} = h_3 = h_1 + h_2 \\ RP_1 &= M_{01} = -\frac{1}{2}(-e_4 - q\tilde{e}_{-4} + e_2 + e_{-2}) \\ RP_2 &= M_{02} = \frac{1}{2}(e_4 - q\tilde{e}_{-4} + e_2 - e_{-2}) \\ RP_3 &= M_{03} = -\frac{i}{\sqrt{2}}(q^{-\frac{i}{2}} e_3 - q^{\frac{i}{2}} q\tilde{e}_{-3}) \end{aligned} \tag{4.15}$$

For $q = 1$, the operators $M_i, i = 1, 2, 3$ generate rotations and L_i generate Lorentz boosts. In view of the contraction we have introduced the notation ($P_0 = P_4$)

$$RP_\mu = M_{0\mu} \quad \mu = 1, \cdots, 4 \tag{4.16}$$

Contraction means $R \to \infty$, so that the P_μ's become the translation operators, since for $q = 1$ this limit corresponds to the Poincaré algebra (see Eq. (1.6)).

However, for $q \neq 1$, we run into trouble with the co-product. For example

$$\begin{aligned} \Delta(e_2) &= e_2 \otimes q^{\frac{1}{2}h_2} + q^{-\frac{1}{2}h_2} \otimes e_2 \\ &= e_2 \otimes q^{\frac{1}{2}(RP_0 - h_1)} + q^{-\frac{1}{2}(RP_0 - h_1)} \otimes e_2 \end{aligned} \tag{4.17}$$

In the limit $R \to \infty$, $\Delta(e_2)$, and therefore $\Delta(L_2)$, become infinite. The same problem arose when the Firenze group [18] tried to obtain the q-deformation of the Euclidean algebra $E(3)$ from the contraction of $U_q(s\ell(2)) \otimes U_q(s\ell(2))$. We follow their solution to the conundrum and put

$$q = e^{\frac{1}{\kappa R}} \tag{4.18}$$

where κ is a new parameter with the dimension of an inverse length, *i.e.* an energy (in the usual units with $\hbar = c = 1$).

We now get, for $R \to \infty$

$$q^{h_3} = e^{\frac{P_0}{\kappa}} \quad ; \quad q^{h_1} = e^{\frac{h_1}{\kappa R}} \to 1 \; . \tag{4.19}$$

From this follows that the rotation subalgebra (4.13) is not deformed. The physical relevance of this result deserves some comment: usual intuition would immediately throw out any deformation of the Poincaré algebra which blew off the local spatial isotropy. The κ–deformation we propose preserves the sacrosanct three–dimensional spatial rotational symmetry but distinguishes the dimension of time, so that the boosts are not what Lorentz imagined.

4.4. κ–POINCARÉ ALGEBRA

The procedure is now obvious. Use the commutation relations (4.1), (4.4) to (4.7), to calculate the κ–Poincaré algebra with the help of the definitions (4.13) to (4.15) and the contraction (4.16), (4.18) and (4.19), taking the limit $R \to \infty$.

The commutation relations simplify considerably with the following non linear change of variables for the deformed boosts :

$$\begin{aligned}
N_1 &= L_1 - \frac{1}{4\kappa}P_2 - \frac{1}{4\kappa}(M_2P_3 + P_3M_2) \\
N_2 &= L_2 + \frac{1}{4\kappa}P_1 + \frac{1}{4\kappa}(M_1P_3 + P_3M_1) \\
N_3 &= L_3 + \frac{1}{2\kappa}P_3 - \frac{1}{4\kappa}(M_1P_2 + P_2M_1 - M_2P_1 - P_1M_2) \\
&= L_3 + \frac{1}{2\kappa}P_3 - \frac{1}{2\kappa}(P_2M_1 - M_2P_1)
\end{aligned} \tag{4.20}$$

A similar change of variables was proposed by Giller *et al.* [53] in order to simplify the κ–Poincaré algebra of [8].

The κ–Poincaré algebra finally reads most appeallingly as [9]

$$\begin{aligned}
&[M_i, M_j] = i\varepsilon_{ijk}M_k \quad i,j,k = 1,2,3 \\
&[M_i, P_j] = i\varepsilon_{ijk}P_k \\
&[M_i, P_0] = 0 \\
&[P_\mu, P_\nu] = 0 \quad \mu,\nu = 0,\cdots,3
\end{aligned} \tag{4.21}$$

the rotation and translation algebra is not deformed.

$$
\begin{aligned}
[M_i, N_j] &= i\varepsilon_{ijk}N_k \\
[P_0, N_k] &= -iP_k \\
[P_i, N_j] &= -i\kappa\delta_{ij}\sinh\frac{P_0}{\kappa} \\
[N_i, N_j] &= -i\varepsilon_{ijk}(M_k\cosh\frac{P_0}{\kappa} - \frac{1}{4\kappa^2}P_k(\vec{P}\vec{M}))
\end{aligned}
\tag{4.22}
$$

the boosts are deformed.

The usual Jacobi identities and the standard reality conditions are satisfied. For $\kappa \to \infty$ one recovers the usual Poincaré algebra.

One verifies that the κ–deformation of the quadratic Casimir, describing the quantum relativistic mass square operator is

$$
\begin{aligned}
C_1 &= 2\kappa^2(\cosh\frac{P_0}{\kappa} - 1) - P_1^2 - P_2^2 - P_3^2 \\
&= (2\kappa\sinh\frac{P_0}{2\kappa})^2 - \vec{P}^2
\end{aligned}
\tag{4.23}
$$

Expanding in power of $\frac{1}{\kappa^2}$ one gets

$$
C_1 = -\vec{P}^2 + P_0^2 + \frac{1}{12\kappa^2}P_0^4 + O(\kappa^{-4}) \tag{4.24}
$$

so that the energy-momentum relation is modified at high energy for fixed κ. This could be checked by experiment, or alternatively, it could be used to give a lower experimental limit on κ (see the conclusions).

The second Casimir can be obtained by introducing the κ–deformed Pauli–Lubanski four–vector

$$
\begin{aligned}
W_0 &= \vec{P}\cdot\vec{M} \\
W_k &= \kappa M_k\sinh\frac{P_0}{\kappa} + \varepsilon_{ijk}P_iN_j
\end{aligned}
\tag{4.25}
$$

so that

$$
C_2 = (\cosh\frac{P_0}{\kappa} - \frac{\vec{P}^2}{4\kappa^2})W_0^2 - \vec{W}^2 \tag{4.26}
$$

commutes with everything in the algebra and it provides us with a good quantum number. This curious expression can be obtained from the one given by Giller *et al.* [53] for the non standard κ -Poincaré algebra [8] by the re placement $\kappa \to i\kappa$.

4.5. κ–POINCARÉ HOPF ALGEBRA

To get the co-product Δ, co-unit ε and antipode S for the quantum Poincaré algebra, we start from $U_q(so(3,2))$ and perform the contraction (4.19). Because of (4.19) and (4.13), Δ, ε and S are trivial for the rotation subalgebra

$$\begin{aligned} \Delta(M_i) &= M_i \otimes I + I \otimes M_i \\ \varepsilon(M_i) &= 0 \\ S(M_i) &= -M_i \end{aligned} \tag{4.27}$$

The same is true for P_0 because, before the contraction, P_0 is proportional to the Cartan generator $h_3 = h_1 + h_2$

$$\begin{aligned} \Delta(P_0) &= P_0 \otimes I + I \otimes P_0 \\ \varepsilon(P_0) &= 0 \\ S(P_0) &= -P_0 \end{aligned} \tag{4.28}$$

From (4.15), (4.10), (4.3) to (4.5), it is clear that the co-product for P_i is non-trivial. After contraction one gets

$$\begin{aligned} \Delta(P_i) &= P_i \otimes \exp(\frac{P_0}{2\kappa}) + \exp(-\frac{P_0}{2\kappa}) \otimes P_i \\ \varepsilon(P_i) &= 0 \\ S(P_i) &= -P_i \end{aligned} \tag{4.29}$$

The expressions for Δ, ε and S of the boost generators N_i follow from (4.20), (4.14), (4.10), (4.3) to (4.5). After some algebra one gets, after contraction :

$$\begin{aligned} \Delta(N_i) &= N_i \otimes \exp(\frac{P_0}{2\kappa}) + \exp(\frac{-P_0}{2\kappa}) \otimes N_i + \\ &+ \frac{1}{2\kappa}\varepsilon_{ijk}[P_j \otimes M_k \exp(\frac{P_0}{2\kappa}) + \exp(-\frac{P_0}{2\kappa})M_j \otimes P_k] \\ \varepsilon(N_i) &= 0 \\ S(N_i) &= -N_i + \frac{3i}{2\kappa}P_i \end{aligned} \tag{4.30}$$

The coproduct and the antipode satisfy the standard reality conditions (see Eq. (3.6)).

In order to prove that the κ–Poincaré algebra is a Hopf algebra one has to show that

1) the co-product is co-associative and it is a homomorphism of the algebra as in (2.27),

$$\begin{aligned} (\Delta \otimes I)\Delta &= (I \otimes \Delta)\Delta \\ \Delta(xy) &= \Delta(x)\Delta(y) \qquad \forall x, y \in \kappa\text{–Poincaré} \end{aligned} \tag{4.31}$$

2) The maps $m, i, \Delta, \varepsilon, S$ satisfy (2.28) :

$$m \circ (S \otimes I) \circ \Delta(x) = m \circ (I \otimes S) \circ \Delta(x) = i \circ \varepsilon(x) \tag{4.32}$$

as well as $(I \otimes \varepsilon) \circ \Delta = I$ and $(\varepsilon \otimes I) \circ \Delta = I$. Conditions (1) and (2) involve some algebra and have been checked both for the sets of operators (M_i, L_i, P_μ) and (M_i, N_i, P_μ) (see Eqs (4.13), (4.14), (4.15) and (4.20)). Hence the two versions of κ–Poincaré algebras given in [8,9] are real Hopf algebras, a standard one [9] and a non -standard one [8].

Up to now we did not find a universal R–matrix, so we cannot decide whether the Hopf algebra is quasi-triangular or not. We have found, nevertheless, R–matrices in the four– and five–dimensional representations of κ–Poincaré[54].

4.6. REALIZATIONS OF κ–POINCARÉ

The following realization of κ–Poincaré using derivatives acting on a commutative spacetime was proposed for spin zero systems by Zaugg [55] and also, independently, by Giller *et al.* [53]; the latter also proposed for arbitrary spin s [56]:

$$\begin{aligned} P_\mu &= -i\frac{\partial}{\partial x^\mu} \\ M_i &= \varepsilon_{ijk} x_j P_k + m_i \\ N_i &= x_0 P_i - \kappa x_i \sinh\frac{P_0}{\kappa} + \exp[\mp\frac{P_0}{2\kappa}] n_i \pm \frac{1}{2\kappa}\varepsilon_{ijk} P_j m_k \end{aligned} \tag{4.33}$$

where $(\vec{m}, \vec{n})$ is a standard finite-dimensional representation of the Lorentz algebra

$$\begin{aligned} [m_i, m_j] &= i\varepsilon_{ijk} m_k \\ [n_i, n_j] &= -i\varepsilon_{ijk} m_k \\ [m_i, n_j] &= i\varepsilon_{ijk} n_k \end{aligned} \tag{4.34}$$

Acting on scalar fields $\phi(\vec{x}, t)$ the κ-deformed boosts N_i are realized putting $m_i = n_i = 0$ [10]

$$N_i \phi(\vec{x}, t) = -i x_0 \frac{\partial}{\partial x_i}\phi(\vec{x}, t) + i\kappa x_i [\phi(\vec{x}, t + \frac{i}{\kappa}) - \phi(\vec{x}, t - \frac{i}{\kappa})] \,. \tag{4.35}$$

This introduces imaginary finite time–differences, which are the trademark for the passage from trigonometric to elliptic functions.

From the Casimir operator C_1 (4.23) and the realization (4.33) one deduces the κ–deformed Klein-Gordon equation

$$\left[\Delta - 2\kappa^2\left(1 - \cos\frac{\partial_t}{\kappa}\right)\right]\phi(\vec{x}, t) = \left[\Delta - \left(2\kappa \sin\frac{\partial_t}{2\kappa}\right)^2\right]\phi(\vec{x}, t) = M_0^2 \phi(\vec{x}, t) \,. \tag{4.36}$$

where Δ is the Laplacian. This should be the starting point of a κ–deformed field theory with regularized Feynman propagators [9].

After setting $P_1 = P_2 = 0$ it is possible [10] to integrate the equation

$$[N_1, P_0] = P_3$$
$$[N_1, P_3] = \kappa \sinh \frac{P_0}{\kappa} \tag{4.37}$$

with the Casimir operator

$$C_1 = 4\kappa^2 \left(\sinh \frac{P_0}{2\kappa}\right)^2 - P_3^2 \equiv M_0^2 \; . \tag{4.38}$$

For $\kappa \to \infty$, we get the usual result, if we first define the initial value

$$P_0(\eta_0) = M \cosh \eta_0$$
$$P_3(\eta_0) = M \sinh \eta_0 \tag{4.39}$$

with

$$P_0^2(\eta) - P_3^2(\eta) = M^2 \tag{4.40}$$

and then use the addition formula for hyperbolic functions to get

$$P_0(\eta) = P_0(\eta_0) \cosh(\eta - \eta_0) + P_3(\eta_0) \sinh(\eta - eta_0)$$
$$P_3(\eta) = P_0(\eta_0) \sinh(\eta - \eta_0) + P_3(\eta_0) \cosh(\eta - eta_0) \tag{4.41}$$

For κ–Poincaré, the rapidity relations (4.39) are replac ed by

$$2\kappa \sinh \frac{P_0}{2\kappa} = M_0 \text{ nc } (u|m)$$
$$P_3 = M_0 \text{ sc } (u|m) \tag{4.42}$$

with

$$u = \sqrt{1 + \frac{M_0^2}{4\kappa^2}}\, \eta_0$$
$$m = \left(1 + \frac{M_0^2}{4\kappa^2}\right)^{-1} \tag{4.43}$$

and η_0 the rapidity. In these expressions, nc(u|m) and sc(u|m) are Jacobi elliptic functions [57]. As one should expect, (4.42) reduces to (4.39) when $\kappa \to \infty$.

The κ–deformation of (4.41) is now obtained by using the addition formula for elliptic functions [57, 10].

One can use the action of the operators (4.33) on $2s+1$ component Weyl spinor fields and calculate their finite κ–Lorentz transformation[10]. For s =1/2 one gets the κ–Dirac equation, which reads as follows [10]:

$$\begin{pmatrix} 0 & \mathcal{P}_{rr'} \\ \mathcal{P}^{-1}_{rr'} & 0 \end{pmatrix} \begin{pmatrix} \Psi^{(+)}_{r'} \\ \Psi^{(-)}_{r'} \end{pmatrix} = \begin{pmatrix} \Psi^{(+)}_{r} \\ \Psi^{(-)}_{r} \end{pmatrix} \qquad r, r' = \pm\frac{1}{2} \tag{4.44}$$

where

$$\mathcal{P}^{\pm 1} = \frac{1}{M_0\sqrt{1+\frac{M_0^2}{4\kappa^2}}}\left[2\kappa \sinh\left(\frac{P_0}{2\kappa}\right) \pm \vec{P}\cdot\vec{\sigma}\right]\left[\cosh\left(\frac{P_0}{2\kappa}\right) \pm \frac{1}{2\kappa}\vec{P}\cdot\vec{\sigma}\right] \tag{4.45}$$

and $\Psi^{(\pm)}$ are two–component Weyl spinors in momentum space.

This agrees with [58].

The κ Dirac equation is therefore:

$$\left[(\kappa \sinh\left(\frac{P_0}{\kappa}\right) + \frac{1}{2\kappa}\vec{P}^2)\gamma_0 + \exp\left(\frac{P_0}{2\kappa}\right)\vec{P}\vec{\gamma} - M_0\sqrt{1+\frac{M_0^2}{4\kappa^2}}\right]\Psi = 0 \tag{4.46}$$

γ_0 and $\vec{\gamma}$ are the usual Dirac matrices. The reader is invited to check that the four-component spinor Ψ satisfies the κ-Klein–Gordon equation. Introducing the electromagnetic field by the minimal substitution $P_\mu \to P_\mu - eA_\mu$ one finds [10] that the κ–corrections to the fine structure of the energy levels of the hydrogen atom are of order κ^{-2}, whereas the "classical" g-factor of the electron becomes

$$g = 2[1 + \frac{m}{\kappa}] + 0(\kappa^{-2}) \, . \tag{4.47}$$

with m the mass of the electron. This value of course does not include radiative corrections, but it signals the crucial modification of our usual space-time concepts arising from a new version of relativistic invariance, in full agreement with the hamiltonian description of quantum field theory which clearly distinguishes between time and space.

5. Conclusions

The mathematical theory of the q–deformation of semisimple Lie algebras is well developed [15, 16, 50]. The deformed commutation relations in the Chevalley basis look rather simple, the new feature being exponentials of the Cartan generators (see Eq. (2.50)). They have to be supplemented by the highly non-linear q-Serre relations ((2.68),(2.74)), except for $s\ell(2)$. These relation s are best expressed in terms of the

q–adjoint operator whose general definition is rather involved ((2.51),(2.55)). The more cumbersome Cartan–Weyl basis allows a much simpler interpretation of the q–Serre relations ((2.77)). This basis is also more convenient, although not always sufficient, for defining a "physical" basis ((4.13)).

The deformed co-product is non symmetric ((2.59)). This of course has its influence on the tensor product of representations([25, 2, 35, 59] and references therein). The real forms of q–algebras follow a pattern similar to that of classical Lie algebras. However, the reality of the co-product imposes new restrictions so that equivalent Lie algebras differ by the allowed values of q once they get deformed (see end of paragraph 3.2).

There exists up to now no general theory for the quantum deformation of non semisimple Lie algebras. A certain number of examples have been given (see the introduction). The deformation of inhomogeneous algebras (e.g. the semi–direct product of rotations and translations) seems particularly interesting. Three main directions are being explored. 1) Subalgebras of semisimple Lie algebras [20, 21]. 2) Differential calculus on noncommutative space [23]. 3) Contraction of semisimple Lie algebras [18, 8, 9].

In these lectures we have followed this last track, in particular the quantum deformation of the Poincaré algebra, obtained by contraction of the q–deformed simple algebra $so(3,2)$[8, 9]. The final expression is a rather simple example of a Hopf algebra ((4.21),(4.22),(4.27) to (4.30)). Abstracting some general features from this result, taking also into account quantum E(3) [18], one may hope to find a canonical way to quantise inhomogeneous Lie algebras without going through the cumbersome procedure of compactification. We now list some features which could have a more general signification.

One of the generators (P_0) plays the role of a Cartan generator in the coproduct ((4.29),(4.30)). As a consequence, the deformation parameter q is replaced by the dimensionful parameter κ ((4.19)) [18, 8, 9]. The rotation subalgebra remains classical ((4.21)(4.27)). The deformation of the boosts l eads to new expressions for the two Casimir operators ((4.23),(4.26)).

The algebra can be realized in terms of partial derivatives on a commuting manifold, supplemented with finite–dimensional representations of the classical Lorentz algebra for arbitrary spin ((4.33),(4.34)). We can apply these operators on a field and thus easely get the κ–Klein–Gordon equation ((4.36)). Because P_0 is in the exponential one gets a finite time difference equation ((4.35)).

We have been able to integrate the deformed boosts for finite rapidities. The usual hyperbolic functions are replaced by elliptic functions ((4.42)). It has also been possible to give the explicit relation between Weyl wave functions at rest and Weyl wave functions boosted to the momentum p, for any spin s [10]. This allows to obtain the Wigner representations for the quantum Poincaré. One can also deduce the κ covariant Dirac equation ((4.44) to (4.46)).

An open problem is to formulate a field theory with interactions. Another interesting topic are the mathematical and physical properties of finite time difference equations. We have mainly worked in the momentum representation. The structure of space–time compatible with quantum Poincaré invariance is not yet understood[10].

The physical applications of q-semisimple algebras are numerous, espe cially for two-dimensional problems : integrable systems, conformal field theory, statistical models.

The Poincaré, or inhomogeneous Lorentz invariance, is fundamental for physics. It is worthwhile to compare the standard relativistic algebra with another consistent invariance algebra, depending on a dimensionful parameter κ, which has the Poincaré invariance as a limiting case.

The first item to compare is the quadratic Casimir operator C_1 ((4.24)). Fr om κ-Poincaré one gets a dispersion relation for light in vacuum and a modified relation between energy and momentum for massive particles [60]. An amusing effect appears if one calculates the modified classical partition function : one finds a limiting Hagedorn temperature proportional to κ analogous to the one obtained in string theory [54]. A more mondane piece of data is the measurement of the energy and velocity of 15 to 20 GeV electrons and γ–rays at SLAC [61, 62]. Taking into account the experimental errors, one finds a lower limit for κ of about 10^4 GeV. A more stringent test, because the energies are much higher, comes from astrophysical data concerning the detection time of 10^5 to 10^6 GeV γ -rays from the pulsar in Hercules X–1. This requires a value of κ larger than at least 10^{12} GeV [63, 64].

Once a κ–field theory with interactions is available, one may hope that the dimensionful parameter κ will serve as a natural cut–off. Further tests are then possible. One may also speculate on the modification of space–time at the Planck mass. An unsolved problem is the pertinence of a nonsymmetric coproduct. Further mathematical work and new physical insight are needed. It is a meager consolation that this difficulty does not arise in two–dimensional systems, because, (un)fortunately, the real world has four dimensions.

Acknowledgments

This work was partially supported by the Swiss National Science Foundation. I thank Jerzy Lukierski and Philippe Zaugg for judicious advice and Martí Ruiz–Altaba for his help in considerably improving these notes.

References

[1] L. Faddeev and L. Takhtajan. Preprint Université Paris VI, 1985; In : *Lect. Notes in Physics* **246** (1986) 166-179.

[2] L. Alvarez-Gaumé, C. Gomez and G. Sierra. *Phys. Lett.* **220** (1989) 142; *Nucl. Phys.* **B319** (1989) 155; **B330** (1990) 347.

[3] C. Gomez and G. Sierra. *Phys. Lett.* **255B** (1991) 51.

[4] C. Ramírez, H. Ruegg and M. Ruiz-Altaba. *Nucl. Phys.* **B364** (1991) 195.

[5] V. Pasquier. *Comm. Math. Phys.* **118** (1988) 355; V. Pasquier and H. Saleur. *Nucl. Phys.* **B330** (1990) 523.

[6] M.T. Batchelor, L. Mezinescu, R.I. Nepomechie and V. Rittenberg. *J. Phys. A : Math. Gen.* **23L** (1990) 141.

[7] S. Meljanac, M. Milekovic' and S. Pallua. *J. Phys. A : Math. Gen.* **21** (1991) 581.

[8] J. Lukierski, A. Nowicki, H. Ruegg and V.N. Tolstoy. *Phys. Lett.* **B264** (1991) 331.

[9] J. Lukierski, A. Nowicki and H. Ruegg. *Phys. Lett.* **B293** (1992) 34 4.

[10] J. Lukierski, H. Ruegg and W. Rühl. *From κ–Poincaré Algebra to κ–Lorentz Quasigroup : A Deformation of Relativistic Symmetry*, Kaiserslautern Preprint KL-TH-92/22, Dec. 1992.

[11] C.W. Misner, K.S.Thorne and J.A.Wheeler. *Gravitation*, Freeman and Company, San Francisco, 1973.

[12] E. Celeghini, R. Giachetti, E. Sorace and M. Tarlini. *Phys. Lett.* **B280** (1992) 180.
F. Bonechi, E. Celeghini, R. Giachetti, E. Sorace and M. Tarlini. *Phys. Rev. Lett.* **68** (1992) 3718.
F. Bonechi, E. Celeghini, R. Giachetti, E. Sorace and M. Tarlini. *Quantum Galilei Group as Symmetry of Magnons,* University of Florence Preprint, DFF 156/3/92.

[13] L. Faddeev, N. Reshetikhin and L. Takhtajan. *Alg. Anal.* **1** (1989) 178. [*Leningrad Math. Journ.* **1** (1990) 193].

[14] L. Woronowicz. *Comm. Math. Phys.* **111** (1987) 613.

[15] W.G. Drinfeld. "Quantum Groups", Proc. Internat. Congress of Mathematics (Berkeley, USA, 1986); p. 70.

[16] M. Jimbo. *Lett. Math. Phys.* **10** (1985) 63 and **11** (1986) 247.

[17] M. Chaichian and P. Kulish. *Phys. Lett.* **B234** (1990) 72.

[18] E. Celeghini, R. Giachetti, E. Sorace and M. Tarlini. *J. Math. Phys.* **31** (1990) 2548; *J. Math. Phys.* **32** (1991) 1155; *J. Math. Phys.* **32** (1991) 1159.

[19] D. Arnaudon and A. Chakrabarti. *Phys. Lett.* **B255** (1991) 242.
D. Arnaudon. *Comm. Math. Phys.* **134** (1990) 523.
A. Chakrabarti. *J. Math. Phys.* **32** (1991) 1227.

[20] V. Dobrev. *Canonical q-Deformations of Noncompact Lie (Super) Algebra,* Göttingen Preprint July 1991; *q-Deformations of Noncompact Lie (Super-) Algebras,* Trieste Preprint IC/92/13, January 1992, to be published in the Proceedings of II-nd Wigner Symposium (Clausthal, July 1991).

[21] J. Lukierski and A. Nowicki, *Phys. Lett.* **B279** (1992) 299.

[22] Yu.T. Manin. "Quantum Groups and Noncommutative Geometry", Publ. Centre de Recherches Math., Univ. Montreal, 1989.

[23] O. Ogievetsky, W.B. Schmidtke, J. Wess and B. Zumino. *Comm. Math. Phys.* **150** (1992) 495.

[24] P.P. Kulish and N. Reshetikhin. *J. Sov. Math.* **23** (1983) 2435. [Russian original in : *Zapiski Nauch Semin LOMI* **101** (1981) 101].

[25] A.N. Kirillov and N. Reshetikhin. LOMI preprint E-9-88 (1988). In : *Infinite Dimensional Lie Algebras and Groups,* Marseille 1988 Meeting, V. Kac (ed.), World Scientific, 1989.

[26] H.-D. Doebner, J.D. Hennig and W. Lücke. *Lecture Notes in Physics* **370** (1990) 29-63.

[27] S. Majid. "Quasitriangular Hopf Algebras and Yang-Baxter Equations ", to appear in *Int. J. Mod. Phys.* **A5** (1990) 1.

[28] S.M. Khoroshkin and V.N. Tolstoy. *Comm. Math. Phys.* **141** (1991) 599.

[29] R. Floreanini, D.A. Leites and L. Vinet. *Lett. Math. Phys.* **23** (1991) 127.

[30] J.E. Humphrey. *Introduction to Lie Algebras and Representation Theory,* Springer, New York, 1972, p. 99.

[31] V.G. Kac. *Infinite Dimensional Lie Algebras,* Birkhäuser, Boston, 1983.

[32] M. Rosso. *Comm. Math. Phys.* **124** (1989) 307.

[33] P. Truini and V.S. Varadarajan. *Lett. Math. Phys.* 26 (1992) 53; *Quantization of Reductive Lie Algebras : Construction and Universality,* Genoa Preprint GEF-Th-13/1992.

[34] F.N. Jackson. *Q.J. Pure. Appl. Math.* **41** (1910) 143.

[35] H. Ruegg. *J. Math. Phys.* **31** (1990) 1085.

[36] N. Burroughs. *Comm. Math. Phys.* **127** (1990) 109.

[37] V.N. Tolstoy. *Lecture Notes in Physics* **370** (1990) 188.

[38] R. Gilmore."*Lie Groups, Lie Algebras and Some of Their Applications*",John Wiley, New York 1974.

[39] E. Twietmeyer. *Lett. Math. Phys.* **24** (1992) 49.

[40] J. Lukierski, A. Nowicki and H. Ruegg. *Phys. Lett.* **B271** (1991) 321.

[41] M. Scheunert. Bonn University preprint Bonn, HE-92-13.

[42] M. Mozrzymas. Bordeaux University preprint LPTB 92-2, February 1992.

[43] B. Durhuus, H.P. Jakobsen and R. Nest. *Nucl. Phys. (Proc. Suppl.)* **B666** (1991) 1.

[44] G. Mack and V. Schomerus. *Nucl. Phys.* **B370** (1992) 185.

[45] M. Flato, L.K. Hadjiivanov and I.T. Todorov.*Quantum Deformations of Singletons and of Free Zero-Mass Fields,* Université de Bourgogne and Bulgarian Academy of Sciences preprint, Dec. 1992.

[46] M. Chaichian, J. de Azcarraga and P. Presnajder. *Phys. Lett.* **B291** (1992) 441

[47] T. Masuda, K. Mimachi, Y. Nakagani, M. Noumi, Y. Saburi and K. Ueno. *Lett. Math. Phys.* **19** (1990) 187.

[48] N. Bourbaki. *Groupes et algèbres de Lie* , Chap. 4, 5 et 6, Hermann, Paris 1968.

[49] J. Lukierski, A. Nowicki and H. Ruegg. Wroclaw Univ. preprint, ITP UWr 810/92, July 1992, to be published in the Proceedings of Karpacz Winter School (February 1992); Elsevier Science Publishers B.V.,Amsterdam,1993.

[50] P. Podles and S.L. Woronowicz. *Comm. Math. Phys.* **130** (1990) 381

[51] U. Carow-Watamura, M.Schlieker, M. Scholl and S. Watamura, *Z. Phys.* **C4 8** (1990) 159.

[52] O. Ogievetsky, W.B. Schmidke, J. Wess and B. Zumino. *Lett. Math. Phys.* **23** (199 1) 233.

[53] S. Giller, P. Kosinski, M. Majewski, P. Maslanka and J. Kunz. *Phys. Lett.* **B286** (1992) 57.

[54] M. Ruiz-Altaba. Private communication.

[55] Ph. Zaugg. Private communication.

[56] S. Giller, P. Kosinski, M. Majewski, P. Maslanka and J. Kunz. *On q-covariant wave functions,* University of Lodz preprint, August 1992.

[57] M. Abramowitz and I.A. Stegun. *Handbook of Mathematical Functions,* National Bureau of Standards, Applied Mathematics Series **55**, Washington 1972.

[58] A. Nowicki, E. Sorace and M. Tarlini. "The Quantum Deformed Dirac Equation from the κ-Poincaré Algebra", INFN Preprint DFF 177/12/92 Firenze, Dec. 1992.

[59] M. Gould and L.C. Biedenharn. *J. Math. Phys.* **33** (1992) 3613.

[60] J. Lukierski, A. Nowicki and H. Ruegg. *Quantum Poincaré Algebra and some Physical Consequences,* Geneva University Preprint UGVA-DPT 1992/12-795 Dec. 1992, to be published in the Proceedings of the International Symposium "Symmetry VI, Bregenz, 2.-7.08.92, B. Gruber ed., Plenum Press.

[61] Z.G.T. Guiragossian *et al.*. *Phys. Rev. Lett.* **34** (1975) 335.

[62] D.W. MacArthur. *Phys. Rev.* **A33** (1986) 1.

[63] T.J. Haines *et al.*. *Phys. Rev.* **D41** (1990) 692.

[64] G. Domokos. *Astrophysical Limits on the Deformation of the Poincaré Group,* The Johns Hopkins University Preprint, JHU-TIPAC-920027, Oct. 1992.

QUANTUM ALGEBRAS AND QUANTUM GROUPS IN q-SPECIAL FUNCTION THEORY

Roberto Floreanini
Istituto Nazionale di Fisica Nucleare, Sezione di Trieste
Dipartimento di Fisica Teorica, Università di Trieste
Strada Costiera 11, 34014 Trieste, Italy

Luc Vinet (*)
Laboratoire de Physique Nucléaire
and
Centre de Recherches Mathématiques
Université de Montréal
Montréal, Canada H3C 3J7

Abstract

The quantum algebra and quantum group interpretation of q-special functions is reviewed. Taking the algebra $\mathcal{U}_q(sl(2))$ as example, we shall see how its representation theory can be used to make advances in the study of the q-hypergeometric series ${}_2\phi_1(a, b; c; q, z)$.

The connection between quantum algebras and q-special functions is now well-established. As in standard Lie theory,[1] these functions arise as matrix elements of certain operators in the algebra generators and also as basis vectors of the corresponding representation spaces. This algebraic setting naturally leads to generating relations, orthogonality properties and addition formulas involving the q-special functions.[2-13]

Here, we shall illustrate the power of this "group-theoretic" interpretation by examining the relation between the quantum algebra $\mathcal{U}_q(sl(2))$ and the q-hypergeometric series ${}_2\phi_1(a, b; c; q, z)$.[6,9,11] Though we mainly work within the simpler quantum algebra framework, we shall also make connection with the corresponding "dual" approach, based on the quantum group $SL_q(2)$. Indeed, matrix

(*) Supported in part by the National Sciences and Engineering Research Council (NSERC) of Canada and the Fonds FCAR of Québec.

L. A. Ibort and M. A. Rodríguez (eds.), Integrable Systems, Quantum Groups, and Quantum Field Therapy 83–94.

elements of corepresentations of $SL_q(2)$ are also seen to involve the function ${}_2\phi_1$.[14] We shall show that the algebra and the group settings are completely equivalent, so that results obtained in one approach can be rephrased in the other.[9]

Finally, using the quantum algebraic interpretation, we shall give explicit examples of generating functions and summation formulas involving the q-hypergeometric function ${}_2\phi_1$.[11] These relations should first be looked at as identities between formal power series; it could happen that they converge only over a finite radius or only when the series terminate.

In the quantum algebra interpretation of the q-hypergeometric function ${}_2\phi_1$, an important role is played by the following q-analogs of the exponential function[15]

$$e_q(z) = \sum_{n=0}^{\infty} \frac{1}{(q;q)_n} z^n = \frac{1}{(z;q)_\infty}, \qquad |z| < 1, \tag{1a}$$

$$E_q(z) = \sum_{n=0}^{\infty} \frac{q^{\frac{1}{2}n(n-1)}}{(q;q)_n} z^n = (-z;q)_\infty, \tag{1b}$$

where $(a;q)_\alpha = (a;q)_\infty/(aq^\alpha;q)_\infty$ is the q-shifted factorial (a and α being arbitrary complex numbers), with $(a;q)_\infty = \prod_{k=0}^{\infty}(1-aq^k)$, $|q|<1$. Note that $e_q(z)\,E_q(-z) = 1$, and that $\lim_{q\to 1^-} e_q\big(z(1-q)\big) = \lim_{q\to 1^-} E_q\big(z(1-q)\big) = e^z$. We shall denote by T_z the q-dilatation operator which acts as $T_z\,\varphi(z) = \varphi(qz)$, on functions of the variable z; out of it, the q-difference operators

$$D_z^+ = z^{-1}(1-T_z), \tag{2a}$$

$$D_z^- = z^{-1}(1-T_z^{-1}), \tag{2b}$$

are constructed. Observe that $\frac{1}{(1-q)}D_z^+ \to d/dz$ and $\frac{1}{(1-q^{-1})}D_z^- \to d/dz$ as $q \to 1$, and that the q-exponentials obey

$$D_z^+\, e_q(\lambda z) = \lambda e_q(\lambda z), \tag{3a}$$

$$D_z^-\, E_q(\lambda z) = -q^{-1}\lambda E_q(\lambda z), \tag{3b}$$

where λ is a complex parameter. The basic hypergeometric series ${}_r\phi_s$ is defined by[15]

$$\begin{aligned} &{}_r\phi_s(a_1, a_2, \ldots, a_r; b_1, \ldots, b_s; q; z) \\ &\quad = \sum_{n=0}^{\infty} \frac{(a_1;q)_n (a_2;q)_n \ldots (a_r;q)_n}{(q;q)_n (b_1;q)_n \ldots (b_s;q)_n} \left[(-1)^n q^{\frac{n(n-1)}{2}}\right]^{1+s-r} z^n, \end{aligned} \tag{4}$$

with $q \neq 0$ when $r > s+1$. Since $(q^{-m};q)_n = 0$, for $n = m+1, m+2, \ldots$, the series ${}_r\phi_s$ terminates if one of the numerator parameters $\{a_i\}$ is of the form q^{-m} with $m = 0, 1, 2 \ldots$, and $q \neq 0$. By the ratio test, when $0 < |q| < 1$, the ${}_r\phi_s$

series converges absolutely for all z if $r \le s$, and for $|z| < 1$ if $r = s+1$. This series also converges absolutely when $|q| > 1$ and $|z| < |b_1 b_2 \ldots b_s|/|a_1 a_2 \ldots a_r|$. It diverges for $z \neq 0$ when $0 < |q| < 1$ and $r > s+1$, and when $|q| > 1$ and $|z| > |b_1 b_2 \ldots b_s|/|a_1 a_2 \ldots a_r|$, unless it terminates. In the following, we shall concentrate on the case

$$
{}_2\phi_1(a, b; c; q, z) = \sum_{n=0}^{\infty} \frac{(a;q)_n (b;q)_n}{(q;q)_n (c;q)_n} \, z^n \,, \qquad |z| < 1 \,, \tag{5}
$$

in terms of which various matrix elements will be expressed. Notice that as $c \to q^{1-m}$, with m a positive integer, this function satisfies the following limit relation[6]

$$
\frac{1}{(q;q)_{-m}} \, {}_2\phi_1(a, b, q^{1-m}; q; z) = z^m \, {}_2\phi_1\big(aq^m, bq^m, q^{m+1}; q, z\big) \, \frac{(a;q)_m (b;q)_m}{(q;q)_m} \,. \tag{6}
$$

Furthermore, in the following we shall always assume $|q| < 1$.

The quantum universal enveloping algebra $\mathcal{U}_q(sl(2))$ is the Hopf algebra generated by the elements k, k^{-1}, e and f satisfying the relations[16,17]

$$
k\,e\,k^{-1} = q^{1/2} e \,, \qquad k\,f\,k^{-1} = q^{-1/2} f \,, \qquad [e, f] = \frac{k^2 - k^{-2}}{q^{1/2} - q^{-1/2}} \,, \tag{7}
$$

and $k\,k^{-1} = k^{-1}k = 1$. The coproduct $\Delta : \mathcal{U}_q(sl(2)) \to \mathcal{U}_q(sl(2)) \otimes \mathcal{U}_q(sl(2))$, antipode $S : \mathcal{U}_q(sl(2)) \to \mathcal{U}_q(sl(2))$ and counit $\varepsilon : \mathcal{U}_q(sl(2)) \to \mathbf{C}$ are defined by:

$$
\begin{aligned}
&\Delta(k) = k \otimes k \,, && \Delta(e) = e \otimes k + k^{-1} \otimes e \,, && \Delta(f) = f \otimes k + k^{-1} \otimes f \,, \\
&S(k) = k^{-1} \,, && S(e) = -q^{1/2}\, e \,, && S(f) = -q^{-1/2}\, f \,, \\
&\varepsilon(k) = 1 \,, && \varepsilon(e) = 0 \,, && \varepsilon(f) = 0 \,.
\end{aligned} \tag{8}
$$

The algebra $\mathcal{U}_q(sl(2))$ has a Poincaré-Birkhoff-Witt basis given by: $e^\mu k^\rho f^\nu$, with $\rho \in \mathbf{Z}$ and $\mu, \nu \in \mathbf{N}$.

We now introduce a left $\mathcal{U}_q(sl(2))$-module $V^{(\lambda, m_0)} = \bigoplus_{j \in I} \mathbf{C}\xi_j$, where $I = \{i |\ i = m_0 + n,\ n \in \mathbf{Z}\}$, and λ, m_0 are complex numbers.[9,14] $V^{(\lambda, m_0)}$ is infinite dimensional, unless $\lambda + m_0$ and $\lambda - m_0$ are both positive integers. The corresponding representation is characterized by the following action of the generators on the basis vectors ξ_j, $j \in I$:

$$
\begin{aligned}
k\,\xi_j &= q^{-j/2}\, \xi_j \,, \\
e\,\xi_j &= q^{(1-2\lambda)/4}\, \frac{1 - q^{\lambda+j}}{1-q}\, \xi_{j-1} \,, \\
f\,\xi_j &= q^{(1-2\lambda)/4}\, \frac{1 - q^{\lambda-j}}{1-q}\, \xi_{j+1} \,.
\end{aligned} \tag{9}
$$

Given any $a \in \mathcal{U}_q(sl(2))$, its matrix elements $W_{ij}(a)$ in this representation are defined by

$$a\,\xi_j = \sum_{i\in \mathbf{Z}+m_0} \xi_i\, W_{ij}(a)\ ; \tag{10}$$

it clearly follows that

$$W_{ij}(ab) = \sum_{k\in \mathbf{Z}+m_0} W_{ik}(a)\, W_{kj}(b)\ . \tag{11}$$

Analogy with ordinary Lie theory[2] suggests to consider the following element in the completion of $\mathcal{U}_q(sl(2))$:

$$U(\alpha,\beta,\gamma) = E_q(\alpha\, e)\, e_q(\beta\, f)\, k^{\gamma}\ , \tag{12}$$

where α, β and γ are complex parameters. Indeed, set $k = q^{-h/2}$ and $q = e^{-\eta}$. In the limit $q \to 1^-$, $\eta \to 0^+$, the defining relations (7) become those of $sl(2)$: $[h,e] = -e$, $[h,f] = f$, $[e,f] = -2h$, and $U\big((1-q)\alpha, (1-q)\beta, 2\gamma/\eta\big)$ goes into the $SL(2)$ group element $e^{\alpha e} e^{\beta f} e^{\gamma h}$.

The matrix elements of $U(\alpha,\beta,\gamma)$ on $V^{(\lambda,m_0)}$ turn out to be expressible in terms of the function ${}_2\phi_1(a,b;c;q,z)$. In fact, with the help of (1), using (9) and identities involving q-shifted factorials, it is straightforward to show that

$$\begin{aligned} W_{ij}\Big(U(\alpha,\beta,\gamma)\Big) = {} & q^{-\gamma j/2} \left(\beta\, \frac{q^{(1-2\lambda)/4}}{1-q}\right)^{i-j} \frac{\left(q^{\lambda-i+1};q\right)_{i-j}}{(q;q)_{i-j}} \\ & \times {}_2\phi_1\left(q^{\lambda+i+1}, q^{i-\lambda}; q^{i-j+1}; q, -\alpha\beta\, \frac{q^{(1-2i)/2}}{(1-q)^2}\right)\ , \quad \text{if } i-j \geq 0\ , \end{aligned} \tag{13a}$$

$$\begin{aligned} W_{ij}\Big(U(\alpha,\beta,\gamma)\Big) = {} & q^{-\gamma j/2}\, q^{(j-i)(j-i-1)/2} \left(\alpha\, \frac{q^{(1-2\lambda)/4}}{1-q}\right)^{j-i} \frac{\left(q^{\lambda+i+1};q\right)_{j-i}}{(q;q)_{j-i}} \\ & \times {}_2\phi_1\left(q^{\lambda+j+1}, q^{j-\lambda}; q^{j-i+1}; q, -\alpha\beta\, \frac{q^{(1-2i)/2}}{(1-q)^2}\right)\ , \quad \text{if } i-j \leq 0\ , \end{aligned} \tag{13b}$$

with $i, j \in \mathbf{Z} + m_0$. This establishes most simply the connection between the basic hypergeometric series ${}_2\phi_1$ and $\mathcal{U}_q(sl(2))$. Notice that by using the limiting formula (6), the above two expressions for $W_{ij}\Big(U(\alpha,\beta,\gamma)\Big)$ are valid irrespective of the sign of $i-j$.

Before using (13) to obtain properties for the function ${}_2\phi_1$, let us make contact with the quantum group approach[18–20,14] and indicate in particular how the results described so far enable one to recover the matrix elements of the $SL_q(2)$ corepresentation given in Ref.[14].

Let $\mathcal{A}$ be the space dual to $\mathcal{U}_q(sl(2))$. To introduce the coordinate ring $A(SL_q(2))$ of the quantum group $SL_q(2)$, consider the fundamental representation $X:\ \mathcal{U}_q(sl(2)) \to \mathcal{E}nd\,\mathbf{C}^2$, defined by

$$X(k) = \begin{pmatrix} q^{1/4} & 0 \\ 0 & q^{-1/4} \end{pmatrix} , \qquad X(e) = \begin{pmatrix} 0 & 1 \\ 0 & 0 \end{pmatrix} , \qquad X(f) = \begin{pmatrix} 0 & 0 \\ 1 & 0 \end{pmatrix} . \tag{14}$$

Since X is a representation, we have $X(ab) = X(a)X(b)$, for any $a, b \in \mathcal{U}_q(sl(2))$. The matrix elements of X, that we shall denote as

$$X = \begin{pmatrix} x & u \\ v & y \end{pmatrix} , \tag{15}$$

define mappings $\mathcal{U}_q(sl(2)) \to \mathbf{C}$, via the evaluation map $a \to X(a)$; hence x, u, v and y are elements of $\mathcal{A}$. We shall call $A(SL_q(2))$ the subalgebra of $\mathcal{A}$ generated by these coordinate elements. This algebra is seen to inherit from $\mathcal{U}_q(sl(2))$ a Hopf structure. The coproduct $\Delta_{\mathcal{A}}:\ \mathcal{A} \to \mathcal{A} \otimes \mathcal{A}$, antipode $S_{\mathcal{A}}:\ \mathcal{A} \to \mathcal{A}$ and counit $\varepsilon_{\mathcal{A}}:\ \mathcal{A} \to \mathbf{C}$ are defined by

$$\Delta_{\mathcal{A}}(X)\,(a \otimes b) = X(ab) , \qquad a, b \in \mathcal{U}_q(sl(2)) , \tag{16}$$

and

$$S_{\mathcal{A}}(X) = \begin{pmatrix} y & -q^{1/2}u \\ -q^{-1/2}v & x \end{pmatrix} , \qquad \varepsilon_{\mathcal{A}}(X) = \begin{pmatrix} 1 & 0 \\ 0 & 1 \end{pmatrix} . \tag{17}$$

In terms of matrix elements, the definition (16) is equivalent to $\Delta_{\mathcal{A}}(X_{rs}) = \sum_{t=1,2} X_{rt} X_{ts}$, $r, s = 1, 2$, with the understanding that $(\Phi \otimes \Psi)(a \otimes b) = \Phi(a)\Psi(b)$ for $\Phi, \Psi \in \mathcal{A}$, $a, b \in \mathcal{U}_q(sl(2))$. The product $\mu_{\mathcal{A}}:\ \mathcal{A} \otimes \mathcal{A} \to \mathcal{A}$ is determined by

$$\mu_{\mathcal{A}}(\Phi \otimes \Psi)(a) \equiv \Phi\Psi(a) = \Phi \otimes \Psi(\Delta(a)) . \tag{18}$$

With these definitions one finds that

$$\begin{aligned} x^L u^M v^N \left(e^\mu k^\rho f^\nu\right) &= q^{-\frac{1}{4}\rho(\mu+\nu-L)-\frac{1}{4}L(\mu+\nu)-\frac{1}{4}(\mu(\mu-1)+\nu(\nu-1))+\frac{1}{2}(\mu-M)(\nu-N)} \\ &\qquad \times \frac{(q;q)_L}{(q;q)_{\mu-M}\,(q;q)_{L-\mu+M}} \frac{(q;q)_\mu\,(q;q)_\nu}{(1-q)^{\mu+\nu}} , \\ &\qquad \text{for } \mu - M = \nu - N, \quad M \le \mu \le M + L , \\ &= 0 , \qquad \text{otherwise} , \end{aligned} \tag{19a}$$

and

$$u^M v^N y^L \left(e^\mu k^\rho f^\nu\right) = q^{-\frac{1}{4}\rho(\mu+\nu+L)-\frac{1}{4}L(\mu+\nu)-\frac{1}{4}(\mu(\mu-1)+\nu(\nu-1))} \times \frac{(q;q)_\mu\,(q;q)_\nu}{(1-q)^{\mu+\nu}}\,\delta_{\mu,M}\,\delta_{\nu,N} , \tag{19b}$$

with M, N, and L nonnegative integers. Given these evaluations on the Poincaré-Birkhoff-Witt basis elements of $\mathcal{U}_q(sl(2))$, it is immediate to see that the coordinate elements obey the following commutation relations

$$q^{1/2}xu = ux \quad q^{1/2}xv = vx \quad q^{1/2}uy = yu \quad q^{1/2}vy = yv \quad uv = vu \ , \tag{20}$$

and

$$\det_q X = xy - q^{-1/2}uv = yx - q^{1/2}uv = \mathbf{1}_{\mathcal{A}} \ . \tag{21}$$

The matrix elements W_{ij} defined in (10) can also be viewed as elements of $\mathcal{A}$; they indeed provide linear mappings from $\mathcal{U}_q(sl(2))$ into $\mathbf{C}$, $a \to W_{ij}(a)$, $a \in \mathcal{U}_q(sl(2))$. Introducing an infinite matrix W, with elements W_{ij}, the composition relation (11) can be rewritten in the form

$$\Delta_{\mathcal{A}} W = W \otimes W \ , \tag{22}$$

and we conclude that W defines a corepresentation of $SL_q(2)$.

Starting from the formulas (13), the evaluation of W_{ij} on $U(\alpha, \beta, \gamma)$, it is now possible to derive the analytic expressions for the elements W_{ij} of W in terms of the coordinates x, u, v and y of $\mathcal{A}$. In fact, the element $U(\alpha,\beta,\gamma) = E_q(\alpha e)\, e_q(\beta f)\, k^\gamma$ generates the complete basis of $\mathcal{U}_q(sl(2))$:

$$(-1)^\mu q^{\mu(2-\rho)/2} \left(\left(D_\alpha^-\right)^\mu \left(D_\beta^+\right)^\nu U(\alpha,\beta,\rho) \right) \Big|_{\alpha=\beta=0} = e^\mu\, k^\rho\, f^\nu \ , \tag{23}$$

and therefore an arbitrary element of $\mathcal{A}$ is completely specified once its evaluation on $U(\alpha,\beta,\gamma)$ is given. Since $x^L u^M v^N$ and $u^M v^N y^L$ span $A(SL_q(2))$, with the help of the formulas (19), the expressions for W_{ij} can be easily abstracted from the matrix elements $W_{ij}(U(\alpha,\beta,\gamma))$ given before. The details can be found in Ref.[9]. One obtains

$$W_{ij} = q^{(i-j)(\lambda-j)} \frac{\left(q^{\lambda+i+1};q\right)_{j-i}}{(q;q)_{j-i}} \, {}_2\phi_1\!\left(q^{\lambda+j+1}, q^{j-\lambda}; q^{j-i+1}; q, -q^{1/2}uv\right) u^{j-i} y^{i+j} \ , \qquad \text{if } i-j \le 0 \ , \ i+j \ge 0 \ , \tag{24a}$$

$$W_{ij} = q^{(i-j)(\lambda+i)} \frac{\left(q^{\lambda+i+1};q\right)_{j-i}}{(q;q)_{j-i}} \, x^{-(i+j)} u^{j-i} \, {}_2\phi_1\!\left(q^{\lambda-i+1}, q^{-i-\lambda}; q^{j-i+1}; q, -q^{1/2}uv\right) \qquad \text{if } i-j \le 0 \ , \ i+j \le 0 \ , \tag{24b}$$

$$W_{ij} = q^{(j-i)(\lambda-i)} \frac{\left(q^{\lambda-i+1};q\right)_{i-j}}{(q;q)_{i-j}} \, {}_2\phi_1\!\left(q^{\lambda+i+1}, q^{i-\lambda}; q^{i-j+1}; q, -q^{1/2}uv\right) v^{i-j} y^{i+j} \ , \qquad \text{if } i-j \ge 0 \ , \ i+j \ge 0 \ , \tag{24c}$$

$$W_{ij} = q^{(j-i)(\lambda+j)} \frac{(q^{\lambda-i+1};q)_{i-j}}{(q;q)_{i-j}} x^{-(i+j)} v^{i-j} \, {}_2\phi_1\Big(q^{\lambda-j+1}, q^{-j-\lambda}; q^{i-j+1}; q, -q^{1/2}uv\Big) \quad \text{if } i-j \geq 0\,,\ i+j \leq 0\,. \tag{24d}$$

These four cases need to be distinguished in order for the elements W_{ij} to be analytic in x, u, v and y. The matrix elements (24) were computed in a different way in Ref.[14]. The quantum algebra derivation, that we have skeched, is straightforward and simpler, and explicitly shows the equivalence between the quantum group and the quantum algebra approach to q-special functions.

For simplicity, up to now all our considerations were based on the specific element (12) of $\mathcal{U}_q(sl(2))$. Off course, other combinations of little and big exponentials can also be used. For instance, take

$$\widetilde{U}(\alpha,\beta,\gamma) = E_q(\beta\, f)\, e_q(\alpha\, e)\, k^\gamma\ ; \tag{25}$$

its matrix elements in the representation (9) can be easily worked out and one explicitly finds $(i, j \in \mathbf{Z} + m_0)$

$$W_{ij}\Big(\widetilde{U}(\alpha,\beta,\gamma)\Big) = q^{-\gamma j/2}\, q^{(i-j)(i-j-1)/2} \left(\beta\, \frac{q^{(1-2\lambda)/4}}{1-q}\right)^{i-j} \frac{(q^{\lambda-i+1};q)_{i-j}}{(q;q)_{i-j}} \times {}_2\phi_1\Big(q^{\lambda-j+1}, q^{-j-\lambda}; q^{i-j+1}; q, -\alpha\beta\, \frac{q^{(1+2i)/2}}{(1-q)^2}\Big)\,, \quad \text{if } i-j \geq 0\,, \tag{26a}$$

$$W_{ij}\Big(\widetilde{U}(\alpha,\beta,\gamma)\Big) = q^{-\gamma j/2} \left(\alpha\, \frac{q^{(1-2\lambda)/4}}{1-q}\right)^{j-i} \frac{(q^{\lambda+i+1};q)_{j-i}}{(q;q)_{j-i}} \times {}_2\phi_1\Big(q^{\lambda-i+1}, q^{-i-\lambda}; q^{j-i+1}; q, -\alpha\beta\, \frac{q^{(1+2i)/2}}{(1-q)^2}\Big)\,, \quad \text{if } i-j \leq 0\,. \tag{26b}$$

The case of $e_q(\alpha\, e)\, E_q(\beta\, f)\, k^\gamma$ has been considered in Ref.[9], while the matrix elements of operators in $\mathcal{U}_q(sl(2))$ involving two little or two big q-exponentials can be expressed in terms of the q-hypergeometric series ${}_3\phi_1$ and ${}_2\phi_2$, and will not be discussed here. Using the explicit expressions (13) and (26) for the matrix elements of the operators $U(\alpha,\beta,\gamma)$ and $\widetilde{U}(\alpha,\beta,\gamma)$, one can now obtain various identities involving the q-hypergeometric function ${}_2\phi_1$.

As a first example, let us work out an orthogonality relation involving two ${}_2\phi_1$ functions. By recalling that $e_q(z)\, E_q(-z) = 1$ and setting $\gamma = 0$, one sees that

$$U(\alpha,\beta,0)\, \widetilde{U}(-\alpha,-\beta,0) = \mathbf{1}\ ; \tag{27}$$

acting on ξ_j, one then finds $(i, j \in \mathbf{Z} + m_0)$

$$\delta_{i-j,0} = \sum_{l \in \mathbf{Z}+m_0} W_{il}\Big(U(\alpha,\beta,0)\Big)\, W_{lj}\Big(\widetilde{U}(-\alpha,-\beta,0)\Big)\ . \tag{28}$$

Insert now for the matrix elements $W_{il}(U)$ and $W_{lj}(\widetilde{U})$ the expression (13b) and the one that it is obtained from (26a) with the use of the following transformation rule for the ${}_2\phi_1$ series,[15]

$$
{}_2\phi_1(q^a, q^b; q^c; q, z) = (q^{a+b-c}z; q)_{c-a-b}\, {}_2\phi_1(q^{c-a}, q^{c-b}; q^c; q, q^{a+b-c}z)\,. \qquad (29)
$$

After some simplifications and the redefinition $z = -\alpha\beta\, q^{1/2}/(1-q)^2$, one finally arrives at the following relation[11]

$$
\begin{aligned}
\delta_{i-j,0} = & \frac{q^{(i^2+j^2)/2}\,(q/z;q)_j}{(q^{\lambda+1};q)_i\,(q^{-\lambda};q)_j} \sum_{l\in\mathbf{Z}+m_0} q^{l(l-1)/2}\big(-zq^{-i-j}\big)^l\, \frac{(z;q)_l\,(q^{\lambda+1};q)_l\,(q^{-\lambda};q)_l}{(q;q)_{l-j}\,(q;q)_{l-i}} \\
& \times {}_2\phi_1\big(q^{\lambda+l+1}, q^{l-\lambda}; q^{l-i+1}; q, zq^{-i}\big)\, {}_2\phi_1\big(q^{\lambda+l+1}, q^{l-\lambda}; q^{l-j+1}; q, zq^{-j}\big)\,.
\end{aligned} \qquad (30)
$$

To get generating relations for the ${}_2\phi_1$, one first notices that it is possible to give a one-variable model for the representation (9), where the generators are expressed as q-difference operators in the complex variable z acting on the space of all linear combinations of the functions z^n, $n \in \mathbf{Z}$. Indeed, by taking

$$
\begin{aligned}
k &= q^{-m_0/2}\, T_z^{-1/2}\,, \\
e &= q^{(1-2\lambda)/4}\left(\frac{1}{1-q} D_z^+ + \frac{1-q^{\lambda+m_0}}{1-q}\,\frac{1}{z}\,T_z\right), \\
f &= q^{(1-2\lambda)/4}\left(\frac{z^2}{1-q} D_z^- + \frac{1-q^{\lambda-m_0}}{1-q}\, z\, T_z^{-1}\right),
\end{aligned} \qquad (31)
$$

and $\xi_j = z^n$, $j = m_0 + n$, $n \in \mathbf{Z}$, for the basis vectors, one can check that the relations (9) are satisfied. Let us now act directly with the operator $U(\alpha,\beta,\gamma)$, with e, f and k expressed as in (31), on z^{j-m_0}. With some manipulations, one finds that this action can be expressed in terms of a ${}_2\phi_0$ q-hypergeometric series,

$$
\begin{aligned}
U(\alpha,\beta,\gamma)\, z^{j-m_0} = & \; z^{j-m_0}\, q^{-j\gamma/2}\left(-\frac{\alpha}{z}\,\frac{q^{(1-2\lambda)/4}}{1-q}; q\right)_{\lambda+j} \\
& \times {}_2\phi_0\left(-\frac{\alpha}{z}\,\frac{q^{j+(1+2\lambda)/4}}{1-q}, q^{j-\lambda}; q, \beta z\, \frac{q^{-j+(1+2\lambda)/4}}{1-q}\right).
\end{aligned} \qquad (32)
$$

Since the series ${}_2\phi_0(a, b; q, z)$ does not converge, unless it terminates or $z = 0$, the action of $U(\alpha,\beta,\gamma)$ on the module $V^{(\lambda,m_0)}$ is ill-defined in this model, unless $V^{(\lambda,m_0)}$ is finite-dimensional or $\beta = 0$. Nevertheless, by proceeding formally one can obtain a generating relation for the q-hypergeometric function ${}_2\phi_1$. Recall the definition (10) for the matrix elements of $U(\alpha,\beta,\gamma)$, and insert (32) for the l.h.s., while in the r.h.s. substitute for $W_{ij}\Big(U(\alpha,\beta,\gamma)\Big)$ the result (13a). After using the transformation formula (29), set $\gamma = 0$, $j = m_0$, $x = -\alpha\beta\, q^{m_0+1/2}/(1-q)^2$ and $y = -(1-q)q^{(2\lambda+3)/4}\, z/\alpha$, to get[11]

$$(xq^{-2m_0};q)_{2m_0}\,(q/y;q)_{m_0+\lambda}\;{}_2\phi_0\big(q^{m_0+\lambda+1}/y, q^{m_0-\lambda};q, xyq^{-2m_0-1}\big)$$
$$= \sum_{l\in\mathbf{Z}} \frac{(q^{m_0-\lambda};q)_l}{(q^{2m_0+1}/x;q)_l\,(q;q)_l}\, y^l\;{}_2\phi_1\big(q^{\lambda-m_0+1}, q^{-m_0-\lambda}; q^{l+1}; q, x\big)\,. \tag{33}$$

The l.h.s. of this relation between formal series yields the ${}_2\phi_1$ as coefficients when expanded in powers of y. If $\lambda + m_0$ and $\lambda - m_0$ are both positive numbers, so that $V^{(\lambda,m_0)}$ is finite-dimensional, the relation (33) becomes an identity between polynomial functions. Note that here again the summand in the r.h.s. is well defined for all integers l, thanks to the limiting relation (6). An equivalent generating formula can be similarly obtained starting with the operator $\widetilde{U}(\alpha,\beta,\gamma)$.

To get further properties of the q-hypergeometric function ${}_2\phi_1$, observe that the matrix elements $W_{ij}\big(U(\alpha,\beta,\gamma)\big)$ and $W_{ij}\big(\widetilde{U}(\alpha,\beta,\gamma)\big)$ themselves define models of the module $V^{(\lambda,m_0)}$. For simplicity, set $\gamma = 0$. Then for each element $a \in \mathcal{U}_q(sl(2))$ define the operator $\pi(a)$ acting on the variables α, β such that $\pi(a)\,U(\alpha,\beta,0) = U(\alpha,\beta,0)\,a$. Upon operating on ξ_j, one obtains

$$\pi(a)\,W_{ij}\big(U(\alpha,\beta,0)\big) = \sum_{l\in\mathbf{Z}+m_0} W_{il}\big(U(\alpha,\beta,0)\big)\,W_{lj}(a)\,, \tag{34}$$

that is, the functions $W_{ij}\left(U(\alpha,\beta,0)\right)$ transform like the vectors ξ_j. From the properties (3) of the q-exponentials it is easy to construct the operators π for k, e and f, and thus for any $a \in \mathcal{U}_q(sl(2))$ by composition. For instance, one finds that

$$\pi^{(l)}(k) = q^{-l/2}\,T_\alpha^{-1/2}\,T_\beta^{1/2}\,, \tag{35a}$$

$$\pi^{(l)}(e) = -qD_\alpha^- - \frac{q^{1/2}}{(1-q)^2}\,\beta\left(q^l T_\alpha T_\beta^{-1} - q^{-l}T_\alpha^{-1}\right), \tag{35b}$$

$$\pi^{(l)}(f) = D_\beta^+\,, \tag{35c}$$

acting on the basis vectors

$$\xi_j^{(l)}(\alpha,\beta) = W_{lj}\big(U(\alpha,\beta,0)\big)\,, \qquad j,l \in \mathbf{Z}+m_0\,, \tag{36}$$

obey the same commutation relations as k, e and f. Similar results are obtained using $W_{ij}\big(\widetilde{U}(\alpha,\beta,0)\big)$.

Since we have a two-variable model of the module $V^{(\lambda,m_0)}$, from the general definition (10) one can write, recalling (36),

$$\widetilde{U}(\alpha,\beta,0)\;\xi_j^{(l)}(x,y) = \sum_{i\in\mathbf{Z}+m_0} W_{li}\big(U(x,y,0)\big)\;W_{ij}\big(\widetilde{U}(\alpha,\beta,0)\big)\,, \tag{37}$$

where the model independent matrix elements $W_{ij}\big(\widetilde{U}(\alpha,\beta,0)\big)$ are still given by (26). This allows deriving addition formulas for the q-hypergeometric series ${}_2\phi_1$: one just needs to evaluate explicitly the l.h.s. of (37), *i.e.* to compute directly the action of $\widetilde{U}(\alpha,\beta,0)$ on the basis functions $\xi_j^{(l)}(x,y)$, when e and f are realized as in (35), α and β being replaced by x and y.

We shall not give here the details of this evaluation (see Ref.[11]), but only collect the final results. In the case $\alpha = 0$, one finds the following generating relation

$$\frac{(wq^{j-\lambda};q)_{l-j}\,(q^{\lambda-l+1};q)_{l-j}}{(q;q)_{l-j}}\;{}_3\phi_2\big(q^{\lambda+l+1},q^{l-\lambda},wq^{l-\lambda};q^{l-j+1},0;q,z\big)$$
$$=\sum_{k=0}^{\infty}\frac{(q^{\lambda-l+1};q)_{l-j-k}\,(q^{j-\lambda};q)_k}{(q;q)_{l-j-k}\,(q;q)_k}\,w^k\;{}_2\phi_1\big(q^{\lambda+l+1},q^{l-\lambda};q^{l-j-k+1};q,z\big)\,,\quad |z|<1\,, \tag{38}$$

while for $\alpha\beta \neq 0$, one has

$${}_2\phi_1\big(q^{\lambda+l+1},q^{l-\lambda};0;q,z\big)\;{}_2\phi_0\big(q^{\lambda-j+1},q^{-\lambda-j};q,w/q\big)=\sum_{k\in\mathbf{Z}}\frac{(-1)^k q^{k(k+1)/2}}{(q;q)_{l-j-k}\,(q;q)_k}$$
$$\times\;{}_2\phi_1\big(q^{\lambda+l+1},q^{l-\lambda};q^{l-j-k+1};q,z\big)\;{}_2\phi_1\big(q^{\lambda-j+1},q^{-j-\lambda};q^{k+1};q,wq^k\big)\,. \tag{39}$$

Concerning the convergence of this last result, the remarks made after Eq.(33) also apply. In particular, when the module $V^{(m_0,\lambda)}$ is finite-dimensional, (39) becomes an identity among polynomials.

Indeed, take $m_0 = 0$ and λ a positive integer. Recalling the definition of the little q-Jacobi polynomials,[15]

$$p_n(z;a,b;q)={}_2\phi_1(q^{-n},abq^{n+1};aq;q,qz)\,, \tag{40}$$

and setting for simplicity $l=j$, $m=\lambda+j$ and $n=\lambda-j$, from (39) one gets

$$c_n\big(q^{m+1};-q^n/z;q\big)\;\mathcal{P}_m\big(wq^{-m};q^{n+1};q\big)$$
$$=\sum_{k=-m}^{n}q^{k(k-n)}\begin{bmatrix}n\\k\end{bmatrix}\frac{(q^{m+1};q)_k}{(q;q)_k}\,z^k\;p_{n-k}\big(z;q^k,q^{n-m};q\big)\;p_m\big(wq^k;q^k,q^{n-m-k};q\big) \tag{41}$$

The c_n are q-Charlier polynomials,[15]

$$c_n(x;a;q)={}_2\phi_1(q^{-n},x;0;q,-q^{n+1}/a)\,, \tag{42}$$

while the polynomials $\mathcal{P}_n$ are defined by

$$\mathcal{P}_n(z;a;q)=\sum_{k=0}^{n}\begin{bmatrix}n\\k\end{bmatrix}(a;q)_k\,z^k\,, \tag{43}$$

with

$$\begin{bmatrix} n \\ k \end{bmatrix} = \frac{(q;q)_n}{(q;q)_k (q;q)_{n-k}} . \tag{44}$$

They satisfy the following three-term recursion relation

$$\mathcal{P}_{n+1}(z;a;q) = \left[1 + z(1 - aq^n)\right] \mathcal{P}_n(z;a;q) - z\,(1 - q^n)\,\mathcal{P}_{n-1}(z;a;q) , \tag{45}$$

and reduce to the Rogers-Szegö polynomials[15,21-22] in the limit $a \to 0$.

References

1. Miller, W., *Lie Theory and Special Functions*, (Academic Press, New York, 1968)
2. Agarwal, A.K., Kalnins, E.G. and Miller, W., Canonical equations and symmetry techniques for q-series, SIAM J. Math. Anal. **18**, 1519-1538 (1987)
3. Floreanini, R. and Vinet, L., q-Orthogonal polynomials and the oscillator quantum group, Lett. Math. Phys. **22**, 45-54 (1991)
4. Floreanini, R. and Vinet, L., The metaplectic representation of $su_q(1,1)$ and the q-Gegenbauer polynomials, J. Math. Phys. **33**, 1358-1363 (1992)
5. Floreanini, R. and Vinet, L., q-Conformal quantum mechanics and q-special functions, Phys. Lett. B **277**, 442-446 (1992)
6. Floreanini, R. and Vinet, L., Quantum algebras and q-special functions, Ann. of Phys., to appear
7. Floreanini, R. and Vinet, L., Addition formulas for q-Bessel functions, University of Montreal-preprint, J. Math. Phys., to appear
8. Floreanini, R. and Vinet, L., Representations of quantum algebras and q-special functions, Proceedings of the *II International Wigner Symposium*, Dobrev, V. and Scherer, W., eds., (Springer-Verlag, Berlin, 1992), to appear
9. Floreanini, R. and Vinet, L., On the quantum group and quantum algebra approach to q-special functions, University of Montreal-preprint, UdeM-LPN-TH86, 1992
10. Floreanini, R. and Vinet, L., Generalized q-Bessel functions, University of Montreal-preprint, UdeM-LPN-TH87, 1992
11. Floreanini, R. and Vinet, L., Using quantum algebras in q-special function theory, University of Montreal-preprint, UdeM-LPN-TH90, 1992
12. Kalnins, E.G., Manocha, H.L. and Miller, W., Models of q-algebra representations: I. Tensor products of special unitary and oscillator algebras, J. Math. Phys. **33**, 2365-2383 (1992)

13. Kalnins, E.G, Miller, W., and Mukherjee, S., Models of q-algebra representations: the group of plane motions, University of Minnesota preprint, 1992

14. Masuda, T., Mimachi, K., Nakagami, Y., Noumi, M., Saburi, Y. and Ueno, K., Unitary representations of the quantum group $SU_q(1,1)$: structure of the dual space of $\mathcal{U}_q(sl(2))$, Lett. Math. Phys. **19**, 187-194 (1990); Unitary representations of the quantum group $SU_q(1,1)$: II-matrix elements of unitary representations and the basic hypergeometric functions, *ibid.* **19**, 195-204 (1990)

15. Gasper, G. and Rahman, M., *Basic Hypergeometric Series*, (Cambridge University Press, Cambridge, 1990)

16. Drinfel'd, V.G., Quantum groups, in: *Proceedings of the International Congress of Mathematicians*, Berkeley (1986), vol. **1**, pp. 798-820, (The American Mathematical Society, Providence, 1987)

17. Jimbo, M., A q-difference analogue of $U(g)$ and the Yang-Baxter equation, Lett. Math. Phys. **10**, 63-69 (1985); A q-analogue of $U(gl(N+1))$, Hecke algebra and the Yang-Baxter equation, *ibid.* **11**, 247-252 (1986)

18. Vaksman, L.L. and Soibelman, Ya.S., Algebra of functions of the quantum group $SU(2)$, Funct. Anal. Appl. **22**, 1-14 (1988)

19. Masuda, T., Mimachi, K., Nakagami, Y., Noumi, M. and Ueno, K., Representations of the quantum group $SU_q(2)$ and the little q-Jacobi polynomials, J. Funct. Anal. **99**, 357-386 (1991)

20. Koornwinder, T.H., Representations of the twisted $SU(2)$ quantum group and some q-hypergeometric orthogonal polynomials, Nederl. Akad. Wetensch. Proc. Ser. **A92**, 97-117 (1989)

21. Carlitz, L., Some polynomials related to theta functions, Annali di Matematica Pura ed Applicata (4) **41**, 359-373 (1955)

22. Carlitz, L., Some polynomials related to theta functions, Duke Math. J. **24**, 521-527 (1957)

LECTURES ON TOPOLOGICAL QUANTUM FIELD THEORY

DANIEL S. FREED

Department of Mathematics
University of Texas at Austin

What follows are lecture notes about Topological Quantum Field Theory. While the lectures were aimed at physicists, the content is highly mathematical in its style and motivation. The subject of Topological Quantum Field Theory is young and developing rapidly in many directions. These lectures are not at all representative of this activity, but rather reflect particular interests of the author.

Our theme is *locality and gluing laws*. The idea that global quantities can be computed from local formulas is an old one in topology and geometry. So too are formulas which express the change in a global invariant when we change a space locally. One should think of the Mayer-Vietoris sequence in homology theory [Sp] as an example. So too the local index theorem [ABP]. In knot theory one has the skein relations [K]. The Casson invariant obeys a surgery formula, and some recent work exploits formulas which calculate the change of Donaldson's invariant under certain surgeries [GM]. These examples illustrate that the gluing laws are not simply a theoretical nicety, but rather a useful tool for computation.

The idea of these lectures, then, is to explore the origin of locality in field theory from first principles. Since this notion relies on Feynman's path integral in the quantum theory, we restrict ourselves to a very simple toy model in which the path integral is a finite sum. Thus we avoid all of the analytical difficulties usually encountered in quantum theory. The penalty we pay is that this toy model has little real mathematical or physical interest. The payoff is that it nicely illustrates some of the main ideas, and allows us to explore some more subtle features. There is also locality and gluing laws in classical field theory, and in Lecture #4 we discuss a particularly illuminating case.

The author is supported by NSF grant DMS-8805684, an Alfred P. Sloan Research Fellowship, a Presidential Young Investigators award, and by the O'Donnell Foundation. He also received generous hospitality from the Geometry Center at the University of Minnesota while these notes were completed.

These are notes from 5 lectures given at the GIFT Summer School on Recent Problems in Mathematical Physics which was held in Salamanca, Spain, June, 1992. *Quisiera expresar mi agradecimiento a Alberto Ibort y Miguel Rodríguez por haberme invitado a dar estas charlas y por su magnífica hospitalidad en Salamanca.*

L. A. Ibort and M. A. Rodríguez (eds.), Integrable Systems, Quantum Groups, and Quantum Field Therapy 95–156.

Here is a brief summary of the lectures. In Lecture #1 we give a rough description of the definition and properties of a lagrangian field theory. We state them as "axioms", though they should not be taken as a rigid mathematical system like Euclidean geometry. The detailed axioms are slightly different in different theories. Our intent is only to coerce the reader into formulating familiar theories in terms which will help him or her relate to the more mathematical theories which follow. Lectures #2 and #3 discuss gauge theory with finite gauge group, our toy model. Lecture #4 is an introduction to the rich geometry of classical Chern-Simons theory. Lecture #5 is a report on some computer computations designed to convince skeptics that it all works!

A more complete treatment of the topics in Lectures #2–#5 may be found in several papers of the author listed in the bibliography. That fact alone should dispel any illusion of impartiality or completeness of these lectures!

Lecture #1: The Basic Structure of Field Theories

Today we begin with a general discussion of basic features of field theories, both on the classical and quantum levels. We only focus on the very basic features which interest us in our discussion of a special type of field theories: *topological* field theories. Missing is any discussion of local analytic properties, specifics about the Lorentz metric, etc. Rather, following Segal [S] and Atiyah [A], we list a general set of axioms which field theories are meant to satisfy. They resemble the Steenrod axioms for homology theory [Sp] much more closely than the Wightman axioms for field theory [GJ]. By no means are they meant to capture all properties of field theory, just as one could hardly do intricate calculations in homology theory just knowing the axioms. But they do provide a focus for our discussion, and they emphasize the properties of interest. We present them in the context of general field theories, and illustrate them with some non-topological examples. We encourage the reader to think about these axioms in familiar examples of field theories (and also mechanical systems, which are the $0+1$ dimensional case of a field theory). There is a much more detailed and precise treatment of these axioms in [Q1].

Fields and the Action.

We speak of a $(d+1)$-dimensional field theory: d space dimensions and 1 time dimension. So $d = 0, 1, \dots$ is the dimension of space—it is fixed at the beginning. I will always use 'X' to denote a spacetime—some sort of $d+1$ dimensional manifold—and 'Y' to denote a space—some sort of d dimensional manifold. By "sort" of manifold we mean manifold with extra structure. This can take the form of topological data, like an orientation or spin structure, or differential geometric data, like a metric or conformal structure. Usual field theories are often done in $X = \mathbb{R}^{d+1}$ and $Y = \mathbb{R}^d$, or perhaps in some open subsets of these Euclidean spaces. A key feature of topological theories is that X and Y are allowed to be *curved manifolds*. The spacetimes X are more fundamental than the spaces Y, so I'll talk now only about spacetimes. In traditional field theory one sometimes considers $X = S^{d+1}$, a $(d+1)$-dimensional sphere, or $X = S^1 \times \cdots \times S^1$, a $(d+1)$-dimensional torus. These are then thought of as compactifications of $\mathbb{R}^{d+1}$. But we want to consider more complicated spaces.

We generally restrict to *compact* spacetimes, both with and without boundary.

Definition. A manifold is called *closed* if it is compact and has no boundary.

This is a different use of 'closed' from the 'open' and 'closed' sets of topology. Most field theories are defined on manifolds endowed with a *metric*, traditionally of Lorentz signature. The *topological* theory we consider in Lectures #2–#5 (Chern-Simons Theory) does not use a metric—in fact, the spacetimes have no continuous parameters. However, our first example today, the σ-model, is of the traditional type; it is defined using a metric. We use Riemannian metrics, so do Euclidean field theory.

σ-model.

Usually a field theory begins with some fixed data, independent of any spacetimes, fields, etc. Here this is:

$d = 0, 1, \ldots$	dimension of the theory minus one
(M, h)	a *Riemannian* manifold with metric h

These data are fixed once and for all. M is not to be confused with the spacetimes, which are

(X^{d+1}, g) a compact, oriented Riemannian manifold of dimension $d+1$ with metric g.

Now to have a field theory we need fields! These are some sort of function on the spacetime, though the function may be vector-valued or even more twisted. For example, in gauge theory the field is a connection on the spacetime, though it is not immediately obvious how to think of this as a function on spacetime. In the σ-model the fields are:

$$\mathcal{C}_X = \text{Map}\,(X, M)$$

These are the fields on the spacetime X. It is useful to think about the assignment

$$X \longmapsto \mathcal{C}_X$$

which attaches a space of fields to every spacetime. In this case a field is a smooth map $\phi \colon X \to M$. The other key ingredient is the *action*, which is a function on the space of fields:

$$S_X \colon \mathcal{C}_X \to \mathbb{R}. \tag{1}$$

In the typical case this function has real values, but in later lectures when we consider the Chern-Simons theory this will have to be modified. The σ-model action is:

$$S_X(\phi) = \int_X |d\phi|^2 \, d\,\text{vol}_X\,. \tag{2}$$

It might look more familiar if we introduce local coordinates x^i on X and y^α on M. Then ϕ looks like $y^\alpha = y^\alpha(x^1, \ldots, x^{d+1})$ and

$$S_X(\phi) = \int_X \frac{\partial y^\alpha}{\partial x^i} \frac{\partial y^\beta}{\partial x^j} h_{\alpha\beta}\, g^{ij} \sqrt{\det(g_{ij})}\, dx^1 \ldots dx^{d+1}.$$

We require X compact to avoid convergence problems.

Formal Axioms.

What I want to observe here are formal properties of the fields and action, properties which carry over to any field theory. We will not discuss the σ-model any further; we only introduced the σ-model action here to illustrate the general structure of field theories. So these axioms apply to *any* field theory whose spacetimes are compact oriented Riemannian manifolds. They need only slight modifications, depending on the particular types of spacetimes in the theory, to apply to any field theory.

Functoriality Axiom. *Suppose $f: (X', g') \to (X, g)$ is an orientation preserving isometry. Thus $f: X' \to X$ is a diffeomorphism of manifolds which preserves the orientation, and $f^*(g) = g'$. We allow the possibility $X = X'$ and $g = g'$; then f is an isometry of (X, g). Think of these as symmetries of spacetimes. Then f induces a map of fields*

$$f^*: \mathcal{C}_X \to \mathcal{C}_{X'}$$

and the action is preserved:

$$S_{X'}(f^*\phi) = S_X(\phi), \qquad \phi \in \mathcal{C}_X.$$

This is the axiom which we need to modify depending on the type of theory. Generally, the map f should be a diffeomorphism which preserves whatever extra geometric and topological structures the spacetimes carry.

Exercise 1. Check this for the σ-model action (2). In local coordinates it is the change of variable formula for the integral.

Exercise 2. If $X = \mathbb{R}^4$ with the Minkowski metric, what are the isometries $f: X \to X$? What does the Functoriality Axiom tell us about the action on Minkowski space?

Exercise 3. A compact 1-manifold which is connected is diffeomorphic to either a circle or a closed interval. Now determine the possible metrics on the manifold up to isometry. (Hint: Construct invariants of the metric which are unchanged by isometry.)

Exercise 4. Suppose $d = 0$. Then the spacetime X is an interval or a circle, and a field is a path in M. Interpret the action (2). Is the length of the path a possible action?

Orientation Axiom. *If $-X$ denotes X with the opposite orientation, then*

$$\mathcal{C}_{-X} = \mathcal{C}_X$$

and

$$S_{-X}(\phi) = -S_X(\phi), \qquad \phi \in \mathcal{C}_X. \tag{3}$$

This is ultimately an expression of the *unitarity* of the theory, and may not hold in nonunitary theories. In the simple theory we discuss in Lecture #2 the spacetimes do not even carry an orientation.

The next axiom explains how to express the fields and the action for nonconnected spacetimes in terms of the fields and action for connected spacetimes.

Additivity Axiom. *Suppose X_1, X_2 are spacetimes and $X = X_1 \sqcup X_2$ the disjoin union. Then*

$$\mathcal{C}_X \cong \mathcal{C}_{X_1} \times \mathcal{C}_{X_2}.$$

Let $\phi_1 \in \mathcal{C}_{X_1}$, $\phi_2 \in \mathcal{C}_{X_2}$ and $\phi = \phi_1 \sqcup \phi_2 \in \mathcal{C}_X$. Then

$$S_X(\phi) = S_{X_1}(\phi_1) + S_{X_2}(\phi_2). \tag{4}$$

Exercise 5. Verify (3) and (4) for the σ-model.

The final axiom is a generalization of the Additivity Axiom, and is the central idea of these lectures. I call it the Gluing Axiom—it expresses the idea that the action and the fields are *local.*

Gluing Axiom. *Let X be a spacetime and $Y \hookrightarrow X$ an oriented codimension one submanifold. Cut along Y to obtain a new spacetime X^{cut} (see Figure 1). Then there is an "equalizer diagram"*

$$\mathcal{C}_X \to \mathcal{C}_{X^{\text{cut}}} \rightrightarrows \mathcal{C}_Y. \tag{5}$$

If $\phi \in \mathcal{C}_X$ with $\phi^{\text{cut}} \in \mathcal{C}_{X^{\text{cut}}}$ the corresponding field on X^{cut}, then

$$S_X(\phi) = S_{X^{\text{cut}}}(\phi^{\text{cut}}). \tag{6}$$

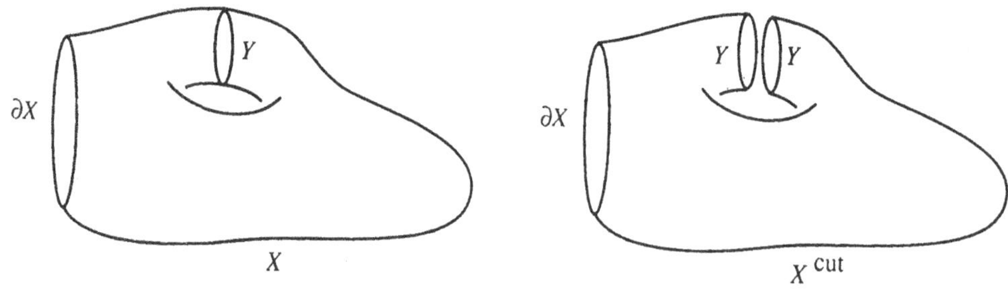

Figure 1: Cutting a manifold along a codimension one submanifold

What does (5) mean? It means that a field on X is a field on $X^{\rm cut}$ which has the same boundary values on the two components of $\partial X^{\rm cut}$ corresponding to Y. (Incidentally, this is very similar to the defining property of a *sheaf*.) What (6) means is that the action on X can just as well be compared on $X^{\rm cut}$. To simplify notation we will often picture cutting and pasting as in Figure 2, where we cut a closed manifold into two pieces. Then what the preceding means in that figure is that if we have fields $\phi_L \in \mathcal{C}_{X_L}$ and $\phi_R \in \mathcal{C}_{X_R}$ which agree on Y, then there is a field $\phi \in \mathcal{C}_X$ which agrees with these on each piece, and the action is

$$S_X(\phi) = S_{X_L}(\phi_L) + S_{X_R}(\phi_R). \tag{7}$$

We will see these pictures many times throughout these lectures.

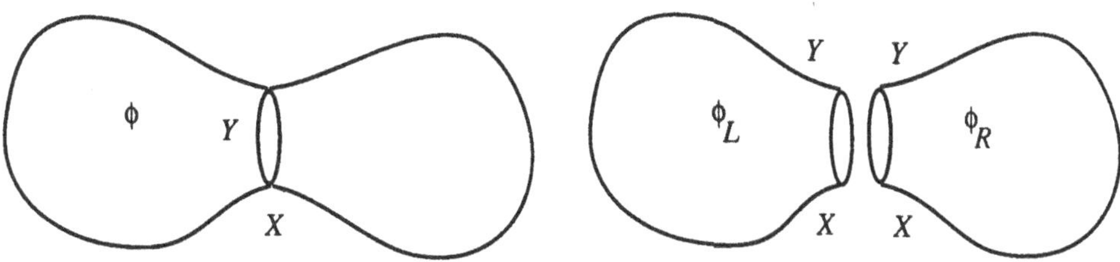

Figure 2: Cutting a closed manifold into two pieces

Exercise 6. Derive (7) from (6). You will need to use the Additivity Axiom as well.

Exercise 7. Verify the Gluing Law for the σ-model action. Check the special case of $d = 0$.

The four axioms are characteristic properties of the classical action of a field theory. We do not mean the above to be a rigorous statement of the axioms—we will have to modify them anyway in the Chern-Simons Theory. Nor do we wish to develop the subject axiomatically. Rather, we wish to isolate and emphasize these properties.

Topological Actions.

A *topological* field theory is defined on spacetimes with no continuous parameters, i.e., no differential geometric data like a (background) metric. The spacetimes may carry discrete topological data like an orientation or spin structure. Of course, the fields may still have continuous parameters.

We give a typical example of the form of a topological action. This is a simple variant of the topological actions which arise in practice. The fixed data for this theory is:

$d = 0, 1, \ldots$	dimension of the theory minus one
M	a smooth manifold
$\omega \in \Omega_M^{d+1}$	a $(d+1)$-form on M

Again, M is not to be confused with the spacetimes, which we now list together with the fields and the action.

Spacetimes: X a compact oriented $(d+1)$-manifold.

Fields: $\mathcal{C}_X = \mathrm{Map}(X, M)$

Action: If $\phi\colon X \to M$ then the action is

$$S_X(\phi) = \int_X \phi^* \omega. \tag{8}$$

So this theory is a sort of topological version of the σ-model in M, though it is *not* what is usually referred to as the Topological Sigma Model [W2].

Exercise 8. Verify the axioms stated above for this action. What is the proper statement of the Functoriality Axiom?

Exercise 9.

a) Suppose that ω is *closed*, i.e., $d\omega = 0$. Show that if $\partial X = 0$ (X is closed), then $S_X(\phi)$ only depends on the homotopy class of ϕ.
b) If X is not closed, then $S_X(\phi)$ varies as ϕ moves through a homotopy. Nonetheless, we can reinterpret the action as an invariant of the homotopy class. Try to do this by explaining the dependence of the action on ϕ. (Hint: The action only depends on the boundary values of ϕ.)

The Chern-Simons action may be considered as a variant of (8), but the symbols 'ω' and '$\int$' have to be radically reinterpreted.

Quantum Theory.

The path integral quantization requires a measure μ_X on $\mathcal{C}_X$ which satisfies certain properties which we do not specify here. Then for X *closed* we define the *partition function* as the *path integral*

$$Z_X = \int_{\mathcal{C}_X} e^{iS_X(\phi)}\, d\mu_X(\phi). \tag{9}$$

Note that Z_X is a complex number.

The big mathematical problem is that in many examples of interest μ_X has not been rigorously defined. This has been a big subject of investigation. Also, in

Euclidean field theory the 'i' is replaced by '-1' and we integrate $e^{-S_X(\phi)}$ instead of $e^{iS_X(\phi)}$. You can add a parameter $\hbar$ if you like. Here my concern is only the formal properties, which for now do not include the properties as $\hbar \to 0$. We return to that *perturbation theory* in Lecture #5 for the Chern-Simons theory. Here we only consider formal properties of the exact path integral.

If $\partial X \neq \phi$ then Z_X is a function of a field on the boundary ∂X. So if $\psi \in \mathcal{C}_{\partial X}$ let

$$\mathcal{C}_X(\psi) = \{\phi \in \mathcal{C}_X : \partial\phi = \psi\},$$

That is, $\mathcal{C}_X(\psi)$ consists of fields on X whose boundary value is ψ. Then

$$Z_X(\psi) = \int_{\mathcal{C}_X(\psi)} e^{iS_X(\phi)}\, d\mu_X(\phi). \tag{10}$$

The function Z_X belongs to a vector space $E(\partial X)$ of "certain" functions on $\mathcal{C}_{\partial X}$— this is the *quantum Hilbert space*. The construction of $E(\partial X)$ is the subject of canonical quantization. We discuss it in the Chern-Simons theory in Lecture #4, where we will explain a little about what 'certain' means above.

As an example, consider the (Euclidean) σ-model for $d = 0$. The classical action in this theory is the energy of a path on a Riemannian manifold M. The quantum field theory is the usual quantum mechanics on M. A "space" in this theory, which might occur as (part of) the boundary of a spacetime, is simply a point. The quantum Hilbert space is

$$E(pt) = \mathcal{F}(M) \tag{11}$$

the space of functions on M. Think of smooth functions; here I won't bother with technicalities about exactly what functions on M we ought to consider. A spacetime which is connected is either a closed interval or a circle, endowed with a metric. For the interval $X_t = [0, t]$ we obtain $Z_{X_t}(x, y) = p_t(x, y)$ depending on points $x, y \in M$, which comprise the fixed field on the boundary $\partial([0, t]) = \{t\} \cup -\{0\}$. (The minus sign is for the orientation.) This p_t for the Euclidean theory is the *heat kernel* of M, the probability density which measures the amount of heat which flows from x to y in time t. If $H: \mathcal{F}(M) \to \mathcal{F}(M)$ is the Laplace operator we can write this heat operator as

$$Z_{[0,t]} = e^{-tH} : \mathcal{F}(M) \to \mathcal{F}(M) \tag{12}$$

The fundamental property of *gluing* is illustrated by:

Figure 3: $e^{-(t_1+t_2)H} = e^{-t_1 H} e^{-t_2 H}$

Another case of gluing is the following:

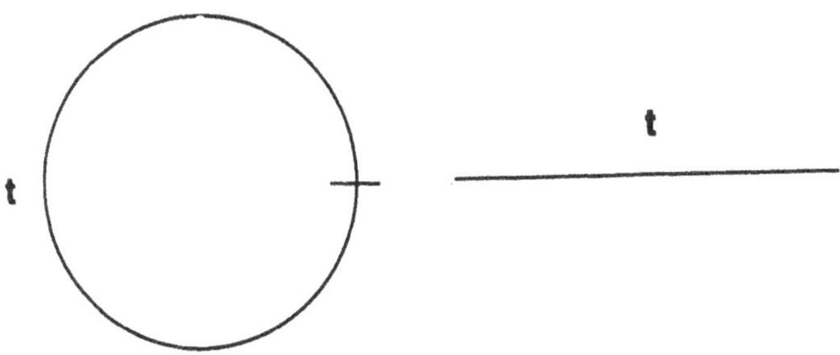

Figure 4: $Z_{S^1(t)} = \operatorname{Tr} e^{-tH}$

That is, the partition function for the circle is the trace of the heat kernel, depending on the length t of the circle.

Axioms for Quantum Theory.

The properties of the quantum Hilbert spaces $E(Y)$ associated to *any* space Y (d-manifold + extra structure) and the generalized partition functions (path integrals) Z_X associated to a spacetime X (($d+1$)-manifold + extra structure) mirror those mentioned above for the classical action. Indeed, they can be made to formally follow from them if one assumes the existence of a measure on the space of fields with appropriate properties. These axioms were stated by Segal [S] for conformal field theory and Atiyah [A] for topological theories. But they are quite general—the only difference between different types of theories is the structure of the manifold and so the symmetries in the Functoriality Axiom.

¡*Repetimos*! A field theory is formulated on spacetimes X^{d+1} and spaces Y^d with certain extra structure (orientation, metric, ...) depending on the type of theory. There may also be some fixed data. Our spacetimes and spaces are compact. The basic objects in the quantum theory are the *quantum Hilbert space* and the *path integral*.

Quantum Hilbert Space: $\quad Y \longmapsto E(Y) \qquad (\partial Y = \phi)$

Path Integral: $\quad X \longmapsto Z_X \in E(\partial X)$

The path integral over a closed manifold is also called the *partition function*. The term 'correlation function' is often used when the boundary is nontrivial, though we tend to (mis)use the term 'partition function' instead. The Hilbert spaces and path integrals satisfy properties analogous to those of the classical theory.

Functoriality Axiom. *If $f\colon Y' \to Y$ is a diffeomorphism preserving the extra structure (orientation, metric, ...) then there is an induced map*

$$f_*\colon E(Y') \to E(Y).$$

If $f\colon X' \to X$ is a diffeomorphism preserving the extra structure with induced boundary map $\partial f\colon \partial X' \to \partial X$ then

$$(\partial f)_*(Z_{X'}) = Z_X.$$

This says that symmetries of the spaces are implemented on the Hilbert spaces, and that symmetries of the spacetimes preserve the path integrals. In many quantum field theories, including the Chern-Simons theory, this is not correct as it stands—the symmetries only act *projectively*. This means that the composition of symmetries acts as the composition of the individual actions up to an overall phase. In other terms, it means that a *central extension* of the symmetries acts on the Hilbert spaces. Often this is used to disqualify a quantum field theory as unrealistic. The study of these central extensions and projective factors goes under the name *anomalies*, and has been the subject of much mathematical investigation and understanding.

Orientation Axiom.

$$E(-Y) \cong \overline{E(Y)} \qquad \textit{(conjugate Hilbert space)} \tag{13}$$
$$Z_{-X} = \overline{Z_X}.$$

Multiplicativity Axiom.

$$E(Y_1 \sqcup Y_2) \cong E(Y_1) \otimes E(Y_2) \qquad \textit{(tensor product)} \tag{14}$$
$$Z_{X_1 \sqcup X_2} = Z_{X_1} \otimes Z_{X_2} \tag{15}$$

The multiplicativity axiom expresses the behavior of the quantum Hilbert spaces and path integrals under disjoint union of spaces and spacetimes.

The final, and for us the most important axiom, is a generalization of (15) which tells how the path integral behaves when we glue manifolds along an oriented closed codimension one submanifold, as depicted in Figure 5.

Gluing Law. *Suppose $Y \hookrightarrow X$ is a closed oriented codimension one submanifold, and X^{cut} the spacetime obtained by cutting along Y. Then we have the path integrals $Z_X \in E(\partial X)$ and*

$$Z_{X^{\text{cut}}} \in E(\partial X^{\text{cut}}) = E(\partial X \sqcup Y \sqcup -Y) \cong E(\partial X) \otimes E(Y) \otimes \overline{E(-Y)}.$$

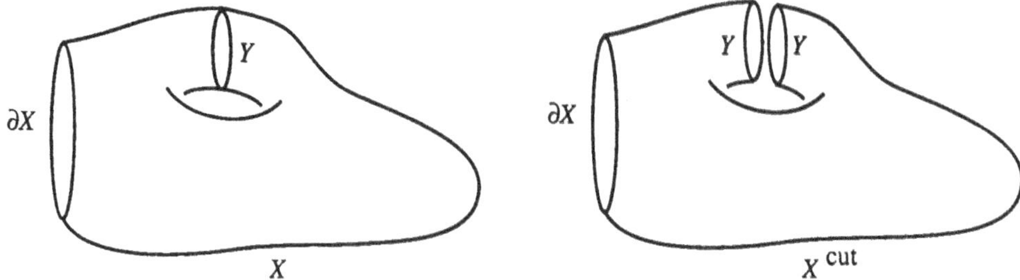

Figure 5: Cutting a manifold along a codimension one submanifold

Now there is a trace

$$\mathrm{Tr}_Y : E(\partial X^{\mathrm{cut}}) \cong E(\partial X) \otimes \underbrace{E(Y) \otimes \overline{E(Y)}}_{\text{contract using inner product}} \longrightarrow E(\partial X).$$

The axiom asserts that the path integrals satisfy

$$Z_X = \mathrm{Tr}_Y\, Z_{X^{\mathrm{cut}}} \tag{16}$$

Exercise 10. Show that quantum mechanics (11) and (12) satisfies these axioms.

Exercise 11. Think of other (relativistic) field theories in higher dimensions and try to see what properties of these the axioms capture.

Exercise 12.
a) What does the gluing axiom say about $Z_{[0,t]}$ in a *topological* $0+1$ dimensional theory?
b) More generally, compute $Z_{[0,1]\times Y^d}$ in a $d+1$ dimensional topological theory.
c) Also, compute $Z_{S^1\times Y^d}$ in a $d+1$ dimensional topological theory.

LECTURE #2: GAUGE THEORY WITH A FINITE GAUGE GROUP

Let G be a *finite* group. For example, G could be the cyclic group $\mathbb{Z}/n\mathbb{Z}$ of n elements, or the symmetric group S_n of $n!$ elements. There is no restriction on G. We construct, in arbitrary dimensions, a gauge theory with structure group G. It is surely the simplest quantum field theory there is, and is an excellent toy model to illustrate certain features. It exhibits the algebraic and geometric structure we are interested in, but has no analytic difficulties whatsoever: all integrals in the theory are finite sums. It also illustrates some basic properties of gauge theory, especially the role of symmetries and reducible connections.

This lecture is based on joint work with Frank Quinn [FQ]. The theory discussed here first appeared in a paper of Dijkgraaf and Witten [DW]. There is related unpublished work by Segal, Kontsevich, and Bernstein/Kazhdan. More recent work [F5] examines finer algebraic structures in this theory, in particular the connection to *quantum groups*, which is a topic of many other lectures at this Summer School.

Fields.

Let X be a spacetime. For this theory we only need a "bare" $(d+1)$-manifold, assumed compact. No orientation, metric, etc. is needed. The space of fields is

(17)
$$\mathcal{C}_X = \left\{ \begin{matrix} P \\ \downarrow \\ X \end{matrix} : P \text{ is a } \textit{principal (Galois, regular) covering space} \text{ with structure group } G. \right\}$$

In other terms, P is a principal bundle with structure group G. Thus G acts freely on P and the quotient is $P/G = X$. We always take G to act on the *right*.

As an example, consider $G = \mathbb{Z}/3\mathbb{Z}$. If $X = pt$ then picture is

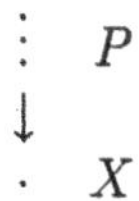

with G cyclically permuting the 3 points. If $X = S^1$ then there are 3 possibilities, up to isomorphism, as illustrated in Figure 6. The nontrivial coverings are pictured as pieces of helices, but the endpoints are meant to be identified. Topologically, the total space P in these covers is a circle.

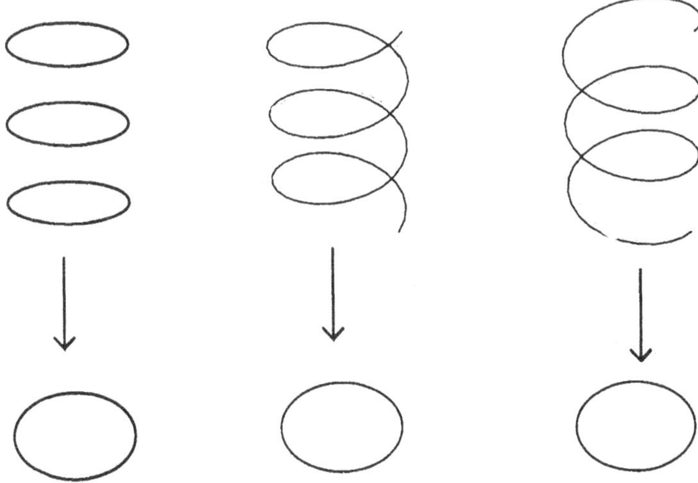

Figure 6: Triple Coverings of S^1

As opposed to the σ-model, where we have a nice space of fields, the fields in (17) do not technically form a topological space or even a set. (The collection of all of anything is not a set.) But there is a notion of *isomorphic* fields, and the fields in our theory fall into a finite number of isomorphism classes.

Definition. An *isomorphism* φ from P' to P is a map $\varphi\colon P' \to P$ which commutes with the G action and such that the induced map on the quotient X is the identity.

This means that for $p' \in P'$ and $g \in G$ we have $\varphi(p' \cdot g) = \varphi(p') \cdot g$, where the first '$\cdot$' indicates the G action on P' and the second '$\cdot$' the G action on P. Any such map induces a map $\overline{\varphi}\colon X \to X$, and we restrict our isomorphism to have $\overline{\varphi} = \mathrm{id}$. More generally, we can consider diagrams

$$\begin{array}{ccc} P' & \xrightarrow{\varphi} & P \\ \downarrow & & \downarrow \\ X' & \xrightarrow{\bar{\varphi}} & X \end{array}$$

They enter when we consider symmetries of the spacetimes as well as symmetries of the fields.

So we get a picture of $\mathcal{C}_X$ which we schematically render in Figure 7. Each point is a field and each arrow is a symmetry of fields, that is, an isomorphism.

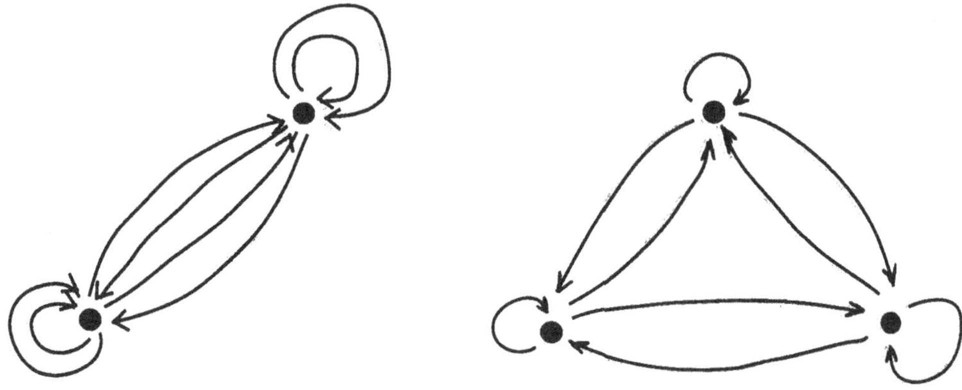

Figure 7: Configuration space for gauge theory with finite gauge group

We can compose two arrows if the second one starts where the first one ends. Notice that every object has an identity *automorphism*. (An *automorphism* is an isomorphism which starts and ends at the same place). Also, every arrow has an inverse. The arrows which start and end at an object P form a *group* called $\operatorname{Aut} P$, the automorphism group of P. There is a finite number of arrows from any point to any other. $\operatorname{Aut} P$ is also called the group of *gauge transformations* of P.

We remark that this picture applies to the space of (gauge) fields in any gauge theory, except that there are now continuous parameters. The technical word for this structure is a *groupoid*. See [F1,§1] for a discussion.

The fields P and P' are *equivalent* ($P \cong P'$) or *isomorphic* if there is an arrow between them. Let

$$\overline{\mathcal{C}_X} = \text{ set of equivalence classes of fields on } X.$$

What makes gauge theories with finite gauge group so simple is that $\overline{\mathcal{C}_X}$ is a finite set if X is compact.

Exercise 13. Determine $\overline{\mathcal{C}_X}$ for $X = pt$. For $X = S^1$.

Exercise 14. Show that $\mathcal{C}_X$ satisfies the formal axioms listed in Lecture #1. Does $\overline{\mathcal{C}_X}$ satisfy these axioms? (Notice that there is no orientation axiom in this theory.)

Exercise 15. Show that $\overline{\mathcal{C}_X}$ is a finite set for any compact manifold X.

Suppose X is connected. Fix a *basepoint* $x \in X$ and $p \in P_x$, where P_x is the fiber of P at x. Then a field $P \to X$ determines a map

$$\text{loops at } \; x \to G$$

by taking the *holonomy* around the loop using the basepoint p. (See Figure 8.) Any loop at x lifts uniquely to a *path* in P which starts at p and ends at some p' in the fiber P_x of P over x. The holonomy $h \in G$ is the unique h satisfying $p' = p \cdot h$. The holonomy only depends on the homotopy class of the loop.

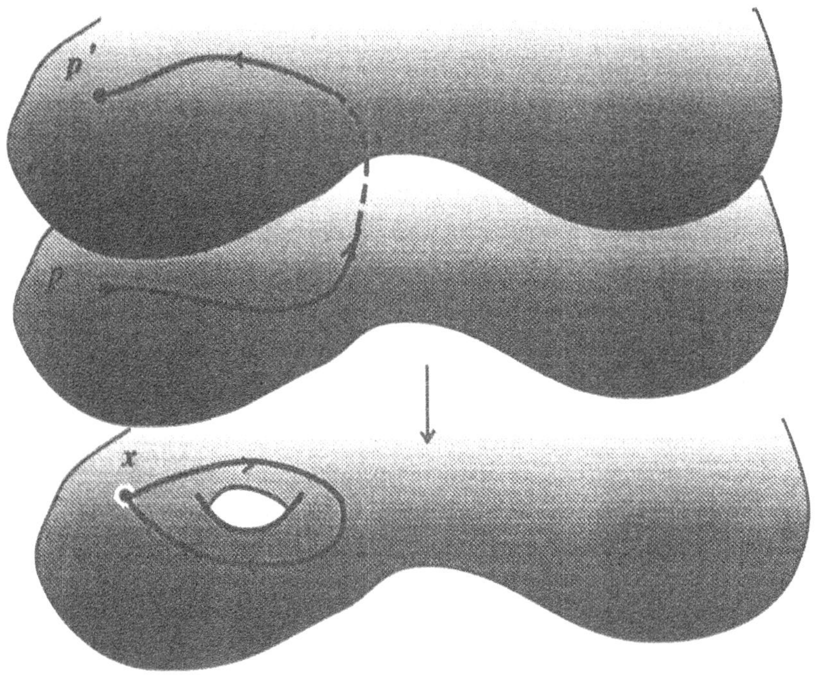

Figure 8: Definition of holonomy

Exercise 16.
a) Check this last assertion.

b) Show that the map $\pi_1(X, x) \to G$ defined by holonomy is a homomorphism of groups.
c) If γ is this homomorphism, and we change the basepoint p and/or x, then the new homomorphism is $g\gamma g^{-1}$ for $g \in G$.

Exercise 16 holds more generally for any *flat connection* with arbitrary gauge group.

The above exercise shows that there is an isomorphism (of sets)

$$\overline{\mathcal{C}_X} \cong \operatorname{Hom}(\pi_1(x), G)/G \tag{18}$$

if X is connected. (We omit the irrelevant basepoint.)

In Exercise 14 it is important to observe that the fields $\mathcal{C}_X$ are *local*, i.e., they can be cut and glued.

Action.

After so much work just to define fields, we can relax—we define the action to be identically zero!

$$S_X(P) = 0 \quad \text{for all} \quad P \in \mathcal{C}_X.$$

There are other possibilities, but this requires using the cohomology of the classifying space of the group, or equivalently group cohomology. These "twisted theories" are in essence constructed through a modification of (8) and the exercises which follow. See [FQ], [Q1], [Q2] for details.

So the classical action in our toy theory is completely trivial! It's surprising that we can learn anything from such an action. The fact that we will demonstrates the rich structure already present in the symmetries of the fields.

Now on to the quantization.

Measure.

Since $\mathcal{C}_X$ is discrete, a measure is just an assignment of a mass to each field P. The appropriate choice is

$$\mu_X(P) = \frac{1}{\#\operatorname{Aut} P}. \tag{19}$$

The '#' sign counts the number of elements in a set. Notice that if $P \cong P'$ then $\operatorname{Aut} P \cong \operatorname{Aut} P'$. So μ passes to a measure on the quotient $\overline{\mathcal{C}_X}$.

We did not state axioms for the measure in Lecture #1. But we do need to know how the measure behaves under gluing. Fix a bundle $Q \to Y$. We consider the gluing map

$$g_Q\colon \overline{\mathcal{C}_{X^{\text{cut}}}(Q \sqcup Q)} \to \overline{\mathcal{C}_X(Q)} \tag{20}$$

from (equivalence classes of) bundles over X^{cut} whose restriction to each copy of Y is Q to (equivalence classes of) bundles over X. (This needs to be made precise

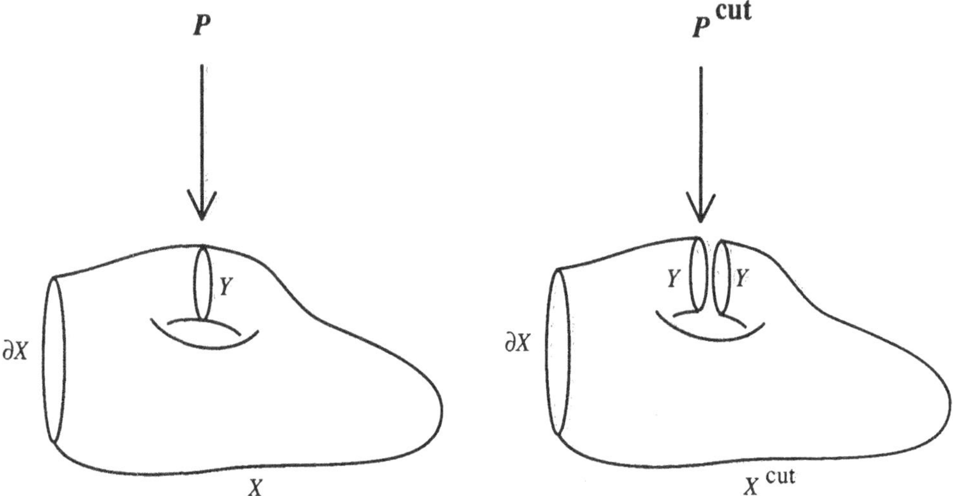

Figure 9: Gluing bundles

in a careful way.) The key formula is that for any $[P] \in \overline{\mathcal{C}_X}$ whose restriction to Y is isomorphic to Q,

$$\mu_X([P]) = \mathrm{vol}\big(g_Q^{-1}([P])\big) \cdot \mu_X(Q) \tag{21}$$

We use $[P]$ to denote the equivalent classes of bundles containing P.

As an illustration consider gluing the two ends of an interval together to obtain a circle. Then Q is the trivial bundle

$$\begin{array}{cc} \vdots & Q \\ \downarrow & \\ \cdot & pt \end{array}$$

Now the equivalence classes of bundles over an interval are in $1:1$ correspondence with the elements of G, since we are meant to have isomorphisms of the restriction to each endpoint with the trivial bundle, so can identify the holonomy, as in Figure 10. Finally, the equivalence classes of bundles over the circle are in $1:1$ correspondence with the *conjugacy classes* in G. This happens since when we glue the two ends of the interval together we are supposed to forget the basepoint and the trivialization, so we can only determine the holonomy up to conjugacy. Therefore, for this particular gluing (20) specializes to the map

$$g\colon G \to \mathrm{Conj}(G)$$

which assigns to each $h \in G$ its conjugacy class $[h]$. Then (21) reduces to a standard formula in group theory:

$$\#[h] \cdot \#C_h = \#G. \tag{22}$$

Here C_h is the *centralizer* of h, i.e., the subgroup of elements of G which commute with h, and h is an arbitrary element in the conjugacy class $[h]$.

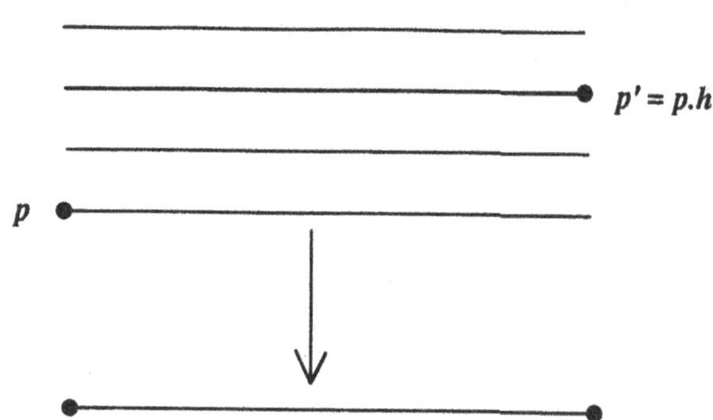

Figure 10: Bundle with basepoints over the interval

Quantization.

Given the fields, the action, and the measure we can carry out the quantization using (9) and (10). Keep in mind that the set of equivalence classes of bundles is *finite*, so we are dealing with finite measure spaces. In other words, the "path integral" in this case is a *finite* sum. Also, note that since there are symmetries on the fields, we carry out the path integral on the equivalence classes of fields. Note too that the space of equivalence classes is *not* the quotient of a set by a group action. Rather, the picture looks like Figure 7.

That said, the quantum Hilbert space is

$$E(Y) = L^2(\overline{\mathcal{C}_Y}, \mu_Y)$$

and the path integral on a *closed* manifold is

$$Z_X = \int_{\overline{\mathcal{C}_X}} e^{iS_X([P])}\, d\mu_X([P]) = \mathrm{vol}(\overline{\mathcal{C}_X}). \tag{23}$$

If X has boundary, then for each $Q \in \mathcal{C}_{\partial X}$ we let $\overline{\mathcal{C}_X}(Q)$ be the set of bundles over X whose restriction to ∂X is Q, up to isomorphisms which are the identity on ∂X. If X has no closed components, then elements of $\overline{\mathcal{C}_X}(Q)$ have no automorphisms. Define

$$Z_X(Q) = \int_{\overline{\mathcal{C}_X}(Q)} e^{iS_X([P])}\, d\mu_X([P]) = \mathrm{vol}\big(\overline{\mathcal{C}_X}(Q)\big). \tag{24}$$

This is the standard procedure: we integrate over fields which satisfy a boundary condition. It is easy to check that for equivalent fields $Q \cong Q'$ on Y, we have $Z_X(Q) = Z_X(Q')$. So Z_X is a function on the set of equivalence classes of fields, that is, it is an element of $E(\partial X)$.

Theorem. *Equations (22)–(24) define a topological quantum field theory.*

That is, these definitions satisfy the axioms stated in Lecture #1.

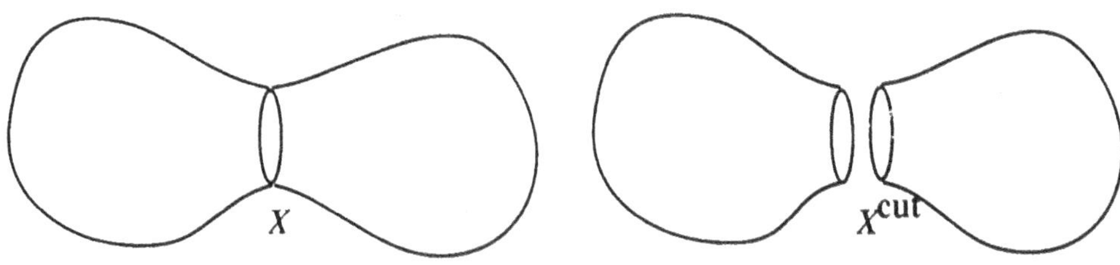

Figure 11: A simple gluing

We don't write a complete proof, but do illustrate the gluing law in case $\partial X = \emptyset$ (Figure 11). (As usual, this simplification is only for ease of notation.) Let $\{Q\}$ be a set of representations of $\overline{\mathcal{C}_Y}$. Recall the gluing map (20). Then

$$\begin{aligned} Z_X &= \mathrm{vol}(\overline{\mathcal{C}_X}) \\ &= \sum_{Q\in\{Q\}} \int_{\overline{\mathcal{C}_X}(Q)} d\mu_X([P]) \\ &= \sum_{Q\in\{Q\}} \left\{ \int_{\overline{\mathcal{C}_{X^{\mathrm{cut}}}}(Q\sqcup Q)} d\mu_{X^{\mathrm{cut}}}([P^{\mathrm{cut}}]) \right\} \cdot \mu_Y(Q) \qquad \text{by (21))} \\ &= \sum_{Q\in\{Q\}} Z_{X^{\mathrm{cut}}}(Q\sqcup Q)\mu_Y(Q) \\ &= \mathrm{Tr}_Y\, Z_{X^{\mathrm{cut}}}. \end{aligned}$$

Exercise 17. Carefully verify the theorem above.

Consequence of the Axioms.

We note a few elementary properties of topological quantum field theories in general. First, let I be a closed interval (it doesn't matter which—they are all diffeomorphic) and Y a space (d-manifold). Consider the product spacetime $I \times Y$. It is a spacetime which is globally a product of time and space. (We write time first for reasons of orientation.) Its boundary is $-Y \sqcup Y$. So using the orientation and multiplicativity axioms (13) and (14) we identify $Z_{I\times Y}$ as a map

$$Z_{I\times Y}\colon E(Y) \to E(Y)$$

Now the gluing law applied to

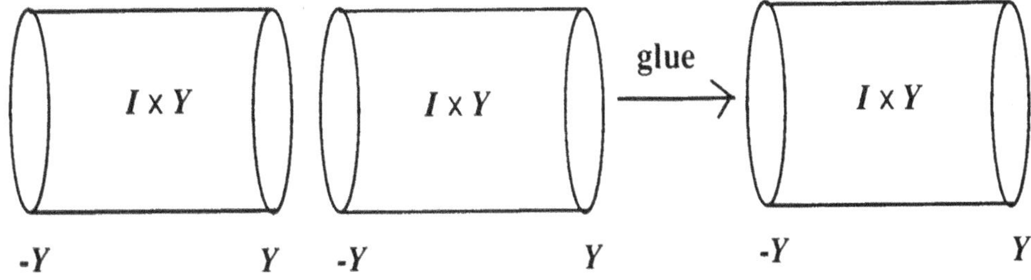

Figure 12: Gluing cylinders

implies

(25) $$(Z_{I\times Y})^2 = Z_{I\times Y}.$$

We may as well assume a nondegeneracy condition:

(26) $$Z_{I\times Y} = \mathrm{id}.$$

In general, (25) only implies that $Z_{I\times Y}$ is a projection operator on $E(Y)$. But only the states in the image ever appear in the theory, so it is hardly a loss of generality to deal with theories which are nondegenerate in the sense that (26) holds.

Similarly, the gluing law applied to

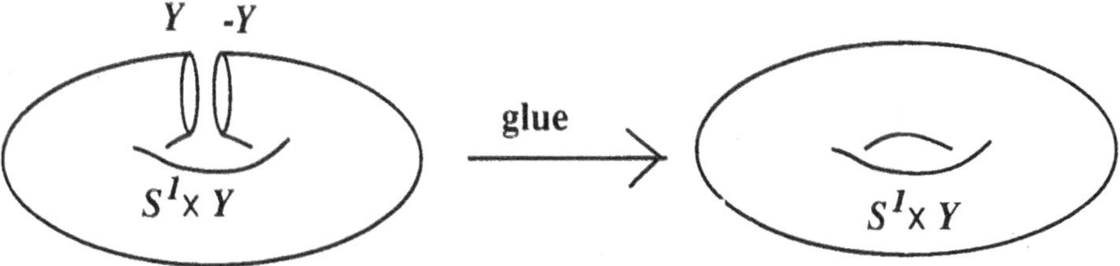

Figure 13: Gluing the ends of a cylinder

shows that

(27) $$Z_{S^1\times Y} = \dim E(Y)$$

in a nondegenerate theory.

Exercise 18.

a) One of the axioms states that a diffeomorphism $f: Y' \to Y$ induces a linear map $f_*: E(Y') \to E(Y)$. Show that (in a nondegenerate theory) if f_0 is homotopic to f_1, then $(f_0)_* = (f_1)_*$.
b) Suppose that we have a 2+1 dimensional theory. Show that $E(S^1 \times S^1)$ carries a representation of $SL_2(\mathbf{Z})$, the group of 2×2 matrices with integer entries and positive determinant which are invertible and whose inverse also has integer entries. (If there are no orientations the appropriate group is $GL_2(\mathbf{Z})$.)

$0+1$ Dimensional Theory.

There is not much to say here. The basic vector space is $E(pt)$ and if the theory is nondegenerate, then by (26)

$$Z_I = \text{identity} : E(pt) \to E(pt).$$

By (27), $Z_{S^1} = \dim E(pt)$.

In the theory constructed from a finite group $\overline{\mathcal{C}_{pt}}$ has one element, the equivalence class of the trivial bundle. So $E(pt)$ is a 1 dimensional vector space. Now $\overline{\mathcal{C}_{S^1}}$ is in $1:1$ correspondence with the conjugacy classes, so (27) implies

$$\sum_{\substack{\text{conjugacy} \\ \text{classes } [h]}} \frac{1}{\#C_h} = 1.$$

In fact, this is true and is equivalent to (22).

$1+1$ Dimensional Theory.

We first discuss $1+1$ dimensional topological theories in general. Our discussion is not meant to be precise, but in a given example one can usually make it so. This material is well-known.

There is a basic vector space in the theory

$$E = E(S^1).$$

The disk D^2 gives an element in this vector space

$$Z_{D^2} \in E.$$

Recall that E is also assumed to be endowed with an inner product. This really is related to the orientation, but we will ignore that here. Also, for simplicity we ignore the conjugation here and treat E as a *real* vector space. The other crucial ingredient is the path integral over the "pair of pants" pictured in Figure 14.

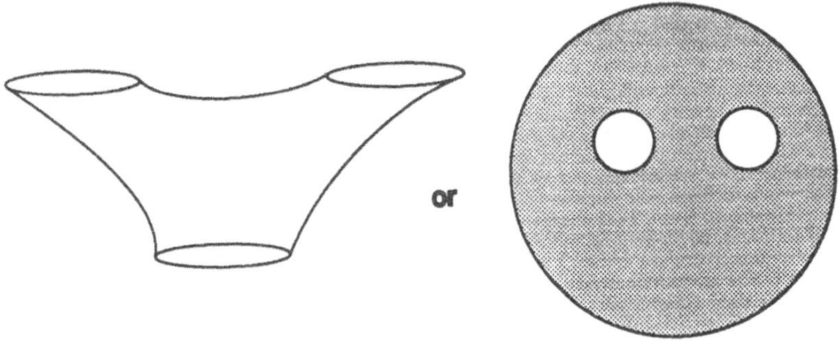

Figure 14: Pair of pants

We identify this path integral (using the inner product to identify $E \cong E^*$) as a map

$$m \colon E \otimes E \to E.$$

Proposition. *The map m is a multiplication which makes E a commutative, associative algebra with identity element Z_{D^2}. Furthermore, m is compatible with the inner product in the sense that*

$$a \otimes b \otimes c \longmapsto \big(m(a,b), c\big), \qquad a, b, c \in E,$$

is totally symmetric in a, b, c.

This is easy to see with pictures, and is proved using the Functoriality Axiom and the gluing law. For example, to see that Z_{D^2} is the identity map, glue the disk to the pair of pants to obtain a cylinder:

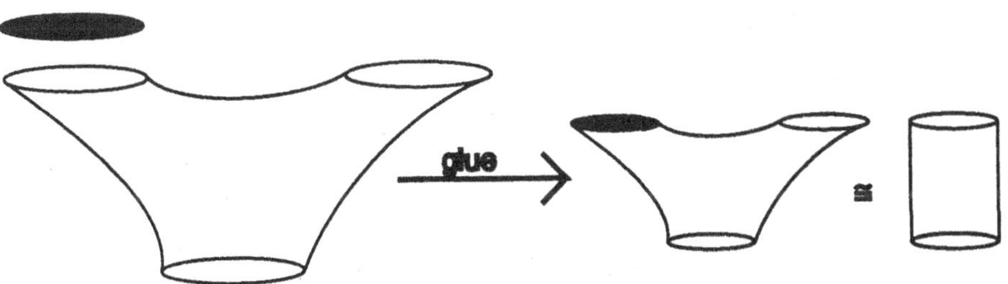

Figure 15: Multiplication by the identity element

By (26) the path integral over the cylinder is the identity map. Associativity is the equivalence of the two diagrams

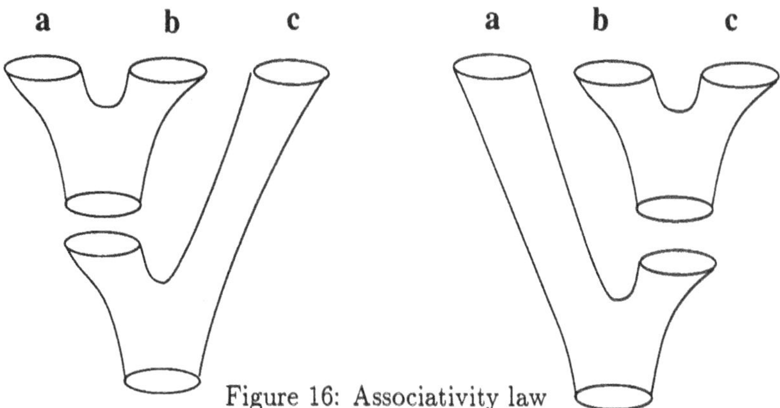

Figure 16: Associativity law

after gluing. The other assertions follow from the self-diffeomorphisms of the pair of pants (Figure 14) which permute the boundary components.

Exercise 19. Prove carefully the assertions in the proposition.

In addition, using the inner product one can show that E is a *semisimple* algebra. This means that there is an isomorphism

$$E \cong \mathbb{C} \times \cdots \mathbb{C}$$

to a product of copies of $\mathbb{C}$. In other words, there exist elements $d_i \in E$, $i = 1, \ldots, \dim E$ such that

$$d_i d_j = \begin{cases} 0 & i \neq j\,; \\ d_j & i = j. \end{cases} \tag{28}$$

These elements are unique up to order.

Exercise 20.
a) Express the identity element of E in terms of d_i.
b) Show that the d_j are orthogonal.

Up to isomorphism the algebra E and inner product are determined by the numbers

$$\lambda_i^2 = (d_i, d_i) = |d_i|^2.$$

Furthermore, the partition function of a surface Σ_g of genus g

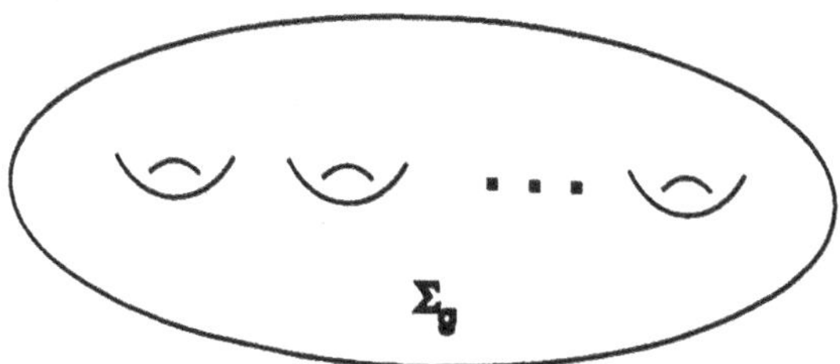

Figure 17: Surface of genus g

is computable in terms of the λ_i^2 as

$$Z_{\Sigma_g} = \sum_i (\lambda_i^2)^{1-g}. \tag{29}$$

Exercise 21. Prove formula (29).

In the context of "conformal blocks", which we will meet in the next lecture, (29) is called the *Verlinde formula*. Here it is presented abstractly for $1+1$ dimensional topological field theories.

We apply this general discussion to the $1+1$ dimensional gauge theory with finite gauge group G. Since $\overline{\mathcal{C}_{S^1}}$ is in $1:1$ correspondence with the conjugacy classes in G, we identify

$$\begin{aligned} E(S^1) &\cong \{\text{central functions on } G\} \\ &\cong \{\text{functions } f\colon G \to \mathbb{C} \text{ invariant under conjugation}\} \end{aligned}$$

Some computation shows that the inner product is

$$(f_1, f_2) = \frac{1}{\#G} \sum_{g\in G} f_1(g)\overline{f_2(g)}. \tag{30}$$

Exercise 22. Derive (30) from the measure (19) and the definition (22).

Finally, we can do the path integral (24) over the pair of pants (Figure 14) to determine the multiplication. It turns out to be the convolution

$$(f_1 * f_2)(g) = \sum_{g_1 g_2 = g} f_1(g_1) f_2(g_2). \tag{31}$$

Exercise 23. Derive (31).

Now we can apply the *Schur orthogonality relations* of finite group theory to determine the special elements d_i in (28). They are

$$d_i = \frac{\dim V_i}{\#G} \cdot \chi_i,$$

where $V_1, \ldots, V_s$ are the *irreducible* representations of G and χ_i is the character of V_i. From (29) we conclude

$$Z_{\Sigma_g} = \sum_i \left(\frac{\dim V_i}{\#G} \right)^{2-2g}. \tag{32}$$

A cute formula.

As a simple application of the ideas in this lecture, and as a further illustration of the gluing law, we use (32) to derive a nice formula which counts the representations of surface groups in finite groups. Although it may hold some intrinsic interest, we present it here as a toy illustration of how the ideas in topological quantum field theory, particularly the gluing law, can be used to derive a result in a field external to quantum field theory.

To begin, recall from (23) that

$$Z_{\Sigma_g} = \mathrm{vol}(\overline{\mathcal{C}_{\Sigma_g}}).$$

Let $\Gamma_g = \pi_1 \Sigma_g$ be the surface group. Then from (18) and (19) we have

$$Z_{\Sigma_g} = \sum_{[\gamma] \in \mathrm{Hom}(\Gamma_g, G)/G} \frac{1}{\#C_\gamma}$$

where $\gamma \colon \Gamma_g \to G$ is a homomorphism, $[\gamma]$ its conjugacy class in $\mathrm{Hom}(\nabla_g, G)/G$, and $C_\gamma \subset G$ the stabilizer of γ in $\mathrm{Hom}(\Gamma_g, G)$ under the conjugation action of G. So

$$\begin{aligned} Z_{\Sigma_g} &= \sum_{[\gamma] \in \mathrm{Hom}(\Gamma_g, G)/G} \frac{1}{\#C_\gamma} \\ &= \sum_{\gamma \in \mathrm{Hom}(\Gamma_g, G)} \frac{1}{\#[\gamma]} \frac{1}{\#C_\gamma} \\ &= \frac{1}{\#G} \# \mathrm{Hom}(\Gamma_g, G). \end{aligned} \tag{33}$$

Combining (33) and (32) we find our desired "cute formula":

$$\# \mathrm{Hom}(\Gamma_g, G) = (\#G)^{2g-1} \sum_i (\dim V_i)^{2-2g}. \tag{34}$$

Recall that $V_1, \ldots, V_s$ is a representative list of irreducible representations of G. Equation (34) can be proved using standard group theoretical methods, though it is not obvious. Here we have derived it using the gluing laws of quantum field theory.

Exercise 24. Check (34) for abelian groups G.

Exercise 25. Check (34) for $g = 0$ and $g = 1$.

LECTURE #3: 2 + 1 DIMENSIONAL THEORIES WITH FINITE GAUGE GROUP

The structure of 2+1 dimensional theories has been investigated by Reshetikhin/Tura Kontsevich, Walker, Alvarez-Gaume, Sierra, Gomez, and many more—I am not sure of exact references. One gets the impression from these investigations that eventually there is an equivalence

$$\left\{\begin{array}{l}2+1\text{ dimensional TQFT with}\\ \text{stronger gluing laws than}\\ \text{stated in these lectures}\end{array}\right\} \longleftrightarrow \left\{\begin{array}{l}\text{“braided monoidal categories”}\\ \text{of a certain type, or some}\\ \text{sort of quantum group data}\end{array}\right\}$$

(Geometry) (Algebra)

Much attention has been devoted to constructing the geometry from the algebra. For example, the work of Reshetikhin/Turaev [RT], Walker [Wa] give precise theorems about this. Here we discuss the reverse, showing how some algebraic structures emerge out of the gluing laws of the TQFT. Rather than work in some abstract axiomatic setting, we consider the gauge theory with finite structure group G that we discussed in Lecture #2. The algebraic structures we encounter are closely related to the symmetries of the fields.

For the material in this lecture I refer to my paper [FQ] with Quinn. I also refer to Quinn's lectures [Q1], which go into more detail about the algebraic structures which arise. My recent work [F5] derives the algebraic structure (including the modular tensor categories and quantum groups) directly from the classical action.

Surfaces with boundary..

In the 2 + 1 dimensional theory the "spaces" are 2 dimensional. The new point is to consider cutting and pasting (gluing) of surfaces, that is, to extend the theory to surfaces with boundary. Recall that for *closed* surfaces Y we constructed a quantum Hilbert space $E(Y)$ in (22). Our goal is to extend this construction to compact surfaces with boundary, and then to derive a gluing law for this extension.

For a closed surface Y the quantum Hilbert space $E(Y)$ consists of functions on $\overline{\mathcal{C}_Y}$, the set of *equivalence classes* of bundles over Y. If we want to allow surfaces with boundary, define quantum Hilbert spaces attached to these surfaces, and glue such surfaces along boundary components, then this gluing must induce a gluing on *equivalence classes* of bundles. On the other hand, gluing is *not* well-defined on equivalence classes of bundles.

Example (in lower dimensions): Up to isomorphism there is only one bundle over the interval I—the trivial bundle. But if we glue the two ends of I to form a circle

S^1, then there are many possible bundles over S^1 obtained by gluing the trivial bundle.

To make gluing well-defined on equivalence classes we introduce *basepoints*. Thus a surface Y with boundary is meant to have a basepoint y *on each component of* ∂Y as in Figure 18. Then we define

$$C'_Y = \{\langle Q; q_1, \dots, q_k\rangle : Q \to Y \text{ is a principal } G \text{ bundle and } q_i \in Q_{y_i}.\}$$

Here Q_{y_i} is the fiber of Q over the basepoint y_i.

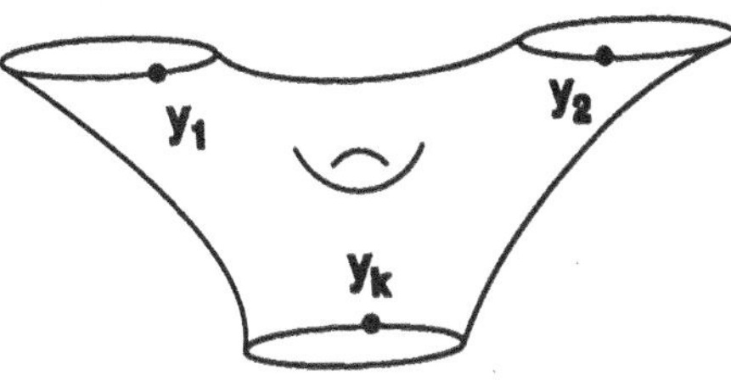

Figure 18: Surface with basepoints

Example (in lower dimensions): A bundle over I with basepoints looks like

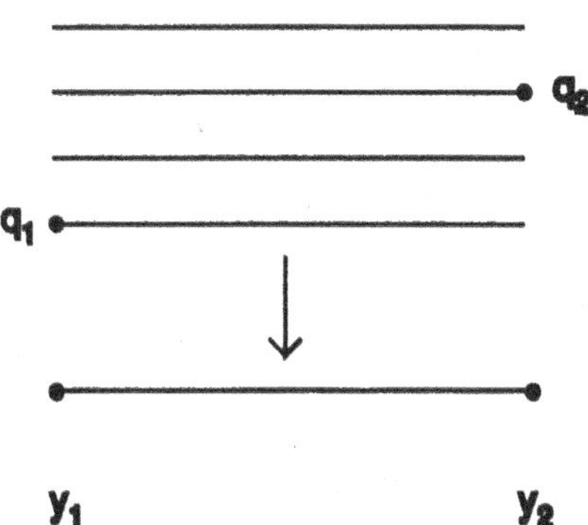

Figure 19: A principal G bundle over the interval with basepoints

Now there is no ambiguity about the gluing—we identify q_1 and q_2 when we glue. If $q_2 = q_1 \cdot h$ for $h \in G$, then the holonomy of the glued bundle is the conjugacy class of h.

There is again a notion of isomorphism for elements of C'_Y—now bundle maps are required to preserve the basepoints. The set of equivalence classes (still finite) is denoted $\overline{C'_Y}$. As we did in the example above, we can define C'_Y for all compact manifolds Y, not just for surfaces.

Exercise 26.

a) Determine $\overline{C'_I}$ for an interval I.

b) Determine $\overline{C'_{D^2}}$ for the disk.

c) Determine $\overline{C'_{I\times S^1}}$ for the cylinder.

d) Determine $\overline{C'_Y}$ for the pair of points Y.

e) Determine $\overline{C'_Y}$ for the surface $Y = \Sigma_g$ – disk.

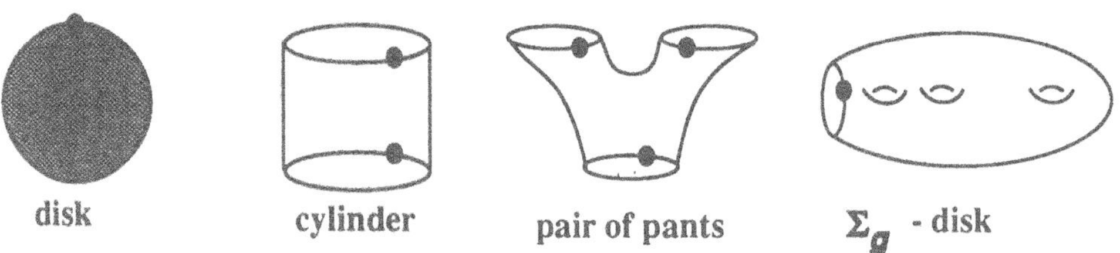

Figure 20: Elementary surfaces with basepoints

Exercise 27. Suppose S is a *connected* manifold with basepoint $s \in S$. Let $R \to S$ and $R' \to S'$ be *isomorphic* G bundles with basepoints $r \in R_S$, $r' \in R'_S$ in the fibers over s. Prove that there is a *unique* isomorphism $R' \to R$ which maps r' to r.

This last exercise implies that elements of C'_Y do *not* have nontrivial automorphisms. In other words, the picture of C'_Y is

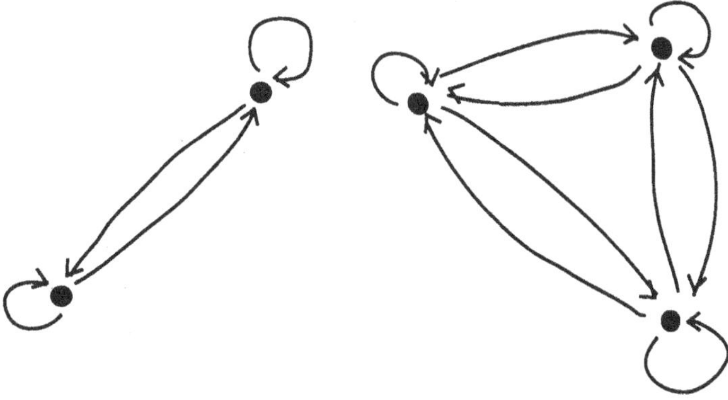

Figure 21: The configuration space for bundles with basepoints

There is a *unique* arrow in each direction between isomorphic bundles. This is why gluing is well-defined on the quotient.

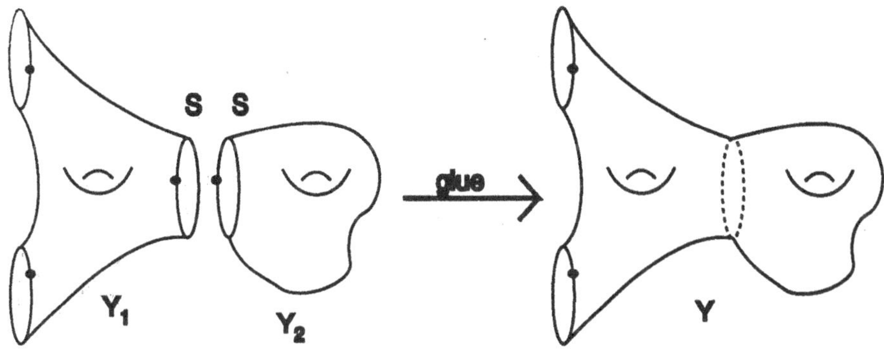

Figure 22: Gluing surfaces and bundles with basepoints

Now a bundle over S^1 *with basepoint* has a definite holonomy $h \in G$. (Recall that without the basepoint the holonomy is determined only up to conjugacy.) So in Figure 22, if $Q_1 \in \mathcal{C}'_{Y_1}$ and $Q_2 \in \mathcal{C}'_{Y_2}$, then we can glue Q_1 to Q_2 along S if the holonomies agree, and this gluing is well-defined on equivalence classes. Denote the glued bundle by $Q_1 \circ Q_2$. Note we can glue along any components of a surface with boundary—for example, as pictured in Figure 23. As usual, we use the picture in Figure 22 only to simplify the notation. So the gluing operation is well defined as a map

$$\begin{array}{ccccc} \overline{\mathcal{B}_{Y_1,Y_2}} & \subset & \overline{\mathcal{C}'_{Y_1}} & \times & \overline{\mathcal{C}'_{Y_2}} \\ & \searrow g & & & \\ & & \overline{\mathcal{C}'_{Y}} & & \end{array}$$

where $\mathcal{B}_{Y_1,Y_2}$ is the set of pairs $\langle Q_1, Q_2 \rangle$ (with basepoints) whose holonomies agree along S, and $\overline{\mathcal{B}_{Y_1,Y_2}}$ is the set of equivalence classes of such pairs of bundles.

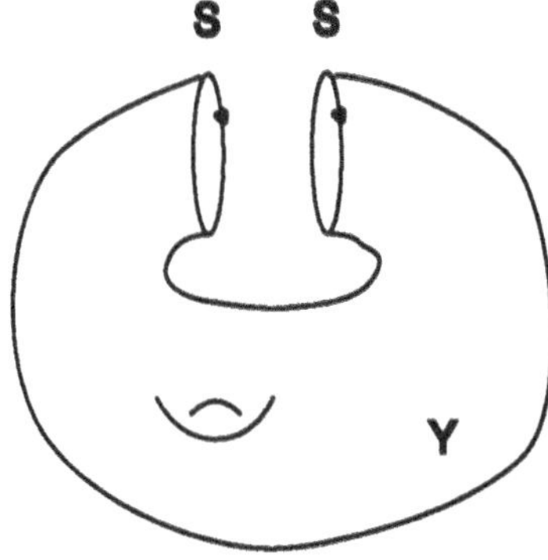

Figure 23: Another gluing of pointed surfaces

The cylinder.

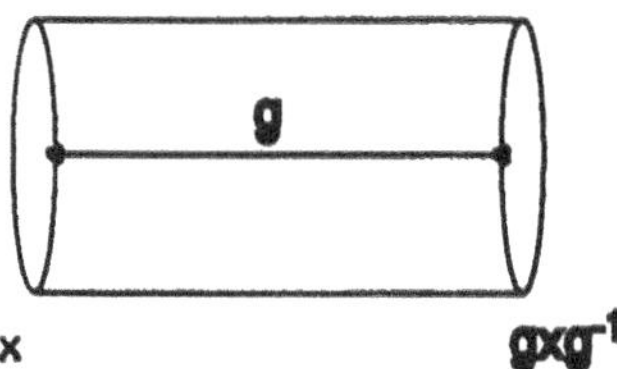

Figure 24: Pointed bundle over the cylinder

Consider the cylinder $I \times S^1$. If $Q \to I \times S^1$ is a bundle with basepoints, then we can determine group elements $g, x \in G$ where g is the parallel transport from one end to the other, and x is the holonomy around the left end. The holonomy around the right end is gxg^{-1} (Figure 24). Now consider the gluing

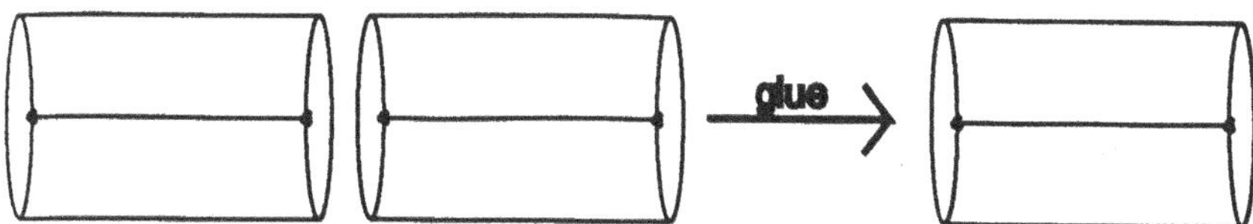

Figure 25: Gluing pointed cylinders

If we have bundles Q_1, Q_2 over the cylinders on the left, we can glue if the holonomies in middle agree. So if Q_i is given by group elements $\langle x_i, g_i \rangle$, then we can glue if and only if $x_2 = g_1 x_1 g_1^{-1}$.

We picture the bundle $\langle x, g \rangle$ as an arrow

Figure 26: Schematic representation of a pointed bundle over the pointed cylinder

So the set of $(\#G)^2$ equivalence classes of bundles over $I \times S^1$ (always with basepoints) is the set of arrows in the picture:

Figure 27: Equivalence classes of pointed bundles over the cylinder

(This is a picture for the symmetric group S_3, in fact.) Note that at the identity e sits a copy of G, and more generally the set of arrows from x to x is the *centralizer* subgroup $C_x \subset G$. This is the subgroup of elements of G which commute with x. The set of all arrows is a *groupoid*.

Exercise 28. Draw this picture for an abelian group G.

Quantization.

The measure we use on $\overline{C'_Y}$ for any Y is still (19), but now understood with basepoints. If Y has no closed components, then there are no nontrivial automorphisms, and the measure is identically 1.

The generalization of (22) is

$$E(Y) = L^2(\overline{C'_Y}, \mu_Y)$$

for any compact surface Y (with basepoints).

Let's first consider the cylinder. We've already seen that $\overline{C'_{I\times S}}$ is a groupoid. This differs from a group in that composition is not always possible. Now, the multiplication

$$m\colon H \times H \longrightarrow H$$

on an arbitrary *group* H determines by duality a *coalgebra* structure

$$\Delta\colon \mathcal{F}(H) \longrightarrow \mathcal{F}(H \times H) \cong \mathcal{F}(H) \otimes \mathcal{F}(H)$$

on the vector space of functions $\mathcal{F}(H)$ on the group. (The last isomorphism needs to be thought about if H is not a finite group.) The same holds for our groupoid of fields on the cylinder.

Proposition. *Let $A = E(I \times S^1)$. Then A is a coalgebra. That is, there is a comultiplication*

$$\Delta\colon A \to A \otimes A$$

and a counit

$$\varepsilon\colon A \to \mathbb{C}$$

There is also a "compatible" inner product, the L^2 inner product. We give the formulas, writing an equivalence class of bundles on $I \times S^1$ as a pair of group elements $\langle g, x\rangle$. Suppose $a \in A$. So $a\colon \overline{C'_{I\times S^1}} \to \mathbb{C}$. An element of $A \otimes A$ is a function on $\overline{C'_{I\times S^1}} \times \overline{C'_{I\times S^1}}$. The coproduct is

$$\Delta a\big(\langle x_1, g_1\rangle, \langle x_2, g_2\rangle\big) = \begin{cases} a(\langle x_1, g_2 g_1\rangle & \text{if } x_2 = g_1 x_1 g_1^{-1}; \\ 0, & \text{if } x_2 \neq g_1 x_1 g_1^{-1}. \end{cases}$$

$$\varepsilon(a) = \sum_x a(\langle x, e\rangle)$$

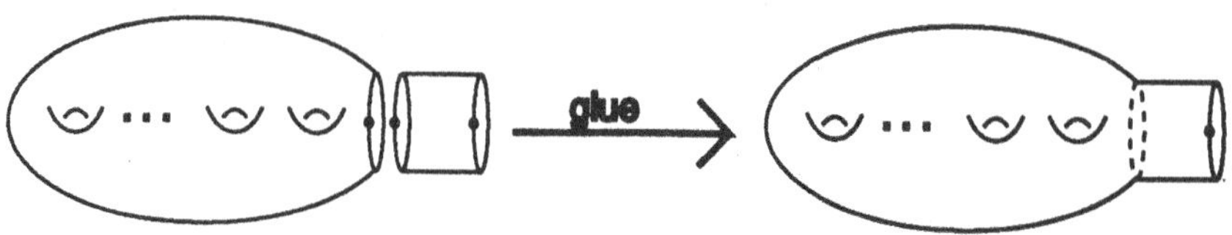

Figure 28: Gluing a pointed cylinder onto a pointed surface

Gluing Laws.

Now consider an arbitrary compact surface Y. By gluing a cylinder onto the boundary (Figure 28) we obtain an "action" of the groupoid $\overline{\mathcal{C}'_{I\times S^1}}$ on $\overline{\mathcal{C}'_Y}$. In the quantum picture this translates to

Proposition. *$E(Y)$ is a unitary comodule for the coalgebra A.*

A *comodule* is a *corepresentation*, dual to a representation of an algebra. Finally, the gluing law states that in the picture

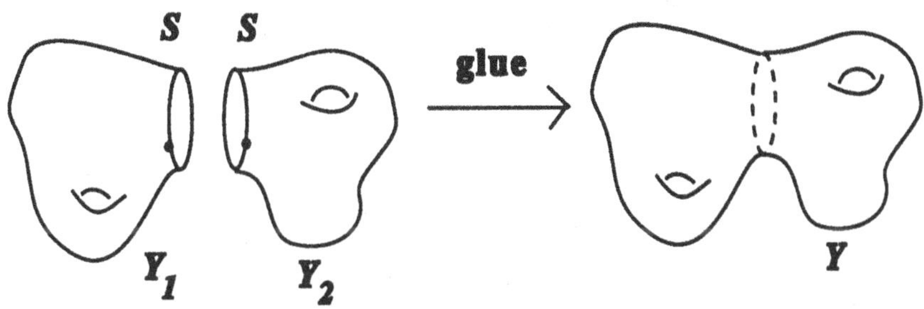

Figure 29: Gluing pointed surfaces

we have

$$E(Y) \cong \frac{1}{\#G}\, E(Y_1) \boxtimes_A E(Y_2).$$

Here '$\boxtimes_A$' is the *cotensor product*, which is dual to the tensor product of modules of algebras. The factor $1/\#G$ is for the inner product.

Conformal Blocks.

This picture can be made more concrete, and so related to standard notions in conformal field theory. Briefly, the coalgebra A is *semisimple*, so there is a finite number of irreducible corepresentations up to isomorphism, say $E_1, \dots, E_N$. Thus for any compact surface Y we can decompose the comodule (corepresentation) $E(Y)$ into irreducibles:

$$E(Y) \cong \bigoplus_\lambda E(Y, \lambda) \otimes E_\lambda$$

where now the $E(Y, \lambda)$ are *vector spaces*. The λ which label the irreducible representations are called the *labels*, and we think of assigning the vector space $E(Y, \lambda)$ to the surface Y with boundary labeled by λ. This is called a *conformal block* in the physics literature; Graeme Segal [S] uses the term *modular functor*. Whatever the language, the gluing law takes the form

$$E(Y) \cong \bigoplus_{\lambda} \frac{\dim E_\lambda}{\#G} \cdot E(Y_1, \lambda) \otimes E(Y_2, \lambda). \tag{35}$$

The factor refers to the inner product structure; it multiplies the inner product on the Hilbert space which follows.

Exercise 29. Show that the definitions

$$\tilde{E}(S^1) = E(S^1 \times S^1)$$
$$\tilde{Z}_Y = \dim E(Y)$$

define a $1+1$ dimensional TQFT. The associated algebra is called the *Verlinde algebra*.

Exercise 30. By the axioms of TQFT, specifically the Functoriality Axiom, the symmetries of $S^1 \times S^1$ act on $E(S^1 \times S^1)$. This gives a representation of $SL_2(\mathbb{Z})$ on $E(S^1 \times S^1)$. Compute this representation.

Exercise 31. Show that the factor in (35) is S^1_λ, where S is the matrix $\begin{pmatrix} 0 & 1 \\ -1 & 0 \end{pmatrix}$ in $SL_2(\mathbb{Z})$, 1 is the trivial representation, and λ is the given label λ. The matrix element S^1_λ refers to the action of S on $E(S^1 \times S^1)$, as computed in Exercise 30.

Tensor Products.

I just mention that the vector space $E(P)$ of the pair of pants

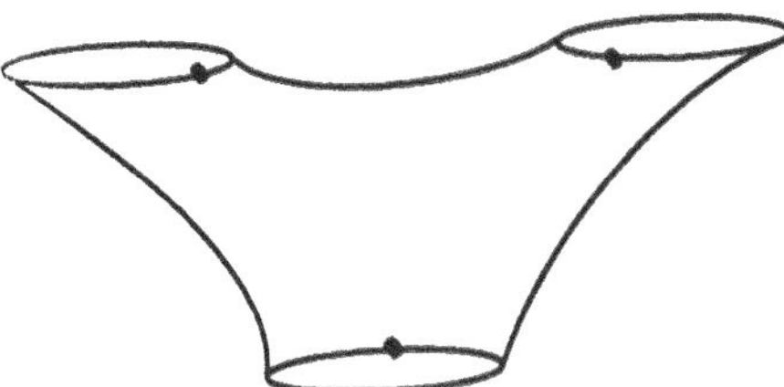

Figure 30: Pointed pair of pants

plays a special role. Namely it is an $(A \times A \times A)$-comodule since there are 3 boundary components. Then if E_1 and E_2 are A-comodules we can form a new A-module $E_1 \otimes E_2$ by taking a sort of double cotensor product over A:

$$E_1 \otimes E_2 = E_1 \boxtimes_A E(P) \boxtimes_A E_2.$$

This leads to a Hopf algebra structure on A, or more precisely to a Hopf algebra whose underlying coalgebra is isomorphic to A. Hopf algebras (and quasi-Hopf algebras) appeared in connection with finite groups in [DPR]. From the classical action one first gets to the modular tensor categories, etc., which have been studied recently. The complete story is in in [F5].

LECTURE #4: CLASSICAL SU_2 CHERN-SIMONS THEORY

We move now to the Chern-Simons gauge theory with gauge group SU_2, a *topological* theory in $2+1$ dimensions. The action is no longer trivial, and the space of equivalence classes of fields is not finite, nor even discrete. So new geometric and analytic ideas now enter, and indeed I would say that they are far from understood at this writing. In this lecture we discuss the *classical theory*; in the next lecture we turn to the quantum theory.

Classical mechanics and classical field theory have long been studied by mathematicians, certainly dating back to Newton's time. Over the years it has inspired many mathematical developments, notably the calculus of variations and symplectic geometry (corresponding to the Lagrangian and Hamiltonian pictures of the theory). More recently, classical gauge theory in physics was discovered to fit neatly into the independently developed mathematical theory of connections and curvature. So it is no surprise that the Chern-Simons lagrangian (see [CS] for the original reference in the mathematics literature) can be understood within our present-day mathematics. In fact, a careful study is illuminating, and I refer to my series of papers in progress [F1], [F2], and [F3] for a detailed development. A survey of these ideas appears in my lecture in the Colloquium connected to this Summer School [F4]. Here we outline some of the basic ideas for the SU_2 case.

Fixed Data.

The dimension of the theory is $d = 2$; this is a $2+1$ dimensional theory. Our gauge group is the Lie group

$$G = SU_2 = \left\{ \begin{pmatrix} \alpha & \beta \\ -\bar{\beta} & \bar{\alpha} \end{pmatrix} : \alpha, \beta \in \mathbb{C},\ |\alpha|^2 + |\beta|^2 = 1 \right\}.$$

As a manifold G is diffeomorphic to the 3-sphere S^3. We also need to fix a bilinear form on the Lie algebra

$$\mathfrak{g} = \mathfrak{su}_2 = \left\{ \begin{pmatrix} ix & \alpha \\ -\bar{\alpha} & -ix \end{pmatrix} : x \in \mathbb{R},\ \alpha \in \mathbb{C} \right\}.$$

This bilinear form is

$$\langle a, b \rangle = \frac{-k}{8\pi^2} \operatorname{Tr}(ab), \qquad a, b \in \mathrm{SU}_2, \tag{36}$$

for some integer

$$k = 1, 2, \ldots$$

Note that (36) is invariant under conjugation. The reason for this normalization is the following. Let θ be the Maurer-Cartan 1-form on SU_2. If $g \in SU_2$ is written as a matrix, we write

$$\theta = g^{-1} dg$$

which is a matrix valued 1-form, and the values are elements of $\mathfrak{su}_2$, i.e., traceless skew-hermitian matrices.

Exercise 32. Write $g = \begin{pmatrix} \alpha & \beta \\ -\bar{\beta} & \bar{\alpha} \end{pmatrix}$ and write a formula for θ in terms of α, β.

Our normalization of (36) is chosen so that

$$\int_{SU_2} \frac{k}{24\pi^2} \operatorname{Tr}(\theta^3) \qquad \text{is an integer.} \tag{37}$$

Exercise 33. Verify (37).

Fields.

Our spacetimes are compact *oriented* 3-manifolds X. The fields are *gauge fields* or *connections*:

$$\mathcal{C}_X = \{SU_2 \text{ connections } \Theta \text{ over } X\}.$$

What is a connection? First of all it lives on a principal G bundle $P \to X$. So P is a manifold on which G acts (on the right) *freely* with quotient X. This is the same definition as in Lecture #2, where G was a finite group. But now SU_2 has no nontrivial topology in dimensions 0 and 1, and some elementary topology proves that there exists an isomorphism $P \cong X \times G$, where $X \times G$ is the *trivial* G bundle with G action

$$\begin{aligned} (X \times G) \times G &\longrightarrow (X \times G) \\ \langle x, g\rangle \times h &\longmapsto \langle x, gh\rangle. \end{aligned}$$

Another way to say this is to say that there exists a section

$$p: X \to P$$

of the principal bundle $\pi: P \to X$; that is, $\pi \circ p = id_X$.

Since all SU_2 bundles are trivial, we might think of fixing once and for all $P = X \times G$. However, as we saw in the theory with finite gauge group, this does not exhibit all of the symmetry when we glue bundles. Recall that an *isomorphism* of bundles is a map

$$\varphi: P' \to P$$

such that

$$\varphi(p' \cdot g) = \varphi(p')g$$

for all $g \in G$, $p' \in P'$ and φ induces the identity map on the quotients $P'/G = P/G = X$. If $p' = p$ then φ is called a gauge transformation. We denote the group of gauge transformations by $\mathcal{G}_P$.

Exercise 34. Show that the group of gauge transformations of the trivial bundles $X \times G$ can be identified with the group of maps $\mathrm{Map}(X, G)$ operating by *left* multiplication.

Exercise 35.

a) If X is a closed, oriented, and connected 3-manifold, use the 3-form (37) to show how to distinguish different homotopy classes of maps $X \to G$, i.e., different components of $\mathrm{Map}(X, G)$.

b) The set of components $\pi_0\mathcal{G}$ of any group $\mathcal{G}$ has a natural group structure. Show that if X is a closed, oriented, and connected 3-manifold, then

$$\pi_0 \, \mathrm{Map}(X, G) \cong \mathbb{Z}.$$

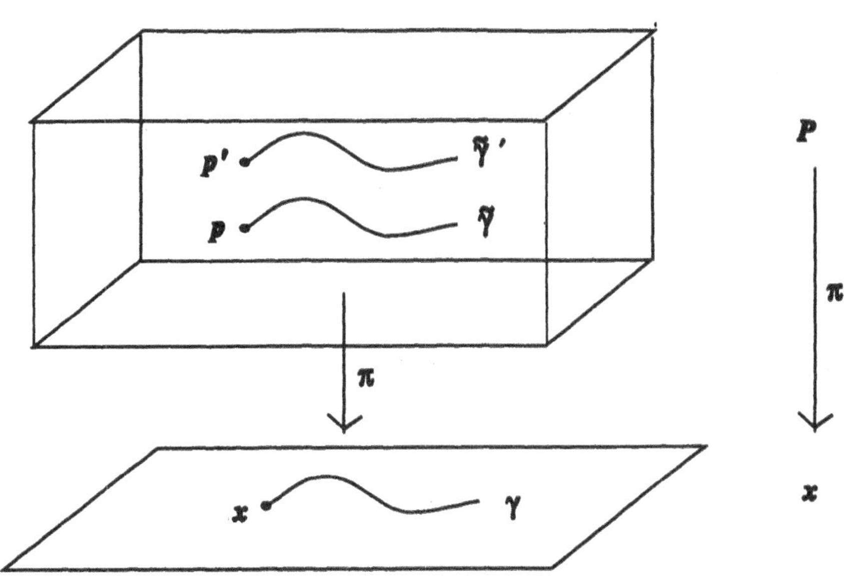

Figure 31: Horizontal lifts in a principal bundle with connection

Geometrically, a *connection* Θ is a structure on P which enables us to lift paths on X in a G-invariant way. (Figure 31) That is, given a path γ in X starting at $x \in X$, and a point $p \in P_x$ in the fiber of P over x, then there is a lifted path $\tilde{\gamma}$ in P starting at p so that $\pi(\tilde{\gamma}) = \gamma$. Of course such paths exist; the connection tells us how to choose one. If $p' = p \cdot g$ is another point of P_x, then the lifted path $\tilde{\gamma}'$ starting at p' must equal $\tilde{\gamma} \circ g$.

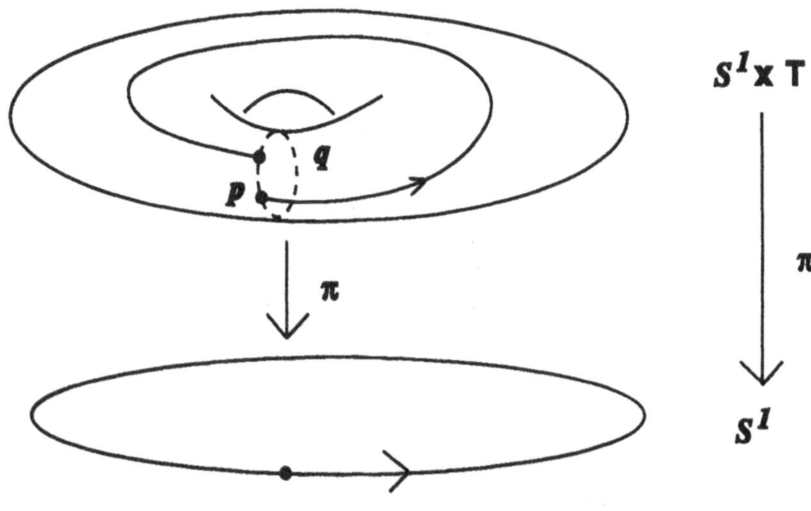

Figure 32: Holonomy of a connection on a $\mathbb{T}$ bundle over S^1

We will return presently to the analytic expression for a connection.

Chern-Simons Action on S^3.

As a warm-up we first look at a geometric definition of holonomy for circle bundles with connection. It suffices to consider the trivial bundle $S^1 \times \mathbb{T} \xrightarrow{\pi} S^1$. Here S^1 is the circle as a manifold and $\mathbb{T}$ is the circle group, also called U_1:

$$\mathbb{T} = \{e^{i\theta} : \theta \in \mathbb{R}\} = \{\lambda \in \mathbb{C} : |\lambda| = 1\}.$$

Now a connection on this bundle gives us a way to lift the base to a *path* in $S^1 \times \mathbb{T}$, which is not necessarily closed, if we give an initial point p. Let q be the terminal point. Now in the fiber over x, here identified with $\mathbb{T}$, we can measure the directed length from p to q. Suppose the total length of $\mathbb{T}$ is 1; then this directed length ℓ is only well-defined up to addition of integers, since we can go around the circle any number of times to measure the directed length. So $e^{2\pi i \ell} \in \mathbb{T}$ is well-defined, and this is the holonomy. (See Figure 32)

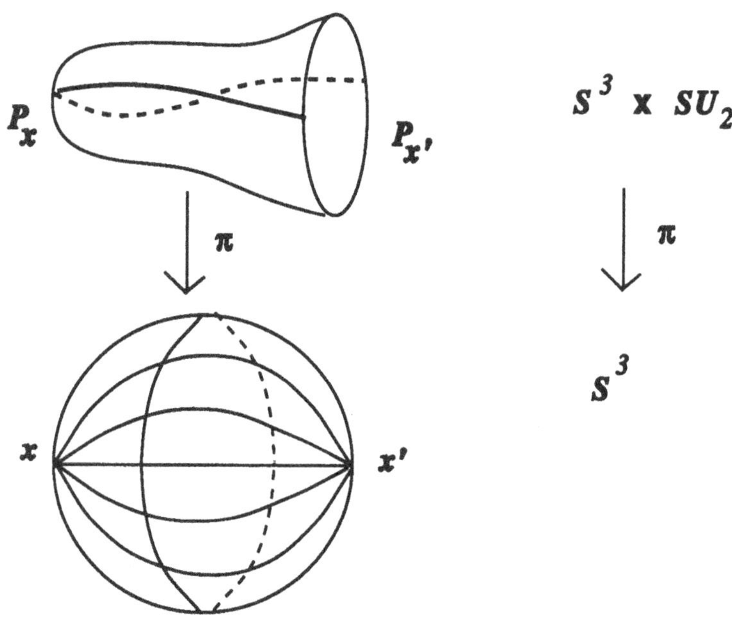

Figure 33: Holonomy of a connection on an SU_2 bundle over S^3

For an SU_2 connection over S^3 the Chern-Simons action is defined similarly. Consider the trivial bundle $P = S^3 \times SU_2 \xrightarrow{\pi} S^3$ with a connection Θ. Fix a point x in S^3 and let x' be the antipodal point. Also, fix $p \in P_x$. Then lift all of the semi-great circles from x to x' starting at p. (See Figure 33.) The lifted paths terminate in the fiber $P_{x'}$ which we identify with SU_2. (Recall this is diffeomorphic to a 3-sphere.) These terminal points fill out a 2-sphere in $P_{x'}$. In fact, the semi-great circles are parametrized by the equatorial S^2, so the terminal points comprise a map $\tau\colon S^2 \to P_{x'}$. Let R be an oriented region in $P_{x'}$ whose boundary is the image of τ (Figure 34). Then the Chern-Simons action is

$$S_X(\Theta) = \int_R \frac{k}{24\pi^2} \operatorname{Tr}(\theta^3). \tag{38}$$

Because of the condition (37), if we change the region R, then (38) changes by an integer. So the quantity

$$e^{2\pi i S_X(\Theta)} \in \mathbb{T}$$

is well-defined. Notice that (38) is just the Wess-Zumino-Novikov-Witten action for the map τ. Already we see a relationship between-Simons and Wess-Zumino-Novikov-Witten. We can think of this relationship as a *transgression*, which is not a sin, but rather a relationship in a fiber bundle between the geometry of the fiber and the geometry of the base (with a shift in dimension). Here the 2 dimensional geometry of the fiber is related to the 3 dimensional geometry of the base.

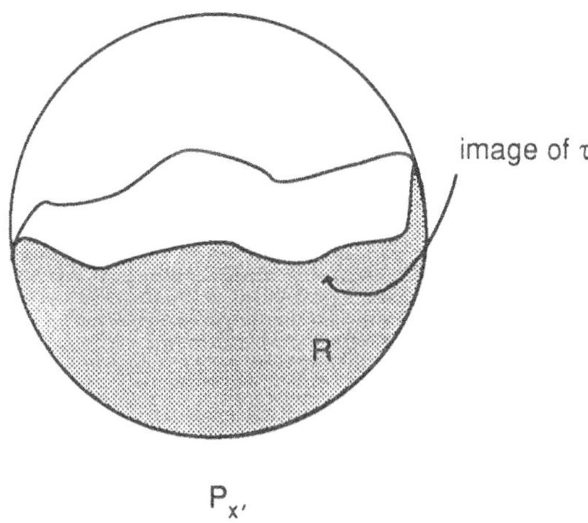

Figure 34: Bounding the image of τ

Chern-Simons Action on closed spacetimes.

Now let Θ be a connection on a bundle $\pi\colon P \to X$, where X is an arbitrary closed, oriented 3-manifold. We now need analytic formulas to define the Chern-Simons action. Rather than give an abstract treatment of connections in general, which you can find in my paper [F1], let me express this in a form often found in the physics literature. Thus suppose $p\colon X \to P$ is a *section* of $P \xrightarrow{\pi} X$, i.e., $\pi \circ p = id_X$. Set

$$p^*\Theta = A = A_i\, dx^i$$

relative to local coordinates x^1, x^2, x^3 on X. Here $A_i = A_i(x)$ are functions on X with values in the Lie algebra $\mathfrak{g}$. If p' is a different section, then we can write

$$p' = p \cdot g \tag{39}$$

for some function $g\colon X \to G$. We have

$$\begin{aligned} p^*\Theta = A^1 &= g^{-1}Ag + g^{-1}dg \\ &= \left(g^{-1}A_i\, g + g^{-1}\frac{\partial g}{\partial x^i}\right) dx^i. \end{aligned}$$

The *curvature* Ω is a $\mathfrak{g}$-valued 2-form on P constructed from the connection Θ, which recall is a $\mathfrak{g}$-valued 1-form on P:

$$\Omega = d\Theta + \frac{1}{2}[\Theta \wedge \Theta].$$

The formula on X, induced by a section $p\colon X \to P$, is perhaps more familiar:

$$p^*\Omega = F = dA + A^2 = \frac{1}{2}F_{ij}\, dx^i \wedge dx^j = \frac{1}{2}\left(\frac{\partial A_j}{\partial x^i} - \frac{\partial A_i}{\partial x^j} + [A_i, A_j]\right) dx^i \wedge dx^j.$$

Exercise 36. If p' is another section as in (39), show

$$p'^*\Omega = F' = gFg^{-1}.$$

The *Chern-Simons form* α is a 3-form which lives on P:

$$\alpha = \alpha(\Theta) = \langle\Theta \wedge \Omega\rangle - \frac{1}{6}\langle\Theta \wedge [\Theta \wedge \Theta]\rangle.$$

Again, the form on the base may look more familiar:

$$\begin{aligned} p^*\alpha &= \frac{-k}{8\pi^2}\,\mathrm{Tr}(A \wedge F - \frac{1}{3}A^3) \\ &= \frac{-k}{8\pi^2}\,\mathrm{Tr}(\frac{1}{2}A_iF_{jk} - \frac{1}{3}A_iA_jA_k)\,\varepsilon^{ijk}\, dx^1 \wedge dx^2 \wedge dx^3. \end{aligned}$$

Exercise 37.

a) Compute $d\alpha$ in case X has dimension ≥ 4.
b) Compute $(p')^*\alpha$ for the section (39) in terms of $p^*\alpha$.

Define an action which now depends on the section p:

$$S_X(\Theta, p) = \int_X p^*\alpha(\Theta). \tag{40}$$

We would like to get rid of the dependence on p, so we investigate how S_X changes when we change p. Using Exercise 37b we see that for the section p' (39):

$$S_X(\Theta, p') = S_X(\Theta, p) + \int_X d\langle g^{-1}Ag \wedge g^{-1}dg\rangle - \frac{1}{6}\int_X \langle g^{-1}dg \wedge [g^{-1}dg \wedge g^{-1}dg]\rangle. \tag{41}$$

The second term vanishes by Stokes' Theorem, if $\partial X = 0$. The third term can be rewritten as

$$\int_X \frac{k}{24\pi^2}\operatorname{Tr}\big((g^{-1}dg)^3\big). \tag{42}$$

If X is closed, this is an integer by (37). So (40) is independent of p up to an integer. Hence

$$e^{2\pi i S_X(\Theta)} \in \mathbb{T}, \qquad X \text{ closed}, \quad \Theta \in \mathcal{C}_X,$$

is well-defined.

Exercise 38. Check that if $\varphi\colon P' \to P$ is an isomorphism, then

$$e^{2\pi i S_X(\varphi^*\Theta)} = e^{2\pi i S_X(\Theta)}$$

In particular, taking $P' = P$ we see that the Chern-Simons action is gauge invariant.

Exercise 39. Check how the action behaves under orientation reversal and disjoint union.

Notice that our action does not quite fit the axioms of Lecture #1, in particular (1). The value is not in $\mathbb{R}$ but rather in $\mathbb{R}/\mathbb{Z}$, or upon exponentiating, in $\mathbb{T}$. There is a more drastic modification of these axioms for manifolds with boundary.

Chern-Simons Action on spacetimes with boundary.

Suppose X is a compact oriented 3-manifold, possibly with nonempty boundary ∂X. Let Θ be a connection over X. We would like to compute the action by (40), but now we find a more complicated dependence on the trivializing section

p. Namely, (41) still holds since it is a local formula. But now the second term is (Stokes' Theorem):

$$\begin{aligned}\int_X d\langle g^{-1}Ag \wedge g^{-1}dg\rangle &= \int_{\partial X} \langle g^{-1}Ag \wedge g^{-1}dg\rangle \\ &= \int_{\partial X} \frac{-k}{8\pi^2} \operatorname{Tr}(g^{-1}Ag \wedge g^{-1}dg) \\ &= \int_{\partial X} \frac{-k}{8\pi^2} \operatorname{Tr}\left(g^{-1}A_i \frac{\partial g}{\partial y^j}\right) \varepsilon^{ij}\, dy^1 \wedge dy^2\end{aligned}$$

Also, by the usual arguments for the Wess-Zumino-Novikov-Witten term, the third term (42) just depends on the restriction ∂g of g to ∂X—call it $W_{\partial X}(\partial g)$—*if we consider it up to integers.* Again we see the Wess-Zumino-Novikov-Witten action in $1+1$ dimensions related to Chern-Simons in $2+1$. Set

$$c_{\partial X}(a,g) = \exp\left(2\pi i \int_{\partial X} \frac{-k}{8\pi^2} \operatorname{Tr}(g^{-1}ag \wedge g^{-1}dg) + W_{\partial X}(g)\right)$$

where a is a 1-form on ∂X with values in $\mathfrak{g}$ and $g\colon \partial X \to G$. Then the preceding is summarized by

$$e^{2\pi i S_X(\Theta, p\cdot g)} = c(p^*\partial\Theta, g)e^{2\pi i S_X(\Theta,p)}. \tag{43}$$

Both sides are elements of $\mathbb{T}$, or if we like elements of unit norm in $\mathbb{C}$—that is, complex numbers of norm one.

So how do we make sense of the action? It certainly is not independent of the trivializing section $p\colon X \to P$. But the variation depends on the boundary values $\partial p\colon \partial X \to \partial P$ of p and the restriction $\partial\Theta$ of Θ to the boundary ∂X. Let $S_{\partial P}$ be the space of all sections of $\partial P \to \partial X$. Then set

$$L_{\partial X,\partial\Theta} = L_{\partial\Theta} = \{f\colon S_{\partial P} \to \mathbb{C} : f(q\cdot g) = c(q^*\partial\Theta, g)\cdot f(q)\} \text{ for all } g\colon \partial X \to G.$$

In other words, $L_{\partial\Theta}$ is the set of functions from sections of $\partial P \to \partial X$ to $\mathbb{C}$ which transform according to (43). This is a one dimensional complex vector space with an inner product, i.e., a *hermitian line*.

Exercise 40. Verify this last assertion.

Then equation (43) states that the action makes sense as an element of this hermitian line (of unit norm):

$$e^{2\pi i S_X(\Theta)} \in L_{\partial\Theta}.$$

So again we must modify our axioms for an action. Now the action is *exponentiated*. For a closed spacetime it is an element of $\mathbb{T}$ (or $\mathbb{C}$). For a nonclosed spacetime it is an element of a complex line.

The construction of this Chern-Simons line applies to any connection η an a G bundle $Q \to Y$ over a *closed* oriented 2-manifold Y—we obtain a line $L_{Y,\eta} = L_\eta$.

Exercise 41. Verify the following properties of the lines L_η.

a) (*Functoriality*) If $\psi \colon Q' \to Q$ is a bundle isomorphism for bundles Q', Q over Y, and η is a connection on Y, then there is an induced isometry

$$\psi_* \colon L_{\psi^*\eta} \to L_\eta. \tag{44}$$

b) (*Orientation*) There is an isometry

$$L_{-Y,\eta} \cong \overline{L_{Y,\eta}}.$$

c) (*Disjoint Union*) There is an isometry

$$L_{\eta_1 \sqcup \eta_2} \cong L_{\eta_1} \otimes L_{\eta_2}$$

if η_1, η_2 are fields (connections) on Y_1, Y_2.

Exercise 42. State and derive analogous properties of the Chern-Simons action. These should be analogous to the corresponding axioms in Lecture #1.

Gluing Law.

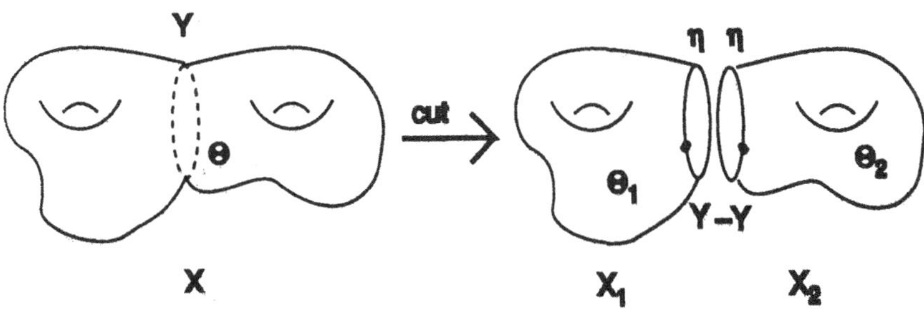

Figure 35: Gluing connections

The locality of a usual sort of action is the gluing law (6), (7). In Chern-Simons this gets modified to accommodate the different sort of action that we have. Also, because of the exponential it looks like a multiplicativity property rather than an additivity property. Figure 35 illustrates a typical gluing. A closed 3-manifold X is cut along Y and a connection Θ cut into two connections Θ_1, Θ_2 which both restrict to η on Y. Then

$$e^{2\pi i S_{X_1}(\Theta_1)} \in L_{Y,\eta}, \qquad e^{2\pi i S_{X_2}(\Theta_2)} \in L_{-Y,\eta} \cong \overline{L_{Y,\eta}}.$$

The gluing axiom is

$$e^{2\pi i S_X(\Theta)} = \left(e^{2\pi i S_{X_1}(\Theta_1)}, e^{2\pi i S_{X_2}(\Theta_2)}\right) \tag{45}$$

where the inner product takes place in $L_{Y,\eta}$.

Exercise 43. Demonstrate (45).

Classical Solutions.

We leave as an exercise the derivation of the variation of the action

$$\delta S_X(\Theta) = 2\int_X \frac{-k}{8\pi^2}\operatorname{Tr}(F \wedge \delta A)$$

and the resultant Euler-Lagrange equations for the classical solutions:

$$F = 0. \tag{46}$$

(If $\partial X \neq \emptyset$ then we must assume that Θ is fixed on ∂X.) So a solution is a *flat connection*, i.e., a connection with zero curvature. Let

$$\mathcal{M}_X = \text{ space of flat connections on } X \text{ up to equivalence.}$$

As with the finite group case in Lecture #2 (see (18)), where all connections are flat, it is true that

$$\mathcal{M}_X \simeq \operatorname{Hom}(\pi_1 X, G)/G.$$

This may be a finite set (see the Brieskorn spheres in Lecture #5) or it may have continuous parameters.

Hamiltonian Theory.

The Hamiltonian theory concerns spacetimes which are products of space and time, or better paths of fields in a space Y. Thus we study $X = [0,\infty) \times Y$. (The fact that this particular X is noncompact won't bother us.) Fix a G bundle $Q \to Y$, which is of course trivializable. Let

$$\mathcal{A}_Q = \text{ space of connections on } Q.$$

Then up to gauge equivalence any connection on $[0,\infty)\times Q\to[0,\infty)\times Y$ is a path in $\mathcal{A}_Q$, as we see using the "temporal gauge". We want to see what the action is on such paths, but now on closed paths.

Let η_t, $0\le t\le 1$, be a path of connections on Q. We view this as a connection η on the product bundle $[0,1]\times Q\to[0,1]\times Y$ which has no dt component, i.e., in the temporal gauge. So we can consider the action $e^{2\pi i S_{[0,1]\times Y}(\eta)}$. It is an element of the line attached to the boundary connections $\eta_0\sqcup\eta_1$. Because the orientation on the left of the cylinder is opposite the orientation on the right,

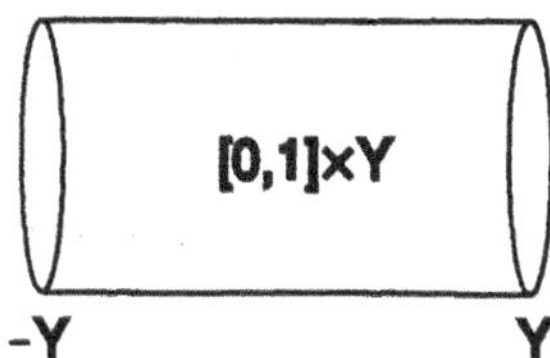

Figure 36: $[0,1]\times Y$

the action is a linear map

$$e^{2\pi i S_{[0,1]\times Y}(\eta)}: L_{\eta_0}\to L_{\eta_1}. \tag{47}$$

We picture this in Figure 37. The lines L_η fit together into a complex line bundle

$$L_Q\to\mathcal{A}_Q \tag{48}$$

and the action gives us a way of lifting paths to act on L_Q. The gluing law for the action implies that these maps compose properly when we glue paths together (Figure 38). One can show that this means that (47) *is the parallel transport of a connection on the bundle* (48). This is the link between the Lagrangian and Hamiltonian Theories. One computes the curvature of this connection to be

$$\omega(\delta_1 A,\delta_2 A)=-2\int_Y\frac{-k}{8\pi^2}\,\mathrm{Tr}(\delta_1 A\wedge\delta_2 A)$$

where $\delta_i A$ are infinitesimal variations of the gauge field, i.e., tangent vectors in $\mathcal{A}_Q$. This curvature is a *symplectic form* on $\mathcal{A}_Q$.

Again: The action on a path in the space of fields on space is the parallel transport for a connection on that space of fields. If the lagrangian is nondegenerate, its curvature is a symplectic form. Many mathematical treatments of Hamiltonian mechanics begin with this symplectic structure. Here we derive it from a Lagrangian theory. Also, (44) says that the action of gauge transformations $\mathcal{G}_Q$ on $\mathcal{A}_Q$ lifts to the line bundle L_Q. This lift is also constructed directly from the Chern-Simons action. So not only does the Hamiltonian Theory involve a symplectic manifold

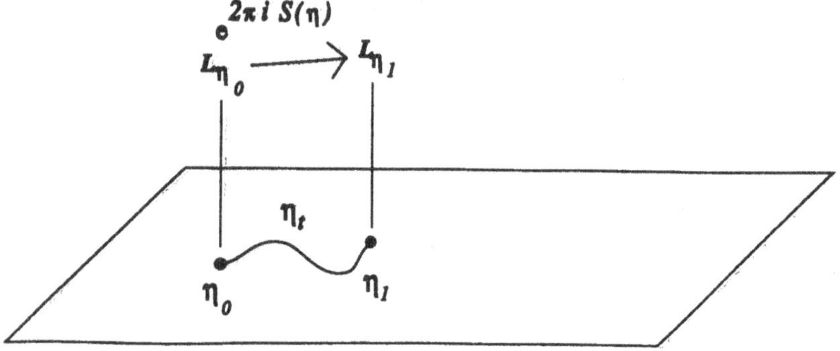

Figure 37: The action as parallel transport

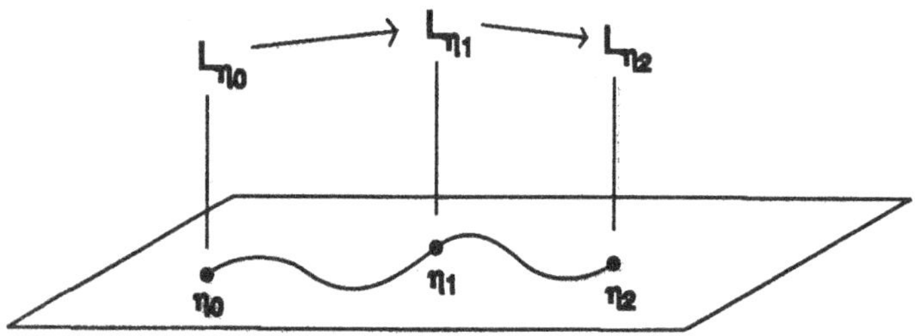

Figure 38: Composing paths and parallel transports

$\mathcal{A}_Q$ (phase space), but the action produces a line bundle $L_Q \to \mathcal{A}_Q$ with a connection whose curvature is the symplectic form. And, as mentioned, the gauge transformations $\mathcal{G}_Q$ lift to act on L_Q.

From this lifted action we can compute a *momentum map* for the action of $\mathcal{G}_Q$. Phase space reduction consists of restricting to the zero momentum subspace and then dividing out by $\mathcal{G}_Q$. Geometrically, the zero momentum space is the subspace of $\mathcal{A}_Q$ over which the connection on L_Q drops to the quotient $\mathcal{A}_Q/\mathcal{G}_Q$. This is the space of *flat* connections, and so we obtain a quotient line bundle with connections

$$L_Y \to \mathcal{M}_Y.$$

Here $\mathcal{M}_Y$ is the set of flat connections on Y up to equivalence. This *Chern-Simons line bundle* and connection enters the quantization of the theory.

This construction of the line bundle appears in [RSW].

Beyond the Hamiltonian Theory.

Recall from Lecture #3 that we obtain a more intricate algebraic structure by considering (in that simple quantum theory) surfaces with boundary. it is natural to ask whether in this classical theory, and ultimately in its quantized version, one can profitably deal with surfaces with boundary. The answer is yes, though the story which develops is quite complicated. It involves strange algebraic objects called *gerbes*. We avoided them in Lecture #3 by introducing basepoints; that can be done here as well [F1,§4–§5]. The full story will be presented in [F2] and [F3]. There we will finally understand the classical Chern-Simons action (for arbitrary gauge group) as the integral of the lagrangian over spacetime, just as ordinary actions, such as the harmonic map action(2), are integrals of a lagrangian over spacetime. Furthermore, one dimension lower we see the Chern-Simons line as the integral of the lagrangian over space. And this continues to lower dimensions and more complicated integrations. (See [F4] for a survey.) But in these topological theories the lagrangian and the integral are somewhat exotic.

Lecture #5: Quantum SU_2 Chern-Simons Theory: Computer Calculations

Although present-day mathematics is well-suited to describe *classical* field theory, and the Chern-Simons theory in particular, the *quantum* field theory most likely requires new mathematical ideas. Certainly there is a large literature on the mathematics of quantum field theory, and there is much nontrivial progress towards understanding many aspects. But I think it is fair to say that the mechanism of quantization is still by and large a mystery. There is plenty of evidence to suggest that quantum field theory will lead to new developments in mathematics; indeed, it already has. Certainly mathematics has also contributed to quantum field theory as well. But, as I say, I believe there is more mystery here than light.

In this final lecture we abandon our theoretical tools and turn to computations in the quantum Chern-Simons theory. This is joint work with Bob Gompf, and is based on an influential paper of Witten [W1]. Witten considers the Chern-Simons theory for SU_2 at "level" k, as discussed in the previous lecture. (He also treats other gauge groups, but we stick to SU_2.) He does two things:

(i) Gives a prescription for computing the *exact* path integral $Z_X(k)$, depending on the level k;

(ii) Gives an *asymptotic* formula for $Z_X(k)$ as $k \to \infty$ by standard perturbation theory techniques.

The exact solution in (i) is, I believe, quite rare in quantum field theory; usually one only has the perturbation theory (ii). The asymptotic formula (see (53) below) involves more or less standard topological invariants of the closed oriented 3-manifold X and representations of its fundamental group. Witten asserts that the exact formula for $Z_X(k)$ is a topological invariant, and this has been borne out by the work of Reshetikhin/Turaev [RT], Walker [Wa] and others. They start with complicated algebraic data and construct the exact solution. Recall that this is the reverse of what we did in Lecture #3. From the special properties of the algebraic data they can prove the properties of quantum field theory (functoriality, gluing, etc.). In particular, they prove that $Z_X(k)$ *is* a topological invariant. While this is very important work, and an important step towards understanding Chern-Simons theory, it should be pointed out that:

(i) This work never mentions the Chern-Simons action.

(ii) Locality is not immediately apparent from this algebraic data. Better, I don't know of a notion of "locality" in their algebraic data which would suggest why one gets out a field theory.

(iii) They do not discuss the perturbative expansion, and it is not at all clear how the topological invariants involved in that expansion (see (53)) emerge from this algebraic data.

I only mean these comments to indicate that there is a long way to go towards understanding locality and quantum field theory in a manner usable in a variety of geometric and topological problems. They are not at all a criticism of this beautiful work, which is an important step towards a more complete understanding these theories.

We do computations of $Z_X(k)$ for some relatively simple 3-manifolds X—*lens spaces* $X = L(p,q)$ and *Brieskorn spheres* $X = \sum(p,q,r)$. We will see the desired properties of the exact solution in our computations, and we will also see that the partition function is a nontrivial effective topological invariant, at least in these simple examples. We will also compute $Z_X(k)$ for large values of k and compare with the asymptotic formula predicted by perturbation theory. We find excellent *numerical* agreement. For lens spaces this is now understood in *analytic* terms (i.e., proved!), but for Brieskorn spheres the analytic results are only qualitative.[1] (See the theses of Lisa Jeffrey [J] and Stavros Garoufalidis [G].) The agreement between the exact solution and the perturbation theory, though expected from formal arguments, is another mystery about the path integral which we don't really understand, and which hints at new geometric structures.

Another discussion of these results from a physics perspective is contained in [FG2]. The complete results are contained in [FG1]. Since these references should be fairly accessible and readily available, our discussion here is brief.

Exact Solution.

Let $e^{2\pi i S_X(\Theta)}$ denote the Chern-Simons action constructed in the last lecture *for* $k = 1$; then the action for arbitrary k is $e^{2\pi i k S_X(\Theta)}$. An important point to remember is this: *We do not have measures available on the space of fields*, so cannot quantize directly using measure theory as we did in Lecture #2. Of course, we can assume the existence of a measure $d\mu_X$ on the space of fields, assume that the measure has all the good properties we need, and write a *formal* path integral

$$Z_X(k) = \int_{\mathcal{C}_X} e^{2\pi i k S_X(\Theta)}\, d\mu_X(\Theta), \qquad X^3 \text{ closed},$$

and suppose that we have a map

$$Y \rightsquigarrow E(Y), \qquad Y^2 \text{ closed}, \tag{49}$$

from closed 2-manifolds to finite dimensional Hilbert spaces such that these maps satisfy the axioms of Lecture #1. (In fact, there is an *anomaly* which enters, which leads to certain *central extensions* and to the requirement that our spaces and spacetimes have a little more structure. We refer to Witten's paper [W1] for this and for more about the entire theory.) We also have the path integral over 3-manifolds with boundary, and we assume that we also have the refined theory involving surfaces with boundary as discussed in Lecture #3.

[1] As we write Lev Rozansky is making great strides towards an analytic understanding.

Now, of course, Witten does not merely assume all of this and go on. There are definite criteria which indicate whether a quantum theory makes sense, and what modifications may have to be made (e.g., anomalies). Also, the procedure for producing the quantum Hilbert spaces (49) is well-studied—it is usually called *canonical quantization* and is the Hamiltonian picture of quantum theory. In this case Witten also appeals to *geometric quantization* for the construction of the quantum Hilbert space. There is much mathematical work towards justifying this construction (for example [ADW], [H]). Further, Witten noticed that the construction fits into *conformal field theory*, more specifically *rational conformal field theory*, and so he was able to borrow results from there to compute with the $E(Y)$. This is also one entree into the theory (discussed in Lecture #3) involving surfaces with boundary. In any case, we simply observe here that the quantum Chern-Simons theory in $2+1$ dimensions is related to the quantum Wess-Zumino-Novikov-Witten (topological) theory in $1+1$ dimensions, just as the corresponding classical theories are related.

Just a word about the quantization. Recall the line bundle with connection

$$L_Y \to \mathcal{M}_Y \tag{50}$$

over the space of flat connections on Y. If Y is given a *complex* structure, then $\mathcal{M}_Y$ and L_Y inherit a complex structure, and the quantization is

$$E(Y) = \mathrm{Hol}(\mathcal{M}_Y; L_Y)$$

the space of holomorphic sections of (50). To see that this is well-defined we must check that it is independent of the complex structure. (In fact, only the underlying *projective* space is independent, since there is an anomaly.) This is like quantizing a classical system with variables p, q by the Hilbert space $\mathrm{Hol}(q + p\sqrt{-1})$ of holomorphic functions of the complex variable $q + p\sqrt{-1}$. Other choices are possible and typical—for example, L^2 functions of q or L^2 functions of p. One of the main problems in quantizing is to identify all of the different quantizations, at least projectively, and that is what the mathematical work referred to in the previous paragraph does for the Chern-Simons theory.

These spaces have led to a solution of some mathematical old problems about $\mathcal{M}_Y$, which are of interest in the theory of Riemann surfaces.

The torus.

One can identify the vector space

$$E(S^1 \times S^1) \cong \mathbb{C}\{e_1, \dots, e_{k+1}\}$$

as the span of certain vectors $e_1, \dots, e_{k+1}$. (This canonical basis can be deduced from the refined gluing laws analogous to those in Lecture #3.) The axioms of topological field theory say that the orientation-preserving diffeomorphisms of $S^1 \times S^1$

should act on $E(S^1 \times S^1)$, and the action only depends on homotopy classes of diffeomorphisms (see Exercise 18). That is, $SL_2(\mathbb{Z})$ acts on $E(S^1 \times S^1)$. From conformal field theory, and ultimately from the theory of Kač-Moody Lie algebras [KM], one concludes that the generators

$$T = \begin{pmatrix} 1 & 1 \\ 0 & 1 \end{pmatrix} \quad \text{and} \quad S = \begin{pmatrix} 0 & -1 \\ 1 & 0 \end{pmatrix}$$

act on $E(S^1 \times S^1)$ via the matrices

$$\widetilde{T}^{\alpha}_{\beta} = e^{2\pi i (h_\alpha - c/24)} \delta^{\alpha}_{\beta} \quad ; \quad \widetilde{S}^{\alpha}_{\beta} = \sqrt{\frac{2}{k+2}} \sin\left(\frac{\alpha\beta\pi}{k+2}\right)$$

where

$$c = \frac{3k}{k+2} \quad ; \quad h_\alpha = \frac{\alpha^2 - 1}{4(k+1)}.$$

These are the explicit formulas we use in our computations below.

Other considerations from conformal field theory show

$$\dim E(S^2) = 1. \tag{51}$$

Surgery Formulas.

So canonical quantization, or rather the link between the path integral and canonical quantization, together with some results from conformal field theory and Kač-Moody Lie algebras (or better topological Wess-Zumino-Novikov-Witten Theory) are one part of the exact solution. One should note that this makes heavy use of *symmetry*. The other part of the exact solution we need is derived directly from the set of axioms (see Lecture #1) that the path integral is meant to satisfy.

We need a starting point, and this is

$$Z_{S^1 \times S^2} = 1, \tag{52}$$

which follows from (51) and (27). The other main tool is the Gluing Law(16), from which we derive the change of the path integral under *surgery*.

Let $T \subset S^1 \times S^2$ be a solid torus as shown in Figure 39. The complement $X = (S^1 \times S^2) - T$ is also a solid torus. We have

$$S^1 \times S^2 = X \cup T$$

glued along the common boundary $\partial T \cong S^1 \times S^1$, which is an ordinary torus. This factors the partition function (52) into

$$1 = (Z_X, Z_T)_{E(S^1 \times S^1)}.$$

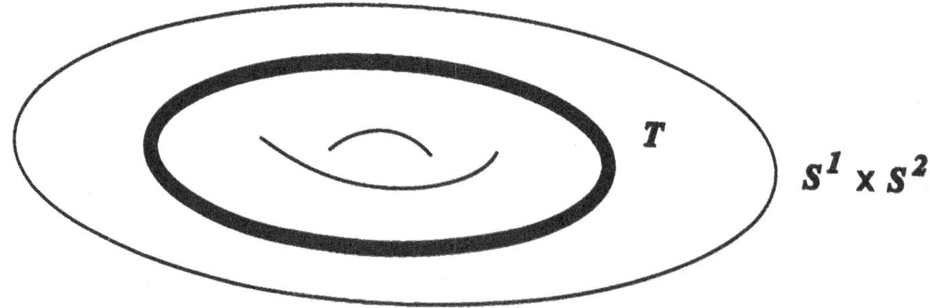

Figure 39: Writing $S^1 \times S^2$ as the union of 2 solid tori

If we glue X and T together by a diffeomorphism f of $\partial T \cong S^1 \times S^1$, then the Functoriality Axiom and the Gluing Law imply that

$$Z_{X \cup_f T} = (Z_X, f_* Z_T)_{E(S^1 \times S^1)},$$

where $X \cup_f T$ is the new 3-manifold obtained by this surgery. If

$$f = \begin{pmatrix} q & r \\ p & s \end{pmatrix} \in SL_2\mathbb{Z}$$

then the resulting manifold is the *lens space* $L(p,q)$. A similar surgery on 3 parallel solid tori in $S^1 \times S^2$ yields the *Brieskorn sphere* $\sum(p,q,r)$.

This is a sketch of the procedure given by Witten, and used by Gompf and me to compute some of the exact partition functions which we compute below.

Perturbation Theory.

In this theory one should think of $k = 1/\hbar$, except that $\hbar$ is now "quantized" to be the reciprocal of an integer. So here the perturbative limit $\hbar \to 0$ is $k \to \infty$. The leading order behavior is determined around the critical points of the classical action, which are the (equivalence classes of) flat connections $A \in \mathcal{M}_X$ (see (46)). The perturbation theory then gives a formula in terms of topological invariants of the flat connection A. For simplicity we state it when the space $\mathcal{M}_X$ is finite and consists entirely of *irreducible* connections. This holds for $X = \sum(p,q,r)$, for example. The asymptotic formula is

$$(53) \qquad Z_X(k) \sim \frac{1}{2} e^{-3\pi i/4} \sum_{A \in \mathcal{M}_X} \sqrt{\tau_X(A)}\, e^{-2\pi i I_A/4} e^{2\pi i (k+2) S_X(A)}$$

where

$$T_X(A) = \text{Reidemeister torsion of } X \text{ with coefficients twisted by } A;$$
$$I_A = \text{a certain ``spectral flow''. It is an integer determined modulo 8.}$$

The ***Reidemeister torsion*** is a topological invariant introduced a long time ago to distinguish various lens spaces [R], [Fr]. It appears here in the form of *analytic torsion*, which was introduced by Ray and Singer [RS] as a certain combination of determinants of Laplace operators. The equality of the analytic torsion and Reidemeister torsion was established by Cheeger [C] and Müller [M]. The *spectral flow* is a topological invariant introduced by Atiyah/Patodi/Singer [APS]. Note the shift $k \to k+2$ in the action, which occurs in various guises in different accounts of the Chern-Simons theory. We emphasize that (53) is derived (by Witten—see also [FG1,§1]) using standard Gaussian integrals and gauge fixing, together with a mathematical theorem which identifies a certain combination of determinants as the Reidemeister torsion.

The Computations.

We present only a few representative computations here. More complete results can be found in [FG1].

We gave surgery descriptions of the lens space $L(p,q)$ and the Brieskorn sphere $\Sigma(p,q,$ above. (Here p,q and p,q,r are pairwise relatively prime.) Now we give another description of these spaces simply to help the reader picture them. For the lens space, we start with the 3-sphere S^3 thought of as the set of points in $\mathbb{C}^2 = \{\langle z,w\rangle\}$ of unit norm. Let the cyclic group $\mathbb{Z}/p\mathbb{Z}$ act on $\mathbb{C}^2$ by letting the generator be the transformation

$$\langle z,w\rangle \longmapsto \langle \omega z, \omega^q w\rangle,$$

where $\omega = e^{2\pi i/p}$. This restricts to a *free* action on S^3 if p and q are relatively prime. The quotient of S^3 by this action is the lens space $L(p,q)$. In other words, there is a $p:1$ covering $S^3 \to L(p,q)$. Notice that $L(2,1)$ is the familiar projective space $\mathbb{RP}^3$.

For the Brieskorn sphere $\Sigma(p,q,r)$ with p,q,r pairwise relatively prime, begin with the complex surface Z defined by the equation

$$z_1^p + z_2^q + z_3^r = 0$$

in $\mathbb{C}^3$. Let S^5 denote the unit 5-sphere in $\mathbb{C}^3$. Then

$$\Sigma(p,q,r) = Z \cup S^5.$$

Table 1: Asymptotic values of the Witten invariant for $L(12,5)$

k	exact value		asymptotic value		ratio	
191	−0.032080	+ 0.018579 *i*	−0.032268 +	0.018630 *i*	0.994936	− 0.001350 *i*
192	0.101535		0.101535		1.000000	
193	−0.087706	− 0.087706 *i*	−0.087706 −	0.087706 *i*	1.000000	
194	−0.050274	+ 0.087616 *i*	−0.050508 +	0.087482 *i*	0.999996	− 0.002671 *i*
195	0.119247	+ 0.068636 *i*	0.119199 +	0.068819 *i*	0.999642	− 0.001328 *i*
196		− 0.174078 *i*		−0.174078 *i*	1.000000	
197	−0.118550	+ 0.068653 *i*	−0.118598 +	0.068473 *i*	1.000351	− 0.001316 *i*
198	0.050227	+ 0.086471 *i*	0.050000 +	0.086603 *i*	0.999997	− 0.002618 *i*
199	0.086387	− 0.086387 *i*	0.086387 −	0.086387 *i*	1.000000	
200	−0.099503		−0.099503		1.000000	

Our first data in Table 1 is for the lens space $L(12,5)$. For some fairly large values of k we display both the exact value of the path integral and the asymptotic value, as predicted by (53). (In fact, (53) must be modified to account for the fact that flat SU_2 connections on $L(12,5)$ are *reducible*.) This perturbative formula works out to be

$$\begin{aligned} Z_{L(12,5)}(k) \sim \frac{i}{2}\frac{\sqrt{2}}{\sqrt{12\cdot(k+2)}}\Big\{ & 4\sin^2(\frac{2}{12}\pi)e^{2\pi i(\frac{5}{12})(k+2)} \\ & - 4\sin^2(\frac{4}{12}\pi)e^{2\pi i(\frac{8}{12})(k+2)} \\ & + 4\sin^2(\frac{6}{12}\pi)e^{2\pi i(\frac{9}{12})(k+2)} \\ & - 4\sin^2(\frac{4}{12}\pi)e^{2\pi i(\frac{8}{12})(k+2)} \\ & + 4\sin^2(\frac{2}{12}\pi)e^{2\pi i(\frac{5}{12})(k+2)}\Big\}. \end{aligned}$$

Notice the different signs, which come from the spectral flow term. The numerical agreement observed in Table 1 has now been verified analytically [J].

Table 2: Asymptotic values of the Witten invariant for $\Sigma(2,3,17)$

k	exact value	asymptotic value	ratio
141	$0.607899 + 0.102594\,i$	$0.596099 + 0.151172\,i$	$0.999182 - 0.081285\,i$
142	$-0.104966 - 0.151106\,i$	$-0.094614 - 0.157913\,i$	$0.997181 - 0.067244\,i$
143	$0.123614 - 0.139016\,i$	$0.132261 - 0.128045\,i$	$1.007707 - 0.075491\,i$
144	$-0.612014 + 0.038199\,i$	$-0.614913 - 0.008261\,i$	$0.994271 - 0.075479\,i$
145	$-0.291162 - 0.132171\,i$	$-0.281928 - 0.153204\,i$	$0.993986 - 0.071336\,i$
146	$-0.413944 + 0.674785\,i$	$-0.465909 + 0.642185\,i$	$0.994797 - 0.077144\,i$
147	$0.400490 - 0.286350\,i$	$0.419276 - 0.254325\,i$	$1.001116 - 0.075706\,i$
148	$-0.091879 + 0.669230\,i$	$-0.143660 + 0.661309\,i$	$0.995194 - 0.077257\,i$
149	$0.946786 - 0.263649\,i$	$0.962119 - 0.191329\,i$	$0.999048 - 0.075356\,i$
150	$-0.024553 - 0.058313\,i$	$-0.021860 - 0.059113\,i$	$1.002906 - 0.044484\,i$

Our second example is the Brieskorn sphere $\Sigma(2,3,17)$. This example is better since, except for the trivial connection, all flat SU_2 connections are irreducible and isolated points in the space of flat connections (up to equivalence). In fact, there are exactly 6 of them! So the asymptotic formula (53) works out to be

$$\begin{aligned} Z_{\Sigma(2,3,17)}(k) \sim \sqrt{\frac{2}{17}}\, e^{-3\pi i/4}\Big\{ & \sin(\frac{\pi}{17})e^{-\pi i(k+2)/204} \\ & + \sin(\frac{4\pi}{17})e^{-169\pi i(k+2)/204} \\ & - \sin(\frac{8\pi}{17})e^{-217\pi i(k+2)/204} \\ & + \sin(\frac{3\pi}{17})e^{-145\pi i(k+2)/204} \\ & + \sin(\frac{2\pi}{17})e^{-361\pi i(k+2)/204} \\ & - \sin(\frac{7\pi}{17})e^{-49\pi i(k+2)/204} \Big\}. \end{aligned} \tag{54}$$

As you can see from Table 2, the exact solution is indeed asymptotic to (54), at least numerically.

References

[A] M. F. Atiyah, *Topological quantum field theory*, Publ. Math. Inst. Hautes Etudes Sci. (Paris) **68** (1989), 175–186.

[ABP] M. F. Atiyah, R. Bott, V. K. Patodi, *On the heat equation and the index theorem*, Invent. math. **19** (1973), 279–330.

[ADW] S. Axelrod, S. Della Pietra, E. Witten, *Geometric quantization of Chern-Simons gauge theory*, J. Diff. Geo. **33** (1991), 787-902.

[APS] M. F. Atiyah, V. K. Patodi, I. M. Singer, *Spectral asymmetry and Riemannian geometry. III*, Math. Proc. Cambridge Philos. Soc. **79** (1976), 71–99.

[C] J. Cheeger, *Analytic torsion and the heat equation*, Ann. Math. **109** (1979), 259–322.

[CS] S. S. Chern, J. Simons, *Characteristic forms and geometric invariants*, Ann. Math. **99** (1974), 48–69.

[DW] R. Dijkgraaf, E. Witten, *Topological gauge theories and group cohomology*, Commun. Math. Phys. **129** (1990), 393–429.

[DPR] R. Dijkgraaf, V. Pasquier, P. Roche, *Quasi-quantum groups related to orbifold models*, Nuclear Phys. B. Proc. Suppl. **18B** (1991), 60–72.

[F1] D. S. Freed, *Classical Chern-Simons theory, Part 1*, Adv. Math. (to appear).

[F2] D. S. Freed, *Higher line bundles* (in preparation).

[F3] D. S. Freed, *Classical Chern-Simons theory, Part 2* (in preparation).

[F4] D. S. Freed, *Locality and integration in topological field theory*, Proceedings of the XIX International Colloquium on Group Theoretical Methods in Physics, Ciemat (to appear).

[F5] D. S. Freed, *Higher algebraic structures and quantization* (preprint, 1992).

[FG1] D. S. Freed, R. E. Gompf, *Computer calculation of Witten's 3-manifold invariant*, Commun. Math. Phys. **141** (1991), 79–117.

[FG2] D. S. Freed, R. E. Gompf, *Computer tests of Witten's Chern-Simons theory against the theory of three-manifolds*, Phys. Rev. Lett. **66** (1991), 1255–1258.

[FQ] D. S. Freed, F. Quinn, *Chern-Simons theory with finite gauge group*, Commun. Math. Phys. (to appear).

[Fr] W. Franz, *Über die Torsion einer Überdeckung*, J. Reine Angew. Math. **173** (1935), 245–254.

[G] S. Garoufalidis, *Relations among 3-manifold invariants* (University of Chicago Ph.D. thesis, 1992).

[GJ] J. Glimm, A. Jaffe, *Quantum Physics. A functional integral point of view.*, Second edition, Springer-Verlag, New York-Berlin, 1987.

[GM] R. Gompf, R. Mrowka, *Irreducible four-manifolds need not be complex* (preprint, 1991).

[H] N. Hitchin, *Flat connections and geometric quantization*, Commun. Math. Phys. **131** (1990), 347–380.

[J] L. C. Jeffery, *On some aspects of Chern-Simons gauge theory* (Oxford Univ. D.Phil. thesis, 1991).

[K] L. H. Kauffman, *Knots and Physics*, World Scientific, 1991.

[KM] V. G. Kac, M. Wakimoto, Adv. Math. **70** (1988), 156.

[M] W. Müller, *Analytic torsion and R-torsion of Riemannian manifolds*, Adv. Math. **28** (1978), 233–305.

[Q1] F. Quinn, *Lectures on axiomatic topological quantum field theory* (preprint, 1992).

[Q2] F. Quinn, *Topological foundations of topological quantum field theory* (preprint, 1991).

[R] K. Reidemeister, *Homotopieringe und Linsenräume*, Hamburger Abhandl **11** (1935), 102–109.

[RS] D. B. Ray, I. M. Singer, *R-torsion and the laplacian on Riemannian manifolds*, Adv. Math. **7** (1971), 145–210.

[RSW] T. R. Ramadas, I. M. Singer, J. Weitsman, *Some comments on Chern-Simons gauge theory*, Commun. Math. Phys. **126** (1989), 409.

[RT] N. Y. Reshetikhin, V. G. Turaev, *Ribbon graphs and their invariants derived from quantum groups*, Commun. Math. Phys. **127** (1990), 1–26.

[S] G. Segal, *The definition of conformal field theory* (preprint).

[Sp] E. H. Spanier, *Algebraic Topology*, Springer-Verlag, New York, 1981.

[W1] E. Witten, *Quantum field theory and the Jones polynomial*, Commun. Math.

Phys. **121** (1989), 351–399.

[W2] E. Witten, *Topological sigma models*, Commun. Math. Phys. **118** (1988), 411.

[Wa] K. Walker, *On Witten's 3-manifold invariants* (preprint, 1991).

DEPARTMENT OF MATHEMATICS, UNIVERSITY OF TEXAS, AUSTIN, TX 78712
E-mail address: dafr@math.utexas.edu

CANONICAL QUANTUM GRAVITY AND THE PROBLEM OF TIME [1] [2]

C.J. Isham

Blackett Laboratory, Imperial College
South Kensington, London SW7 2BZ, United Kingdom

ABSTRACT

The aim of this paper is to provide a general introduction to the problem of time in quantum gravity. This problem originates in the fundamental conflict between the way the concept of 'time' is used in quantum theory, and the role it plays in a diffeomorphism-invariant theory like general relativity. Schemes for resolving this problem can be sub-divided into three main categories: (I) approaches in which time is identified before quantising; (II) approaches in which time is identified after quantising; and (III) approaches in which time plays no fundamental role at all. Ten different specific schemes are discussed in this paper which also contain an introduction to the relevant parts of the canonical decomposition of general relativity.

CONTENTS

[1]Lectures presented at the NATO Advanced Study Institute "Recent Problems in Mathematical Physics", Salamanca, June 15–27, 1992.

[2]Research supported in part by SERC grant GR/G60918.

L. A. Ibort and M. A. Rodríguez (eds.), Integrable Systems, Quantum Groups, and Quantum Field Therapy 157–287.

1 INTRODUCTION

1.1 Preamble

These notes are based on a course of lectures given at the NATO Advanced Summer Institute "Recent Problems in Mathematical Physics", Salamanca, June 15–27, 1992. The notes reflect part of an extensive investigation with Karel Kuchař into the problem of time in quantum gravity. An excellent recent review is Kuchař (1992*b*), to which the present article is complementary to some extent. In particular, my presentation is slanted towards the more conceptual aspects of the problem and, as this a set of lecture notes (rather than a review paper proper), I have also included a fairly substantial technical introduction to the canonical theory of general relativity. However, there is inevitably a strong overlap with many portions of Kuchař's paper, and I am grateful to him for permission to include this material plus a number of ideas that have emerged in our joint discussions. Therefore, the credit for any good features in the present account should be shared between us; the credit for the mistakes I claim for myself alone.

1.2 Preliminary Remarks

The problem of 'time' is one of the deepest issues that must be addressed in the search for a coherent theory of quantum gravity. The major conceptual problems with which it is closely connected include:

- the status of the concept of *probability* and the extent to which it is *conserved*;
- the status of the associated concepts of *causality* and *unitarity*;
- the time-honoured debate about whether quantum gravity should be approached via a *canonical*, or a *covariant*, quantisation scheme;
- the extent to which *spacetime* is a meaningful concept;
- the extent to which classical *geometrical* concepts can, or should, be maintained in the quantum theory;
- the way in which our *classical world* emerged from some primordial quantum event at the big-bang;
- the whole question of the *interpretation* of quantum theory and, in particular, the domain of applicability of the conventional Copenhagen view.

The prime source of the problem of time in quantum gravity is the invariance of classical general relativity under the group Diff($\mathcal{M}$) of diffeomorphisms of the spacetime manifold $\mathcal{M}$. This stands against the simple Newtonian picture of a fixed time parameter, and tends to produce quantisation schemes that apparently lack any fundamental notion of time at all. From this perspective, the heart of the problem is contained in the following questions:

1. How should the notion of time be re-introduced into the quantum theory of gravity?

2. In particular, should attempts to identify time be made at the classical level, *i.e.*, before quantisation, or should the theory be quantised first?

3. Can 'time' still be regarded as a fundamental concept in a quantum theory of gravity, or is its status purely phenomenological? If the concept of time is not fundamental, should it be replaced by something that is: for example, the idea of a *history* of a system, or *process*, or an *ordering structure* that is more general than that afforded by the conventional idea of time?

4. If 'time' is only an approximate concept, how reliable is the rest of the quantum-mechanical formalism in those regimes where the normal notion of time is not applicable? In particular, how closely tied to the concept of time is the idea of probability? This is especially relevant in those approaches to quantum gravity in which the notion of time emerges only *after* the theory has been quantised.

In addition to these questions—which apply to quantum gravity in general—there is also the partly independent issue of the applicability of the concept of time (and, indeed, of quantum theory in general) in the context of quantum cosmology. Of particular relevance here are questions of (i) the status of the Copenhagen interpretation of quantum theory (with its emphasis on the role of measurements); and (ii) the way in which our present classical universe, including perhaps the notion of time, emerged from the quantum origination event.

A key ingredient in all these questions is the realisation that the notion of time used in conventional quantum theory is grounded firmly in Newtonian physics. Newtonian time is a fixed structure, external to the system: a concept that is manifestly incompatible with diffeomorphism-invariance and also with the idea of constructing a quantum theory of a truly closed system (such as the universe itself). Most approaches to the problem of time in quantum gravity [3] seek to address this central issue by identifying an *internal* time which is defined in terms of the system itself, using either the gravitational field or the matter variables that describe the material

[3]A similar problem arises when discussing thermodynamics and statistical physics in a curved spacetime. A recent interesting discussion is Rovelli (1991*d*) which makes specific connections between this problem of time and the one that arises in quantum gravity.

content of the universe. The various schemes differ in the way such an identification is made and the point in the procedure at which it is invoked. Some of these techniques inevitably require a significant reworking of the quantum formalism itself.

1.3 Current Research Programmes in Quantum Gravity

An important question that should be raised at this point is the relation of the problem of time to the various research programmes in quantum gravity that are currently active. The primary distinction is between approaches to quantum gravity that start with the *classical* theory of general relativity (or a simple extension of it) to which some quantisation algorithm is applied, and schemes whose starting point is a *quantum* theory from which classical general relativity emerges in some low-energy limit, even though that theory was not one of the initial ingredients. Most of the standard approaches to quantum gravity belong to the former category; superstring theory is the best-known example of the latter.

Some of the more prominent current research programmes in quantum gravity are as follows (for recent reviews see Isham (1985, 1987, 1992) and Alvarez (1989)).

- *Quantum Gravity and the Problem of Time.* This subject—the focus of the present paper—dates back to the earliest days of quantum gravity research. It has been studied extensively in recent years—mainly within the framework of the canonical quantisation of the classical theory of general relativity (plus matter)—and is now a significant research programme in its own right. A major recent review is Kuchař (1992*b*). Earlier reviews, from somewhat different perspectives, are Barbour & Smolin (1988) and Unruh & Wald (1989). There are also extensive discussions in Ashtekar & Stachel (1991) and in Halliwell, Perez-Mercander & Zurek (1992); other relevant literature will be cited at the appropriate points in our discussion.

- *The Ashtekar Programme.* A major technical development in the canonical formalism of general relativity was the discovery by Ashtekar (1986, 1987) of a new set of canonical variables that makes general relativity resemble Yang-Mills theory in several important respects, including the existence of non-local physical observables that are an analogue of the Wilson loop variables of non-abelian gauge theory (Rovelli & Smolin 1990, Rovelli 1991*a*, Smolin 1992); for a recent comprehensive review of the whole programme see Ashtekar (1991). This is one of the most promising of the non-superstring approaches to quantum gravity and holds out the possibility of novel non-perturbative techniques. There have also been suggestions that the Ashtekar variables may be helpful in resolving the problem of time (for example, in Ashtekar (1991)[pages 191-204]). The Ashtekar programme is not discussed in these notes, but it is a significant development and has important implications for quantum gravity research in general.

- *Quantum Cosmology*. This subject was much studied in the early days of quantum gravity and has enjoyed a renaissance in the last ten years, largely due to the work of Hartle & Hawking (1983) and Vilenkin (1988) on the possibility of constructing a quantum theory of the creation of the universe. The techniques employed have been mainly those of canonical quantisation, with particular emphasis on minisuperspace (*i.e.*, finite-dimensional) approximations. The problem of time is central to the subject of quantum cosmology and is often discussed within the context of these models, as are a number of other conceptual problems expected to arise in quantum gravity proper. A major recent review is Halliwell (1991*a*) which also contains a comprehensive bibliography (see also Halliwell (1990) and Halliwell (1992*a*)).

- *Low-Dimensional Quantum Gravity*. Studies of gravity in $1+1$ and $2+1$ dimensions have thrown valuable light on many of the difficult technical and conceptual issues in quantum gravity, including the problem of time (for example, Carlip (1990, 1991)). The reduction to lower dimensions produces major technical simplifications whilst maintaining enough of the flavour of the $3+1$-dimensional case to produce valuable insights into the full theory. Lower-dimensional gravity also has direct physical applications. For example, idealised cosmic strings involve the application of gravity in $2+1$ dimensions, and the theory in $1+1$ dimensions has applications in statistical mechanics. The subject of gravity in $2+1$ dimensions is reviewed in Jackiw (1992*b*) whilst recent developments in $1+1$ dimensional gravity are reported in Jackiw (1992*a*) and Teitelboim (1992) (in the proceedings of this Summer School).

- *Semi-Classical Quantum Gravity*. Early studies of this subject (for a review see Kibble (1981)) were centered on the equations

$$G_{\alpha\beta}(X,\gamma] = \langle\psi|T_{\alpha\beta}(X;\gamma,\hat{\phi}]|\psi\rangle \tag{1.3.1}$$

where the source for the classical spacetime metric γ is an expectation value of the energy-momentum tensor of quantised matter $\hat{\phi}$. More recently, equations of this type have been developed from a WKB-type approximation to the quantum equations of canonical quantum gravity, especially the Wheeler-DeWitt equation (§5.1.5). Another goal of this programme is to provide a coherent foundation for the construction of quantum field theory in a fixed background spacetime: a subject that has been of enduring interest since Hawking's discovery of quantum-induced radiation from a black hole.

The WKB approach to solving the Wheeler-DeWitt equation is closely linked to a semi-classical theory of time and will be discussed in §5.4.

- *Spacetime Structure at the Planck Length.* There is a gnostic subculture of workers in quantum gravity who feel that the structure of space and time may undergo radical changes at scales of the Planck length. In particular, the idea surfaces repeatedly that the continuum spacetime picture of classical general relativity may break down in these regions. Theories of this type are highly speculative but could have significant implications for the question of 'time' which, as a classical concept, is grounded firmly in continuum mathematics.

- *Superstring Theory.* This is often claimed to be the 'correct' theory of quantum gravity, and offers many new perspectives on the intertwining of general relativity and quantum theory. Of particular interest is the recurrent suggestion that there exists a minimal length, with the implication that many normal spacetime concepts could break down at this scale. It is therefore unfortunate that the current approaches to superstring theory are mainly perturbative in character (involving, for example, graviton scattering amplitudes) and are difficult to apply directly to the problem of time. But the question of the implications of string theory is an intriguing one, not least because of the very different status assigned by the theory to important spacetime concepts such as the diffeomorphism group.

It is noteable that almost all studies of the problem of time have been performed in the framework of 'conventional' canonical quantum gravity in which attempts are made to apply quantisation algorithms to the field equations of classical general relativity. However, the resulting theory is well-known to be perturbatively non-renormalisable and, for this reason, much of the work on time has used finite-dimensional models that are free of ultraviolet divergences. Therefore, it must be emphasised that many of the specific problems of time are *not* connected with the pathological short-distance behaviour of the theory, and appear in an authentic way in these finite-dimensional model systems .

Nevertheless, the non-renormalisability is worrying and raises the general question of how seriously the results of the existing studies should be taken. The Ashtekar programme may lead eventually to a finite and well-defined 'conventional' quantum theory of gravity, but quite a lot is being taken on trust. For example, any addition of Riemann-curvature squared counter-terms to the normal Einstein Lagrangian would have a drastic effect on the canonical decomposition of the theory and could render irrelevant much of the discussion involving the Wheeler-DeWitt equation.

In practice, most of those who work in the field seem to believe that, whatever the final theory of quantum gravity may be (including superstrings), enough of the conceptual and geometrical structure of classical general relativity will survive to ensure the relevance of most of the general questions that have been asked about the meaning and significance of 'time'. However, it should not be forgotten that the question of what constitutes a conceptual problem—such as the nature of time—

often cannot be decided in isolation from the technical framework within which it is posed. Therefore, it is feasible that when, for example, superstring theory is better understood, some of the conceptual issues that appear now to be genuine problems will be seen to be the result of asking ill-posed questions, rather than reflecting a fundamental problem in nature itself.

1.4 Outline of the Paper

We begin in §2 by discussing the status of time in general relativity and conventional quantum theory, and some of the *prima facie* problems that arise when attempts are made to unite these two, somewhat disparate, conceptual structures. As we shall see, approaches to the problem of time fall into three distinct categories—those in which time is identified before quantising, those in which time is identified after quantising, and 'timeless' schemes in which no fundamental notion of time is introduced at all. A brief description is given in §2 of the principle schemes together with the major difficulties that are encountered in their implementation.

Most approaches to the problem of time involve the canonical theory of general relativity, and §3 is devoted to this topic. Section §4 deals with the approaches to the problem of time that involve identifying time before quantising, while §5 addresses those schemes in which the gravitational field is quantised first. Then follows an account in §6 of the schemes that avoid invoking time as a fundamental category at all. As we shall see, attempts of this type become rapidly involved in general questions about the interpretation of quantum theory and, in particular, of the domain of applicability of the traditional Copenhagen approach. The paper concludes with a short summary of the current situation and some speculations about the future.

1.5 Conventions

The conventions and notations to be used are as follows. A Lorentzian metric γ on a four-dimensional spacetime manifold $\mathcal{M}$ is assumed to have signature $(-1,1,1,1)$. Lower case Greek indices refer to coordinates on $\mathcal{M}$ and take the values $0,1,2,3$. We adopt the modern differential-geometric view of coordinates as local *functions* on the manifold. Thus, if X^α, $\alpha = 0\ldots3$ is a coordinate system on $\mathcal{M}$ the *values* of the coordinates of a point $Y \in \mathcal{M}$ are the four numbers $X^\alpha(Y)$, $\alpha = 0\ldots3$. However, by a familiar abuse of notation, sometimes we shall simply write these numbers as Y^α. Lower case Latin indices refer to coordinates on the three-manifold Σ and range through the values $1,2,3$.

The symbol $f : X \to Y$ means that f is a map from the space X into the space Y. In the canonical theory of gravity there are a number of objects F that are functions simultaneously of (i) a point x in a finite-dimensional manifold (such as Σ or $\mathcal{M}$);

and (ii) a function $G : X \to Y$ where X, Y can be a variety of spaces. It is useful to adopt a notation that reflects this property. Thus we write $F(x, G]$ to remind us that G is itself a function. If F depends on n points $x_1, \ldots x_n$ in a finite-dimensional manifold, and m functions $G_1, \ldots G_m$, we shall write $F(x_1, \ldots x_n; G_1, \ldots G_m]$.

2 QUANTUM GRAVITY AND THE PROBLEM OF TIME

2.1 Time in Conventional Quantum Theory

The problem of time in quantum gravity is deeply connected with the special role assigned to temporal concepts in standard theories of physics. In particular, in Newtonian physics, time—the parameter with respect to which change is manifest—is *external* to the system itself. This is reflected in the special status of time in conventional quantum theory:

1. Time is not a physical observable in the normal sense since it is not represented by an operator. Rather, it is treated as a background parameter which, as in classical physics, is used to mark the evolution of the system. In particular, it provides the parameter t in the time-dependent Schrödinger equation

$$i\hbar \frac{d\psi_t}{dt} = \widehat{H}\psi_t. \tag{2.1.1}$$

   This special property of time applies both to non-relativistic quantum theory and to relativistic particle dynamics and quantum field theory. It is the reason why the meaning assigned to the time-energy uncertainty relation $\delta t \, \delta E \geq \frac{1}{2}\hbar$ is quite different from that pertaining to, for example, the position and the momentum of a particle.

2. This view of time is related to the difficulty of describing a truly *closed* system in quantum-mechanical terms. Indeed, it has been cogently argued that the only physical states in such a system are *eigenstates* of the Hamiltonian operator, whose time evolution is essentially trivial (Page & Wooters 1983). Of course, the ultimate closed system is the universe itself.

3. The idea of events happening at a single time plays a crucial role in the technical and conceptual foundations of quantum theory:

   - The notion of a *measurement* made at a particular time is a fundamental ingredient in the conventional Copenhagen interpretation. In particular, an *observable* is something whose value can be measured at a fixed time.

On the other hand, a 'history' has no direct physical meaning except in so far as it refers to the outcome of a sequence of time-ordered measurements.

- One of the central requirements of the *scalar product* on the Hilbert space of states is that it be conserved under the time evolution (2.1.1). This is closely connected to the unitarity requirement that probabilities always sum to one.
- More generally, a key ingredient in the construction of the Hilbert space for a quantum system is the selection of a complete set of observables that are required to *commute* at a fixed value of time.

4. These ideas can be extended to systems that are compatible with special relativity: one simply replaces the unique time system of Newtonian physics with the set of relativistic inertial reference frames. The quantum theory can be made independent of a choice of frame if it carries a unitary representation of the Poincaré group. In the case of a relativistic quantum field theory, this is closely related to the requirement of microcausality, *i.e.*,

$$[\widehat{\phi}(X), \widehat{\phi}(X')] = 0 \tag{2.1.2}$$

for all spacetime points X and X' that are spacelike separated.

The background Newtonian time appears explicitly in the time-dependent Schrödinger equation (2.1.1), but it is pertinent to note that such a time is truly an abstraction in the sense that no *physical* clock can provide a precise measure of it (Unruh & Wald 1989). For suppose there is some quantum observable T that can serve as a 'perfect' physical clock in the sense that, for some initial state, its observed values increase monotonically with the abstract time parameter t. Since $\widehat{T}$ may have a continuous spectrum, let us decompose its eigenstates into a collection of normalisable vectors $|\tau_0\rangle, |\tau_1\rangle, |\tau_2\rangle \ldots$ such that $|\tau_n\rangle$ is an eigenstate of the projection operator onto the interval of the spectrum of $\widehat{T}$ centered on τ_n. Then to say that T is a perfect clock means:

1. For each m there exists an n with $n > m$ and $t > 0$ such that the probability amplitude for $|\tau_m\rangle$ to evolve to $|\tau_n\rangle$ in Newtonian time t is non-zero (*i.e.*, the clock has a non-zero probability of running forwards with respect to the abstract Newtonian time t). This means that

$$f_{mn}(t) := \langle\tau_n|U(t)|\tau_m\rangle \neq 0 \tag{2.1.3}$$

where $U(t) := e^{-it\widehat{H}/\hbar}$.

2. For each m and for all $t > 0$, the amplitude to evolve from $|\tau_m\rangle$ to $|\tau_n\rangle$ vanishes if $m > n$ (*i.e.*, the clock never runs backwards).

Unruh & Wald (1989) show that these conditions are incompatible with the physical requirement that the energy of the system be positive. This follows by studying the function $f_{mn}(t)$, $m > n$, in (2.1.3) for complex t and with $m > n$. Since $\widehat{H}$ is bounded below, f_{mn} is holomorphic in the lower-half plane and hence cannot vanish on any open real interval unless it vanishes identically for all t whose imaginary part is less than, or equal to, zero. However, the requirement that the clock never runs backwards is precisely that $f_{mn}(t) = 0$ for all $t > 0$, and hence $f_{mn}(t) = 0$ for all real t. But then, if $m < n$,

$$f_{mn}(t) = \langle \tau_n | U(t) | \tau_m \rangle = \langle \tau_m | U(t)^\dagger | \tau_n \rangle^* = f^*_{nm}(-t) \tag{2.1.4}$$

which, as we have just shown, vanishes for all $t > 0$. Hence the clock can never run forwards in time, and so perfect clocks do not exist. This means that any physical clock always has a small probability of sometimes running backwards in abstract Newtonian time.

An even stronger requirement on T is that it should be a 'Hamiltonian time observable' in the sense that

$$[\,\widehat{T}, \widehat{H}\,] = i\hbar. \tag{2.1.5}$$

This implies at once that $U(t)|T\rangle = |T+t\rangle$ where $\widehat{T}|T\rangle = T|T\rangle$, which is precisely the type of behaviour that is required for a perfect clock. However, it is well-known that self-adjoint operators satisfying (exponentiable) representations of (2.1.5) necessarily have spectra equal to the whole of $\mathbb{R}$, and hence (2.1.5) is manifestly incompatible with the requirement that $\widehat{H}$ be a positive operator.

This inability to represent abstract Newtonian time with any genuine physical observable is a fundamental property of quantum theory. As we shall see in §4.3, it is reflected in one of the major attempts to solve the problem of time in quantum gravity.

2.2 Time in a Diff($\mathcal{M}$)-invariant Theory

The special role of time in quantum theory suggests strongly that it (or, more generally, the system of inertial reference frames) should be regarded as part of the *a priori* classical background that plays such a crucial role in the Copenhagen interpretation of the theory. But this view becomes highly problematic once general relativity is introduced into the picture; indeed, it is one of the main sources of the problem of time in quantum gravity.

One way of seeing this is to focus on the idea that the equations of general relativity transform covariantly under changes of spacetime coordinates, and physical results are meant to be independent of choices of such coordinates. However, 'time' is frequently regarded as a coordinate on the spacetime manifold $\mathcal{M}$ and it might be expected therefore to play no fundamental role in the theory. This raises several important questions:

1. If time is indeed merely a coordinate on $\mathcal{M}$—and hence of no direct physical significance—how does the, all-too-real, property of change emerge from the formalism?

2. Like all coordinates, 'time' may be defined only in a local region of the manifold. This could mean that:

   - the time variable t is defined only for some finite range of values for t; or
   - the subset of points t=const may not be a complete three-dimensional submanifold of the spacetime.

   How are such global problems to be handled in the quantum theory?

3. Is this view of the coordinate nature of time compatible with normal quantum theory, based as it is on the existence of a universal, Newtonian time?

The global issues are interesting, but they are not central to the problem of time and therefore in what follows I shall assume the topology of the spacetime manifold $\mathcal{M}$ to be such that global time coordinates can exist. If $\mathcal{M}$ is equipped with a Lorentzian metric γ, such a global time coordinate can be regarded as the parameter in a foliation of $\mathcal{M}$ into a one-parameter family of spacelike hypersurfaces: the hypersurfaces of 'equal-time'. This useful picture will be employed frequently in our discussions.

A slightly different way of approaching these issues is to note that the equations of general relativity are covariant with respect to the action of the group $\mathrm{Diff}(\mathcal{M})$ of diffeomorphisms (*i.e.*, smooth and invertible point transformations) of the spacetime manifold $\mathcal{M}$. We shall restrict our attention to diffeomorphisms with compact support, by which I mean those that are equal to the unit map outside some closed and bounded region of $\mathcal{M}$. Thus, for example, a Poincaré-group transformation of Minkowski spacetime is not deemed to belong to $\mathrm{Diff}(\mathcal{M})$. This restriction is imposed because the role of transformations with a non-trivial action in the asymptotic regions of $\mathcal{M}$ is quite different from those that act trivially. It must also be emphasised that $\mathrm{Diff}(\mathcal{M})$ means the group of active point transformations of $\mathcal{M}$. This should not be confused with the pseudo-group which describes the relations between overlapping pairs of coordinate charts (although of course there is a connection between the two).

Viewed as an active group of transformations, the diffeomorphism group $\mathrm{Diff}(\mathcal{M})$ is analogous in certain respects to the gauge group of Yang-Mills theory. For example, in both cases the groups are associated with field variables that are non-dynamical, and with a canonical formalism that entails constraints on the canonical variables. However, in another, and very important sense the two groups are quite different.

Yang-Mills transformations occur at a fixed spacetime point whereas the diffeomorphism group moves points around. Invariance under such an active group of transformations robs the individual points in $\mathcal{M}$ of any fundamental ontological significance. For example, if ϕ is a scalar field on $\mathcal{M}$ the value $\phi(X)$ at a particular point $X \in \mathcal{M}$ has no invariant meaning. This is one aspect of the Einstein 'hole' argument that has featured in several recent expositions (Earman & Norton 1987, Stachel 1989). It is closely related to the question of what constitutes an *observable* in general relativity—a surprisingly contentious issue that has generated much debate over the years and which is of particular relevance to the problem of time in quantum gravity. In the present context, the natural objects that are manifestly Diff($\mathcal{M}$)-invariant are spacetime integrals like, for example,

$$F[\gamma] := \int_{\mathcal{M}} d^4X\,(-\det\gamma(X))^{\frac{1}{2}}\,R^{\alpha\beta\gamma\delta}(X,\gamma]\,R_{\alpha\beta\gamma\delta}(X,\gamma]. \tag{2.2.1}$$

Thus 'observables' of this type are intrinsically non-local.

These implications of Diff($\mathcal{M}$)-invariance pose no real difficulty in the classical theory since once the field equations have been solved the Lorentzian metric on $\mathcal{M}$ can be used to give meaning to concepts like 'causality' and 'spacelike separated', even if these notions are not invariant under the action of Diff($\mathcal{M}$). However, the situation in the quantum theory is very different. For example, whether or not a hypersurface is spacelike depends on the spacetime metric γ. But in any quantum theory of gravity there will presumably be some sense in which γ is subject to quantum fluctuations. Thus causal relationships, and in particular the notion of 'spacelike', appear to depend on the quantum state. Does this mean that 'time' also is state dependent?

This is closely related to the problem of interpreting a microcausality condition like (2.1.2) in a quantum theory of gravity. As emphasised by Fredenhagen & Haag (1987), for most pairs of points $X, X' \in \mathcal{M}$ there exists at least one Lorentzian metric with respect to which they are *not* spacelike separated, and hence, in so far as all metrics are 'virtually present' (for example, by being summed over in a functional integral) the right hand side of (2.1.2) is never zero. This removes at a stroke one of the bedrocks of conventional quantum field theory!

The same argument throws doubts on the use of canonical commutation relations: with respect to what metric is the hypersurface meant to be spacelike? This applies in particular to the equation $[\widehat{g}_{ab}(x), \widehat{g}_{cd}(x')] = 0$ which, in a conventional reading, would imply that the canonical configuration variable g (a metric on a three-dimensional manifold Σ) can be measured simultaneously at two points x, x' in Σ. This difficulty in forming a meaningful interpretation of commutation relations also renders dubious any attempts to find a quantum gravity analogue of the powerful C^*-algebra approach to conventional quantum field theory.

These rather general problems concerning time and the diffeomorphism group might be resolved in several ways. One possibility is to restrict attention to spacetimes

$(\mathcal{M}, \gamma)$ that are asymptotically flat. It is then possible to define *asymptotic* quantities, including time and space coordinates, using the values of fields in the asymptotic regions of $\mathcal{M}$. This works because the diffeomorphism group transformations have compact support and hence quantities of this type are manifestly invariant. In such a theory, time evolution would be meaningful only in these regions and could be associated with the appropriate generator of some sort of asymptotic Poincaré group. DeWitt used the concept of an asymptotic observable in his seminal investigations of covariant quantum gravity (DeWitt 1965, 1967a, 1967b, 1967c) .

However, it is not clear to what extent this really solves the problem of time. An asymptotic time might suffice for calculations of graviton-graviton scattering amplitudes, but it is difficult to see how it could be used in the quantum cosmological situations to which so much attention has been devoted in recent years. One can also question the extent to which such a notion of time can be related operationally to what could be measured with the aid of real physical clocks. This is particularly relevant in the light of the discussion below of the significance of internal time coordinates.

In any event, the present paper is concerned mainly with the problem of time in the context of compact three-spaces, or of non-compact spaces but with boundary data playing no fundamental role. In theories of this type, the problem of time and the spacetime diffeomorphism group might be approached in several different ways.

- General relativity could be forced into a Newtonian framework by assigning special status to some particular foliation of $\mathcal{M}$. For example, a special background metric γ could be introduced to act as a source of preferred reference frames and foliations. However, the action of a diffeomorphism on $\mathcal{M}$ generally maps a hypersurface into a different one, and hence insisting on preserving the special foliation generates a type of symmetry breaking in which the $\mathrm{Diff}(\mathcal{M})$ invariance is reduced to the group of transformations that leave the foliation invariant. If the foliation is tied closely to the properties of the metric γ, this group is likely to be just the group of isometries (if any) of γ.

  There is an interesting point at issue at here. Some people (for example, certain process philosophers) might argue that, since we do in fact live in one particular universe, we should be free to exploit whatever special characteristics it might happen to have. Thus, for example, we could use the background 3^0K radiation to define some quasi-Newtonian universal time associated with the Robertson-Walker homogeneous cosmology with which it is naturally associated. [4] On the other hand, theoretical physicists tend to want to consider all *possible* universes under the umbrella of a single theoretical structure. Thus there is not

[4] An interesting recent suggestion by Valentini (1992) is that a preferred foliation of spacetime could arise from the existence of nonlocal hidden-variables.

much support for the idea of focussing on a special foliation associated with a contingent feature of the actual universe in which we happen to find ourselves.

- At the other extreme, one might seek a new interpretation of quantum theory in which the concept of time does not appear at all (or, at the very least, plays no fundamental role). Our familiar notion of temporal evolution will then have to 'emerge' from the formalism in some phenomenological way. As we shall see, some of the most interesting approaches to the problem of time are of this type.
- Events in $\mathcal{M}$ might be identified using the positions of physical particles. For example, if ϕ is a scalar field on $\mathcal{M}$ the value of the field where a particular particle *is*, is a Diff($\mathcal{M}$)-invariant number and is hence an observable in the sense discussed above. This idea can be generalised in several ways. In particular, the value of ϕ could be specified at that event where some collection of fields takes on a certain set of values (assuming there is a unique such event). These fields, or 'internal coordinates', might be specified using distributions of matter, or they could be part of the gravitional field complex itself. Almost all of the existing approaches to the problem of time involve ideas of this type.

This last point emphasises the important fact that a physical definition of time requires more than just saying that it is a local coordinate on the spacetime manifold. For example, in the real world, time is often measured with the aid of a spatially-localised physical clock, and this usually means the proper time along its worldline. This quantity is Diff($\mathcal{M}$)-invariant if the diffeomorphism group is viewed as acting simultaneously on the points in $\mathcal{M}$ (and hence on the world line) and the spacetime metric γ. More precisely, if the beginning and end-points of the world-line are labelled using internal coordinates, the proper time along the geodesic connecting them is an intrinsic property.

The idea of labelling spacetime events with the aid of physical clocks and spatial reference frames is of considerable importance in both the classical and the quantum theory of relativity. However, in the case of the quantum theory the example of proper time raises another important issue. The calculation of the value of an interval of proper time involves the spacetime metric γ, and therefore it has a meaning only *after* the equations of motion have been solved (and of course these equations include a contribution from the energy-momentum tensor of the matter from which the clock is made). This causes no problem in the classical theory, but difficulties arise in any theory in which the geometry of spacetime is subject to quantum fluctuations and therefore has no fixed value. There is an implication that time may become a quantum operator: a problematic concept that is not part of standard quantum theory .

2.3 Approaches to the Problem of Time

Most approaches to resolving the problem of time in quantum gravity agree that 'time' should be identified in terms of the internal structure of the system rather than being regarded as any sort of external parameter. Their differences lie in:

- the way in which such an identification is made;
- whether it is done before or after quantisation;
- the degree to which the resulting entity resembles the familiar time of conventional classical and quantum physics;
- the role it plays in the final interpretation of the theory.

In turn, these differences are closely related to the general question of how to handle the constraints on the canonical variables that are such an intrinsic feature of the canonical theory of general relativity (§3).

As discussed in Kuchař (1992*b*), the various approaches to the problem of time in quantum gravity can be organised into three broad categories:

I The first class of schemes are those in which an internal time is identified as a functional of the canonical variables, and the canonical constraints are solved, *before* the system is quantised. The aim is to reproduce something like a normal Schrödinger equation with respect to this choice of time. This category is the most conservative of the three since its central assumption is that the construction of *any* coherent quantum-theoretical structure requires something resembling the external, classical time of standard quantum theory.

II In the second type of scheme the procedure above is reversed and the constraints are imposed at a quantum level as restrictions on allowed state vectors and with time being identified only *after* this step. The states can be written as functionals $\Psi[g]$ of a three-geometry $g_{ab}(x)$ (the basic configuration variable in the theory) and the most important of the operator constraints is a functional differential equation for $\Psi[g]$ known as the Wheeler-DeWitt equation (5.1.17). The notion of time has to be recovered in some way from the solutions to this central equation. The key feature of approaches of this type is that the final probabilistic interpretation of the theory is made only *after* the identification of time. Thus the Hilbert space structure of the final theory may be related only very indirectly (if at all) to that of the quantum theory with which the construction starts.

III The third class of scheme embraces a variety of methods that aspire to maintain the timeless nature of general relativity by avoiding any specific conception of time in the quantum theory. Some start, as in II, by imposing the constraints at a quantum level, others proceed along somewhat different lines, but they all agree in espousing the view that it is possible to construct a technically coherent, and *conceptually complete*, quantum theory (including the probabilistic interpretation) without needing to make any direct reference to the concept of time which, at most, has a purely phenomenological status. It is this latter feature that separates these schemes from those of category II.

These three broad categories can be further sub-divided into the following ten specific approaches to the problem of time.

I Tempus ante quantum

1. *The internal Schrödinger interpretation.* Time and space coordinates are identified as specific functionals of the gravitational canonical variables, and are then separated from the dynamical degrees of freedom by a canonical transformation. The constraints are solved classically for the momenta conjugate to these variables, and the remaining physical (*i.e.*, 'non-gauge') modes of the gravitational field are then quantised in a conventional way, giving rise to a Schrödinger evolution equation for the physical states.
2. *Matter clocks and reference fluids.* This is an extension of the internal Schrödinger interpretation in which matter variables coupled to the geometry are used to label spacetime events. They are introduced in a special way aimed at facilitating the handling of the constraints that yield the Schrödinger equation.
3. *Unimodular gravity.* This is a modification of general relativity in which the cosmological constant λ is considered as a dynamical variable. A 'cosmological' time is identified as the variable conjugate to λ, and the constraints yield the Schrödinger equation with respect to this time. This approach can be treated as a special case of a reference fluid.

II Tempus post quantum

1. *The Klein-Gordon interpretation.* The Wheeler-DeWitt equation is considered as an infinite-dimensional analogue of the Klein-Gordon equation for a relativistic particle moving in a fixed background geometry. The probabilistic interpretation of the theory is based on the Klein-Gordon norm with the hope that it will be positive on some appropriate subspace of solutions to the Wheeler-DeWitt equation.
2. *Third quantisation.* The problems arising from the indefinite nature of the scalar product of the Klein-Gordon interpretation are addressed by suggesting that the solutions $\Psi[g]$ of the Wheeler-DeWitt equation are to

be turned into operators. This is analogous to the second quantisation of a relativistic particle whose states are described by the Klein-Gordon equation.

3. *The semi-classical interpretation.* Time is deemed to be a meaningful concept only in some semi-classical limit of the quantum gravity theory based on the Wheeler-DeWitt equation. Using a form of WKB expansion, the Wheeler-DeWitt equation is approximated by a conventional Schrödinger equation in which the time variable is extracted from the state $\Psi[g]$. The probabilistic interpretation arises only at this level.

III Tempus nihil est

1. *The naïve Schrödinger interpretation.* The square $|\Psi[g]|^2$ of a solution $\Psi[g]$ of the Wheeler-DeWitt equation is interpreted as the probability density for 'finding' a spacelike hypersurface of $\mathcal{M}$ with the geometry g. Time enters as an internal coordinate function of the three-geometry, and is represented by an operator that is part of the quantisation of the complete three-geometry.
2. *The conditional probability interpretation.* This can be regarded as a sophisticated development of the naïve Schrödinger interpretation whose primary ingredient is the use of conditional probabilities for the results of a pair of observables A and B. This is deemed to be correct even in the absence of any proper notion of time; as such, it is a modification of the conventional quantum-theoretical formalism. In certain cases, one of the observables is regarded as defining an instant of time (*i.e.*, it represents a physical, and therefore imperfect, clock) at which the other variable is measured [5]. Dynamical evolution is then equated to the dependence of these conditional probabilities on the values of the internal clock variables.
3. *Consistent histories approach.* This is based on a far-reaching extension of normal quantum theory to a form that does not require the conventional Copenhagen interpretation. The main ingredient is a precise prescription from within the formalism itself that says when it is, or is not, meaningful to ascribe a probability to a *history* of the system. The extension to quantum gravity involves defining the notion of a 'history' in a way that avoids having to make any direct reference to the concept of time. In its current form, the scheme culminates in the hope that functional integrals over spacetime fields may be well-defined, even in the absence of a conventional Hilbert space structure.

[5]Or, perhaps, 'has a value'; the language used reflects the extent to which one favours operational or realist interpretations of quantum theory. In practice, quantum schemes of type III are particularly prone to receive a 'many-worlds' interpretation.

4. *The frozen time formalism.* Observables in quantum gravity are declared to be operators that commute with all the constraints, and are therefore constants of the motion. Attempts are made to show that, although 'timeless', such observables can nevertheless be used to give a picture of dynamical evolution.

2.4 Technical Problems With Time

In the context of either of the first two types of scheme—I constrain before quantising, or II quantise before constraining—a number of potential technical problems can be anticipated. Some of the most troublesome are as follows (see Kuchař (1992*b*)).

- *The ultra-violet divergence problem.* Both schemes involve complicated classical functions of fields defined at the same spatial point. The perturbative non-renormalisability of quantum gravity suggests that the operator analogues of these expressions are very ill-defined. This pathology has no direct connection with most aspects of the problem of time but it throws a big question mark over some of the techniques used to tackle that problem.

- *The operator-ordering problem.* Horrendous operator-ordering difficulties arise when attempts are made to replace the classical constraints and Hamiltonians with operator equivalents. These difficulties cannot easily be separated from the ultra-violet divergence problem.

- *The global time problem.* Experience with non-abelian gauge theories suggests the existence of global obstructions to making the crucial canonical transformations that untangle the physical modes of the gravitational field from internal spacetime coordinates.

- *The multiple-choice problem.* The Schrödinger equation based on one particular choice of internal time may give a different quantum theory from that based on another. Which, if any, is correct? Can these different quantum theories be seen to be part of an overall scheme that *is* covariant? A similar problem can be anticipated with the identification of time in the solutions of the Wheeler-DeWitt equation.

- *The Hilbert space problem.* Schemes of type I have the big advantage of giving a natural inner product that is conserved with respect to the internal time variables. This leads to a straightforward interpretative framework. In the alternative constraint quantisation schemes the situation is quite different. The Wheeler-DeWitt equation is a second-order functional differential equation, and as such presents familiar problems if one tries to construct a genuine, positive-definite inner product on the space of its solutions. This is the 'Hilbert space problem'.

- *The spatial metric reconstruction problem.* The classical separation of the canonical variables into physical and non-physical parts can be inverted and, in particular, the metric g_{ab} can be expressed as a functional of the dynamical and non-dynamical modes. The 'spatial metric reconstruction problem' is whether something similar can be done at the quantum level. This is part of the general question of the extent to which classical geometrical properties can be, or should be, preserved in the quantum theory.

- *The spacetime problem.* If an internal space or time coordinate is to operate within a conventional spacetime context, it is necessary that, viewed as a function on $\mathcal{M}$, it be a *scalar* field; in particular, it must not depend on any background foliation of $\mathcal{M}$. However, the objects used in the canonical approach to general relativity are functionals of the *canonical* variables, and there is no *prima facie* reason for supposing they will satisfy this condition. The spacetime problem consists in finding functionals that do have this desirable property or, if this is not possible, understanding how to handle the situation and what it means in spacetime terms.

- *The problem of functional evolution.* This affects both approaches to the canonical theory of gravity. In the scheme in which time is identified before quantisation, the problem is the possible existence of anomalies in the algebra of the local Hamiltonians that are associated with the generalised Schrödinger equation. Any such anomaly would render this equation inconsistent. In quantisation schemes of type II (and in the appropriate schemes of type III), the worry is potential anomalies in the quantum version of the canonical analogue of the Lie algebra of the spacetime diffeomorphism group Diff($\mathcal{M}$). In both cases, the consistency of the classical evolution (guaranteed by the closing properties of the appropriate algebras) is lost.

In theories of category III, time plays only a secondary role, and therefore, with the exception of the first two, most of the problems above are not directly relevant. However, analogues of several of them appear in type III schemes, which also have additional difficulties of their own. Further discussion is deferred to the appropriate sections of the notes.

3 CANONICAL GENERAL RELATIVITY

3.1 Introductory Remarks

Most of the discussion of the problem of time in quantum gravity has been within the framework of the canonical theory whose starting point is a three-dimensional manifold Σ which serves as a model for physical space. This is often contrasted with the so-called 'spacetime' (or 'covariant') approaches in which the basic entity is a four-dimensional spacetime manifold $\mathcal{M}$. The virtues and vices of these two different approaches have been the subject of intense debate over the years and the matter is still far from being settled. Not surprisingly, the problem of time looks very different in the two schemes. Some of the claimed advantages of the canonical approach to quantum gravity are as follows.

1. The spacetime approaches often employ formal functional-integral techniques in which a number of serious difficulties are swept under the carpet as problems concerned with the integration measure, the contour of integration *etc.* On the other hand, canonical quantisation is usually discussed in an operator-based framework, with the advantage that problems appear in a more explicit and, perhaps, tractable way.

2. The development of quantisation techniques that do not depend on a background metric seems to be easier in the canonical framework. This is particularly relevant to the problem of time.

3. For related reasons, canonical quantisation is better suited for discussing quantum cosmology, spacetime singularities and similar topics.

4. Canonical methods tend to place more emphasis on the geometrical structure of general relativity. In particular, it is easier to address the issue of the extent to which such structure is, or should be, maintained in the quantum theory.

5. Many of the deep conceptual problems in quantum gravity are more transparent in a canonical approach. This applies in particular to the problem of time and the, not unrelated, general question of the domain of applicability of the interpretative framework of conventional quantum theory.

It should be emphasised that some of these advantages arise only in comparison with weak-field perturbation theory, and this does not mean that canonical methods are intrinsically superior to those in which spacetime fields are used from the outset. For example, there could exist *bona fide* approaches to quantum gravity that involve the calculation of a functional integral like

$$Z = \int \mathcal{D}[g]\, e^{iS(g)} \tag{3.1.1}$$

using methods other than weak-field perturbation theory. A particularly interesting example is the consistent histories approach to the problem of time discussed in §6.2.

3.2 Quantum Field Theory in a Curved Background

3.2.1 The Canonical Formalism

In approaching the canonical theory of general relativity it is helpful to begin by considering briefly the canonical quantisation of a scalar field ϕ propagating in a background Lorentzian metric γ on a spacetime manifold $\mathcal{M}$. The action for the system is

$$S[\phi] = -\int_{\mathcal{M}} d^4X\,(-\det\gamma(X))^{\frac{1}{2}}\left(\frac{1}{2}\gamma^{\alpha\beta}(X)\,\partial_\alpha\phi(X)\,\partial_\beta\phi(X) + V(\phi(X))\right) \tag{3.2.1}$$

where $V(\phi)$ is an interaction potential for ϕ (which could include a mass term).

This system has a well-posed classical Cauchy problem only if $(\mathcal{M},\gamma)$ is a globally-hyperbolic pair (Hawking & Ellis 1973). In particular, this means that $\mathcal{M}$ is topologically equivalent to the Cartesian product $\Sigma\times\mathbb{R}$ where Σ is some spatial three-manifold and $\mathbb{R}$ is a global time direction. To acquire the notion of dynamical evolution we need to *foliate* $\mathcal{M}$ into a one-parameter family of embeddings $\mathcal{F}_t:\Sigma\to\mathcal{M}$, $t\in\mathbb{R}$, of Σ in $\mathcal{M}$ that are spacelike with respect to the background Lorentzian metric $\gamma_{\alpha\beta}$; the real number t can then serve as a time parameter. To say that an embedding $\mathcal{E}:\Sigma\to\mathcal{M}$ is *spacelike* means that the pull-back $\mathcal{E}^*(\gamma)$—a symmetric rank-two covariant tensor field on Σ—has signature $(1,1,1)$ and is positive definite. We recall that the components of $\mathcal{E}^*(\gamma)$ are [6]

$$(\mathcal{E}^*(\gamma))_{ab}(x) := \gamma_{\alpha\beta}(\mathcal{E}(x))\,\mathcal{E}^\alpha{}_{,a}(x)\,\mathcal{E}^\beta{}_{,b}(x) \tag{3.2.2}$$

on the three-manifold Σ.

The canonical variables are defined on Σ and consist of the scalar field ϕ and its conjugate variable π (a scalar density) which is essentially the time-derivative of ϕ. Classically these constitute a well-defined set of Cauchy data and satisfy the Poisson bracket relations

$$\{\phi(x),\phi(x')\} = 0 \tag{3.2.3}$$
$$\{\pi(x),\pi(x')\} = 0 \tag{3.2.4}$$
$$\{\phi(x),\pi(x')\} = \delta(x,x'). \tag{3.2.5}$$

The Dirac δ-function is defined such that the smeared fields satisfy

$$\{\phi(f),\pi(h)\} = \int_\Sigma d^3x\,f(x)\,h(x) \tag{3.2.6}$$

[6] The symbol $\mathcal{E}^\alpha(x)$ means $X^\alpha(\mathcal{E}(x))$ where X^α, $\alpha=0,1,2,3$ is a coordinate system on $\mathcal{M}$. The quantity defined by the left hand side of (3.2.2) is independent of the choice of such a system.

where the test-functions f and h are respectively a scalar density and a scalar [7] on the three-manifold Σ.

The dynamical evolution of the system is obtained by constructing the Hamiltonian $H(t)$ in the usual way from the action (3.2.1) and the given foliation of $\mathcal{M}$. The resulting equations of motion are

$$\frac{\partial\phi(x,t)}{\partial t} = \{\phi(x), H(t)\} \tag{3.2.7}$$

$$\frac{\partial\pi(x,t)}{\partial t} = \{\pi(x), H(t)\}. \tag{3.2.8}$$

Note that these give the evolution with respect to the time parameter associated with the specified foliation. However, the physical fields can be evaluated on any spacelike hypersurface, and this should not depend on the way the hypersurface happens to be included in a particular foliation. Thus it should be possible to write the physical fields as functions $\phi(x,\mathcal{E}]$, $\pi(x,\mathcal{E}]$ of $x \in \Sigma$ and functionals of the embedding functions $\mathcal{E}$. Indeed, the Hamiltonian equations of motion (3.2.7–3.2.8)) are valid for all foliations and imply the existence of four functions $h_\alpha(x,\mathcal{E}]$ of the canonical variables such that $\phi(x,\mathcal{E}]$ and $\pi(x,\mathcal{E}]$ satisfy the functional differential equations

$$\frac{\delta\phi(x,\mathcal{E}]}{\delta\mathcal{E}^\alpha(x')} = \{\phi(x,\mathcal{E}], h_\alpha(x',\mathcal{E}]\} \tag{3.2.9}$$

$$\frac{\delta\pi(x,\mathcal{E}]}{\delta\mathcal{E}^\alpha(x')} = \{\pi(x,\mathcal{E}], h_\alpha(x',\mathcal{E}]\} \tag{3.2.10}$$

that describe how these fields change under an infinitesimal deformation of the embedding $\mathcal{E}$.

3.2.2 Quantisation of the System

The formal canonical quantisation of this system follows the usual rule of replacing Poisson brackets with operator commutators. Thus (3.2.3–3.2.5) become

$$[\,\widehat{\phi}(x), \widehat{\phi}(x')\,] = 0 \tag{3.2.11}$$

$$[\,\widehat{\pi}(x), \widehat{\pi}(x')\,] = 0 \tag{3.2.12}$$

$$[\,\widehat{\phi}(x), \widehat{\pi}(x')\,] = i\hbar\,\delta(x,x') \tag{3.2.13}$$

which can be made rigorous by smearing and exponentiating in the standard way.

The next step is to choose an appropriate representation of this operator algebra. One might try to emulate elementary wave mechanics by taking the state space of

[7]We are using a definition of the Dirac delta function $\delta(x,x')$ that is a scalar in x and a scalar density in x'. Thus, if f is a scalar function on Σ we have, formally, $f(x) = \int_\Sigma d^3x'\,\delta(x,x')f(x')$.

the quantum field theory to be a set of functionals Ψ on the topological vector space E of all classical fields, with the canonical operators defined as

$$(\widehat{\phi}(x)\Psi)[\phi] := \phi(x)\Psi[\phi] \tag{3.2.14}$$

$$(\widehat{\pi}(x)\Psi)[\phi] := -i\hbar\frac{\delta\Psi[\phi]}{\delta\phi(x)}. \tag{3.2.15}$$

Thus the inner product would be

$$\langle\Psi|\Phi\rangle = \int_E d\mu[\phi]\,\Psi^*[\phi]\,\Phi[\phi] \tag{3.2.16}$$

and, for a normalised function Ψ,

$$\mathrm{Prob}(\phi \in B; \Psi) = \int_B d\mu[\phi]\,|\Psi[\phi]|^2 \tag{3.2.17}$$

is the probability that a measurement of the field on Σ will find it in the subset B of E.

This analysis can be made rigorous after smearing the fields with functions from an appropriate test function space. It transpires that the support of a state functional Ψ is typically on *distributions*, rather than smooth functions, so that the inner product is really

$$\langle\Psi|\Phi\rangle = \int_{E'} d\mu[\phi]\,\Psi^*[\phi]\,\Phi[\phi] \tag{3.2.18}$$

where E' denotes the topological dual of E. Furthermore, there is no infinite-dimensional version of Lebesgue measure, and hence if (3.2.15) is to give a self-adjoint operator it must be modified to read

$$(\widehat{\pi}(x)\Psi)[\phi] := -i\hbar\frac{\delta\Psi[\phi]}{\delta\phi(x)} + i\rho(x)\Psi[\phi] \tag{3.2.19}$$

where $\rho(x)$ is a function that compensates for the weight factor in the measure $d\mu$ used in the construction of the Hilbert space of states $L^2(E', d\mu)$.

The dynamical evolution of the system can be expressed in the Heisenberg picture as the commutator analogue of the Poisson bracket relations (3.2.7–3.2.8). Alternatively, one can adopt the Schrödinger picture in which the time evolution of state vectors is given by [8]

$$i\hbar\frac{\partial\Psi(t,\phi]}{\partial t} = H(t;\widehat{\phi},\widehat{\pi}\,]\,\Psi(t,\phi] \tag{3.2.20}$$

[8] As usual, the notation $H(t;\widehat{\phi},\widehat{\pi}]$ must be taken with a pinch of salt. The classical fields $\phi(x)$ and $\pi(x)$ can be replaced in the Hamiltonian $H(t;\phi,\pi]$ with their operator equivalents $\widehat{\phi}(x)$ and $\widehat{\pi}(x)$ only after a careful consideration of operator ordering and regularisation.

or, if the classical theory is described by the functional equations (3.2.9–3.2.10), by the functional differential equations

$$i\hbar \frac{\delta \Psi[\mathcal{E}, \phi]}{\delta \mathcal{E}^{\alpha}(x)} = h_{\alpha}(x; \mathcal{E}, \widehat{\phi}, \widehat{\pi}\,]\, \Psi[\mathcal{E}, \phi]. \tag{3.2.21}$$

However, note that the steps leading to (3.2.20) or (3.2.21) are valid only if the inner product on the Hilbert space of states is t (resp. $\mathcal{E}$) independent. If not, a compensating term must be added to (3.2.20) (resp. (3.2.21)) if the scalar product $\langle \Psi_t | \Phi_t \rangle_t$ (resp. $\langle \Psi_{\mathcal{E}} | \Phi_{\mathcal{E}} \rangle_{\mathcal{E}}$) is to be independent of t (resp. the embedding $\mathcal{E}$). This is analogous to the quantum theory of a particle moving on an n-dimensional Riemannian manifold Q with a time-dependent background metric $g(t)$. The natural inner product

$$\langle \psi_t | \phi_t \rangle_t := \int_Q d^n q \, (\det g(q,t))^{\frac{1}{2}} \, \psi_t^*(q) \, \phi_t(q) \tag{3.2.22}$$

is not preserved by the naïve Schrödinger equation

$$i\hbar \frac{d\psi_t}{dt} = \widehat{H}(t) \psi_t \tag{3.2.23}$$

because the time derivative of the right hand side of (3.2.22) acquires an extra term coming from the time-dependence of the metric $g(t)$. In the Heisenberg picture the weight function $\rho(q)$ becomes time-dependent.

The major technical problem in quantum field theory on a curved background $(\mathcal{M}, \gamma)$ is the existence of infinitely many unitarily inequivalent representations of the canonical commutation relations (3.2.11–3.2.13). If the background metric γ is static, the obvious step is to use the timelike Killing vector to select the representation. This gives rise to a consistent one-particle picture of the quantum field theory. In other special, but non-static, cases (for example, a Robertson-Walker metric) there may be a 'natural' choice for the representations that gives rise to a picture of particle creation by the background metric, and for genuine astrophysical applications this may be perfectly adequate.

The real problems arise if one is presented with a generic metric γ, in which case it is not at all clear how to proceed. A minimum requirement is that the Hamiltonians $\widehat{H}(t)$, or the Hamiltonian densities $\widehat{h}_{\alpha}(x, \mathcal{E}\,]$, should be well-defined. However, there is an unpleasant possibility that the representations could be t (resp. $\mathcal{E}$) dependent, and in such a way that those corresponding to different values of t (resp. $\mathcal{E}$) are unitarily inequivalent, in which case the dynamical equations (3.2.20) (resp. (3.2.21)) are not meaningful. This particular difficulty can be overcome by using a C^*-algebra approach, but the identification of physically-meaningful representations remains a major problem.

3.3 The Arnowitt-Deser-Misner Formalism

3.3.1 Introduction of the Foliation

Let us consider now how these ideas extend to the canonical formalism of general relativity itself. The early history of this subject was grounded in the seminal work of Dirac (1958a,1958b) and culminated in the investigations by Arnowitt, Deser and Misner (1959a, 1959b, 1960a, 1960b, 1960c, 1960d, 1961a, 1961b, 1962). These original studies involved the selection of a specific coordinate system on the spacetime manifold, and some of the global issues were thereby obscured. A more geometrical, global approach was developed by Kuchař in a series of papers Kuchař (1972, 1976a, 1976b, 1976c, 1977, 1981a) and this will be adopted here. The treatment will be fairly cursory since the main aim of this course is to develop the conceptual and structural aspects of the problem of time. A more detailed account of the technical issues can be found in the forthcoming work Isham & Kuchař (1994).

The starting point is a four-dimensional manifold $\mathcal{M}$ and a three-dimensional manifold Σ that play the roles of physical spacetime and three-space respectively. The space Σ is assumed to be compact; if not, some of the expressions that follow must be augmented by surface terms. Furthermore, the topology of $\mathcal{M}$ is assumed to be such that it can be foliated by a one-parameter family of embeddings $\mathcal{F}_t : \Sigma \to \mathcal{M}$, $t \in \mathbb{R}$, of Σ in $\mathcal{M}$ (of course, then there will be many such foliations). As in the case of the scalar field theory, this requirement imposes a significant *a priori* topological limitation on $\mathcal{M}$ since (by the definition of a foliation) the map $\mathcal{F} : \Sigma \times \mathbb{R} \to \mathcal{M}$, defined by $(x,t) \mapsto \mathcal{F}(x,t) := \mathcal{F}_t(x)$, is a diffeomorphism of $\Sigma \times \mathbb{R}$ with $\mathcal{M}$.

Since $\mathcal{F}$ is a diffeomorphism from $\Sigma \times \mathbb{R}$ to $\mathcal{M}$, its inverse $\mathcal{F}^{-1} : \mathcal{M} \to \Sigma \times \mathbb{R}$ is also a diffeomorphism and can be written in the form

$$\mathcal{F}^{-1}(X) = (\sigma(X), \tau(X)) \in \Sigma \times \mathbb{R} \tag{3.3.1}$$

where $\sigma : \mathcal{M} \to \Sigma$ and $\tau : \mathcal{M} \to \mathbb{R}$. The map τ is a global *time function* and gives the natural time parameter associated with the foliation in the sense that $\tau(\mathcal{F}_t(x)) = t$ for all $x \in \Sigma$. However, from a physical point of view such a definition of 'time' is rather artificial (how would it be measured?) and is a far cry from the notion of time employed in the construction of real clocks. This point is not trivial and we shall return to it later.

Note that for each $x \in \Sigma$ the map $\mathcal{F}_x : \mathbb{R} \to \mathcal{M}$ defined by $t \mapsto \mathcal{F}(x,t)$ is a curve in $\mathcal{M}$ and therefore has a one-parameter family of tangent vectors on $\mathcal{M}$, denoted $\dot{\mathcal{F}}_x(t)$, whose components are $\dot{\mathcal{F}}_x^\alpha(t) = \partial\mathcal{F}^\alpha(x,t)/\partial t$. The flow lines of the ensuing vector field (known as the *deformation* vector [9] of the foliation) are illustrated in Figure 1 which also shows the normal vector n_{t_1} on the hypersurface $\mathcal{F}_{t_1}(\Sigma) \subset \mathcal{M}$.

[9] The deformation vector is a hybrid object in the sense that if x is a point in the three-space

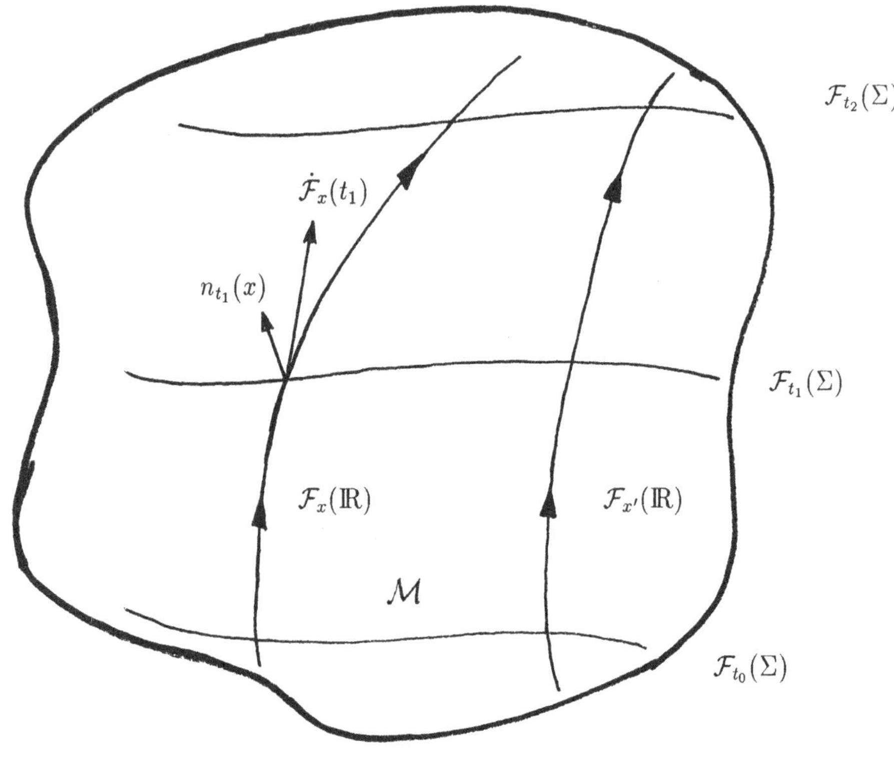

Figure 1: The flow lines of the foliation of $\mathcal{M}$

In general, if $\mathcal{E} : \Sigma \to \mathcal{M}$ is a spacelike embedding, the *normal* vector field n to $\mathcal{E}$ is defined by the equations

$$n_\alpha(x, \mathcal{E}\,]\, \mathcal{E}^\alpha{}_{,a}(x) \;=\; 0 \qquad (3.3.2)$$

$$\gamma^{\alpha\beta}(\mathcal{E}(x))\, n_\alpha(x, \mathcal{E}\,]\, n_\beta(x, \mathcal{E}\,] \;=\; -1 \qquad (3.3.3)$$

for all $x \in \Sigma$. Equation (3.3.2) defines what it means to say that n is *normal* to the hypersurface $\mathcal{E}(\Sigma)$, while (3.3.3) is a normalisation condition on n and emphasises that this vector is timelike with respect to the Lorentzian metric on $\mathcal{M}$. The minus sign on the right hand side of (3.3.3) reflects our choice of signature on $(\mathcal{M}, \gamma)$ as $(-1, 1, 1, 1)$.

In considering the dynamical evolution, a particularly interesting quantity is the functional derivative of $g_{ab}(x, \mathcal{E}\,]$ with respect to $\mathcal{E}$ projected along the normal vector n. A direct calculation (Hojman, Kuchař & Teitelboim 1976, Kuchař 1974) shows

Σ the vector $\dot{\mathcal{F}}_x(t)$ lies in the tangent space $T_{\mathcal{F}(x,t)}\mathcal{M}$ at the point $\mathcal{F}(x,t)$ in the four-manifold $\mathcal{M}$. Such an object is best regarded as an element of the space $T_{\mathcal{F}_t}\mathrm{Emb}(\Sigma, \mathcal{M})$ of vectors tangent to the infinite-dimensional manifold $\mathrm{Emb}(\Sigma, \mathcal{M})$ of embeddings of Σ in $\mathcal{M}$ at the particular embedding $\mathcal{F}_t$. This way of viewing things is quite useful technically but I shall not develop it further here.

that

$$n^\alpha(x,\mathcal{E}]\frac{\delta}{\delta\mathcal{E}^\alpha(x)}g_{ab}(x',\mathcal{E}] = -2K_{ab}(x,\mathcal{E}]\,\delta(x,x') \tag{3.3.4}$$

where K is the *extrinsic curvature* of the hypersurface $\mathcal{E}(\Sigma)$ defined by

$$K_{ab}(x,\mathcal{E}] := -{}^4\nabla_\alpha n_\beta(x,\mathcal{E}]\,\mathcal{E}^\alpha{}_{,a}(x)\,\mathcal{E}^\beta{}_{,b}(x) \tag{3.3.5}$$

where ${}^4\nabla_\alpha n_\beta(x,\mathcal{E}]$ denotes the covariant derivative obtained by parallel transporting the cotangent vector $n(x,\mathcal{E}] \in T^*_{\mathcal{E}(x)}\mathcal{M}$ along the hypersurface $\mathcal{E}(\Sigma)$ using the Lorentzian four-metric γ on $\mathcal{M}$. It is straightforward to show that K is a symmetric tensor, *i.e.*, $K_{ab}(x,\mathcal{E}]{=}K_{ba}(x,\mathcal{E}]$.

3.3.2 The Lapse Function and Shift Vector

The next step is to decompose the deformation vector into two components, one of which lies along the hypersurface $\mathcal{F}_t(\Sigma)$ and the other of which is parallel to n_t. In particular we can write

$$\dot{\mathcal{F}}^\alpha(x,t) = N(x,t)\,\gamma^{\alpha\beta}(\mathcal{F}(x,t))\,n_\beta(x,t) + N^a(x,t)\,\mathcal{F}^\alpha{}_{,a}(x,t) \tag{3.3.6}$$

where we have used $n_\beta(x,t)$ instead of the more clumsy notation $n_\beta(x,\mathcal{F}_t]$. The quantities $N(t)$ and $N^a(t)$ are known respectively as the *lapse* function and the *shift* vector associated with the embedding $\mathcal{F}_t$.

Note that, like the normal vector, the lapse and shift depend on both the spacetime metric γ and the foliation. This relationship can be partially inverted with, for a fixed foliation, the lapse and shift functions being identified as parts of the metric tensor γ. This can be seen most clearly by studying the 'pull-back' $\mathcal{F}^*(\gamma)$ of γ by the foliation $\mathcal{F}:\Sigma\times\mathbb{R}\to\mathcal{M}$ in coordinates X^α, $\alpha = 0\ldots3$, on $\Sigma\times\mathbb{R}$ that are adapted to the product structure in the sense that $X^{\alpha=0}(x,t) = t$, and $X^{\alpha=1,2,3}(x,t) = x^{a=1,2,3}(x)$ where x^a, $a = 1,2,3$ is some coordinate system on Σ. The components of $\mathcal{F}^*(\gamma)$ are

$$(\mathcal{F}^*\gamma)_{00}(x,t) = N^a(x,t)\,N^b(x,t)\,g_{ab}(x,t) - (N(x,t))^2 \tag{3.3.7}$$

$$(\mathcal{F}^*\gamma)_{0a}(x,t) = N^b(x,t)\,g_{ab}(x,t) \tag{3.3.8}$$

$$(\mathcal{F}^*\gamma)_{ab}(x,t) = g_{ab}(x,t) \tag{3.3.9}$$

where $g_{ab}(x,t)$ is shorthand for $g_{ab}(x,\mathcal{F}_t]$ and is given by

$$g_{ab}(x,t) := (\mathcal{F}_t^*\,\gamma)_{ab}(x) = \gamma_{\alpha\beta}(\mathcal{F}(x,t))\,\mathcal{F}^\alpha{}_{,a}(x,t)\,\mathcal{F}^\beta{}_{,b}(x,t). \tag{3.3.10}$$

We see from (3.3.6) that the lapse function and shift vector provide information on how a hypersurface with constant time parameter t is related to the displaced hypersurface with constant parameter $t+\delta t$ as seen from the perspective of an enveloping spacetime. More precisely, for a given Lorentzian metric γ on $\mathcal{M}$ and foliation

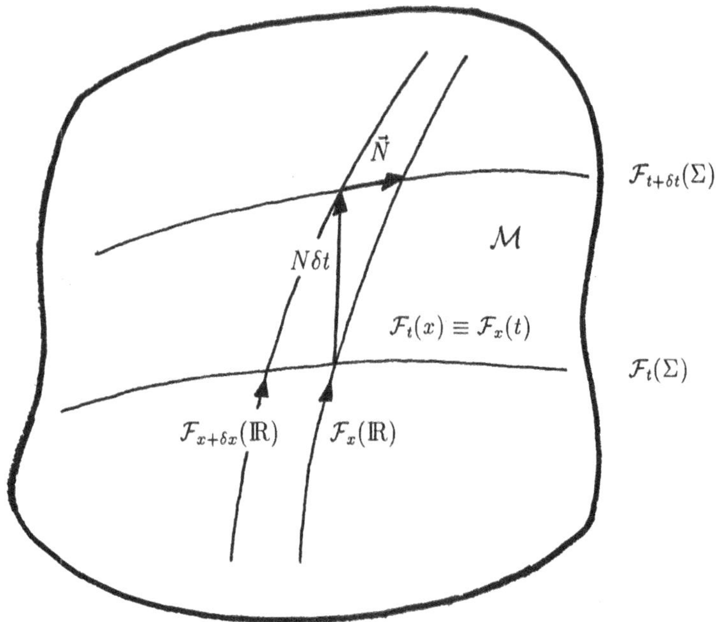

Figure 2: The lapse function and shift vector of a foliation

$\mathcal{F} : \Sigma \times \mathrm{I\!R} \to \mathcal{M}$, the lapse function specifies the proper time separation $\delta_\perp \tau$ between the hypersurfaces $\mathcal{F}_t(\Sigma)$ and $\mathcal{F}_{t+\delta t}(\Sigma)$ measured in the direction normal to the first hypersurface:

$$\delta_\perp \tau(x) = N(x,t)\delta t. \tag{3.3.11}$$

The shift vector $\vec{N}(x)$ determines how, for each $x \in \Sigma$, the point $\mathcal{F}_{t+\delta t}(x)$ in $\mathcal{M}$ is displaced with respect to the intersection of the hypersurface $\mathcal{F}_{t+\delta t}(\Sigma)$ with the normal geodesic drawn from the point $\mathcal{F}_t(x) \equiv \mathcal{F}_x(t)$. If this intersection point can be obtained by evaluating $\mathcal{F}_{t+\delta t}$ at a point on Σ with coordinates $x^a + \delta x^a$ then

$$\delta x^a(x) = -N^a(x,t)\delta t \tag{3.3.12}$$

as illustrated in Figure 2.

It is clear from (3.3.7–3.3.8) that the shift vector and lapse function are essentially the γ_{0a} and γ_{00} parts of the spacetime metric. It was in terms of this coordinate language that the original ADM approach was formulated. As we shall see shortly, the Einstein field equations do not lead to any dynamical development for these variables.

3.3.3 The Canonical Form of General Relativity

The canonical analysis of general relativity proceeds as follows. We choose some reference foliation $\mathcal{F}^{\text{ref}} : \Sigma \times \mathrm{I\!R} \approx \mathcal{M}$ and consider first the situation in which $\mathcal{M}$

carries a *given* Lorentzian metric γ that satisfies the vacuum Einstein field equations $G_{\alpha\beta}(X,\gamma] = 0$ and is such that each leaf $\mathcal{F}_t^{\text{ref}}(\Sigma)$ is a hypersurface in $\mathcal{M}$ that is spacelike with respect to γ. The problem is to find a set of canonical variables for this system and an associated set of first-order differential equations that determine how the variables evolve from one leaf to another and whose solution recovers the given metric γ. Of course, in practice this procedure is used in the situation in which γ is unknown and is to be determined by solving the Cauchy problem using given Cauchy data.

The Riemannian metric on Σ defined by (3.3.10) plays the role of the basic configuration variable. The rate of change of g_{ab} with respect to the label time t is related to the extrinsic curvature (3.3.5) K for the embeddings $\mathcal{F}_t^{\text{ref}} : \Sigma \to \mathcal{M}$ by

$$K_{ab}(x,t) = \frac{1}{2N(x,t)}(-\dot{g}_{ab}(x,t) + L_{\vec{N}} g_{ab}(x,t)) \tag{3.3.13}$$

where $L_{\vec{N}} g_{ab}$ denotes the Lie derivative [10] of g_{ab} along the shift vector field $\vec{N}$ on Σ.

The key step in deriving the canonical form of the action principle is to pull-back the Einstein Lagrangian density by the foliation $\mathcal{F}^{\text{ref}} : \Sigma \times \mathbb{R} \to \mathcal{M}$ and express the result as a function of the extrinsic curvature, the metric g, and the lapse vector and shift function. This gives

$$\begin{aligned}(\mathcal{F}^{\text{ref}})^*(|\gamma|^{\frac{1}{2}} R[\gamma]) &= N|g|^{\frac{1}{2}}(K_{ab}K^{ab} - (K_a^{\ a})^2 + R[g]) \\ &\quad -2\frac{d}{dt}(|g|^{\frac{1}{2}} K_a^{\ a}) - (|g|^{\frac{1}{2}}(K_a^{\ a} N^b - g^{ab} N_{,a}))_{,b}\end{aligned} \tag{3.3.14}$$

where $R[g]$ and $|g|$ denote respectively the curvature scalar and the determinant of the metric on Σ.

The spatial-divergence term vanishes because Σ is assumed to be compact. However, the term with the total time-derivative must be removed by hand to produce a genuine action principle. The resulting 'ADM' action for matter-free gravity can be written as

$$S[g,N,\vec{N}] = \frac{1}{\kappa^2}\int dt \int_\Sigma d^3x\, N|g|^{\frac{1}{2}}(K_{ab}K^{ab} - (K_a^{\ a})^2 + R[g]) \tag{3.3.15}$$

where $\kappa^2 := 8\pi G/c^2$. This action is to be varied with respect to the Lorentzian metric $(\mathcal{F}^{\text{ref}})^*(\gamma)$ on $\Sigma \times \mathbb{R}$, and hence with respect to the paths of spatial geometries, lapse functions, and shift vectors associated via (3.3.7–3.3.9) with $(\mathcal{F}^{\text{ref}})^*(\gamma)$. The extrinsic curvature K is to be thought of as the explicit functional of these variables given by (3.3.13).

---

[10] This quantity is equal to $N_{a|b} + N_{b|a}$ where $|$ means covariant differentiation using the induced metric g_{ab} on Σ.

The canonical analysis now proceeds in the standard way except that, since (3.3.15) does not depend on the time derivatives of N or $\vec{N}$, in performing the Legendre transformation to the canonical action, the time derivative $\partial g_{ab}/\partial t$ is replaced by the conjugate variable p^{ab} but N and $\vec{N}$ are left untouched. The variable conjugate to g_{ab} is

$$p^{ab} := \frac{\delta S}{\delta \dot{g}_{ab}} = -\frac{|g|^{\frac{1}{2}}}{\kappa^2}(K^{ab} - g^{ab}K_c{}^c) \tag{3.3.16}$$

which can be inverted in the form

$$K^{ab} = -\kappa^2 |g|^{-\frac{1}{2}}(p^{ab} - \frac{1}{2}g^{ab}p_c{}^c). \tag{3.3.17}$$

The action obtained by performing the Legendre transformation is

$$S[g,p,N,\vec{N}] = \int dt \int_\Sigma d^3x\,(p^{ab}\dot{g}_{ab} - N\mathcal{H}_\perp - N^a\mathcal{H}_a) \tag{3.3.18}$$

where

$$\mathcal{H}_a(x;g,p] := -2p_a{}^b{}_{|b}(x) \tag{3.3.19}$$

$$\mathcal{H}_\perp(x;g,p] := \kappa^2 \mathcal{G}_{ab\,cd}(x,g]\,p^{ab}(x)\,p^{cd}(x) - \frac{|g|^{\frac{1}{2}}(x)}{\kappa^2}\,R(x,g] \tag{3.3.20}$$

in which

$$\mathcal{G}_{ab\,cd}(x,g] := \frac{1}{2}|g|^{-\frac{1}{2}}(x)(g_{ac}(x)g_{bd}(x) + g_{bc}(x)g_{ad}(x) - g_{ab}(x)g_{cd}(x)) \tag{3.3.21}$$

is the 'DeWitt supermetric' on the space of three-metrics (DeWitt 1967a). The functions $\mathcal{H}_a$ and $\mathcal{H}_\perp$ of the canonical variables (g,p) play a key role in the theory and are known as the *supermomentum* and *super-Hamiltonian* respectively.

It must be emphasised that the variables g, p, N and $\vec{N}$ are treated as independent in varying the action (3.3.18). In particular, $g_{ab}(x,t)$ is no longer viewed as the restriction of a spacetime metric γ to a particular hypersurface $\mathcal{F}_t(\Sigma)$ of $\mathcal{M}$. Rather, $t \mapsto g_{ab}(x,t)$ is any path in the space Riem(Σ) of Riemannian metrics on Σ, and similarly $t \mapsto p^{ab}(x,t)$ (resp. $N(x,t)$, $N^a(x,t)$) is any path in the space of all contravariant, rank-2 symmetric tensor densities (resp. functions, vector fields) on Σ. Consistency with the earlier discussion is ensured by noting that the equation obtained by varying the action (3.3.18) with respect to p^{ab}

$$\dot{g}_{ab}(x,t) = \frac{\delta H[N,\vec{N}]}{\delta p^{ab}(x,t)}, \tag{3.3.22}$$

can be solved algebraically for p^{ab} and, with the aid of (3.3.17), reproduces the relation (3.3.13) between the time-derivative of g_{ab} and the extrinsic curvature. Here, the functional

$$H[N,\vec{N}](t) := \int_\Sigma d^3x\,(N\mathcal{H}_\perp + N^a\mathcal{H}_a) \tag{3.3.23}$$

of the canonical variables (g,p) acts as the Hamiltonian of the system.

Varying the action (3.3.18) with respect to g_{ab} gives the dynamical equations

$$\dot{p}^{ab}(x,t) = -\frac{\delta H[N,\vec{N}]}{\delta g_{ab}(x,t)} \tag{3.3.24}$$

while varying N^a and N leads respectively to

$$\mathcal{H}_a(x;g,p] = 0 \tag{3.3.25}$$

and

$$\mathcal{H}_\perp(x;g,p] = 0 \tag{3.3.26}$$

which are *constraints* on the canonical variables (g,p).

Note that the action (3.3.18) contains no time derivatives of the fields N or $\vec{N}$ which appear manifestly as Lagrange multipliers that enforce the constraints (3.3.26) and (3.3.25). This confirms the status of the lapse function and shift vector as non-dynamical variables that can be specified as *arbitrary* functions on $\Sigma \times \mathbb{R}$; indeed, this must be done in some way before the dynamical equations can be solved (see §3.3.6).

The set of equations (3.3.22), (3.3.24), (3.3.25) and (3.3.26) are completely equivalent to Einstein's equations $G_{\alpha\beta} = 0$ in the following sense (Fischer & Marsden 1979):

- Let γ by any Lorentzian metric on $\mathcal{M}$ that satisfies the vacuum Einstein equations $G_{\alpha\beta}(X,\gamma] = 0$ and let $t \mapsto \mathcal{F}_t : \Sigma \to \mathcal{M}$ be a one-parameter family of spacelike embeddings of Σ in $\mathcal{M}$ that foliates $\mathcal{M}$ with associated lapse function and shift vector given by (3.3.6). Then the family of induced metrics $t \mapsto g(t)$ and momenta $t \mapsto p(t)$ (where $p(t)$ is computed from γ using (3.3.13) and (3.3.16)) satisfies the equations (3.3.22), (3.3.24), (3.3.25) and (3.3.26).
- Conversely, if $t \mapsto \mathcal{F}_t$ is a spacelike foliation of $(\mathcal{M},\gamma)$ such that the evolution and constraint equations above hold, then γ satisfies the vacuum field equations.

In effect, equations (3.3.22) and (3.3.24) reproduce the projections of the Einstein equations that are tangent to the hypersurfaces $t = const.$, while (3.3.25) and (3.3.26) reproduce the normal projections of these equations, $n^\alpha n^\beta G_{\alpha\beta} = 0$ and $n^\alpha G_{\alpha a} = 0$.

3.3.4 The Constraint Algebra

The classical canonical algebra of the system is expressed by the basic Poisson brackets

$$\{g_{ab}(x), g_{cd}(x')\} \;=\; 0 \tag{3.3.27}$$

$$\{p^{ab}(x), p^{cd}(x')\} = 0 \tag{3.3.28}$$
$$\{g_{ab}(x), p^{cd}(x')\} = \delta^c_{(a}\delta^d_{b)}\,\delta(x,x') \tag{3.3.29}$$

which can be used to cast the dynamical equations into a canonical form in which the right hand sides of (3.3.22) and (3.3.24) are $\{g_{ab}(x), H[N,\vec{N}]\}$ and $\{p^{ab}(x), H[N,\vec{N}]\}$ respectively. However, the constraints (3.3.25) and (3.3.26) imply that not all the variables $g_{ab}(x)$ and $p^{cd}(x)$ are independent, and so, as in all systems with constraints, the Poisson bracket relations (3.3.27–3.3.29) need to be used with care.

A crucial property of the canonical formalism is the closure of the Poisson brackets of the super-Hamiltonian and supermomentum, computed using (3.3.27–3.3.29); *i.e.*, the set of constraints is *first class*. This is contained in the fundamental relations (Dirac 1965)

$$\{\mathcal{H}_a(x), \mathcal{H}_b(x')\} = -\mathcal{H}_b(x)\,\partial^{x'}_a\delta(x,x') + \mathcal{H}_a(x')\,\partial^x_b\delta(x,x') \tag{3.3.30}$$
$$\{\mathcal{H}_a(x), \mathcal{H}_\perp(x')\} = \mathcal{H}_\perp(x)\,\partial^x_a\delta(x,x') \tag{3.3.31}$$
$$\{\mathcal{H}_\perp(x), \mathcal{H}_\perp(x')\} = g^{ab}(x)\,\mathcal{H}_a(x)\,\partial^{x'}_b\delta(x,x') - g^{ab}(x')\,\mathcal{H}_a(x')\,\partial^x_b\delta(x,x'). \tag{3.3.32}$$

Using the smeared variables [11]

$$H[N] := \int_\Sigma d^3x\, N(x)\,\mathcal{H}_\perp(x), \quad H[\vec{N}] := \int_\Sigma d^3x\, N^a(x)\,\mathcal{H}_a(x), \tag{3.3.33}$$

where N and $\vec{N}$ are any scalar function and vector field on Σ, these equations can be written as

$$\{H[\vec{N}_1], H[\vec{N}_2]\} = H([\vec{N}_1,\vec{N}_2]) \tag{3.3.34}$$
$$\{H[\vec{N}], H[N]\} = H[L_{\vec{N}}N] \tag{3.3.35}$$
$$\{H[N_1], H[N_2]\} = H[\vec{N}] \tag{3.3.36}$$

where, in (3.3.36), $N^a(x) := g^{ab}(x)(N_1(x)N_2(x)_{,b} - N_2(x)N_1(x)_{,b})$.

The geometrical interpretation of these expressions is as follows.

1. The Lie algebra of the diffeomorphism group Diff(Σ) is generated by the vector fields on Σ with minus the commutator of a pair of vector fields

$$[N_1, N_2]^a := N_1^b\, N^a_{2\,,b} - N_2^b\, N^a_{1\,,b} \tag{3.3.37}$$

playing the role of the Lie bracket. Thus (3.3.34) shows that the map $\vec{N} \to -H[\vec{N}]$ is a homomorphism of the Lie algebra of Diff(Σ) into the Poisson bracket algebra of the theory. A direct calculation of the Poisson brackets of $H[\vec{N}]$ with $g_{ab}(x)$ and $p^{cd}(x)$ confirms the interpretation of $-H[\vec{N}]$ as a generator of spatial diffeomorphisms.

[11] The integrals are well-defined since both $\mathcal{H}_a$ and $\mathcal{H}_\perp$ are densities on Σ.

2. Similarly, $H[N]$ can be interpreted as generating deformations of a hypersurface *normal* to itself as embedded in $\mathcal{M}$. However, note that, unlike the analogous statement for $H[\vec{N}]$, this interpretation applies only *after* the field equations have been solved.

Two important things should be noted about the 'gauge' algebra (3.3.34–3.3.36):

1. It is *not* the Lie algebra of Diff($\mathcal{M}$) even though this was the invariance group of the original theory.

2. The presence of the g^{ab} factor in the right hand side of (3.3.36) means it is *not* a genuine Lie algebra at all.

These two features are closely related since the Dirac algebra (3.3.34–3.3.36) is essentially the Lie algebra of Diff($\mathcal{M}$) *projected* along, and normal to, a spacelike hypersurface. The significance of this in the quantum theory will emerge later.

Even though the Dirac algebra is not a genuine Lie algebra, it still generates an action on the canonical variables (g,p). This is obtained by integrating the infinitesimal changes of the form

$$\delta_{(N,\vec{N})}\, g_{ab}(x) := \{g_{ab}(x), H[N] + H[\vec{N}]\} \tag{3.3.38}$$

$$\delta_{(N,\vec{N})}\, p^{ab}(x) := \{p^{ab}(x), H[N] + H[\vec{N}]\} \tag{3.3.39}$$

for arbitrary infinitesimal smearing functions N and $\vec{N}$. We shall refer to the set of all such trajectories in the phase space $\mathcal{S}$ of pairs (g,p) as the *orbits* of the Dirac algebra. Note that, because of the first-class nature of the constraints $\mathcal{H}_a(x) = 0 = \mathcal{H}_\perp(x)$, the subspace $\mathcal{S}_\mathcal{C} \subset \mathcal{S}$ on which the constraints are satisfied is mapped into itself by the transformations (3.3.38–3.3.39). On this constraint surface, an orbit consists of the set of all pairs (g,p) that can be obtained from spacelike slices of some specific Lorentzian metric γ on $\mathcal{M}$ that satisfies the vacuum Einstein equations.

A peculiar feature of general relativity is that an orbit on the constraint surface includes the *dynamical evolution* of a pair (g,p) with respect to any choice of lapse function and shift vector. Indeed, the dynamical equations (3.3.22) and (3.3.24) are simply a special case of the transformations above. This has an important implication for the notion of an 'observable'. In a system with first-class constraints, an observable is normally defined to be any function on the phase space of the system whose Poisson bracket with the constraints vanishes weakly, *i.e.*, it vanishes on the constraint surface. In the present case, this means that A is an observable if and only if

$$\{\, A, H[N,\vec{N}]\,\} \approx 0 \tag{3.3.40}$$

for all N and $\vec{N}$, where $H[N,\vec{N}] := H[N] + H[\vec{N}]$. Thus any such quantity is a *constant* of the motion with respect to evolution along the foliation associated with

N and $\vec{N}$. We shall deal later with the quantum analogue of this situation which is deeply connected with the problem of time in quantum gravity.

3.3.5 The Role of the Constraints

The constraints (3.3.25–3.3.26) are of major importance in both the classical and the quantum theories of gravity, and it is useful at this point to gather together various results concerning them.

1. The constraints are consistent with the equations of motion (3.3.22) and (3.3.24) in the sense of being automatically maintained in time. For example,

$$\frac{d\mathcal{H}_\perp(x)}{dt} = \{\mathcal{H}_\perp(x), H[N, \vec{N}]\} \tag{3.3.41}$$

   and the right hand side vanishes on the constraint surface $\mathcal{S}_C$ in phase space by virtue of the closing nature of the algebra (3.3.30–3.3.32).

2. The constraints lead to a well-posed Cauchy problem for the dynamical equations (Hawking & Ellis 1973, Fischer & Marsden 1979) once the undetermined quantities N and $\vec{N}$ have been fixed in some way (see below).

3. If a Lorentzian metric γ on the spacetime $\mathcal{M}$ satisfies the vacuum Einstein equations $G_{\alpha\beta}(X,\gamma] = 0$ then the constraint equations (3.3.25–3.3.26) are satisfied on all spacelike hypersurfaces of $\mathcal{M}$. (It is understood that p^{ab} is to be computed from the given spacetime geometry using the definition (3.3.5) of the extrinsic curvature K_{ab} and the relation (3.3.17) between p^{ab} and K^{ab}.)

4. Conversely, let $(\mathcal{M},\gamma)$ be a Lorentzian spacetime with the property that the constraint equations (3.3.25–3.3.26) are satisfied on every spacelike hypersurface. Then γ necessarily satisfies all ten Einstein field equations $G_{\alpha\beta}(X,\gamma] = 0$.

This last result is highly significant since it means the dynamical aspects of the Einstein equations are already contained in the constraints alone. This plays a crucial role in the Dirac quantisation programme (see §5.1) and also in the Hamilton-Jacobi approach to the classical theory.

The proof is rather simple. For any given foliation of $\mathcal{M}$, the Hamiltonian constraints $\mathcal{H}_\perp(x; g(t), p(t)] = 0$ are equivalent to the spacetime equation

$$n^\alpha(x,t)\, n^\beta(x,t)\, G_{\alpha\beta}(\mathcal{F}(x,t),\gamma] = 0 \tag{3.3.42}$$

where n^α is the vector normal to the hypersurface $\mathcal{F}_t(\Sigma)$ at the point $\mathcal{F}(x,t)$ in $\mathcal{M}$. Now, at any given point $X \in \mathcal{M}$, and for any (normalised) timelike vector m in the tangent space $T_X\mathcal{M}$, the foliation can be chosen so that m is the normal vector at that point. Thus, by ranging over all possible foliations, we find that for all $X \in \mathcal{M}$ and for all timelike vectors m at X

$$m^\alpha(X)\, m^\beta(X)\, G_{\alpha\beta}(X, \gamma] = 0. \tag{3.3.43}$$

Now suppose m_1, m_2 is any pair of timelike vectors. Then $m_1 + m_2$ is also timelike and hence $(m_1 + m_2)^\alpha (m_1 + m_2)^\beta\, G_{\alpha\beta} = 0$. However, $m_1^\alpha\, m_1^\beta\, G_{\alpha\beta} = m_2^\alpha\, m_2^\beta\, G_{\alpha\beta} = 0$ and hence, since $G_{\alpha\beta} = G_{\beta\alpha}$, we see that

$$m_1^\alpha\, m_2^\beta\, G_{\alpha\beta} = 0. \tag{3.3.44}$$

Now let σ be any spacelike vector. Then there exist timelike vectors m_1, m_2 such that $\sigma = m_1 - m_2$. It follows from (3.3.44) that for any timelike m,

$$m^\alpha\, \sigma^\beta\, G_{\alpha\beta} = 0. \tag{3.3.45}$$

Finally, a similar result shows that for any pair of spacelike vectors σ_1, σ_2,

$$\sigma_1^\alpha\, \sigma_2^\beta\, G_{\alpha\beta} = 0. \tag{3.3.46}$$

However, any vector $u \in T_p\mathcal{M}$ can be written as the sum of a spacelike and a timelike vector, and hence for all $X \in \mathcal{M}$ and $u, v \in T_X\mathcal{M}$ we have

$$u^\alpha\, v^\beta\, G_{\alpha\beta}[\gamma] = 0, \tag{3.3.47}$$

which means precisely that $G_{\alpha\beta}[\gamma] = 0$.

Note that:

1. The super-Hamiltonian constraints [12] $\mathcal{H}_\perp(x, \mathcal{E}] = 0$ for all spacelike embeddings $\mathcal{E}$ are sufficient by themselves: it is not necessary to impose the supermomentum constraints $\mathcal{H}_a(x, \mathcal{E}] = 0$ in addition.

2. There is a similar result for electromagnetism: if the initial-value equations $\mathrm{div}\vec{E} = 0$ hold in every inertial frame (*i.e.*, on every spacelike hyperplane), the electromagnetic field must necessarily evolve according to the dynamical Maxwell equations. [13]

[12] I have written $\mathcal{H}_\perp(x, \mathcal{E}]$ to emphasise that the variables g_{ab} and p^{ab} that appear in $\mathcal{H}_\perp$ are computed from the given Lorentzian four-metric γ using the spacelike embedding $\mathcal{E} : \Sigma \to \mathcal{M}$.

[13] Karel Kuchař, private communication.

3.3.6 Eliminating the Non-Dynamical Variables

I wish now to return to the canonical equations of motion (3.3.22–3.3.26) and the status of the non-dynamical degrees of freedom. These variables must be removed in some way before the equations of motion can be solved. As emphasised already, these redundant variables include N and $\vec{N}$ which must be fixed before the abstract label time t in the foliation can be related to a physical quantity like proper time. However, it is important to note that these are not the only variables in the theory that are non-dynamical. This is clear from a count of degrees of freedom. The physical modes of the gravitational field should correspond to $2 \times \infty^3$ configuration [14] variables (the two circular polarisations of a gravitational wave; the two helicity states of a graviton), or $4 \times \infty^3$ phase space variables. However, if a specific coordinate system [15] is chosen on Σ then, after eliminating N and $\vec{N}$, we have the $12 \times \infty^3$ variables $(g_{ab}(x), p^{cd}(x))$. In principle, the four constraint equations (3.3.25) and (3.3.26) can be used to remove a further $4 \times \infty^3$ variables, but this still leaves $8 \times \infty^3$, which is $4 \times \infty^3$ too many.

The physical origin of these spurious variables lies in the Diff($\mathcal{M}$)-invariance of the original Einstein action. The orbits of the Dirac algebra on the phase space of pairs (g, p) are similar in many respects to the orbits of the Yang-Mills gauge group on the space of vector-potentials and, as in that case, it is necessary to impose some sort of 'gauge'. This will have the desired effect of eliminating the remaining $4 \times \infty^3$ non-physical degrees of freedom.

The exact way in which these additional non-physical variables are to be identified and removed depends on how the lapse function and shift vector are fixed. The main possibilities are as follows.

1. Set N and $\vec{N}$ equal to specific functions [16] on $\Sigma \times \mathbb{R}$. If these functions are substituted into the dynamical equations (3.3.22) and (3.3.24) they yield four equations of the form $\dot{F}^A(x,t;g(t),p(t)] = 0$, $A = 0,1,2,3$, whose solutions are $F^A(x,t;g(t),p(t)]=f^A(x)$ for any arbitrary set of functions f^A on Σ. Solving these equations enables an extra $4 \times \infty^3$ variables to be removed, as was required.

   This technique can be generalised in various ways. For example, N and $\vec{N}$ can be set equal to specific functionals $N(x,t;g,p]$ and $N^a(x,t;g,p]$ of the canonical variables. Another possibility is to impose conditions on the time derivatives of N and $\vec{N}$ rather than on N and $\vec{N}$ themselves. This is necessary in the context of the so-called 'covariant gauges'.

[14]The notation $n \times \infty^m$ refers to a field complex with n components defined on an m-dimensional manifold.

[15]In most of the discussion in this paper I shall neglect possible global issues such as, for example, the fact that a three-manifold Σ that is compact cannot be covered by a single coordinate system.

[16]To correspond to a proper foliation it is necessary that, for all $(x,t) \in \Sigma \times \mathbb{R}$, $N(x,t) \neq 0$ and, by convention, we require the continuous function N to satisfy $N(x,t) > 0$. This means that proper time increases in the same direction as the foliation time label t; changing the sign of N corresponds to reversing the sign of t.

2. Another option is to start by imposing four conditions

$$F^A(x,t;g(t),p(t)] = 0 \tag{3.3.48}$$

which restrict the paths $t \mapsto (g(t),p(t))$ in the phase space. This method is popular in path-integral approaches to canonical quantisation where it is implemented using a (formal) functional δ-function (see §4.1). The main requirement is that each orbit of the Dirac group contains one, and only one, of these restricted paths on the subspace of the phase space defined by the initial value constraints (3.3.25–3.3.26). As we shall discuss later, there can be global obstructions (in the phase space) to the existence of such functions F^A. Care must also be taken to avoid eliminating genuine physical degrees of freedom in addition to the non-dynamical modes. This potential problem has been discussed at length by Teitelboim (1982, 1983a, 1983b, 1983c, 1983d) .

   Since the equations $F^A(x,t;g(t),p(t)] = 0$ are valid for all times, the time-derivative must also vanish. If the ensuing equations $\dot{F}^A(x,t;g(t),p(t)] = 0$ are substituted into the dynamical equations (3.3.22) and (3.3.24), there results a set of four elliptic partial-differential equations for N and $\vec{N}$ that can be solved (in principle) to eliminate these variables as specific functionals of g and p.

3. The third approach to eliminating the non-dynamical variables will play a major role in our discussions of time. This involves the parametrisation of spacetime points by 'internal' space and time coordinates, defined as the values of various functionals $\mathcal{X}^A(x;g,p]$, $A = 0,1,2,3$ of the canonical variables, which can be set equal to some fixed functions $\chi_t^A(x)$. In effect, this leads to an equation of the type (3.3.48) and has the same technical consequences although the physical interpretation is more transparent.

3.4 Internal Time

3.4.1 The Main Ideas

The idea of specifying spacetime points by values of the fields was emphasised in the canonical theory by Baierlein, Sharp & Wheeler (1962), and is of sufficient importance to warrant a subsection in its own right. As we shall see, this is related to the problem of rewriting (3.3.18) as a true canonical action for just the physical modes of the gravitational field.

The key idea is to introduce quantities $\mathcal{T}(x;g,p]$ and $\mathcal{Z}^a(x;g,p]$, $a = 1,2,3$, that can serve as the time and spatial coordinates for a spacetime event that is associated (in a way yet to be specified) with the point $x \in \Sigma$ and the pair of canonical variables (g,p); the collection of four functions $(\mathcal{T},\mathcal{Z}^a)$ will be written as

$\mathcal{X}^A$, $A = 0,1,2,3$. Strictly speaking, we should worry about the fact that the three-manifold Σ is compact and hence cannot be covered by a single system of coordinates. The same will be true for the spacetime manifold $\Sigma \times \mathbb{R}$, and therefore we should allow for more than one set of the internal spatial functions $\mathcal{Z}^a$. This could be handled in various ways. One option is to work directly with collections of spatial coordinate functions that can cover Σ globally. But this is complicated since the number of coordinate charts used on a manifold is arbitrary provided only that it is greater than, or equal to, the mininum number determined by the manifold topology.

A slightly more elegant (although ultimately equivalent) approach is to realise that, since $\mathcal{M}$ is diffeomorphic to $\Sigma \times \mathbb{R}$, a spacetime event can be labelled by specifying (i) a value for a time variable, and (ii) a *point* in the three-manifold Σ. Thus we could try to extend the scheme above to include functions $\mathcal{Z}(x;g,p]$ whose values lie in Σ, rather than in some region of $\mathbb{R}^3$. When used in this way, I shall refer to Σ as the spatial *label* space and denote it by Σ_l. A particular set of functions $\mathcal{Z}^a(x;g,p]$ can then be associated with each set of local coordinate functions x^a on Σ_l according to the identification $\mathcal{Z}^a(x;g,p] := x^a(\mathcal{Z}(x;g,p])$. I shall not become too involved in this subtlety here since we are mainly concerned with broad principles rather than technical details. However, if desired, the discussion that follows can be generalised to include this concept of a label space.

To see how the idea of internal coordinates is used let us consider first the case of a given Einstein spacetime $(\mathcal{M},\gamma)$ where we wish to use the functionals to locate an event X in $\mathcal{M}$. The key steps are as follows.

1. Fix a reference foliation $\mathcal{F}^{\rm ref} : \Sigma \times \mathbb{R} \to \mathcal{M}$ with spacelike leaves and write its inverse $(\mathcal{F}^{\rm ref})^{-1} : \mathcal{M} \to \Sigma \times \mathbb{R}$ as

$$(\mathcal{F}^{\rm ref})^{-1}(X) := (\sigma^{\rm ref}(X), \tau^{\rm ref}(X)) \in \Sigma \times \mathbb{R} \tag{3.4.1}$$

   where $\sigma^{\rm ref} : \mathcal{M} \to \Sigma$, and where $\tau^{\rm ref} : \mathcal{M} \to \mathbb{R}$ is a global reference-time function on $\mathcal{M}$. This illustrates the general sense in which space and time are not intrinsic spacetime structures but are put into $\mathcal{M}$ by hand. Thus an instant of time, labelled by the number t, is the hypersurface

$$\mathcal{F}^{\rm ref}_t(\Sigma) := \{\mathcal{F}^{\rm ref}_t(x) | x \in \Sigma\} = (\tau^{\rm ref})^{-1}\{t\}, \tag{3.4.2}$$

   and a point in space x is the worldline

$$\mathcal{F}^{\rm ref}_x(\mathbb{R}) := \{\mathcal{F}^{\rm ref}_x(t) | t \in \mathbb{R}\} = (\sigma^{\rm ref})^{-1}\{x\} \tag{3.4.3}$$

   where we recall that $\mathcal{F}^{\rm ref}_t : \Sigma \to \mathcal{M}$ and $\mathcal{F}^{\rm ref}_x : \mathbb{R} \to \mathcal{M}$ are defined by $\mathcal{F}^{\rm ref}_t(x) := \mathcal{F}^{\rm ref}(x,t)$ and $\mathcal{F}^{\rm ref}_x(t) := \mathcal{F}^{\rm ref}(x,t)$ respectively. This is illustrated in Figure 3.

2. Construct the particular leaf of the reference foliation that passes through the given event X. According to the notation above, this embedding of Σ in $\mathcal{M}$

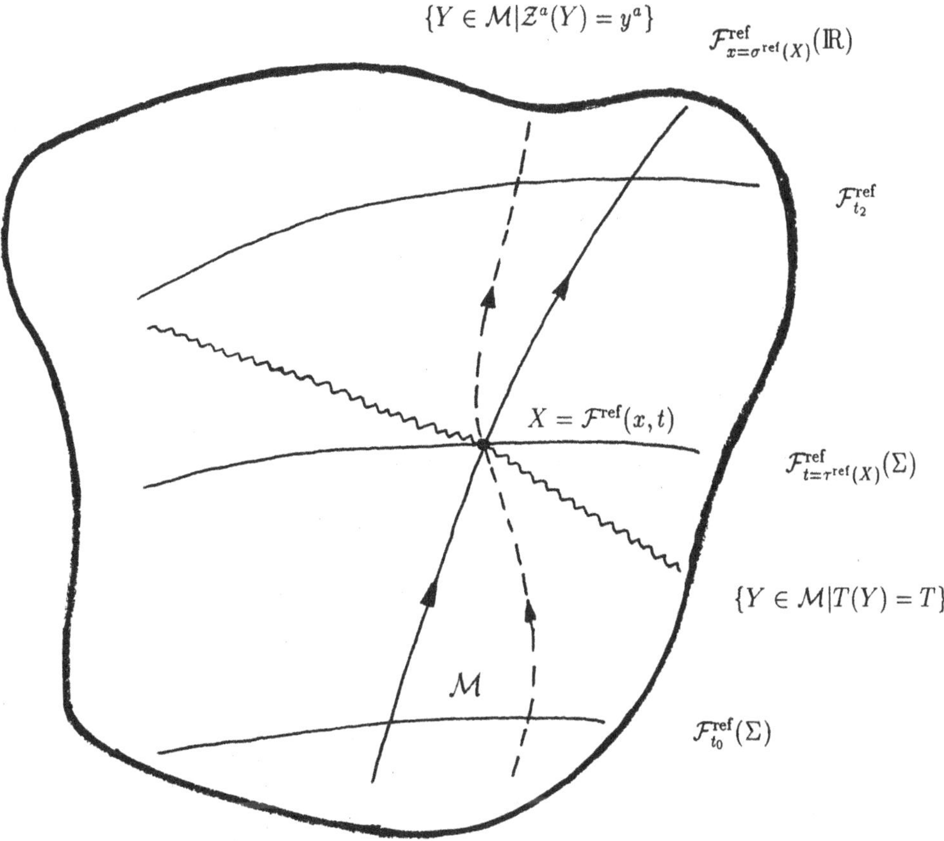

Figure 3: The foliation of $\mathcal{M}$ and the hypersurface $T(Y) = T$ of constant internal time.

is $\mathcal{F}^{\mathrm{ref}}_{t=\tau^{\mathrm{ref}}(X)}$. Let $x \in \Sigma$ be such that $\mathcal{F}^{\mathrm{ref}}_{t=\tau^{\mathrm{ref}}(X)}(x) \equiv \mathcal{F}^{\mathrm{ref}}_x(\tau^{\mathrm{ref}}(X)) = X$, *i.e.*, $x = \sigma^{\mathrm{ref}}(X)$. Compute:

(a) the induced metric $g_{ab}(x, \mathcal{F}^{\mathrm{ref}}_{t=\tau^{\mathrm{ref}}(X)}]$ on Σ using (3.2.2);

(b) the induced momentum $p^{cd}(x, \mathcal{F}^{\mathrm{ref}}_{t=\tau^{\mathrm{ref}}(X)}]$, using the definition (3.3.5) of the extrinsic curvature $K_{ab}(x, \mathcal{F}^{\mathrm{ref}}_{t=\tau^{\mathrm{ref}}(X)}]$ and the relation (3.3.16) between p and K.

3. Use these values of g and p in the definitions of $\mathcal{T}$ and $\mathcal{Z}^a$ to ascribe time and space coordinates to the event X.

In summary, the 'internal time' coordinate of the point $X \in \mathcal{M}$ is defined to be

$$T(X) := \mathcal{T}(\sigma^{\mathrm{ref}}(X); g[\mathcal{F}^{\mathrm{ref}}_{t=\tau^{\mathrm{ref}}(X)}], p[\mathcal{F}^{\mathrm{ref}}_{t=\tau^{\mathrm{ref}}(X)}]], \tag{3.4.4}$$

while the spatial labels are

$$Z^a(X) := \mathcal{Z}^a(\sigma^{\mathrm{ref}}(X); g[\mathcal{F}^{\mathrm{ref}}_{t=\tau^{\mathrm{ref}}(X)}], p[\mathcal{F}^{\mathrm{ref}}_{t=\tau^{\mathrm{ref}}(X)}]]. \tag{3.4.5}$$

Alternatively, one can write

$$T(\mathcal{F}^{\mathrm{ref}}(x,t)) = \mathcal{T}(x; g(t), p(t)] \tag{3.4.6}$$
$$Z^a(\mathcal{F}^{\mathrm{ref}}(x,t)) = \mathcal{Z}^a(x; g(t), p(t)] \tag{3.4.7}$$

where $g(t)$ and $p(t)$ denote the metric and momentum induced from γ on the hypersurface $\mathcal{F}^{\mathrm{ref}}_t(\Sigma)$ of $\mathcal{M}$.

Note that to be consistent such an identification of internal coordinates requires the functions $\mathcal{T}$ and $\mathcal{Z}^a$ to be chosen such that the following two conditions are satisfied with respect to the given Lorentzian metric γ on $\mathcal{M}$:

1. For all values of the constant number $T \in \mathbb{R}$, the 'instant of time' $\{Y \in \mathcal{M} | T(Y) = T\}$ must be a *spacelike* subspace of $(\mathcal{M}, \gamma)$.

2. For all $\vec{y} \in \mathbb{R}^3$, the set $\{Y \in \mathcal{M} | Z^a(Y) = y^a\}$ must be a *timelike* subspace [17] of $(\mathcal{M}, \gamma)$.

[17] Strictly speaking, this will be true only locally if the Z^a are local coordinate functionals. This problem can be overcome by using the labelling space Σ_l and requiring that, for all points $y \in \Sigma_l$, the set $\{Y \in \mathcal{M} | \mathcal{Z}(Y) = y\}$ is timelike.

3.4.2 Reduction to True Canonical Form

In the context of the general canonical theory—where there is no fixed spacetime metric—the internal time and space functionals are used to specify a set of non-dynamical degrees of freedom. This is part of the programme to reduce the theory to true canonical form. The key steps are as follows.

1. Perform a canonical transformation

$$(g_{ab}(x), p^{cd}(x)) \mapsto (\mathcal{X}^A(x), \mathcal{P}_B(x); \phi^r(x), \pi_s(x)) \tag{3.4.8}$$

in which the $12 \times \infty^3$ variables $(g_{ab}(x), p^{cd}(x))$ are mapped into

- the four functions $\mathcal{X}^A(x)$ specifying a particular choice of internal space and time coordinates;
- their four conjugate momenta $\mathcal{P}_B(x)$;
- the two modes $\phi^r(x)$, $r = 1, 2$, which represent the physical degrees of freedom of the gravitational field;
- their conjugate momenta $\pi_s(x)$, $s = 1, 2$.

The statement that $\mathcal{P}_B$ are the momenta conjugate to the four internal coordinate variables $\mathcal{X}^A$ means they satisfy the Poisson bracket relations (computed using the basic Poisson brackets (3.3.27–3.3.29))

$$\{\mathcal{X}^A(x), \mathcal{X}^B(x')\} = 0 \tag{3.4.9}$$
$$\{\mathcal{P}_A(x), \mathcal{P}_B(x')\} = 0 \tag{3.4.10}$$
$$\{\mathcal{X}^A(x), \mathcal{P}_B(x')\} = \delta^A_B\, \delta(x, x') \tag{3.4.11}$$

while the corresponding relations for the 'physical' [18] variables (ϕ^r, π_s) are

$$\{\phi^r(x), \phi^s(x')\} = 0 \tag{3.4.12}$$
$$\{\pi_r(x), \pi_s(x')\} = 0 \tag{3.4.13}$$
$$\{\phi^r(x), \pi_s(x')\} = \delta^r_s\, \delta(x, x'). \tag{3.4.14}$$

It is also assumed that all cross brackets of $\mathcal{X}^A$ and $\mathcal{P}_B$ with ϕ^r and π_s vanish:

$$\begin{aligned} 0 = \{\phi^r(x), \mathcal{X}^A(x')\} = \{\phi^r(x), \mathcal{P}_B(x')\} = \\ \{\pi_s(x), \mathcal{X}^A(x')\} = \{\pi_s(x), \mathcal{P}_B(x')\}. \end{aligned} \tag{3.4.15}$$

[18] To avoid confusion it should be emphasised that ϕ^r and π_s may *not* be observables in the sense of satisfying (3.3.40).

Note that some or all of these relations may need to be generalised to take account of the global topological properties of the function spaces concerned. There may also be global obstructions to some of the steps.

2. Express the super-Hamiltonian and supermomentum as functionals of these new canonical variables and write the canonical action (3.3.18) as

$$S[\phi,\pi,N,\vec{N},\mathcal{X},\mathcal{P}] = \int dt \int_\Sigma d^3x(\mathcal{P}_A\dot{\mathcal{X}}^A + \pi_r\dot{\phi}^r - N\mathcal{H}_\perp - N^a\mathcal{H}_a) \qquad (3.4.16)$$

where all fields are functions of x and t, and where $\mathcal{H}_\perp$ and $\mathcal{H}_a$ are rewritten as functionals of $\mathcal{X}_A$, $\mathcal{P}_B$, ϕ^r and π_s. Note that ϕ^r, π_s, $\mathcal{X}^A$, $\mathcal{P}_B$, N and $\vec{N}$ are all to be varied in (3.4.16) as independent functions. The fields $\mathcal{X}^A$ are interpreted geometrically as defining an embedding of Σ in $\mathcal{M}$ via the parametric equations

$$T = \mathcal{T}(x) \qquad (3.4.17)$$

$$Z^a = \mathcal{Z}^a(x) \qquad (3.4.18)$$

where $\mathcal{T}(x) := \mathcal{X}^0(x)$ and $\mathcal{Z}^a(x) := \mathcal{X}^a(x)$, $a = 1,2,3$. The conjugate variables $P_B(x)$ can be viewed as the energy and momentum densities of the gravitational field measured on this hypersurface. Note however that the spacetime picture that lies behind these interpretations is defined only *after* solving the equations of motion and reconstructing a Lorentzian metric on $\mathcal{M}$.

3. Remove $4 \times \infty^3$ of the $8 \times \infty^3$ non-dynamical variables by solving [19] the constraints $\mathcal{H}_\perp(x) = 0$ and $\mathcal{H}_a(x) = 0$ for the variables $\mathcal{P}_A(x)$ in the form

$$\mathcal{P}_A(x) + h_A(x;\mathcal{X},\phi,\pi] = 0. \qquad (3.4.19)$$

4. Remove the remaining $4 \times \infty^3$ non-dynamical variables by 'deparametrising' the canonical action functional (3.4.16) by substituting into it the solution (3.4.19) of the initial value equations. This gives

$$S[\phi,\pi] = \int dt \int_\Sigma d^3x \,\{\pi_r(x,t)\dot{\phi}^r(x,t) - h_A(x;\chi_t,\phi(t),\pi(t)]\dot{\chi}_t^A(x)\} \qquad (3.4.20)$$

which can be shown to give the correct field equations for the physical fields ϕ^r and π_s. Note that, in (3.4.20), the four quantities $\mathcal{X}^A$ are no longer to be varied but have instead been set equal to some *prescribed* functions χ_t^A of x. This is valid since, after solving the constraints, the remaining dynamical equations of motion give no information about how the variables $\mathcal{X}^A$ evolve in parameter time t. In effect, in terms of the original internal coordinate functionals, we have chosen a set of conditions

$$\mathcal{X}^A(x;g(t),p(t)] = \chi_t^A(x) \qquad (3.4.21)$$

[19]There may be global obstructions to this step.

that restrict the phase-space paths $t \mapsto (g(t), p(t))$ over which the action is to be varied. In this way of looking at things, (3.4.21) can be regarded as additional constraints which, when added to the original constraints, make the entire set second-class.

The lapse function and shift vector play no part in this reduced variational principle. However, it will be necessary to reintroduce them if one wishes to return to a genuine spacetime picture. As discussed in §3.3.6, this can be done by solving the additional set of Einstein equations that are missing from the set generated by the reduced action. These are elliptic partial-differential equations for N and $\vec{N}$.

The equations of motion derived from the reduced action (3.4.20) are those corresponding to the Hamiltonian

$$H_{\mathrm{true}}(t) := \int_\Sigma d^3x\, \dot{\chi}_t^A(x)\, h_A(x; \chi_t, \phi(t), \pi(t)] \tag{3.4.22}$$

and can be written in the form

$$\frac{\partial \phi^r(x,t)}{\partial t} = \{\phi^r(x,t), H_{\mathrm{true}}(t)\}_{\mathrm{red}} \tag{3.4.23}$$

$$\frac{\partial \pi_s(x,t)}{\partial t} = \{\pi_s(x,t), H_{\mathrm{true}}(t)\}_{\mathrm{red}} \tag{3.4.24}$$

where $\{\,,\,\}_{\mathrm{red}}$ denotes the Poisson bracket evaluated using only the physical modes ϕ^r and π_s. Thus, at least formally, the system has been reduced to one that looks like the conventional field theory discussed earlier in section §3.2 and epitomised by the dynamical equations (3.2.7–3.2.8). A particular choice for the four functions χ_t^A (using some coordinate system x^a on Σ) is

$$\chi_t^0(x) = t \tag{3.4.25}$$

$$\chi_t^a(x) = x^a \tag{3.4.26}$$

for which the true Hamiltonian (3.4.22) is just the integral over Σ of h_0. Thus we arrive at the fully-reduced form of canonical general relativity as derived in the work of Arnowitt, Deser and Misner.

3.4.3 The Multi-time Formalism

Although a particular time label t (corresponding to some reference foliation of $\mathcal{M}$ into a one-parameter family of hypersurfaces) has been used in the above, this label is in fact quite arbitrary. This is reflected in the arbitrary choice of the functions χ_t^A in the 'gauge conditions' (3.4.21). Indeed, the physical fields (ϕ^r, π_s) ought to depend only on the hypersurface in $\mathcal{M}$ on which they are evaluated, not on the way in which that hypersurface happens to be included in a particular foliation. Thus it should be possible to write the physical fields $\phi^r(x, \mathcal{X}]$, $\pi_s(x, \mathcal{X}]$ as functions of $x \in \Sigma$

and functionals of the internal coordinates $\mathcal{X}^A$ which are now regarded as *arbitrary* functions of $x \in \Sigma$ rather than being set equal to some fixed set.

This is indeed the case. More precisely, the Hamiltonian equations of motion (3.4.23–3.4.24), which hold for all choices of the functions χ^A, imply that there exist Hamiltonian densities $h_A(x,\mathcal{X}\,]$ such that $\phi^r(x,\mathcal{X}\,]$ and $\pi_s(x,\mathcal{X}\,]$ satisfy the functional differential equations

$$\frac{\delta\phi^r(x,\mathcal{X}\,]}{\delta\mathcal{X}^A(x')} = \{\phi^r(x,\mathcal{X}\,], h_A(x',\mathcal{X}\,]\}_{\text{red}} \tag{3.4.27}$$

$$\frac{\delta\pi_s(x,\mathcal{X}\,]}{\delta\mathcal{X}^A(x')} = \{\pi_s(x,\mathcal{X}\,], h_A(x',\mathcal{X}\,]\}_{\text{red}} \tag{3.4.28}$$

which is a gravitational analogue [20] of the scalar field equations (3.2.9–3.2.10).

This so-called 'bubble-time' or 'multi-time' canonical formalism is of considerable significance and will play an important role in our discussions of the problem of time in quantum gravity (Kuchař 1972).

4 IDENTIFY TIME BEFORE QUANTISATION

4.1 Canonical Quantum Gravity: Constrain Before Quantising

4.1.1 Basic Ideas

We shall now discuss the first of those interpretations of type I in which time is identified as a functional of the geometric canonical variables *before* the application of any quantisation algorithm. These approaches to the problem of time are conservative in the sense that the quantum theory is constructed in an essentially standard way. In particular, 'time' is regarded as part of the *a priori* background structure used in the formulation of the quantum theory. There are two variants of this approach. The first starts with the fully-reduced formalism, and culminates in a quantum version of the Poisson-bracket dynamical equations (3.4.23–3.4.24); the second starts with the multi-time dynamical equations (3.4.27–3.4.28) and ends with a multi-time Schrödinger equation.

In the first approach, the problem is to quantise the first-order action principle (3.3.18) in which the canonical variables $g_{ab}(x)$ and $p^{cd}(x)$ are subject to the constraints $\mathcal{H}_a(x;g,p] = 0 = \mathcal{H}_\perp(x;g,p]$ where $\mathcal{H}_a$ and $\mathcal{H}_\perp$ are given by (3.3.19)

[20] However, note that the quantities $\mathcal{E}^\alpha$ in (3.2.9–3.2.10) are the components of an embedding map $\mathcal{E}:\Sigma\to\mathcal{M}$ with respect some (unspecified) coordinate system on $\mathcal{M}$, whereas the quantities $\mathcal{X}^A$ in (3.4.27–3.4.28) are coordinate functions on $\mathcal{M}$.

and (3.3.20) respectively, and where all non-dynamical variables are removed before quantisation. The account that follows is very brief and does not address the very difficult technical question of whether it is really feasible to construct a consistent quantum theory of gravity in this way by applying some quantisation algorithm to the classical theory of general relativity.

The main steps are as follows.

1. Impose a suitable 'gauge'. As explained in §3.3.6, this can be done in several ways:

- Set the lapse function N and shift vector $\vec{N}$ equal to specific functions (or functionals) of the remaining canonical variables.
- *Or*, impose a set of conditions $F^A(x,t;g,p] = 0$, $A = 0,1,2,3$.
- *Or*, identify a set $\mathcal{X}^A(x;g,p]$, $A = 0,1,2,3$ of internal spacetime coordinates and set these equal to some set of *fixed* functions $\chi_t^A(x)$.

In all three cases, the outcome is that N and $\vec{N}$ drop out of the formalism, as do $8 \times \infty^3$ of the $12 \times \infty^3$ canonical variables.

2. Construct a canonical action (*cf*(3.4.20)) to reproduce the dynamical equations (*cf*(3.4.23–3.4.23)) of the remaining $4 \times \infty^3$ 'true' canonical variables $(\phi^r(x), \pi_s(x))$.

3. Impose canonical commutation relations on these variables and then proceed as in any standard [21] quantum theory. In particular, we obtain a Schrödinger equation of the form

$$i\hbar \frac{\partial \Psi_t}{\partial t} = \widehat{H}_{\mathrm{true}} \Psi_t \tag{4.1.1}$$

where $\widehat{H}_{\mathrm{true}}$ is the quantised version of the Hamiltonian (3.4.22) for the physical modes introduced in the classical Poisson bracket evolution equations (3.4.23–3.4.24).

A variant of this operator approach uses a formal, canonical path-integral to compute expressions like [22]

$$Z = \int \Pi_{x,t} \mathcal{D}g_{ab}(x,t)\, \mathcal{D}p^{cd}(x,t)\, \mathcal{D}N(x,t)\, \mathcal{D}N^a(x,t)\, \delta[F^A]\, |\det\{F^A, \mathcal{H}_B\}|\, e^{\frac{i}{\hbar}S[g,p,N,\vec{N}]} \tag{4.1.2}$$

where $S[g,p,N,\vec{N}]$ is given by (3.3.18), $\mathcal{H}_{A=0}$ denotes $\mathcal{H}_\perp$, and where the gauge conditions $F^A(x,t;g,p] = 0$ are enforced by the functional Dirac delta-function.

[21] This last step is rather problemetic since the reduced phase space obtained by the construction above is topologically non-trivial, and hence the fields $\phi^r(x)$ and $\pi_s(x)$ are only *local* coordinates on this space. This means the naive canonical commutation relations should be replaced by an operator algebra that respects the global structure of the system (Isham 1984).

[22] This is only intended as an example. As defined, Z is just a number, and to produce physical predictions it is necessary to include external sources, or background fields, in the standard way.

Note that because S is a linear functional of N and $\vec{N}$, the functional integrals over these variables in (4.1.2) formally generate functional delta-functions [23] which impose the constraints $\mathcal{H}_\perp(x) = 0 = \mathcal{H}_a(x)$.

4.1.2 Problems With the Formalism

Neither the operator nor the functional-integral programmes are easy to implement, and both are unattractive for a number of reasons. For example:

- The constraints $\mathcal{H}_a(x; g, p] = 0 = \mathcal{H}_\perp(x; g, p]$ cannot be solved in a closed form [24]. Weak-field perturbative methods could be used but these throw little light on the problem of time and inevitably founder on the problem of non-renormalisability.

- The programme violates the geometrical structure of general relativity by removing parts of the metric tensor. This makes it very difficult to explore the significance of spacetime geometry in the quantum theory.

- Choosing a gauge breaks the gauge invariance of the theory and, as usual, the final quantum results must be shown to be independent of the choice. In the functional-integral approach, this is the reason for the functional Jacobian $|\det\{F^A, \mathcal{H}_B\}|$ in (4.1.2). However, this integral is so ill-defined that a proper demonstration of gauge invariance is as hard as it is in the operator approaches. The issue of gauge invariance is discussed at length by Barvinsky (1991).

4.2 The Internal Schrödinger Interpretation

4.2.1 The Main Ideas

The classical dynamical evolution described above employs the single-time Poisson bracket equations of the form (3.4.23–3.4.24) and, in the operator form, leads to the gravitational analogue (4.1.1) of the Schrödinger equation (3.2.20). However, the resulting description of the evolution is in terms of a fixed foliation of spacetime, whereas a picture that is more in keeping with the spirit of general relativity would be one that involves an analogue of the functional-differential equations (3.2.9–3.2.10) and the associated multi-time Schrödinger equation (3.2.21) that describe evolution along the deformation of arbitrary hypersurfaces of $\mathcal{M}$.

[23] There is a subtlety here, since the classical lapse function N is always positive. If the same condition is imposed on the variable N in the functional integral it will not lead to a delta function.

[24] The Ashtekar variables might change this situation as they lead to powerful new techniques for tackling the constraints (Ashtekar 1986, Ashtekar 1987). For two recent developments in this direction see Newman & Rovelli (1992) and Manojlović & Miković (1992).

The internal Schrödinger interpretation refers to such a generalisation in which one quantises the multi-time version of the canonical formalism of classical general relativity. As always in canonical approaches to general relativity, the first step is to introduce a reference foliation $\mathcal{F}^{\text{ref}} : \Sigma \times \mathbb{R} \to \mathcal{M}$ which is used to define the canonical variables. However, it is important to appreciate that this should serve only as an intermediate tool in the development of the theory and, like a basis set in a vector space theory, it should not appear in the final results.

The key steps in constructing the internal Schrödinger interpretation are as follows:

1. Pick a set of classical functions $\mathcal{T}(x; g, p]$, $\mathcal{Z}^a(x; g, p]$ that can serve as internal time and space coordinates.

2. Perform the canonical transformation (3.4.8)
$$(g_{ab}(x), p^{cd}(x)) \mapsto (\mathcal{X}^A(x), \mathcal{P}_B(x); \phi^r(x), \pi_s(x)) \tag{4.2.1}$$
in which $(\phi^r(x), \pi_s(x))$ are identified with the $4 \times \infty^3$ physical canonical modes of the gravitational field.

3. Solve the super-Hamiltonian constraint $\mathcal{H}_\perp(x) = 0$ for the variables $\mathcal{P}_A(x)$ in the form (3.4.19)
$$\mathcal{P}_A(x) + h_A(x; \mathcal{X}, \phi, \pi] = 0. \tag{4.2.2}$$

4. Identify $h_A(x; \mathcal{X}, \phi, \pi]$ as the set of Hamiltonian densities appropriate to a multi-time version of the classical dynamical equations as in (3.4.27–3.4.28).

5. Quantise the system in the following steps:

   (a) Replace the Poisson bracket relations (3.4.12–3.4.14) with the canonical commutation relations
$$[\,\widehat{\phi}^r(x), \widehat{\phi}^s(x')\,] = 0 \tag{4.2.3}$$
$$[\,\widehat{\pi}_r(x), \widehat{\pi}_s(x')\,] = 0 \tag{4.2.4}$$
$$[\,\widehat{\phi}^r(x), \widehat{\pi}_s(x')\,] = i\hbar\delta^r_s\,\delta(x, x'). \tag{4.2.5}$$

   (b) Try to construct quantum Hamiltonian densities $\widehat{h}_A(x)$ by the usual substitution rule in which the fields ϕ^r and π_s in the classical expression $h_A(x; \mathcal{X}, \phi, \pi]$ are replaced by their operator analogues $\widehat{\phi}^r$ and $\widehat{\pi}_s$.

   (c) The classical multi-time evolution equations (3.4.27–3.4.28) can be quantised in one of two ways. The first is to replace the classical Poisson brackets expressions with Heisenberg-picture operator relations. The second is to use a Schrödinger picture in which the constraint equation (4.2.2) becomes the multi-time Schrödinger equation

$$i\hbar \frac{\delta \Psi_{\mathcal{X}}}{\delta \mathcal{X}^A(x)} = h_A(x; \mathcal{X}, \hat{\phi}, \hat{\pi}]\Psi_{\mathcal{X}} \tag{4.2.6}$$

for the state vector $\Psi_{\mathcal{X}}$.

This fundamental equation can be justified from two slightly different points of view. The first is to claim that, since the classical Hamiltonian associated with the internal coordinates $\mathcal{X}^A$ is h_A, the Schrödinger equation (4.2.6) follows from the basic time-evolution axiom of quantum theory, suitably generalised to a multi-time situation. However, note that this step is valid only if the inner product on the Hilbert space of states is $\mathcal{X}$-independent. If not, a compensating term must be added to (4.2.6) if the scalar product $\langle \Psi_{\mathcal{X}}, \Phi_{\mathcal{X}} \rangle_{\mathcal{X}}$ is to be independent of the internal coordinate functions. This is analogous to the situation discussed earlier for the functional Schrödinger equation (3.2.21) for a scalar field theory in a background spacetime.

Another approach to deriving (4.2.6) is to start with the classical constraint equation (4.2.2) and then impose it at the quantum level as a constraint on the allowed state vectors,

$$(\hat{\mathcal{P}}_A(x) + h_A(x; \widehat{X}, \hat{\phi}, \hat{\pi}])\Psi = 0. \tag{4.2.7}$$

The Schrödinger equation (4.2.6) is reproduced if one makes the formal substitution

$$\hat{\mathcal{P}}_A(x) = -i\hbar \frac{\delta}{\delta \mathcal{X}^A(x)}. \tag{4.2.8}$$

This approach has the advantage of making the derivation of the Schrödinger equation look somewhat similar to that of the Wheeler-DeWitt equation. However, once again this is valid only if the scalar product is $\mathcal{X}$-independent. If not, it is necessary to replace (4.2.8) with

$$\hat{\mathcal{P}}_A(x) = -i\hbar \frac{\delta}{\delta \mathcal{X}^A(x)} + F_A(x, \mathcal{X}] \tag{4.2.9}$$

where the extra term $F_A(x, \mathcal{X}]$ is chosen to compensate the $\mathcal{X}$-dependence of the scalar product.

Considerable care must be taken in interpreting (4.2.8) or (4.2.9). These equations might appear to come from requiring the existence of self-adjoint operators $\widehat{\mathcal{X}}^A(x)$ and $\hat{P}_B(x')$ that satisfy canonical commutation relations

$$[\widehat{\mathcal{X}}^A(x), \widehat{\mathcal{X}}^B(x')] = 0 \tag{4.2.10}$$

$$[\hat{\mathcal{P}}_A(x), \hat{\mathcal{P}}_B(x')] = 0 \tag{4.2.11}$$

$$[\widehat{\mathcal{X}}^A(x), \hat{\mathcal{P}}_B(x')] = i\hbar\delta^A_B\, \delta(x, x') \tag{4.2.12}$$

which are the quantised versions of the classical Poisson-bracket relations (3.4.9–3.4.11). However, the example of elementary wave mechanics shows that such an interpretation is misleading. In particular, the $i\hbar\partial/\partial t$ that appears in the time-dependent Schrödinger equation cannot be viewed as an operator defined on the physical Hilbert space of the theory.

4.2.2 The Main Advantages of the Scheme

Many problems arise in the actual implementation of this programme, but let us first list the main advantages.

1. We have argued in the introductory sections that one of the main sources of the problem of time in quantum gravity is the special role played by time in normal physics; in particular, its position as something that is external to the system and part of the *a priori* classical background of the Copenhagen interpretation of quantum theory. The canonical transformation (4.2.1) applied *before* quantisation could be said to cast quantum gravity into the same mould.

2. The relative freedom from interpretational problems is epitomised by the ease with which the theory can be equipped (at least formally) with a Hilbert space structure. In particular, since $(\phi^r(x), \pi_s(x))$ are genuine physical modes of the gravitational field, it is fully consistent with conventional quantum theory to insist that (once suitable smeared) the corresponding operators are *self-adjoint*. This is certainly not the case for the analogous statement in the Dirac, constraint-quantisation, scheme in which all $12 \times \infty^3$ variables $(g_{ab}(x), p^{cd}(x))$ are assigned operator status. It is then consistent to follow the earlier discussion in §3.2 of quantum field theory in a background spacetime and define the quantum states as functionals $\Psi[\phi]$ of the configuration variables $\phi^r(x)$ with the canonical operators $\widehat{\phi}^r(x)$ and $\widehat{\pi}_s(x)$ defined as the analogue of (3.2.14) and (3.2.15) respectively. Of course, the cautionary remarks made earlier about measures on infinite-dimensional spaces apply here too.

3. This Hilbert space structure provides an unambiguous probabilistic interpretation of the theory. Namely, if the normalised state is Ψ, and if a measurement is made of the true gravitational degrees of freedom ϕ^r on the hypersurface represented by the internal coordinates $\mathcal{X}^A$, the probability of the result lying in the subset B of field configurations is (cf(3.2.17))

$$\text{Prob}(\Psi; \phi \in B) = \int_B d\mu(\phi)\, |\Psi[\mathcal{X}, \phi]|^2. \tag{4.2.13}$$

4. The associated inner product also allows the construction of a meaningful notion of a quantum *observable*. This is simply any well-defined operator functional

$\widehat{F} = F[\mathcal{X}, \widehat{\phi}, \widehat{\pi}]$ of the true gravitational variables $\widehat{\phi}^r(x)$, $\widehat{\pi}_s(x)$ and of the internal spacetime coordinates $\mathcal{X}^A(x)$ that is self-adjoint in the inner product. The usual rules of quantum theory allow one to ask, and in principle to answer, questions about the spectrum of $\widehat{F}$ and the probabilistic distributions of the values of the observable F on the hypersurface defined by $\mathcal{X}$.

4.2.3 A Minisuperspace Model

One of the basic questions in this programme is how to select the internal embedding variables and, in particular, the internal time variable $\mathcal{T}(x)$. Three general options have been explored:

- *Intrinsic time* $\mathcal{T}(x, g]$. The internal-time value of a spacetime event is constructed entirely from the internal metric $g_{ab}(x)$ carried by the hypersurface of the reference foliation that passes through the event.
- *Extrinsic time* $\mathcal{T}(x; g, p]$. To identify the time coordinate of an event, one needs to know not only the intrinsic metric on the hypersurface, but also its extrinsic curvature, and hence how the hypersurface would appear from the perspective of a classical spacetime constructed from the pair (g, p).
- *Matter time* $\mathcal{T}(x; g, p, \phi, p_\phi]$. Time is not constructed from geometric data alone but from matter fields ϕ coupled to gravity, or from a combination of matter and gravity fields.

Very little is known about how to choose an internal time in the full theory of quantum gravity although it is noteable that the use of an extrinsic time is natural in the linearised theory and was discussed in some of the original series of ADM papers (Arnowitt, Deser & Misner 1960*a*, Arnowitt, Deser & Misner 1962), and in Kuchař (1970). This type of time appears naturally also in the theory of quantised cylindrical gravitational waves (Kuchař 1971, Kuchař 1973, Kuchař 1992*b*), and in the Ashtekar formalism (Ashtekar 1991).

The full theory is very complicated and most studies have been in the context of so-called *minisuperspace* models in which only a finite number of the gravitational degrees of freedom are invoked in the quantum theory, the remainder being eliminated by the imposition of symmetries on the spatial metric. These models represent homogeneous (but, in general, anisotropic) cosmologies, of which the various Bianchi-type vacuum spacetimes (in which the spatial three-manifolds are assorted three-dimensional Lie groups) have been the most widely studied. The requirement of homogeneity limits the allowed hypersurfaces to a priviledged foliation in which each leaf can be labelled by a single time variable.

Restricting things in this way has the big advantage that the worst quantum field theory problems—in particular, non-renormalisability—do not arise, which means that some of the conceptual issues can be discussed in a technical framework that is mathematically well-defined. The main limitation of these models is the absence of any proper perturbative scheme into which they can be made to fit. At the classical level, models of this type correspond to exact solutions of Einstein's field equations, but this not true in the corresponding quantum theory since, for example, the infinite set of neglected modes presumably possess zero-point fluctuations. Some of these issues have been discussed in detail by Kuchař & Ryan (1989) to which the reader is referred for further information. However, we shall not get too involved with these particular issues here since our main use of minisuperspace models is to illustrate various specific features of the problem of time, and this they do quite well.

A simple example is when the spacetime geometries are restricted to be one of the familiar homogeneous and isotropic Robertson-Walker metrics

$$g_{\alpha\beta}(X)\, dX^\alpha \otimes dX^\beta := -N(t)^2\, dt \otimes dt + a(t)^2 \omega_{ab}(x)\, dx^a \otimes dx^b \tag{4.2.14}$$

where $\omega_{ab}(x)$ is the metric for a three-space of constant curvature k. The case $k = 1$ is associated with Σ being a three-sphere, while $k = 0$ and $k = -1$ correspond to the flat and hyperbolic cases respectively. Note that the lapse vector $\vec{N}$ does not appear in this expression.

Spacetimes of this type produce a non-vanishing Einstein tensor $G_{\alpha\beta}$ and therefore require a source of matter. Much work in recent times has employed a scalar field ϕ for this purpose, and we shall use this here. It can be shown (Kaup & Vitello 1974, Blyth & Isham 1975) that the coupled equations [25] for the variables a, ϕ and N can all be derived from the first-order action [26]

$$S[a, p_a, \phi, p_\phi, N] = \int dt\, (p_a \dot{a} + p_\phi \dot{\phi} - N H_{\text{RW}}) \tag{4.2.15}$$

where p_a, a, p_ϕ, ϕ and N are functions of t and are to be varied independently of each other. The super-Hamiltonian in this model is

$$H_{\text{RW}} := \frac{-p_a^2}{24a} - 6ka + \frac{p_\phi^2}{2a^3} + a^3 V(\phi). \tag{4.2.16}$$

A natural choice for an intrinsic time in the $k = 1$ model is the radius a of the universe. Thus, according to the discussion above, the reduced Hamiltonian for this

[25]There is no specific information in these notes about how the canonical formalism for general relativity is extended to include matter but, generally speaking, nothing very dramatic happens. In particular, although the supermomentum and super-Hamiltonian constraints acquire contributions from the matter variables, the crucial Dirac algebra is still satisfied. However, the details can be quite complicated, especially for systems with internal degrees of freedom; the subject as a whole has been analysed comprehensively in a series of papers by Kuchař (1976a, 1976b, 1976c, 1977).

[26]For simplicity, I have chosen units in which $8\pi G/c^4 = 1$.

system can be found by solving the constraint

$$\frac{-p_a^2}{24a} - 6a + \frac{p_\phi^2}{2a^3} + a^3V(\phi) = 0 \tag{4.2.17}$$

for the variable p_a. Hence

$$h_a = \pm\sqrt{24}\Big(- 6t^2 + \frac{p_\phi^2}{2t^2} + t^4V(\phi)\Big)^{\frac{1}{2}}. \tag{4.2.18}$$

A simple extrinsic time is $t := p_a$ which (in the simple case $V(\phi) = 0$) leads to the Hamiltonian

$$h_{p_a} = \pm\left(\frac{-\frac{1}{12}t^2 \pm (\frac{1}{144}t^4 + 48p_\phi^2)^{\frac{1}{2}}}{24}\right)^{\frac{1}{2}}, \tag{4.2.19}$$

while a natural definition using the matter field is $t := \phi$, which yields

$$h_\phi = \pm(\frac{a^2p_a^2}{12} + 12a^4 - 2a^6V(t))^{\frac{1}{2}}. \tag{4.2.20}$$

4.2.4 Mean Extrinsic Curvature Time

One of the few definitions of internal time that has been studied in depth is one in which spacetime is foliated by hypersurfaces of constant mean extrinsic curvature. The $\mathcal{T}$ functional is

$$\mathcal{T}(x;g,p] := \tfrac{2}{3}|g|^{-\frac{1}{2}}(x)\,p^a{}_a(x) \tag{4.2.21}$$

which is conjugate to

$$\mathcal{P}_\mathcal{T}(x;g,p] := -|g|^{\frac{1}{2}}(x) \tag{4.2.22}$$

in the sense that

$$\{\mathcal{T}(x), \mathcal{P}_\mathcal{T}(x')\} = \delta(x,x'). \tag{4.2.23}$$

The use of (4.2.21) as an internal time variable in the spatially compact case was developed in depth by York and his collaborators (York 1972*b*, York 1972*a*, Smarr & York 1978, York 1979, Isenberg & Marsden 1976). An interesting account of its use in $2+1$ dimensions is Carlip (1990, 1991).

The first step in implementing the quantisation programme is to extend the definitions (4.2.21–4.2.22) to a full canonical transformation of the type (4.2.1). One possible example is

$$(g_{ab}(x), p^{cd}(x)) \mapsto (\mathcal{T}(x), \sigma_{ab}(x), \mathcal{P}_\mathcal{T}(x), \pi^{cd}(x)) \tag{4.2.24}$$

where the 'conformal metric' $\sigma_{ab}(x)$ and its 'conjugate' momentum $\pi^{cd}(x)$ are defined by

$$\sigma_{ab} := |g|^{-\frac{1}{3}}g_{ab}, \tag{4.2.25}$$
$$\pi^{ab} := |g|^{\frac{1}{3}}(p^{ab} - \tfrac{1}{3}p\,g^{ab}). \tag{4.2.26}$$

Note that $\det \sigma_{ab} = 1$ and π^{ab} is traceless, and so each field corresponds to $5 \times \infty^3$ independent variables. [27] The next step therefore should be to remove $3 \times \infty^3$ variables from each by identifying the spatial parts $\mathcal{Z}^a(x; g, p]$ of the embedding variables plus their conjugate momenta. However, since we are concerned primarily with the problem of time, we shall not perform this (rather complex) step but concentrate instead on what is already entailed by the use of the preliminary canonical transformation (4.2.24). The complete analysis is contained in the papers by York *et al* cited above.

In terms of these new variables $\mathcal{T}$, $\mathcal{P}_\mathcal{T}$, σ_{ab} and π^{ab}, the super-Hamiltonian constraint $\mathcal{H}_\perp = 0$ becomes the equation for $\Phi := (-\mathcal{P}_\mathcal{T})^{\frac{1}{6}}$

$$\frac{1}{\kappa^2}\left(\Delta_\sigma \Phi - \tfrac{1}{8} R[g]\right) + \kappa^2\left(\tfrac{1}{8}\pi^{ab}\pi_{ab}\Phi^{-7} - \tfrac{3}{64}\mathcal{T}^2\Phi^5\right) = 0 \tag{4.2.27}$$

where Δ_σ is the Laplacian operator constructed from the metric σ_{ab}. This is a non-linear, elliptic, partial-differential equation which, in principle, can be solved for Φ (and hence for $\mathcal{P}_\mathcal{T} = -|g|^{\frac{1}{2}}$) as a functional of the remaining canonical variables

$$\mathcal{P}_\mathcal{T}(x) + h(x; \mathcal{T}, \sigma, \pi] = 0. \tag{4.2.28}$$

In the quantum theory this leads at once to the desired functional Schrödinger equation

$$i\hbar \frac{\delta \Psi[\mathcal{T}, \sigma]}{\delta \mathcal{T}(x)} = (h(x; \mathcal{T}, \hat{\sigma}, \hat{\pi}]\Psi)[\mathcal{T}, \sigma]. \tag{4.2.29}$$

4.2.5 The Major Problems

As remarked earlier, considerable conceptual advantages accrue from using the internal Schrödinger interpretation of time, and it is therefore unfortunate that the scheme is riddled with severe technical difficulties. These fall into two categories. The first are those quantum-field theoretic problems that are expected to arise in any theory that we know from weak-field perturbative analysis to be non-renormalisable. In particular, the Hamiltonian densities $h_A(x; \mathcal{X}, \phi, \pi]$ are likely to be highly non-linear functions of the fields $\phi^r(x), \pi_s(x)$ and we shall encounter the usual problems when trying to define products of quantum fields evaluated at the same point.

The second class of problems are those that more closely involve the question of time itself and are not directly linked to the non-renormalisability of the theory. Indeed, many such problems arise already in minisuperspace models which, having only a finite number of degrees of freedom, are free from the worst difficulties of

[27] This is reflected in the fact that the Poisson bracket of $\sigma_{ab}(x)$ with $\pi^{cd}(x')$ is proportional to the *traceless* version of the Kronecker-delta. Thus these variables are not a conjugate pair in the strict sense: to get such variables it would be necessary first to solve the second-class constraints $\det \sigma_{ab} = 1$ and $\pi^c_c = 0$. Alternatively, these variables could be employed as part of an over-complete set in a group-theoretical approach to the quantum theory.

quantum field theory. I shall concentrate here mainly on this second type of problem, using the classification mentioned briefly at the end of §2.4; see also Isham (1992) and Kuchař (1992*b*).

The Global Time Problem. It is far from obvious that it is possible to perform a canonical transformation (4.2.1) with the desired characteristics. This raises the following questions:

1. What properties must the functionals $\mathcal{X}^A(x;g,p]$ possess in order to serve as internal spacetime coordinates?
2. If such functions do exist, is it possible to perform a canonical transformation of the type (4.2.1) which is such that:
   (a) the constraints $\mathcal{H}_\perp(x;g,p]=0$ and $\mathcal{H}_a(x;g,p]=0$ can be solved globally (on the phase space) for $\mathcal{P}_A$ in the form (4.2.2); and
   (b) there is a unique such solution?

One example of the first point is the requirement hat the internal space and time coordinates produce a genuine foliation of the spacetime. In particular, if $t\mapsto(g(t),p(t))$ is any curve in the Cauchy data of a solution to the vacuum Einstein equations, we require $d\mathcal{T}(x;g(t),p(t)]/dt>0$. Of course, if γ is static, this condition can never be satisfied, which means that spacetimes of this type do not admit internal time functions. This is hardly surprising, but it does show that we are unlikely to find conditions on $\mathcal{T}$ that are to be satisfied on *all* spacetimes.

The existence of functionals satisfying these conditions in various minisuperspace models has been studied by Hájíček (1986, 1988, 1989, 1990a, 1990b). He showed that these finite-dimensional models generically exhibit global obstructions to the construction of such functionals, and the evidence suggests that the problem may get worse as the number of degrees of freedom increases. See also the comments in Torre (1992).

This is not really surprising since, if χ^A are to serve as proper 'gauge functions', the subspace of the space $\mathcal{S}$ of all (g,p) satisfying the equations $\mathcal{X}^A(x,g,p]=\chi^A(x;g,p]$ (and also the constraints $\mathcal{H}_\perp(x)=0=\mathcal{H}_a(x)$) must intersect each orbit of the Dirac algebra just once. This is analogous to gauge-fixing in Yang-Mills theory and therefore one anticipates the presence of a Gribov phenomenon in the form of obstructions to the construction of such a global gauge (Singer 1978). This possibility arises because the topological structure of the space of physical configurations is non-trivial. However, this structure seems not to have been studied systematically in the full theory of general relativity, and so the subject warrants further investigation.

The Spatial Metric Reconstruction Problem. At a classical level the canonical transformation (3.4.8) can be inverted and, in particular, the metric $g_{ab}(x)$ can be expressed as a functional $g_{ab}(x;\mathcal{X},\mathcal{P},\phi,\pi]$ of the embedding variables $(\mathcal{X}^A,\mathcal{P}_B)$ and the physical degrees of freedom (ϕ^r,π_s). The question is whether something similar can be done at the quantum level. In particular:

- is it possible to make sense of an expression like $g_{ab}(x;\mathcal{X},\widehat{\mathcal{P}},\widehat{\phi},\widehat{\pi}\,]$ with $\widehat{\mathcal{P}}_A$ being replaced by $-\widehat{h}_A$?;
- if so, is there any sense in which this operator looks like an operator version of a Riemannian *metric*?

The first question is exceptionally difficult to answer. Even classically, $g_{ab}(x)$ is likely to be a highly non-linear and non-local function of the physical modes (ϕ^r,π_s), and therefore intractable problems of operator ordering and infinite operator-products can be expected in the quantum theory. One might attempt to answer the second question by showing that, if they can be defined at all, the operators $g_{ab}(x;\mathcal{X},\widehat{\mathcal{P}},\widehat{\phi},\widehat{\pi}\,]$ and $p_a{}^b(x;\mathcal{X},\widehat{\mathcal{P}},\widehat{\phi},\widehat{\pi}\,]$ satisfy the *affine* commutation relations (5.1.6–5.1.8) that arise in the quantisation schemes of type II where all $6\times\infty^3$ modes $g_{ab}(x)$ are afforded operator status (see §5.1).

The Definition of the Operators $\widehat{h}_A(x)$. Even apart from the question of ultra-violet divergences, many problems appear when trying to construct operator equivalents of the Hamiltonian/momentum densities $h_A(x;\mathcal{X},\phi,\pi]$ that arise as the solutions (4.2.2) of the initial-value constraints for the conjugate variables $\mathcal{P}_A(x)$. For example:

1. As mentioned earlier, the solution for $\mathcal{P}_A(x)$ may exist only locally in phase space, and there may be more than one such solution. In the latter case it might be possible to select a particular solution on 'physical grounds' (such as the requirement that the physical Hamiltonian is a positive functional of the canonical variables) but the status of such a step is not clear since it means that certain classical solutions to the field equations are deliberately excluded.

Simple examples of these phenomena can be seen in the Robertson-Walker model (4.2.18–4.2.20). The constraint (4.2.17) is quadratic in the conjugate variable p_a and therefore has the *two* solutions in (4.2.18). Note also that for a number of typical potential functions $V(\phi)$ the expression under the square root in (4.2.18) is negative for sufficiently large t, and the range of such values depends on the values of the canonical variables. Thus even classically the constraint can be solved by a *real* p_a only in a restricted region of phase space and values of t.

2. Even if it does exist, the classical solution for $h_A(x;\mathcal{X},\phi,\pi]$ is likely be a very complicated expression of the canonical variables. Indeed, it may exist only in some implicit sense: a good example is the solution Φ of the elliptic partial-differential equation (4.2.27) that determines $\mathcal{P}_T$ in the case of the mean extrinsic curvature time. The solution may also be a very non-local function of the canonical variables.

These properties of the classical solution pose various problems at the quantum level. For example:

- Operator ordering is likely to be a major difficulty. This is particularly relevant to the problem of functional evolution discussed below.

- The operator that represents a physical quantum Hamiltonian is required to be self-adjoint and positive. In a simple model, the positivity requirement may involve just selecting a particular solution to the constraints, but self-adjointness and positivity are very difficult to check in a situation in which even the classical expression is only an implicit function of the canonical variables.

- As remarked already, the constraint equations may well be algebraic equations for $\mathcal{P}_A$ whose solution therefore involves taking roots of some operator $\widehat{K}$. This can be done with the aid of the spectral theorem provided $\widehat{K}$ is a positive, self-adjoint operator—which takes us back to the preceding problem. If $\widehat{K}$ is *not* positive then even if the quantum Hamiltonian exists it is not self-adjoint, and so the time evolution becomes non-unitary.

3. The solutions to the constraints are likely to be explicit functionals of the internal spacetime coordinate functions, which gives rise to a time-dependent Hamiltonian. This effect, which can be seen clearly in the minisuperspace example, has several implications:

- A time-dependent Hamiltonian means that energy can be fed into, or taken out of, the quantum system. In normal physics, this happens whenever the system is not closed, with the time dependence of the Hamiltonian being determined by the environment to which the system couples. However, a compact three-manifold Σ (the 'universe') has no external environment, and so the time-dependence seems a little odd.

- If $h(t)$ is time-dependent, the Schrödinger equation

$$i\hbar\frac{d\psi_t}{dt} = \widehat{h}(t)\psi_t \tag{4.2.30}$$

does not lead to the simple second-order equation

$$-\hbar^2\frac{d^2\psi_t}{dt^2} = \left(\widehat{h}(t)\right)^2\psi_t \tag{4.2.31}$$

because of the extra term involving the time derivative of $\widehat{h}(t)$. In the case of quantum gravity this means that the functional Schrödinger equation does *not* imply the Wheeler-DeWitt equation. This is not necessarily a bad thing but it does illustrates the potential inequivalence of different approaches to the canonical quantisation of gravity.

- As remarked already, roots of operators can be handled if the object whose root is being taken is self-adjoint and positive. However, in general this will be true only for certain ranges of t (which can depend on the values of the true canonical variables). Thus a proper Hamiltonian operator may exist only for a limited range of time values. A clear example of this is the Robertson-Walker model (4.2.18).

These difficulties in defining the operators $\widehat{h}_A(x)$ may seem collectively to constitute a major objection to the internal Schrödinger interpretation. However, the apparent failure to identify a completely satisfactory set of internal spacetime coordinates is not necessarily a disaster: it might reflect something of genuine physical significance. For example, much work has been done in recent years on quantum theories of the creation of the universe and, in any such theory, something peculiar must necessarily happen to time near the origination point. A good example is the schemes of Hartle & Hawking (1983) and Vilenkin (1988), in both of which there is a sense in which time becomes imaginary. Perhaps the tendency to produce a non-unitary evolution in the internal time variables is a reflection of this effect.

The Multiple Choice Problem. Generically, there is no geometrically natural choice for the internal spacetime coordinates and, classically, all have an equal standing. However, this classical cornucopia becomes a real problem at the quantum level since there is no reason to suppose that the theories corresponding to different choices of time will agree.

The crucial point is that two different choices of internal coordinates are related by a canonical transformation and, in this sense, are classically of equal validity. However, one of the central properties/problems of the quantisation of any non-linear system is that, because of the well-known Van-Hove phenomenon (Groenwold 1946, Van Hove 1951), most classical canonical transformations *cannot* be represented by unitary operators while, at the same time, maintaining the irreducibility of the canonical commutation relations. This means that in quantising a system it is always necessary to select some preferred sub-algebra of classical observables which is to be quantised (Isham 1984). One analogue of this situation in quantum gravity is precisely the dependence of the theory on the choice of internal time. It must be emphasised that this phenomenon is not related to ultra-violet divergences, or other pathologies peculiar to quantum field theories, but arises already in finite-dimensional systems. A good example is the Robertson-Walker model used above to illustrate various types of internal time. The three Hamiltonians (4.2.18), (4.2.19) and (4.2.20) are associated with different quantum theories of the same classical system.

The Problem of Functional Evolution. The problem of functional evolution is concerned with the key question of the consistency of the dynamical evolution generated by the constraints; in particular, the preservation of the constraints as the system evolves in time. Much more than most of the other problems of time, the issue of functional evolution depends critically on our ability to construct quantum gravity as a consistent quantum *field* theory by giving sense to the infinite collection of Hamiltonians and their commutation relations.

At the classical level, there is no problem. The consistency of the original constraints (3.3.25), (3.3.26) with the dynamical equations (3.3.22), (3.3.24) follows from the first-class nature of the constraints, *i.e.*, their Poisson brackets vanish on the constraint subspace by virtue of the Dirac algebra (3.3.30–3.3.32). Similarly, the consistency of the internal time dynamical equations (3.4.27–3.4.28) is ensured by the Poisson bracket relations between $h_A(x)$ with $h_B(x')$. These results mean that the classical evolution of the system from one initial hypersurface to another is independent of the family of hypersurfaces chosen to interpolate between them.

In the internal Schrödinger interpretation, the analogous requirement on the quantum operators is

$$\frac{\delta \widehat{h}_A(x,\mathcal{X}]}{\delta \mathcal{X}^B(x')} - \frac{\delta \widehat{h}_B(x,\mathcal{X}]}{\delta \mathcal{X}^A(x')} + \frac{1}{i\hbar}[\,\widehat{h}_A(x,\mathcal{X}],\widehat{h}_B(x',\mathcal{X}]\,] = 0. \tag{4.2.32}$$

If this fails, the functional Schrödinger equation (4.2.6) breaks down.

It is clear that these conditions can be checked only if the operators concerned are defined properly, which raises the entire gamut of problems in quantum gravity, including those of operator-ordering and ultra-violet divergences. Not surprisingly, very little can be said about this problem in the full theory. The best that can be done is to elucidate some of the issues on sufficiently simple (and hence almost trivial) systems. It is essential however that these systems have an *infinite* number of degrees of freedom, *i.e.*, we have to deal with a genuine *field* theory for the phenomenon to arise at all. In particular, minisuperspace models tell us nothing about this particular problem of time.

One useful example is the functional evolution of a parametrised, massless scalar field propagating on a $1+1$-dimensional flat cylindrical Minkowskian background [28] spacetime: a system that has been studied [29] in some detail by Kuchař (1988, 1989a, 1989b). There is a canonical formulation of this theory which casts the super-Hamiltonian and supermomentum constraints into the same form as those of the midisuperspace model of cylindrical gravitational waves. This links the functional evolution problem in the parametrised field theory with that in quantum gravity proper (Torre 1991).

[28]It should be possible to generalise these results to any parametrised free field theory on a flat four-dimensional background, but this does not seem to have been done.

[29]Similarly, there exists a canonical transformation which casts the super-Hamiltonian and supermomentum constraints of a bosonic string moving in a d-dimensional target space into those of a parametrised theory of $d-2$ independent scalar fields propagating on a two-dimensional Minkowskian background. This enables the functional evolution problem to be posed, and solved, for the bosonic string (Kuchař & Torre 1989, Kuchař & Torre 1991*c*).

In this particular case it has been shown that with the aid of careful regularisation and renormalisation the apparent anomaly in the commutator of the $\widehat{h}_A(x)$ operators can be removed, and hence a consistent quantum evolution attained. However, in the full theory of quantum gravity the problem of functional evolution is particularly difficult to disentangle from the ambiguities generated by ultra-violet divergences, and very little is known about its solution.

The Spacetime Problem. We have seen how in the canonical version of general relativity 'time' is to be viewed as a function variable rather than the single parameter of Newtonian physics. In classical geometrodynamics we are required to locate an event X in an Einstein spacetime $(\mathcal{M}, \gamma)$ using the canonical data on an embedding that passes through X. Given a particular choice $(\mathcal{T}, \mathcal{Z}^a)$ of internal time and space functions, the coordinates associated with the point X were given in (3.4.4) and (3.4.5) as

$$T(X) := \mathcal{T}(\sigma^{\text{ref}}(X); g[\mathcal{F}^{\text{ref}}_{t=\tau^{\text{ref}}(X)}], p[\mathcal{F}^{\text{ref}}_{t=\tau^{\text{ref}}(X)}]], \tag{4.2.33}$$

and

$$Z^a(X) := \mathcal{Z}^a(\sigma^{\text{ref}}(X); g[\mathcal{F}^{\text{ref}}_{t=\tau^{\text{ref}}(X)}], p[\mathcal{F}^{\text{ref}}_{t=\tau^{\text{ref}}(X)}]]. \tag{4.2.34}$$

These internal embedding variables have some peculiar properties. In particular, if a different reference foliation is used, so that the hypersurface $\mathcal{F}^{\text{ref}}_{t=\tau^{\text{ref}}(X)} : \Sigma \to \mathcal{M}$ passing through the same point X in $\mathcal{M}$ is different, then the 'time' value $T(X)$ alloted to the event will generally not be the same. Thus the time of an event depends not just on the event itself, but also on a choice of spatial hypersurface passing through the event.

Viewed from a spacetime perspective, this feature is pathological since coordinates on a manifold are local scalar functions. It is physically highly undesirable since it means that the results and the interpretation of the theory depend on the choice of the reference foliation $\mathcal{F}^{\text{ref}}$. Having to choose a specific background foliation is equivalent to introducing a Newtonian-type universal time parameter, which is something we wish to avoid since there is no natural place in general relativity for such a field-independent reference system. Indeed, our goal is to construct a formalism in which $\mathcal{F}^{\text{ref}}$ drops out entirely from the final result.

This undesirable behaviour of the internal time $\mathcal{T}(x; g, p]$ can be avoided only if it has a vanishing Poisson bracket with the generator of 'tilts' or 'bends' of the hypersurface. In particular, we require (Kuchař 1976*c*, Kuchař 1982, Kuchař 1991*b*)

$$\{\mathcal{T}(x), H[N]\} = 0 \tag{4.2.35}$$

for all test functions N that vanish at a point $x \in \Sigma$. Of course, the same limitation should also be imposed on the internal spatial coordinates $\mathcal{Z}^a(x; g, p]$. The search for internal spacetime coordinates that satisfy (4.2.35) constitutes the *spacetime problem*.

The requirement (4.2.35) is rather strong. It clearly excludes any intrinsic time like $R(x, g]$ that is a local functional of the three-geometry alone. It also excludes many obvious choices of an extrinsic time. For example, the time function (4.2.21) is certainly not a spacetime scalar: when a hypersurface is bent around a given event in a vacuum Einstein spacetime, the mean extrinsic curvature $\frac{2}{3}|g|(x)^{-\frac{1}{2}} g^{ab}(x)\, p_{ab}(x)$ changes, even if the event remains the same. Thus the canonical coordinate (4.2.21) cannot be turned into a coordinate on spacetime.

There do exist functionals $\mathcal{T}(x; g, p]$ of the canonical data that *are* local spacetime scalars. For example, take the square ${}^{(4)}R_{\alpha\beta\gamma\delta}(X, \gamma]\, {}^{(4)}R^{\alpha\beta\gamma\delta}(X, \gamma]$ of the Riemann curvature tensor of a vacuum Einstein spacetime, and reexpress it in terms of the canonical data on a spacelike hypersurface. But note that not all functionals $\mathcal{T}(x; g, p]$ of this type can serve as time functions. Two necessary conditions are:

1. $\{\mathcal{T}(x), \mathcal{T}(x')\} = 0$;

2. for any given Lorentzian metric γ on $\mathcal{M}$ that satisfies the vacuum Einstein field equations, the hypersurfaces of equal $\mathcal{T}$ time must be *spacelike*. (These hypersurfaces can be evaluated using (4.2.33–4.2.34) with any convenient reference foliation $\mathcal{F}^{\mathrm{ref}}$; the answer will be independent of the choice of $\mathcal{F}^{\mathrm{ref}}$ for internal spacetime functionals that are compatible with the spacetime problem.)

However, even if these conditions are satisfied, it is still necessary to split off $\mathcal{T}$ from the rest of the canonical variables by a canonical transformation (3.4.8), and this is by no means a trivial task. In fact, Kuchař and I are not aware of a single concrete example of a decomposition of the canonical variables based on a local scalar time function $\mathcal{T}(x; g, p]$. Of course, another possibility is to construct scalars from *matter* fields in the theory, and this is the topic of the next subsection.

4.3 Matter Clocks and Reference Fluids

4.3.1 The Basic Ideas

We have seen that it is difficult to produce a satisfactory definition of time using only the canonical variables of the gravitational field. In particular, there is nothing in the canonical formalism itself to provide insight into how the internal spacetime functionals $\mathcal{T}(x; g, p]$ and $\mathcal{Z}^a(x; g, p]$ are to be selected. However, in practice, location in time and in space is not performed in this way. Real physical clocks are made of matter with definite properties—an observation that has generated a recent flurry of interest in the idea of 'quantum clocks' with the hope that they may lead to a more tractable approach to the problem of time.

In a sense, the idea of matter clocks has already been implicit in what has been said so far. For example, one of the dynamical variables in the simple minisuperspace model §4.2.3 is the spatially-homogeneous scalar field ϕ, and this can be used to define time. Indeed, (4.2.20) is the Hamiltonian obtained by selecting $t = \phi$ as a classical time variable. However, a 'quantum clock' does not mean an arbitrary collection of particle or matter-field variables, but rather a device whose self-interaction and coupling to the gravitational field are deliberately optimised to serve as a measure of time. The important question is the extent to which the ability to measure (or, more precisely, to define) time using one of these systems is compatible with its realisation as a real physical entity. In particular, it must have a positive energy: a property that, as emphasised in §2, is far from being trivial

Two different types of system are feasible. The first is a point-particle clock that can measure time only at points along its worldline. The second is a cloud of such clocks that fills the space Σ and which can therefore provide a global measure of time and spatial position. The origin of this idea lies in the old classical notion of a reference fluid, and was first applied to quantum gravity in a major way by DeWitt (1962) (see also DeWitt (1967*a*)) who discussed a gravitational analogue of the famous Bohr-Rosenfeld analysis of the measurability of the quantised electromagnetic field (Bohr & Rosenfeld 1933, Bohr & Rosenfeld 1978).

In schemes of this type, the cloud of clocks is regarded as a realistic material medium with a Lagrangian that accurately describes its physical properties. An important question is the precise sense in which the ensuing structure corresponds to a genuine coordinate system on the spacetime manifold. A different approach is to *start* with a fixed set of coordinate conditions imposed on the spacetime metric γ, and then implement them by appending the conditions to the action with a family of Lagrange multipliers. These extra terms in the action are then parametrised (*i.e.*, made invariant under the action of Diff($\mathcal{M}$)) and interpreted as the source terms for a special type of matter. Schemes of this type have their origin in the general problem of understanding if, and how, the full spacetime diffeomorphism group Diff($\mathcal{M}$) (rather than its projections in the form of the Dirac algebra) should be represented in the canonical theory of gravity (Bergmann & Komar 1972, Salisbury & Sundermeyer 1983, Isham & Kuchař 1985*a*, Isham & Kuchař 1985*b*, Kuchař 1986*a*, Lee & Wald 1990).

4.3.2 The Gaussian Reference Fluid

The work of Kuchař and his collaborators has been especially significant and I shall illustrate it here with the example discussed in Kuchař & Torre (1991*a*) of a Gaussian reference fluid. Other examples are harmonic coordinate conditions (Kuchař & Torre 1991*b*, Stone & Kuchař 1992) and the K =*const* slicing condition (Kuchař 1992*a*). Kuchař & Torre (1989) handle the conformal, harmonic and light-cone gauges in the bosonic string in the same way, and the canonical treatment of two-dimensional induced quantum gravity is discussed in Torre (1989).

The main steps in the development of the Gaussian reference fluid are as follows.

1. The Gaussian coordinate conditions are $\gamma^{00}(X) = -1$ and $\gamma^{0a}(X) = 0$, $a = 1,2,3$, but a more covariant-looking expression can be obtained by introducing a set of spacetime functions $\mathcal{T}$, $\mathcal{Z}^a$, $a = 1,2,3$ and imposing the Gaussian conditions in the form

$$\gamma^{\alpha\beta}(X)\mathcal{T}_{,\alpha}(X)\mathcal{T}_{,\beta}(X) = -1 \quad (4.3.1)$$
$$\gamma^{\alpha\beta}(X)\mathcal{T}_{,\alpha}(X)\mathcal{Z}^a{}_{,\beta}(X) = 0. \quad (4.3.2)$$

These should be viewed as a (spacetime-coordinate independent) set of partial differential equations for the functions $(\mathcal{T}, \mathcal{Z}^a)$, a set of whose solutions (there is some arbitrariness) are the Gaussian coordinate functions on $\mathcal{M}$.

2. The conditions (4.3.1–4.3.2) are now added to the dynamical system with the aid of Lagrange multipliers M and M_a, $a = 1,2,3$. The extra term in the action is

$$S_F[\gamma, M, \vec{M}, \mathcal{T}, \vec{\mathcal{Z}}] := \int_{\mathcal{M}} d^4X\,(-\det\gamma)^{\frac{1}{2}}\left(-\tfrac{1}{2}M\left(\gamma^{\alpha\beta}\mathcal{T}_{,\alpha}\,\mathcal{T}_{,\beta}+1\right)+M_a\,\gamma^{\alpha\beta}\mathcal{T}_{,\alpha}\,\mathcal{Z}^a{}_{,\beta}\right). \quad (4.3.3)$$

Note that this action is invariant under coordinate transformations on the a superscript of $\mathcal{Z}^a$ provided that M_a transforms accordingly. This is consistent with the idea advanced in §3.4 that objects like $\mathcal{Z}$ are best thought of as taking their values in a *labelling* three-manifold Σ_l, in which case the superscript a refers to a coordinate system on Σ_l. The Lagrange multiplier M_a is then a hybrid object whose domain is the spacetime manifold $\mathcal{M}$ but whose values lie in the cotangent bundle to Σ_l. [30]

3. The action (4.3.3) is defined covariantly on the spacetime manifold $\mathcal{M}$, and is in 'parametrised' form in the sense that $(\mathcal{T}, \mathcal{Z}^a)$ are regarded as functions that can be varied freely. These four extra variables can be interpreted as describing a material system—the 'Gaussian reference fluid'—that interacts with the gravitational field and has its own energy-momentum tensor. Variation of the action with respect to $\mathcal{T}$ and $\mathcal{Z}^a$ gives the Euler hydrodynamic equations of the reference fluid, which turns out to be a heat-conducting fluid (Kuchař & Torre 1991*a*).

4. A complete canonical analysis shows that the constraints of the full theory have the form

$$\Pi_a(x) + \tilde{\mathcal{H}}_a(x; \mathcal{T}, \mathcal{Z}, g, p] = 0 \quad (4.3.4)$$
$$\Pi_{\mathcal{T}}(x) + \tilde{\mathcal{H}}_\perp(x; \mathcal{T}, \mathcal{Z}, g, g] = 0 \quad (4.3.5)$$

where $\tilde{\mathcal{H}}_a$ and $\tilde{\mathcal{H}}_\perp$ are linear combinations of the usual gravitational supermomentum and super-Hamiltonian (but with coefficients that depend on $\mathcal{T}$, $\mathcal{Z}$ and g), and where $\Pi_a(x)$ and $\Pi_{\mathcal{T}}(x)$ are the momenta conjugate to the variables $\mathcal{Z}^a(x)$ and $\mathcal{T}(x)$ respectively.

---

[30]Thus $\vec{M} : \mathcal{M} \to T^*\Sigma_l$ with $\vec{M}(X) \in T^*_{\mathcal{Z}(X)}\Sigma_l$. In general, if v belongs to the cotangent space $T^*_{\mathcal{Z}(X)}\Sigma_l$ of Σ_l, we interpret the symbol $\mathcal{Z}^a{}_{,\alpha}(X)\,v_a$, $\alpha = 0\ldots3$ as the coordinate representation on $\mathcal{M}$ of the pull-back $(\mathcal{Z}^*v)$ of v to $\mathcal{M}$ by the map $\mathcal{Z} : \mathcal{M} \to \Sigma_l$.

4.3.3 Advantages and Problems

The use of the Gaussian reference fluid has several big advantages over a purely geometrical set of internal coordinate functions. In particular:

- The most obvious property of the constraints (4.3.4–4.3.5) is that they are *linear* in the momenta Π_a and $\Pi_{\mathcal{T}}$ of the matter-field coordinate functions. In the quantum theory, this leads at once to a functional Schrödinger equation of the type (4.2.6), and hence to an uncontentious intepretation of a state $\Psi[\mathcal{T}, \mathcal{Z}, g]$ as the probability amplitude for measuring the three-metric $g_{ab}(x)$ on the hypersurface in $\mathcal{M}$ associated with $(\mathcal{T}, \mathcal{Z}^a)$. Since the constraints are manifestly linear in the reference-fluid momenta, the Hamiltonian densities that appear in the Schrödinger equation are local, explicit functions of the canonical variables. In addition, the Hamiltonian is only quadratic in $p^{ab}(x)$. These are significant advantages over the situation that arises when purely geometrical time and space functions are used.

- The spacetime problem is non-existent because the variables $(\mathcal{T}, \mathcal{Z}^a)$ in the action (4.3.3) are defined from the very start as genuine spacetime scalar fields.

These gains are attractive, and it is therefore unfortunate that they are offset to some extent by several basic problems:

- The energy-momentum tensor of the matter fluid does not satisfy the famous energy-conditions of general relativity, and therefore the system cannot be regarded as physically realisable. On the other hand, if one starts with a physically correct matter system (such as, for example, the scalar field ϕ in the minisuperspace model in §4.2.3) the simple linear dependence on $\Pi_{\mathcal{T}}$ is lost. This suspension between Charybdis and Scylla is arguably an inevitable consequence of the general problem of the non-existence of physical Hamiltonian clocks (§2).

- In classical general relativity, it is well-known that Gaussian coordinate conditions almost invariably breakdown somewhere, and therefore Gaussian coordinates are defined only *locally* on the spacetime manifold $\mathcal{M}$. This should be reflected in the quantum theory at some point, but it is not clear where, or how.

- Even if the technical problems above can be overcome, there still remains the issue of how fundamental are the Gaussian-type coordinate conditions, and with what entities in the real world the associated matter variables are to be identified. This is particularly appropriate in discussions of early universe quantum cosmology where there seems to be no room for a simple 'phenomenological' type of analysis.

4.4 Unimodular Gravity

One rather special example of a reference fluid is associated with the unimodular coordinate condition

$$\det \gamma_{\alpha\beta}(X) = 1 \tag{4.4.1}$$

that has often been used in discussions of classical general relativity. If the procedure outlined above is applied to this particular condition, the ensuing parametrised theory corresponds to the usual theory of general relativity but with a cosmological 'constant' that is a dynamical variable, rather than a fixed constant. The use of this theory as a possible solution to the problem of time has been discussed in several recent papers, and especially by Unruh and Wald (Henneaux & Teitelboim 1989, Unruh 1988, Unruh 1989, Unruh & Wald 1989, Brown & York 1989, Kuchař 1991*a*).

The super-Hamiltonian constraint of this modified theory is

$$\lambda + |g(x)|^{-\frac{1}{2}}\,\mathcal{H}_\perp(x) = 0 \tag{4.4.2}$$

in which what would normally be the cosmological constant λ appears as the momentum conjugate to a variable τ that is identified as a 'cosmological time'. The implication in the quantum theory is that dynamical evolution with respect to τ is described by the family of ordinary Schrödinger equations

$$i\hbar\frac{\partial\Psi(\tau,g]}{\partial\tau} = |g(x)|^{-\frac{1}{2}}\,(\widehat{\mathcal{H}}_\perp(x)\Psi)(\tau,g] \tag{4.4.3}$$

parametrised by the point x in Σ.

The problem is how to interpret such a family of equations. Kuchař studied this question with great care (Kuchař 1991*a*, Kuchař 1992*b*) and showed that the correct dynamical Schrödinger equation is

$$i\hbar\frac{\partial\Psi(\tau,g]}{\partial\tau} = \Big(\int_\Sigma d^3x\,|g|^{\frac{1}{2}}\Big)^{-\frac{1}{2}}\int_\Sigma d^3x\,(\widehat{\mathcal{H}}_\perp(x)\Psi)(\tau,g]. \tag{4.4.4}$$

If this was the only equation satisfied by $\Psi(\tau,g]$ it would be viable to interpret this function as the probability density for the three-metric g at a given value of the time τ. However, the effect of the existence of the family of equations (4.4.3) is that the state vector Ψ must also satisfy the collection of constraints

$$|g(x)|^{-\frac{1}{2}}\,\widehat{\mathcal{H}}_\perp(x))_{,a}\,\Psi(\tau,g] = 0 \tag{4.4.5}$$

where $x \in \Sigma$. But the three-geometry operator does not commute with these constraints, and therefore the interpretation of $\Psi(\tau,g]$ as a probability distribution for g is not tenable.

The geometrical origin of this problem is that the cosmological time measures the four-volume enclosed between two embeddings of the associated internal time functional $\mathcal{T}(x)$ but given one of the embeddings the second is not determined uniquely by the value of τ: two embeddings that differ by a zero four-volume (something that can happen easily in a spacetime with a *Lorentzian* signature) cannot be separated in this way. The extra constraints (4.4.5) are to be interpreted as saying that the theory is independent of this arbitrariness. For further details see the cited papers by Kuchař.

5 IDENTIFY TIME AFTER QUANTISATION

5.1 Canonical Quantum Gravity: Quantise Before Constraining

5.1.1 The Canonical Commutation Relations for Gravity

In approaches to the problem of time of category II, a quantum theory is constructed without solving the constraints, which are then imposed at the quantum level. The identification of 'time' is made *after* this process, and is used to give the final physical interpretation of the theory, particularly the probabilistic aspects. This final structure may be related only loosely to the quantum structure with which the construction started. As we shall see, this leads to a picture of quantum gravity that is radically different from that afforded by the internal Schrödinger interpretation.

The starting point is the operator version of the Poisson-bracket algebra (3.3.27–3.3.29) in the form of the canonical commutation relations

$$[\,\widehat{g}_{ab}(x), \widehat{g}_{cd}(x')\,] \;=\; 0 \tag{5.1.1}$$

$$[\,\widehat{p}^{ab}(x), \widehat{p}^{cd}(x')\,] \;=\; 0 \tag{5.1.2}$$

$$[\,\widehat{g}_{ab}(x), \widehat{p}^{cd}(x')\,] \;=\; i\hbar\, \delta^c_{(a}\delta^d_{b)}\, \delta(x, x') \tag{5.1.3}$$

of operators defined on the three-manifold Σ. Several things should be said about this algebra:

- The classical object $g_{ab}(x)$ is not merely a symmetric covariant tensor: it is also a *metric* tensor, *i.e.*, at each point $x \in \Sigma$ the matrix $g_{ab}(x)$ is invertible with signature $(1,1,1)$. In particular, for any non-vanishing vector-density field v^a (of an appropriate weight) we have

$$g(v \otimes v) := \int_\Sigma d^3x\, v^a(x)\, v^b(x)\, g_{ab}(x) > 0, \tag{5.1.4}$$

  and it is reasonable to require the corresponding quantum operators $\widehat{g}(v \otimes v)$ to satisfy the analogous equations

$$\widehat{g}(v \otimes v) > 0. \tag{5.1.5}$$

It is noteworthy that the canonical commutation relations (5.1.1–5.1.3) are *incompatible* with these geometrical properties of $g_{ab}(x)$ provided that the smeared versions of the operators $\widehat{p}^{cd}(x)$ are self-adjoint. In that case they can be exponentiated to give unitary operators which when acting on $\widehat{g}(v\otimes v)$ show that the spectrum of $\widehat{g}(v\otimes v)$ can take on negative values.

This problem can be partly remedied by replacing (5.1.1–5.1.3) with a set of *affine* relations (Klauder 1970, Pilati 1982, Pilati 1983, Isham 1984, Isham & Kakas 1984*a*, Isham & Kakas 1984*b*, Isham 1992),

$$[\,\widehat{g}_{ab}(x),\widehat{g}_{cd}(x')\,] = 0 \tag{5.1.6}$$
$$[\,\widehat{p}_a{}^b(x),\widehat{p}_c{}^d(x')\,] = i\hbar(\delta^d_a\,\widehat{p}_c{}^b(x) - \delta^b_c\,\widehat{p}_a{}^d(x))\,\delta(x,x') \tag{5.1.7}$$
$$[\,\widehat{g}_{ab}(x),\widehat{p}_c{}^d(x')\,] = i\hbar\,\delta^d_{(a}\widehat{g}_{b)c}(x)\,\delta(x,x'). \tag{5.1.8}$$

At a classical level, the corresponding Poisson brackets are equivalent to the standard canonical relations (3.3.27–3.3.29) with $p_c{}^d(x) := g_{cb}(x)\,p^{bd}(x)$. However, the situation at the quantum level is very different and, for example, there exist many representations of the affine relations (5.1.6–5.1.8) in which the spectrum of the smeared metric operator 'almost' satisfies the operator inequality (5.1.5); 'almost' in the sense that the right hand side is ≥ 0 rather than a strict inequality. This result is helpful, but it provides only a partial resolution of the general question of the extent to which the classical geometrical properties of $g_{ab}(x)$ can be, or should be, captured in the quantum theory. This is how the *spatial metric reconstruction* problem appears in this approach to quantum gravity.

- We are working in the 'Schrödinger representation' in which no time dependence is carried by the canonical variables. As we shall see, there is more to this than meets the eye.

- Equation (5.1.1) (or the affine analogue (5.1.6)) is a form of microcausality. However, the functional form of the constraints is independent of any foliation of spacetime, and therefore it is not clear what this 'microcausal' property means in terms of the usual ideas of an 'equal-time' hypersurface, or indeed how the notion of spacetime structure (as opposed to spatial structure) appears at all. At this stage, the most that can be said is that, whatever the final spacetime interpretation may be, (5.1.1) implies that the points of Σ are to be regarded as spacelike separated.

5.1.2 The Imposition of the Constraints

The key question is how the constraint equations (3.3.25–3.3.26) are to be handled. The essence of the Dirac approach is to impose them as constraints on the physically allowed states in the form

$$\mathcal{H}_a(x;\widehat{g},\widehat{p})\Psi = 0 \tag{5.1.9}$$

$$\mathcal{H}_\perp(x;\widehat{g},\widehat{p})\Psi = 0. \tag{5.1.10}$$

We recall that, in the classical theory, the constraints are *equivalent* to the dynamical equations in the sense that if they are satisfied on all spatial hypersurfaces of a Lorentzian metric γ, then γ necessarily satisfies the Einstein vacuum field equations. This is reflected in the quantum theory by the assumption that the operator constraints (5.1.9–5.1.10) are the *sole* technical content of the theory, *i.e.*, the dynamical evolution equations are *not* imposed as well.

This is closely related to the following fundamental observation. According to the first-order action (3.3.18), the canonical Hamiltonian (3.3.23) associated with general relativity is

$$H[N,\vec{N}](t) = \int_\Sigma d^3x(N(x)\mathcal{H}_\perp(x) + N^a(x)\mathcal{H}_a(x)) \tag{5.1.11}$$

where N and $\vec{N}$ are regarded as external, *c*-number functions. However, (5.1.11) has a rather remarkable implication for the putative Schrödinger equation

$$i\hbar\frac{d}{dt}\Psi_t = \widehat{H}[N,\vec{N}](t)\Psi_t \tag{5.1.12}$$

since if the state Ψ_t satisfies the constraint equations (5.1.9–5.1.10), we see that it has no time dependence at all! Similarly, it is not meaningful to speak of a 'Schrödinger' or a 'Heisenberg' picture since the matrix elements between physical states of a Heisenberg-picture field will be the same as for the field in the Schrödinger picture.

This so-called 'frozen formalism' caused much confusion when it was first discovered since it seems to imply that nothing happens in a quantum theory of gravity. Clearly this is some sort of quantum analogue of the fact that classical observables (*i.e.*, those satisfying (3.3.40)) are constants of the motion. These days, this situation is understood to reflect the absence of any external time parameter in general relativity, and therefore, in particular, the need to discuss the measurement of time with the aid of functionals of the internal variables in the theory. This perspective dominates almost all work on the problem of time in quantum gravity.

5.1.3 Problems with the Dirac Approach

Many problems arise when attempting to implement the Dirac scheme. For example:

1. To what extent can, or should, the classical Poisson-bracket algebra (3.3.30–3.3.32) be maintained in the quantum theory? The constraints (3.3.19) and

(3.3.20) are highly non-linear functions of the canonical variables and involve non-polynomial products of field operators evaluated at the same point. Thus we are lead inevitably to the problems of regularisation, renormalisation, operator ordering, and potential anomalies. This is the form taken by the *functional evolution problem* in this approach to quantisation.

2. It is not clear what properties are expected of the constraint operators $\widehat{\mathcal{H}}_\perp(x)$ and $\widehat{\mathcal{H}}_a(x)$; in particular, should they be self-adjoint? Since one presumably starts with self-adjoint representations of the canonical commutation representations (5.1.1–5.1.3), it is perhaps natural to require self-adjointness for the super-Hamiltonian and supermomentum operators. However, this has been challenged several times and the issue is clearly of significance in discussing the operator-ordering problem (Komar 1979*b*, Komar 1979*a*, Kuchař 1986*c*, Kuchař 1986*b*, Kuchař 1987, Hajicek & Kuchař 1990*a*, Hajicek & Kuchař 1990*b*). The possibility of using a non-hermitian operator can be partly justified by noting that the Hilbert space structure on the space that carries the representation of the canonical algebra may be only distantly related to the Hilbert space structure that ought to be imposed on the *physical* states (*i.e.*, those that satisfy the constraints).

3. More generally, what is the relation between these two Hilbert spaces? This question has important implications for the problem of time.

4. What is meant by an 'observable' in a quantum theory of this type? By analogy with the classical result (3.3.40), one might be tempted to postulate that an operator $\widehat{A}$ defined on the starting Hilbert space corresponds to an observable if
$$[\,\widehat{A}, \widehat{H}[N, \vec{N}]\,] = 0 \tag{5.1.13}$$
for all test functions N and $\vec{N}$. However, it can be argued that it is sufficient for (5.1.13) to be satisfied on the subspace of *solutions* to the Dirac constraints (5.1.9–5.1.10). This is the quantum analogue of the fact that the classical condition (3.3.40) for an observable is a *weak* equality, *i.e.*, it holds only on the subspace of the classical phase space given by the solutions (g,p) to the classical constraints $\mathcal{H}_a(x; g, p] = 0 = \mathcal{H}_\perp(x; g, p]$.

5.1.4 Representations on Functionals $\Psi[g]$

In attempting to find concrete representations of the canonical algebra (5.1.1–5.1.3) it is natural to try an analogue of the quantum scalar field representations (3.2.14) and (3.2.15). Thus the state vectors are taken to be functionals $\Psi[g]$ of Riemannian metrics g on Σ, and the canonical operators are defined as

$$(\widehat{g}_{ab}(x)\Psi)[g] \;:=\; g_{ab}(x)\Psi[g] \tag{5.1.14}$$

$$(\widehat{p}^{cd}(x)\Psi)[g] \ := \ -i\hbar\frac{\delta\Psi[g]}{\delta g_{cd}(x)}. \tag{5.1.15}$$

These equations have been used widely in the canonical approach to quantum gravity although, even when suitably smeared, they do not define proper self-adjoint operators because of the absence of any Lebesgue measure on Riem(Σ) (*e.g.*, Isham (1992)) and, as remarked earlier, they are also incompatible with the positivity requirement (5.1.5).

Let us consider the Dirac constraints (5.1.9–5.1.10) in this representation. From a physical perspective it is easy to see the need for them. Formally, the domain space of the state functionals is Riem(Σ), and to specify a metric $g_{ab}(x)$ at a point $x \in \Sigma$ requires six numbers (the components of the metric in some coordinate system). However, the true gravitational system should have only *two* degrees of freedom per spatial point, and therefore four of the six degrees of freedom need to be lost. This is precisely what is achieved by the imposition of the constraints (5.1.9–5.1.10). As we saw in §4.1, the same counting argument applies if the system is reduced to true canonical form before quantising.

The easiest constraints to handle are those in the first set

$$(\widehat{H}[\vec{N}]\Psi)[g] = 0. \tag{5.1.16}$$

We saw earlier that the classical functions $H[\vec{N}]$ are the infinitesimal generators of the diffeomorphism group of Σ, and the same might be expected to apply here. The key question is whether the operator-ordering problems can be solved so that the algebra (3.3.34) is preserved at the quantum level. In practice, this is fairly straightforward; indeed, one powerful way of *solving* the operator-ordering problem for the $\widehat{H}[\vec{N}]$ generators is to insist that they form a self-adjoint representation of the Lie algebra of Diff(Σ).

The implications of (5.1.16) are then a straightforward analogue of those in conventional Yang-Mills gauge theories. The group Diff(Σ) acts as a group of transformations on the space Riem(Σ) of Riemannian metrics on Σ, with $f \in$ Diff(Σ) sending $g \in$ Riem(Σ) to f^*g. Apart from certain technical niceties, this leads to a picture in which Riem(Σ) is fibered by the orbits of the Diff(Σ) action. Then (5.1.16) implies that the state functional Ψ is constant (modulo possible θ-vacuua effects) on the orbits of Diff(Σ), and therefore passes to a function on the *superspace* Riem(Σ)/Diff(Σ) of Diff(Σ) orbits (Misner 1957, Higgs 1958).

5.1.5 The Wheeler-DeWitt Equation

We must consider now the final constraint $\widehat{\mathcal{H}}_\perp(x)\Psi = 0$. Unlike the constraint $H[\vec{N}]\Psi = 0$, this has no simple group-theoretic interpretation since, as remarked earlier, the presence of the explicit $g^{ab}(x)$ factor on the right hand side of (3.3.32) means that (3.3.30–3.3.32) is not a genuine Lie algebra. Thus the operator-ordering problem becomes much harder. If we choose as a simple example the ordering in which all the p^{cd} variables are placed to the right of the g_{ab} variables, the constraint (5.1.10) becomes

$$-\hbar^2\kappa^2\mathcal{G}_{ab\,cd}(x,g)\frac{\delta^2\Psi[g]}{\delta g_{ab}(x)\,\delta g_{cd}(x)} - \frac{|g|^{\frac{1}{2}}(x)}{\kappa^2}R(x,g)\Psi[g] = 0 \tag{5.1.17}$$

where $\mathcal{G}_{ab\,cd}$ is the DeWitt metric defined in (3.3.21).

Equation (5.1.17) is the famous Wheeler-DeWitt equation (Wheeler 1962, Wheeler 1964, DeWitt 1967*a*, Wheeler 1968). It is the heart of the Dirac constraint quantisation approach to the canonical theory of quantum gravity, and everything must be extracted from it. Needless to say, there are a number of problems and questions that need to be considered. For example:

1. The ordering chosen in (5.1.17) is a simple one but there is no particular reason why it should be correct. One popular alternative is to write the 'kinetic-energy' term as a covariant functional-Laplacian using the DeWitt metric (3.3.21). This part of the operator is then invariant under redefinitions of coordinates on Riem(Σ). Of course, a key issue in discussing the ordering of $\widehat{\mathcal{H}_\perp}(x)$, $x \in \Sigma$, is whether or not these operators are expected to be self-adjoint (the ordering chosen in (5.1.17) is certainly *not* self-adjoint in the scalar product formally associated with the choice (5.1.14–5.1.15) for the canonical operators).

2. The Wheeler-DeWitt equation contains products of functional differential operators evaluated at the same spatial point and is therefore likely to produce $\delta(0)$ singularities when acting on a wide variety of possible state functionals. Thus regularisation will almost certainly be needed.

3. A major question is how to approach the problem of solving the Wheeler-DeWitt equation. One obvious tactic is to deal with it as a functional differential equation *per se*. However, whether or not this is valid depends on the general interpretation of the constraint equations $\widehat{\mathcal{H}_\perp}(x)\Psi = 0$, $x \in \Sigma$. If these equations mean that Ψ is a simultaneous eigenvector of self-adjoint operators $\widehat{\mathcal{H}_\perp}(x)$, $x \in \Sigma$, with eigenvalue 0 then, as with eigenfunction problems for ordinary differential operators, some sort of boundary value conditions need to be imposed on Ψ, and the theory itself is not too informative about what these might be. In practice, there has been a tendency to solve the constraint equation as a functional differential equation without checking that 0 is a genuine eigenvalue (*i.e.*, without worrying about boundary conditions). This can lead to highly misleading results—a fact that is often overlooked, especially in discussions of minisuperspace approximations to the theory.

4. One of the hardest problems is to decide what the Wheeler-DeWitt equation means in physical terms. In particular, the notions of 'time' and 'time-evolution' must be introduced in some way. The central idea is one we have mentioned several

times already: time must be defined as an *internal* property of the gravitational system (plus matter) rather than being identified with some external parameter in the universe. We shall return frequently to this important topic in our analysis of the various approaches to the problem of time in quantum gravity.

5.1.6 A Minisuperspace Example

Further discussion of the Wheeler-DeWitt equation is assisted by having access to the minisuperspace model discussed in §4.2.3. To derive the Wheeler-DeWitt equation for this system we need first to confront some of the problems mentioned above. In particular:

- The classical variable a satisfies the inequality $a \geq 0$. How is this inequality to be implemented in the quantum theory? This is a very simple example of the spatial metric reconstruction problem.
- The classical expression $p_a^2/24a$ in the super-Hamiltonian (4.2.16) will lead to operator-ordering problems in the quantum theory.

The first problem can be tackled in several different ways:

1. Ignore the problem and impose standard commutation relations $[\,\widehat{a}, \widehat{p}_a\,] = i\hbar$ even though we know this leads to a spectrum for $\widehat{a}$ which is the entire real line. This is the minisuperspace analogue of taking the 'naïve' commutation relations (5.1.1–5.1.3). The Hilbert space [31] will be $L^2(\mathbb{R}, da)$ with the operators defined in the usual way as

$$(\widehat{a}\psi)(a) \;:=\; a\psi(a) \tag{5.1.18}$$

$$(\widehat{p}_a\psi)(a) \;:=\; -i\hbar\frac{d\psi(a)}{da}. \tag{5.1.19}$$

   The problem in this approach is to give some physical meaning to the negative values of a.

2. Insist on using the Hilbert space $L^2(\mathbb{R}_+, da)$ of functions that are concentrated on $\mathbb{R}_+$ but keep the definitions (5.1.18–5.1.19). The conjugate momentum $\widehat{p}_a$ is no longer self-adjoint but, nevertheless, it is possible to arrange for the super-Hamiltonian to be a self-adjoint function of $\widehat{a}$, $\widehat{p}_a$, $\widehat{p}_a^\dagger$, $\widehat{\phi}$ and $\widehat{p}_\phi$.

3. Perform a canonical transformation at the classical level to a new variable Ω defined by $a = e^\Omega$. This new variable ranges freely over the entire real line and

[31] This is only that part of the Hilbert space which refers to the a-variable. The full Hilbert space is $L^2(\mathbb{R}^2, da\, d\phi)$.

can therefore be quantised as part of a conventional set of commutation relations using the Hilbert space $L^2(\mathbb{R}, d\Omega)$. The conjugate variable is $p_\Omega := e^{-\Omega} p_a$ and the super-Hamiltonian is

$$H_{\mathrm{RW}} := e^{-3\Omega}\left(\frac{-p_\Omega^2}{24} + \frac{p_\phi^2}{2}\right) - 6ke^{\Omega} + e^{3\Omega}V(\phi). \tag{5.1.20}$$

4. Use the affine relation $[\,\widehat{a}, \widehat{\pi}_a\,] = i\hbar\widehat{a}$ which is the minisuperspace analogue of the full set of affine relations (5.1.6–5.1.8). The affine momentum π_a is related classically to the canonical momentum p_a by $\pi_a := ap_a$, and the super-Hamiltonian becomes

$$H_{\mathrm{RW}} := \frac{1}{a^3}\left(\frac{-\pi_a^2}{24} + \frac{p_\phi^2}{2}\right) - 6ka + a^3V(\phi). \tag{5.1.21}$$

A self-adjoint representation of the affine commutation relations can be defined on the Hilbert space $L^2(\mathbb{R}_+, da/a)$ by

$$(\widehat{a}\psi)(a) \;:=\; a\psi(a) \tag{5.1.22}$$

$$(\widehat{\pi}_a\psi)(a) \;:=\; -i\hbar\, a\frac{d\psi(a)}{da}. \tag{5.1.23}$$

Note that the transformation $a := e^{\Omega}$ sets up an equivalence between this approach and the previous one.

To get a sample Wheeler-DeWitt equation, let us choose method three with the Ω variable satisfying standard commutation relations. Then, ignoring the operator-ordering problem, we get

$$\left(\hbar^2 e^{-3\Omega}\left(\frac{1}{24}\frac{\partial^2}{\partial\Omega^2} - \frac{1}{2}\frac{\partial^2}{\partial\phi^2}\right) - 6ke^{\Omega} + e^{3\Omega}V(\phi)\right)\psi(\Omega, \phi) = 0. \tag{5.1.24}$$

This simple model illustrates many of the features of the full canonical theory of general relativity and will be very useful in what follows. For a recent treatment using path-integral techniques see Linden & Perry (1991).

5.2 The Klein-Gordon Interpretation for Quantum Gravity

5.2.1 The Analogue of a Point Particle Moving in a Curved Spacetime

The key issue now is how the solutions to the Wheeler-DeWitt equation (5.1.17) are to be interpreted. This involves two related questions:

1. What inner product should be placed on the solutions?

2. How is the notion of time evolution to be extracted from the Wheeler-DeWitt equation?

One natural inner product might seem to be [32]

$$\langle\Psi|\Phi\rangle := \int_{\mathrm{Riem}(\Sigma)} \mathcal{D}g\, \Psi^*[g]\,\Phi[g] \tag{5.2.1}$$

since this is the scalar product with respect to which, for example, the canonical operators (5.1.14) and (5.1.15) are self-adjoint (at least formally). Indeed, this is precisely what is used in the so-called 'naïve Schrödinger interpretation' discussed in §6.1. However, the definition (5.2.1) can be applied to *any* functional of the three-metric g, which as we shall see in §6.1 is the cause of considerable difficulties. In the present section we are concerned rather with finding a scalar product that applies only to *solutions* to the Wheeler-DeWitt equation.

The central idea is to explore the analogy between the Wheeler-DeWitt equation (5.1.17) and the Klein-Gordon equation of a particle moving in a curved space with an arbitrary, time-dependent potential (DeWitt 1967*a*). The validity of this analogy can be seen especially clearly in the simple minisuperspace model (5.1.24).

The point-particle model has been discussed in great depth in recent years by Kuchař; here I shall sketch only the main ideas (Kuchař 1991*b*, Kuchař 1992*b*). Consider a relativistic particle of mass M moving in a four-dimensional spacetime $(\mathcal{M},\gamma)$ where γ is a fixed Lorentzian metric. The classical trajectories of the particle in $\mathcal{M}$ are parametrised by an arbitrary real number τ, and the theory is invariant under the reparametrisation $\tau \mapsto \tau'(\tau)$. This invariance leads to the constraint $H(X,P) = 0$ with the super-Hamiltonian

$$H(X,P) := \frac{1}{2M}\gamma^{\alpha\beta}(X)\,P_\alpha\,P_\beta + V(X) \tag{5.2.2}$$

where $V(X)$ is the (positive) potential-energy term.

In the quantum theory, this constraint becomes the Klein-Gordon equation

$$(\gamma^{\alpha\beta}(X)\nabla_\alpha\nabla_\beta + V(X))\Psi(X) = 0 \tag{5.2.3}$$

where a convenient choice has been made for the operator ordering of the kinetic-energy term. The standard interpretation of this equation is based on the pairing between any pair of solutions Ψ, Φ defined by

$$(\Psi,\Phi)_{\mathrm{KG}} := \int_{\mathcal{E}(\Sigma)} d\Sigma_\alpha(X)\,\frac{1}{2i}\gamma^{\alpha\beta}(X)(\Psi(X)^*\overrightarrow{\partial}_\beta\Phi - \Psi(X)^*\overleftarrow{\partial}_\beta\,\Phi(X)) \tag{5.2.4}$$

[32] Here $\mathcal{D}g$ denotes the formal analogue of the Lebesgue measure. No such measure really exists on $\mathrm{Riem}(\Sigma)$ and a more careful discussion would need to take this into account.

where the integral is taken over the hypersurface $\mathcal{E}(\Sigma)$ of $\mathcal{M}$ defined by an embedding $\mathcal{E}:\Sigma\rightarrow\mathcal{M}$ that is spacelike with respect to the background metric γ on $\mathcal{M}$. Note that $d\Sigma_\alpha(X)$ is the directed hypersurface volume-element in $\mathcal{M}$ defined by

$$d\Sigma_\alpha(X) := \epsilon_{\alpha\beta\gamma\delta}\, dX^\beta\wedge dX^\gamma\wedge dX^\delta. \tag{5.2.5}$$

It follows from the Klein-Gordon equation that $\langle\Psi,\Phi\rangle_{\mathrm{KG}}$ is independent of $\mathcal{E}$, which suggests that $\langle\Psi,\Phi\rangle_{\mathrm{KG}}$ might be a suitable choice for a scalar product. However, as things stand, this is not viable since the pairing is *not* positive definite. Indeed

- $\langle\Psi,\Psi\rangle_{\mathrm{KG}}=0$ for all *real* functions Ψ;
- complex solutions to the Klein-Gordon equation exist for which $\langle\Psi,\Psi\rangle_{\mathrm{KG}}<0$.

The standard way of resolving this problem is to look for a timelike vector field U on $\mathcal{M}$ that is a Killing vector for the spacetime metric γ and is also such that the potential V is constant along its flow lines. A natural choice of time function $\tau(X)$ is then the parameter along these flow lines, defined as a solution to the partial differential equation

$$U^\alpha(X)\partial_\alpha\tau(X)=1. \tag{5.2.6}$$

If such a Killing vector exists, it follows at once that the energy $E(X,P)=-P_\tau := -U^\alpha(X)\,P_\alpha$ of the particle is a constant of the motion with $\{E,H\}=0$. On quantisation this becomes

$$[\,\widehat{E},\widehat{H}\,]=0 \tag{5.2.7}$$

which, if all the operators are self-adjoint with respect to the original inner product (5.2.1), means it is possible to find simultaneous eigenstates of

$$\widehat{E} := i\hbar U^\alpha\partial_\alpha. \tag{5.2.8}$$

and the super-Hamiltonian (5.2.2). It is therefore meaningful to select those solutions of the Klein-Gordon equation that have positive energy, and it is straightforward to see that the inner product (5.2.4) is *positive* on such solutions. Furthermore, restricted to such solutions, the Klein-Gordon equation can be shown to be equivalent to a conventional Schrödinger equation using the chosen time parameter. This equation is obtained by factorising the super-Hamiltonian in the form

$$H=(P_\tau+h)(P_\tau-h). \tag{5.2.9}$$

which is equivalent to the constraint

$$P_\tau+h=0 \tag{5.2.10}$$

on the subspace of the phase space of positive-energy solutions. In effect, the Schrödinger equation is obtained by imposing this second constraint as a constraint on allowed state vectors.

The construction above forms the basis for a physically meaningful interpretation of the quantum theory. However, if no suitable Killing vector U exists there is no consistent one-particle quantisation of this theory.

5.2.2 Applying the Idea to Quantum Gravity

In trying to apply these ideas to the Wheeler-DeWitt equation, the key observation is that the DeWitt metric (3.3.21) on Riem(Σ) has a hyperbolic character in which the conformal modes of the metric play the role of time-like directions, *i.e.*, the transformation $\gamma_{ab}(x) \mapsto F(x)\gamma_{ab}(x)$, $F(x) > 0$, is a 'time-like' displacement in Riem(Σ). This suggests that it may be possible to choose some internal time functional $\mathcal{T}(x, g]$ so that the Wheeler-DeWitt equation can be written in the form

$$-\hbar^2\kappa^2\Big(\frac{\delta^2}{\delta\mathcal{T}^2(x)} - \mathcal{F}^{R_1 R_2}(x;\mathcal{T},\sigma]\frac{\delta^2}{\delta\sigma^{R_1}(x)\,\delta\sigma^{R_2}(x)}\Big)\Psi[\mathcal{T},\sigma]$$
$$-\frac{|g|^{\frac{1}{2}}(x;\mathcal{T},\sigma]}{\kappa^2}R(x;\mathcal{T},\sigma]\Psi[\mathcal{T},\sigma] = 0 \qquad (5.2.11)$$

where $\sigma^R(x, g]$, $R = 1, \ldots, 5$ denotes the $5 \times \infty^3$ modes of the metric variables $g_{ab}(x)$ that remain after identifying the $1 \times \infty^3$ internal time modes $\mathcal{T}(x)$.

The starting point is the formal pairing (the analogue of the point-particle expression (5.2.4))

$$\langle\Psi, \Phi\rangle := i\prod_x \int_\Sigma d\Sigma^{ab}(x)\,\Psi^*[\gamma]\Big(\mathcal{G}_{ab\,cd}(x, g]\frac{\overrightarrow{\delta}}{\delta g_{cd}(x)} - \frac{\overleftarrow{\delta}}{\delta g_{cd}(x)}\mathcal{G}_{ab\,cd}(x, g]\Big)\Phi[\gamma] \quad (5.2.12)$$

between solutions Ψ and Φ of the Wheeler-DeWitt equation. The functional integral is over some surface in Riem(Σ) that is spacelike with respect to the DeWitt metric (3.3.21), and $d\Sigma^{ab}(x)$ is the directed surface-element in Riem(Σ) at the point $x \in \Sigma$. Of course, considerable care would be needed to make this expression rigorous. For example, it is necessary to take account of the Diff(Σ)-invariance and then project the inner product down to Riem(Σ)/Diff(Σ), *i.e.*, we must also include the action of the supermomentum constraints $\mathcal{H}_a(x)\Psi = 0$. Note also that the precise form of the scalar product depends on how the operator-ordering problem in the Wheeler-DeWitt equation is solved. However, the essential idea is clear. In particular, the expression (5.2.12) has the important property of being invariant under deformations of the 'spatial' hypersurface in Riem(Σ). This is the quantum-gravity analogue of the requirement in the normal Klein-Gordon equation that the scalar product (5.2.4) be time independent.

In the minisuperspace example, the Wheeler-DeWitt equation (5.1.24) can be simplified by multiplying [33] both sides by $e^{3\Omega}$ to give

$$\left(\hbar^2\Big(\frac{1}{24}\frac{\partial^2}{\partial\Omega^2} - \frac{1}{2}\frac{\partial^2}{\partial\phi^2}\Big) - 6ke^{4\Omega} + e^{6\Omega}V(\phi)\right)\psi(\Omega,\phi) = 0. \tag{5.2.13}$$

The associated scalar product is simply

$$\langle\psi,\phi\rangle := i\int_{\Omega=\text{const}} d\phi\left(\psi^*\frac{\partial\phi}{\partial\Omega} - \phi\frac{\partial\psi^*}{\partial\Omega}\right) \tag{5.2.14}$$

which is conserved in Ω-time by virtue of (5.2.13). Note that (5.2.13) has the anticipated form (5.2.11) for the Wheeler-DeWitt equation expressed using an appropriate internal time $\mathcal{T}(x,g]$.

Unfortunately, the right hand side of (5.2.12) cannot serve as a genuine Hilbert space inner product because, as in the analogous case of the point particle, it is not positive definite. Guided by the point-particle example, one natural way of trying to resolve this problem is to look for a vector field on Riem(Σ) that is a Killing vector for the DeWitt metric (3.3.21) and that scales the potential term $|g|^{\frac{1}{2}}(x)\,R(x,g]$ in an appropriate way. Sadly, Kuchař has shown that Riem(Σ) admits no such vector (Kuchař 1981*a*, Kuchař 1991*b*), and hence there is no possibility of defining physical states as an analogue of the positive-frequency solutions of the normal Klein-Gordon equation (but see Friedman & Higuchi (1989) for the asymptotically-flat case). However, even if such a Killing vector did exist, there are other problems. For example (Kuchař 1992*b*):

- The potential term $|g|^{\frac{1}{2}}(x)\,R(x,g]$ can take on both negative and positive values, whereas the potential $V(X)$ in the point-particle super-Hamiltonian (5.2.2) was required to be positive. Without this condition it is not possible to prove the positivity of the Klein-Gordon scalar product restricted to positive-energy solutions.

- There are good physical reasons for selecting just the positive-energy solutions for the point-particle, but the justification for the analogous step in the gravitational case is not clear. In particular, it is quite legitimate for the geometries along a path in superspace to both expand and contract in volume, and this means a classical solution to Einstein's equations can have either sign of E. Therefore, there is no justification for picking just the positive-frequency modes. Note that this objection applies already to the simple minisuperspace model discussed above with the scalar product (5.2.14).

---

[33] This step is contentious. It is true that the constraint equation $\widehat{\mathcal{H}}_\perp\Psi = 0$ is not formally affected by multiplying on the left by any invertible operator, but this 'renormalisation' of $\widehat{\mathcal{H}}_\perp$ affects the total constraint algebra, and the implications of this need to be considered at some point. Of course, it also affects the hermiticity properties of the original constraint.

- The attempt to construct a Klein-Gordon interpretation of the Wheeler-DeWitt equation entails the selection of some intrinsic time functional $\mathcal{T}(x,g]$. However, as discussed earlier, any such choice will necessarily fall foul of the *spacetime problem* whose resolution requires an internal time to be a functional of the conjugate momenta $p^{cd}(x)$ as well as the metric variables $g_{ab}(x)$. Hence $\mathcal{T}(x,g]$ cannot be interpreted as a genuine spacetime coordinate.

- The feasability of a Klein-Gordon interpretation is dependent on the fact that the classical super-Hamiltonian $\mathcal{H}_\perp(x;g,p]$ is *quadratic* in the momentum variables $p^{ab}(x)$. This property is lost if any powers of the Riemann curvature $R_{\alpha\beta\gamma\delta}(X,\gamma]$ are added to the classical spacetime action of the theory. Expressions of this type are likely to arise as counter-terms in almost any attempt to construct a proper quantisation of the gravitational field (including superstring theory, but excepting the Ashtekar programme in its current form) and it is important to have some feel for how they change the situation. Several of the approaches to the problem of time are sensitive to the precise form of $\mathcal{H}_\perp$, but this is particularly so of the Klein-Gordon interpretation.

The problems above, plus the non-existence of a suitable Killing vector on Riem(Σ), seem to form an immovable block to resolving the Hilbert space problem of how to turn the solutions to the Wheeler-DeWitt equation into a genuine Hilbert space.

5.3 Third Quantisation

There have been several different reactions to the failure of the Klein-Gordon approach to the Wheeler-DeWitt equation. In the case of a relativistic particle with an external spacetime-dependent metric or potential, the impossibility of isolating positive-frequency solutions is connected with a breakdown of the one-particle interpretation of the theory, and the standard resolution is to second-quantise the system by turning the Klein-Gordon wave function into a quantum field.

It has been suggested several times that a similar process might be needed in quantum gravity with $\Psi[g]$ becoming an operator $\hat{\Psi}[g]$ in some new Hilbert space. This procedure is usually called 'third quantisation' since the original Wheeler-DeWitt equation is already the result of a quantum field theory (Kuchař 1981*b*, Coleman 1988, Giddings & Strominger 1988, McGuigan 1988, McGuigan 1989). However, it is unclear what this means, or if the problem of time can really be solved in this way. Some of the many difficulties that arise when trying to implement this programme are as follows.

1. The approach to third quantisation that is closest to conventional quantum field theory involves constructing a Fock space whose 'one-particle' sector is associated with the functionals $\Psi[g]$. But this raises several difficulties:

- What is the analogue of the one-particle Hilbert space which is to form the basis for the Fock-space construction? The difficulty is that the problem of time is closely related to the Hilbert space problem, so where do we start? In the case of a particle in the presence of a spacetime-dependent potential, one common way of resolving this issue is to begin with a well-defined free theory with a proper Hilbert space, and then to regard the interactions with the background as a perturbation that can annihilate and create the quanta of this theory. However, this procedure is dubious if the spacetime dependence comes from the spacetime metric γ itself unless there is some sense in which one can usefully write $\gamma_{\alpha\beta}(X)$ as the sum $\eta_{\alpha\beta} + h_{\alpha\beta}(X)$ of the fixed Minkowskian metric $\eta_{\alpha\beta}$ plus a small perturbation $h_{\alpha\beta}(X)$. This is even more inappropriate in the full quantum-gravity theory since there is no obvious way of writing the DeWitt metric (3.3.21) on Riem(Σ) as a sum of this type.

- If a one-particle Hilbert space *can* be constructed, what is the interpretation of states that are tensor products of the vectors in such a space? In the case of particles, if $|x_1\rangle$ and $|x_2\rangle$ are states corresponding to a particle localised at points x_1 and x_2 respectively, then the tensor product $|x_1\rangle|x_2\rangle$ describes a pair of particles, both of which move in the *same* physical three-space. However, if $|g_1\rangle$ and $|g_2\rangle$ are eigenstates of the metric operator, it is not clear to what the product state $|g_1\rangle|g_2\rangle$ refers. The simplest thing might be to say that g_1 and g_2 are both metrics on the same space Σ, but this has no obvious physical interpretation.

2. A more meaningful interpretation is that g_1 and g_2 are metrics on *different* copies of Σ or, equivalently, a single metric on the disjoint union of two copies of Σ. However this also raises a number of problems:

- The transition from a state $|g\rangle$ to a state $|g_1\rangle|g_2\rangle$ corresponds to a topology change in which Σ bifurcates into two copies of itself. But this is unlikely to be compatible with the Wheeler-DeWitt equation. Are we to look for some non-linear interaction between the operators $\widehat{\Psi}[g]$ that describes such a process; analogous perhaps to what is done in string field theory? Any such term would signify a radical departure from the usual field equations of general relativity.

- A normal Fock-space construction involves Bose statistics but it is not clear what it means physically to say that $|g_1, g_2\rangle$ is *symmetric* in g_1 and g_2. The two copies of Σ are disjoint, and presumably no causal connection can be made between them. So what is the operational significance of a Bose structure? The use of Fermi statistics would be even more bizarre!

- Once the original space Σ has been allowed to bifurcate into a pair of copies of itself it seems logical to extend the topology change to include an *arbitrary* final three-manifold. Thus the scope of the theory is increased enormously.

- What is the analogue of the one-particle Hilbert space which is to form the basis for the Fock-space construction? The difficulty is that the problem of time is closely related to the Hilbert space problem, so where do we start? In the case of a particle in the presence of a spacetime-dependent potential, one common way of resolving this issue is to begin with a well-defined free theory with a proper Hilbert space, and then to regard the interactions with the background as a perturbation that can annihilate and create the quanta of this theory. However, this procedure is dubious if the spacetime dependence comes from the spacetime metric γ itself unless there is some sense in which one can usefully write $\gamma_{\alpha\beta}(X)$ as the sum $\eta_{\alpha\beta} + h_{\alpha\beta}(X)$ of the fixed Minkowskian metric $\eta_{\alpha\beta}$ plus a small perturbation $h_{\alpha\beta}(X)$. This is even more inappropriate in the full quantum-gravity theory since there is no obvious way of writing the DeWitt metric (3.3.21) on Riem(Σ) as a sum of this type.

- If a one-particle Hilbert space *can* be constructed, what is the interpretation of states that are tensor products of the vectors in such a space? In the case of particles, if $|x_1\rangle$ and $|x_2\rangle$ are states corresponding to a particle localised at points x_1 and x_2 respectively, then the tensor product $|x_1\rangle|x_2\rangle$ describes a pair of particles, both of which move in the *same* physical three-space. However, if $|g_1\rangle$ and $|g_2\rangle$ are eigenstates of the metric operator, it is not clear to what the product state $|g_1\rangle|g_2\rangle$ refers. The simplest thing might be to say that g_1 and g_2 are both metrics on the same space Σ, but this has no obvious physical interpretation.

2. A more meaningful interpretation is that g_1 and g_2 are metrics on *different* copies of Σ or, equivalently, a single metric on the disjoint union of two copies of Σ. However this also raises a number of problems:

- The transition from a state $|g\rangle$ to a state $|g_1\rangle|g_2\rangle$ corresponds to a topology change in which Σ bifurcates into two copies of itself. But this is unlikely to be compatible with the Wheeler-DeWitt equation. Are we to look for some non-linear interaction between the operators $\widehat{\Psi}[g]$ that describes such a process; analogous perhaps to what is done in string field theory? Any such term would signify a radical departure from the usual field equations of general relativity.

- A normal Fock-space construction involves Bose statistics but it is not clear what it means physically to say that $|g_1, g_2\rangle$ is *symmetric* in g_1 and g_2. The two copies of Σ are disjoint, and presumably no causal connection can be made between them. So what is the operational significance of a Bose structure? The use of Fermi statistics would be even more bizarre!

- Once the original space Σ has been allowed to bifurcate into a pair of copies of itself it seems logical to extend the topology change to include an *arbitrary* final three-manifold. Thus the scope of the theory is increased enormously.

- It is difficult to see how the bifurcation of space helps with the problem of time. Presumably the Wheeler-DeWitt equation will continue to hold in each disconnected piece of the universe, and then the problem of internal time reappears in each.

3. If $\Psi[g]$ becomes a self-adjoint operator it should correspond to some sort of observable. But what can that be, and how could it be measured?

The idea of third quantisation is intriguing and could lead to a radical change in the way in which quantum gravity is perceived. But, in the light of the comments above, it is difficult to see how it can resolve the problem of time.

5.4 The Semiclassical Approximation to Quantum Gravity

5.4.1 The Early Ideas

Studies of the semiclassical approach to quantum gravity date back to the work of Møller (1962) which was based on the idea that a consistent unification of general relativity and quantum theory might not require quantising the gravitational field itself but only the matter to which it couples. The suggested implementation of this scheme was the system of equations

$$G_{\alpha\beta}(X,\gamma] = \langle\psi|T_{\alpha\beta}(X;\gamma,\widehat{\phi}\,]|\psi\rangle \tag{5.4.1}$$

and

$$i\hbar\frac{d\widehat{\phi}}{dt} = [\,\widehat{H}[\gamma],\widehat{\phi}\,] \tag{5.4.2}$$

in which the source of the gravitational field is the expectation value in some state $|\psi\rangle$ of the energy-momentum tensor of the quantised matter ϕ. The time label in the Heisenberg-picture equation of motion (5.4.2) must be related to the time coordinate used in (5.4.1). If ϕ is field, (5.4.2) would probably be replaced with a set of relativistically-covariant field equations. The hope is that the pair of equations (5.4.1–5.4.2) is *exact* and comprises a consistent, complete solution to the problem of quantum gravity.

The early 1980s saw a renewal of interest in this approach, although the results were not conclusive (see Kibble (1981) for a survey of the situation at that time, and Duff (1981) for a general criticism). One problem is that if the matter is chosen to be a quantised field, the right hand side of (5.4.1) can be defined only after the quantum energy-momentum tensor has been regularised and renormalised (Randjbar-Daemi, Kay & Kibble 1980). This procedure has been much studied in the general context of quantum field theory in a curved background spacetime and is widely agreed to be ambiguous if the metric is non-stationary. Counter-terms arise that involve higher powers of the Riemann curvature and which have a significant effect on the Einstein field equations. For example, there have been claims, and counter claims, that the system is intrinsically unstable (Horowitz & Wald 1978, Horowitz 1980, Horowitz & Wald 1980, Horowitz 1981, Suen 1989*a*, Suen 1989*b*, Simon 1991, Fulling 1990).

From the perspective of conventional quantum theory, the coupled equations (5.4.1) and (5.4.2) are rather peculiar. In particular, different states $|\psi_1\rangle$ and $|\psi_2\rangle$ give rise to different background spacetime geometries γ_1 and γ_2, and so it is difficult to make much sense of the superposition principle. The theory is therefore difficult to interpret since none of the standard quantum-mechanical rules are applicable. Another question is the status of different states $|\psi\rangle$ and metrics γ that satisfy the coupled equations (5.4.1) and (5.4.2). Is one particular state to be selected via, for example, some quantum cosmological theory of the initial state? Or do all possible solutions have some physical meaning?

One might wonder if the whole idea of quantising everything but the gravitational field is simply inconsistent—perhaps along the lines of the famous argument in Bohr & Rosenfeld (1933) which showed that the electromagnetic field *has* to be quantised if it is to couple consistently to the current generated by quantised matter. However, as Rosenfeld himself pointed out, there is no direct analogue for gravity since the proof for electromagnetism involves taking to infinity the ratio e/m of the charge e to the inertial mass m of a test particle—a procedure that is impossible in the gravitational case since the analogue of e is the gravitational mass whose ratio to the inertial mass is fixed by the equivalence principle (Rosenfeld 1963). There have been several attempts since then (*e.g.*, Eppley & Hannah (1977), Page & Geilker (1981)) to clarify the situation but it is still somewhat unclear.

5.4.2 The WKB Approximation to Pure Quantum Gravity

The semi-classical approach has reappeared in recent years in the guise of a Born-Oppenheimer/WKB approximation to quantum gravity, and has been particularly discussed in the context of the problem of time. The starting point is no longer the equations (5.4.1–5.4.2) describing a purely classical spacetime metric coupled to quantum matter. Instead, one begins with the full quantum-gravity theory in the form of the Wheeler-DeWitt equation augmented to incorporate the matter degrees of freedom. A solution $\Psi[g,\phi]$ to this equation is then subject to a WKB-type expansion with the aim of showing that the lowest-order term satisfies a functional *Schrödinger* equation with respect to an internal time function that is determined by the state function Ψ. Thus, to this order of approximation, the second-order Wheeler-DeWitt equation is replaced by a first-order Schrödinger equation, and hence by a system that can be given a probabilistic interpretation using the associated inner product. This is therefore a good example of a type-II scheme: the physical interpretation appears only *after* a time variable has been identified in a preliminary quantum theory.

One aim of this approach is to provide a framework that interpolates between quantum gravity proper and the, better understood, subject of quantum field theory in a fixed background spacetime. For example, there have been several discussions of how the original semiclassical equations (5.4.1) and (5.4.2) arise in this framework. However, since these equations (or, rather, their analogues) are now only *approximate*, some of the problems discussed above disappear or, more precisely, appear in a different light.

As far as the general problem of time in quantum gravity is concerned, the main idea can be summarised by saying that 'time' is only a meaningful concept in a quantum state that has some semi-classical component which can serve to define it. Thus time is an *approximate*, semi-classical concept, and its definition depends on the quantum *state* of the system. In particular, time would have no meaning in a quantum cosmology in which the universe never 'emerges' into a semi-classical region. Thus this approach pays some deference to the general idea that time is part of the classical background assumed in the Copenhagen interpretation of quantum theory.

At a technical level, the starting point is a WKB technique for obtaining an approximate solution to the Wheeler-DeWitt equation for pure geometrodynamics (see Singh & Padmanabhan (1989) for a comprehensive review, and Kuchař (1992*b*) for a recent, and very careful, discussion). This involves looking for a solution in the form

$$\Psi[g] = A[g]e^{iS[g]/\hbar\kappa^2} \tag{5.4.3}$$

where $S[g]$ is real, and where $A[g]$ is a positive, real function of g that is 'slowly varying' in the sense that

$$\hbar\kappa^2 \left|\frac{\delta A[g]}{\delta g_{ab}}\right| \ll \left|A[g]\frac{\delta S[g]}{\delta g_{ab}}\right|. \tag{5.4.4}$$

Inserting (5.4.3–5.4.4) into the Wheeler-DeWitt equation shows that, to lowest order in an expansion in powers of $\hbar\kappa^2 \simeq (L_P)^2$ (where $L_P := (G\hbar/c^3)^{\frac{1}{2}}$ is the Planck length), the phase S satisfies the Hamilton-Jacobi equation

$$\mathcal{G}_{ab\,cd}(x,g]\frac{\delta S[g]}{\delta g_{ab}(x)}\frac{\delta S[g]}{\delta g_{cd}(x)} - |g|^{\frac{1}{2}}(x)R(x,g] = 0 \tag{5.4.5}$$

of classical general relativity (the supermomentum constraints $\widehat{\mathcal{H}}_a(x)\Psi = 0$ have the same implication as before). The amplitude factor A obeys the 'conservation law'

$$\mathcal{G}_{ab\,cd}(x,g]\frac{\delta}{\delta g_{ab}(x)}\left(A^2[g]\frac{\delta S[g]}{\delta g_{cd}(x)}\right) = 0. \tag{5.4.6}$$

It should be noted that the precise form of (5.4.5) and (5.4.6) depends on the choice of operator ordering in the Wheeler-DeWitt equation. I have used a simple ordering in (5.1.17) in which the functional derivatives stand to the right of the DeWitt metric, but one might prefer, for example, a version in which the kinetic energy term is formally invariant under transformations of coordinates on Riem(Σ). This will lead to minor changes in the Hamilton-Jacobi equation (5.4.5) and the conservation law (5.4.6) (to illustrate this point see the analogous equations in Kuchař (1992*b*)).

5.4.3 Semiclassical Quantum Gravity and the Problem of Time

The recent surge of interest in the application of WKB methods to the problem of time began with the important work of Banks (1985). The basic idea is a type of Born-Oppenheimer approach and has been developed further in a number of papers; for example Hartle (1986), Zeh (1986), Zeh (1988), Brout (1987), Brout, Horowitz & Wiel (1987), Brout & Venturi (1989), Englert (1989), Halliwell (1987), Singh & Padmanabhan (1989), Padmanabhan (1989*b*), Kiefer & Singh (1991) and Halliwell (1991*a*).

Banks considered a system of matter fields coupled to gravity for which the Wheeler-DeWitt equation for the combined system can be written as (*cf*(5.1.17))

$$-\hbar^2\kappa^2\mathcal{G}_{ab\,cd}(x,g]\frac{\delta^2\Psi}{\delta g_{ab}(x)\,\delta g_{cd}(x)}[g]-\left(\frac{|g|^{\frac{1}{2}}(x)}{\kappa^2}R(x,g]-\widehat{\mathcal{H}}_m(x,g]\right)\Psi[g]=0 \quad (5.4.7)$$

where $\mathcal{H}_m(x,g]$ is the Hamiltonian density for the matter. The next step is to find solutions to (5.4.7) of the special form

$$\Psi[g,\phi]=A[g]\,\Phi[g,\phi]\,e^{iS[g]/\hbar\kappa^2} \quad (5.4.8)$$

where ϕ denotes the collection of matter variables. The function $\Phi[g,\phi]$ is expanded as a power series [34] in Newton's constant G

$$\Phi[g,\phi]=\psi[g,\phi]+\sum_{n=1}^{\infty}G^n\psi_{(n)}[g,\phi] \quad (5.4.9)$$

which, together with the ansatz (5.4.8), is inserted into the Wheeler-DeWitt equation (5.4.7). Keeping just the lowest-order terms in the expansion shows that the phase factor S satisfies the Hamilton-Jacobi equation (5.4.5) as before. Furthermore, the amplitude factor A can be selected to satisfy the conservation equation (5.4.6) (there is clearly some ambiguity in writing the overall amplitude as a product $A[g]\Phi[g,\phi]$). Finally, the lowest-order term $\psi[g,\phi]$ in (5.4.9) satisfies the *first-order* functional differential equation

$$-2i\hbar\mathcal{G}_{ab\,cd}(x,g]\frac{\delta S[g]}{\delta g_{ab}(x)}\frac{\delta\psi[g,\phi]}{\delta g_{cd}(x)}+(\widehat{\mathcal{H}}_m(x,g]\psi)[g,\phi]=0. \quad (5.4.10)$$

[34] As always, an expansion in a dimensioned constant needs to be handled carefully. The physical expansion parameter will be a dimensionless parameter constructed from G and, for example, some energy scale in the theory.

The key step now is to find a functional $\mathcal{T}(x,g]$ such that

$$2\mathcal{G}_{ab\,cd}(x,g]\frac{\delta S[g]}{\delta g_{ab}(x)}\frac{\delta\mathcal{T}(x',g]}{\delta g_{cd}(x)}=\delta(x,x') \tag{5.4.11}$$

which can serve as an intrinsic time functional for the system. This must be augmented by a set $\sigma^R(x,g]$, $R=1\ldots 5$, of functions on Riem(Σ) that are 'comoving' along the flow lines generated by $\mathcal{T}$ in Riem(Σ) in the sense that

$$2\mathcal{G}_{ab\,cd}(x,g]\frac{\delta S[g]}{\delta g_{ab}(x)}\frac{\delta\sigma^R(x',g]}{\delta g_{cd}(x)}=0. \tag{5.4.12}$$

Then, using the functions $(\mathcal{T},\sigma^R)$ as coordinates on Riem(Σ), (5.4.10) becomes the functional Schrödinger equation

$$i\hbar\frac{\delta\psi[\mathcal{T},\sigma,\phi]}{\delta\mathcal{T}(x)}=(\widehat{\mathcal{H}}_m(x;\mathcal{T},\sigma]\psi)[\mathcal{T},\sigma,\phi], \tag{5.4.13}$$

which is the desired result.

There are various things to note about this construction.

1. The equation (5.4.11) shows clearly how the definition of the time function $\mathcal{T}(x,g]$ depends on the quantum state via the choice of the solution S to the Hamilton-Jacobi equation (5.4.5). This approach to the problem of time in quantum gravity can be traced back to the early ideas of DeWitt (1967*a*) and Misner (1992). It was developed further in Lapchinski & Rubakov (1979) and Banks (1985), and formed a central ingredient in the interpretation of quantum cosmology given in Vilenkin (1989).

2. Padmanabhan (1990) and Greensite (1990, 1991a, 1991b) have suggested an extension of the idea above in which time is defined with respect to the phase of *any* solution of the Wheeler-DeWitt equation, not just one that is in WKB form. Their construction can be regarded as a result of requiring quantum gravity to satisfy the Ehrenfest principle (see also Vink (1992) and Squires (1991)).

3. Only the gravitational field is treated in this semiclassical way. The matter fields are fully quantised, although the probability associated with (5.4.13) is conserved only to the same order in the expansion to which (5.4.13) is valid.

4. The scheme can be extended to include genuine quantum fluctuations of the gravitational field itself (*i.e.*, fluctuations around the background metrics associated with the solution to the Hamilton-Jacobi equation (5.4.5)). Typical examples are the work by Halliwell and Hawking on inhomogeneous perturbations of a homogeneous universe (Halliwell & Hawking 1985, Halliwell 1991*a*), and Vilenkin (1989) who writes the spatial metric $g_{ab}(x)$ as the sum of a classical background $g_{ab}^{\text{class}}(x)$ and a quantum part $\widehat{h}_{ab}(x)$.

5. If S is a *real* solution to the Hamilton-Jacobi equation (5.4.5), the function $\Psi[g] = A[g]\, e^{iS[g]/\hbar\kappa^2}$ oscillates rapidly along paths in superspace. However, imaginary S solutions can also exist, and these produce an exponential type of behaviour. This is often interpreted as indicating a Riemannian rather than Lorentzian spacetime picture and has been much studied in quantum cosmology with the idea that the universe tunnels from an 'imaginary-time' region (Halliwell 1987, Halliwell 1990).

5.4.4 The Major Problems

The WKB approach to the definition of time is interesting, but it raises a number of difficult questions and problems. For example:

1. In so far as one starts with the Wheeler-DeWitt equation, the problems discussed earlier in §5.1.5 apply here too. These include singular operator-products, factor ordering and, in particular, the issue of whether or not the Wheeler-DeWitt equation is to be regarded as a genuine *eigenvalue* equation for a set of self-adjoint operators $\widehat{\mathcal{H}}_{\perp}(x)$, $x \in \Sigma$, defined on the Hilbert space that carries the representation of the canonical operators $(\widehat{g}_{ab}(x), \widehat{p}^{cd}(x))$: *i.e.*, the question of the boundary conditions in Riem(Σ) that are to be used in solving the Wheeler-DeWitt equation.

2. It is not obvious that there exist global solutions on superspace to the defining equation (5.4.11) for the internal time. Indeed, the analysis in Hájíček (1986) of a minisuperspace model suggests otherwise. This is related to the *global time problem* that arises in the internal Schrödinger and Klein-Gordon interpretations of time.

3. The internal time $\mathcal{T}$ defined by (5.4.11) is a functional of the metric $g_{ab}(x)$ only. However, as we argued earlier in the context of the Klein-Gordon interpretation, an intrinsic function of this type will not resolve the *spacetime problem* of constructing a scheme that is independent of the initial choice of a reference foliation.

4. The simple WKB approximation breaks down at the turning points of S, and a more careful treatment is needed near such regions.

5. The elegant first-order form of the Schrödinger equation (5.4.10) is lost at the next order in the WKB approximation. This makes it difficult to assess the proper status of (5.4.10) and to understand how the physics of time changes as one gets nearer to the Planck regime.

6. The WKB ansatz (5.4.8) is only one of a very large number of possible types of solution to the Wheeler-DeWitt equation. Why should it have such a preferred status?

   In particular, why should one not consider *superpositions* of WKB solutions to the Wheeler-DeWitt equation? Indeed, in some of the the original discussions it was assumed that the aim was to construct a coherent superposition of such solutions to produce a quantum state that approximates a single classical spacetime manifold (Gerlach 1969). However, if the state vector is a sum

   $$\Psi[g] := \sum_j A_j[g]\, \Phi_j[g,\phi]\, e^{iS_j[g]/\hbar\kappa^2} \tag{5.4.14}$$

   then each solution S_j of the Hamilton-Jacobi equation will lead to its *own* definition of time. A particularly relevant example is a wave function of the form $e^{iS[g]} + e^{-iS[g]}$ which is real, and is therefore a natural type of semiclassical solution to the Hartle-Hawking ansatz for the wave function of the universe. The two parts of this function are usually said to correspond to expanding and contracting universes respectively (this can mean, for example, their behaviour with respect to an extrinsic time variable like $p^a{}_a(x)$), and it is very difficult to see what such a quantum superposition could mean. The situation is reminiscent of the Schrödinger-cat problem in ordinary quantum theory where a single value for a macroscopic property has to be extracted from a quantum state that is a linear superposition of eigenstates.

7. The alternative is to keep just a single WKB function, but this also raises several difficulties:

   (a) As remarked earlier when discussing the Klein-Gordon inner product (5.2.12), the Wheeler-DeWitt equation is real, and therefore naturally admits real solutions. On the other hand, the time-dependent Schrödinger equation (5.4.13) is intrinsically complex because of the $i = \sqrt{-1}$ in the left hand side. As emphasised by Barbour & Smolin (1988) and Barbour (1990), this means that any attempt to derive the latter from the former will necessarily entail the imposition by hand of some correlation between the real and imaginary parts of Ψ. Keeping to just a single WKB function is an example of this, essentially ad hoc, procedure.

   (b) A quantum cosmologist might respond that this is precisely what is to be expected in a theory that describes the quantum creation of the universe via the medium of a unique solution to the Wheeler-DeWitt equation, and the Vilenkin scheme does indeed produce a solution that is naturally complex (Vilenkin 1989). On the other hand, the Hartle-Hawking ansatz leads to a *real* solution of the equation (although this issue is clouded by ambiguities in deciding what the ansatz really means).

(c) With the aid of Wigner functions and other techniques, Halliwell interprets a single WKB solution as describing a whole *family* of classical spacetimes (Halliwell 1987, Halliwell 1990). In effect, these different classical trajectories in superspace are labelled by the $\sigma^R(x, g]$ functions in (5.4.13). But this again raises the 'Schrödinger-cat' problem in the guise of having to decide how any specific spacetime arises.

8. We mentioned earlier that the WKB scheme has been extended to include quantum fluctuations in the metric. The extreme example of such an extension is to keep classical only those gravitational degrees of freedom that can be used to define internal clocks and spatial reference frames. This requires a split of the gravitational modes akin to that employed in the internal Schrödinger interpretation, and similar difficulties are encountered. In general there is a serious difficulty in deciding which modes of the gravitational field are to remain classical and which are to be subject to quantum fluctuations. There is no obvious physical reason for making such a selection.

9. If time is only a semi-classical concept, the notion of probability—as, for example, in the usual interpretation associated with a Schrödinger equation like (5.4.13)—is also likely to be valid only in some approximate sense. At best, this leaves open the question of how to interpolate between the Schrödinger equation (5.4.13) and the starting, and as yet uninterpreted, Wheeler-DeWitt equation (5.4.7); at worse, it throws doubt on the utility of the entire quantum programme.

5.5 Decoherence of WKB Solutions

5.5.1 The Main Idea in Conventional Quantum Theory

Of the problems listed above, the two that are particularly awkward at the conceptual level are:

1. A single $e^{iS[g]}$ corresponds to *many* classical spacetimes.

2. The lack of any *prima facie* reason for excluding a combination $\sum_i A_i[g]\Phi_i[g,\phi]e^{iS_i[g]}$ of WKB solutions, each term of which gives rise to its *own* intrinsic time function satisfying (5.4.11).

There have been a number of recent claims that these, and related, problems can be solved by invoking the notion of *decoherence*. A particularly useful general review is Zurek (1991).

The idea of decoherence has been developed as part of the general investigation into the foundations of quantum theory that has been growing steadily during the last decade. One reason for this activity has been an increasing awareness of the inadequacy of the traditional Copenhagen interpretation of quantum theory, especially the posited dualism between the classical and quantum worlds, the emphasis on measurement as a primary interpretative category, and the associated invocation of 'reduction of the state vector' as a descriptive process that lies outside the deterministic evolution afforded by the Schrödinger equation. Considerations of this type have been enhanced by attempts to develop physical devices (*e.g.*, SQUIDS) that are of macroscopic size but which are nevertheless expected to exhibit genuine quantum properties.

The other major motivation for the renewed interest in the foundations of quantum theory is the many advances that have taken place in cosmology, especially the realisation that the quantum state of the very early universe could have been responsible for the large-scale properties of the universe we see around us today. But the universe itself is the ultimate closed system, and there can be no external observer to make measurements. In particular, the notion of state-vector reduction is an anathema to most people who work in quantum cosmology.

To see how the idea of decoherence arises, consider a quantum-mechanical system $\mathcal{S}$ and an observable A with eigenstates [35] $|a_1\rangle_S, \ldots |a_N\rangle_S$ corresponding to the eigenvalues $a_1 \ldots a_N$ of the associated self-adjoint operator $\widehat{A}$. Any (normalised) state $|\psi\rangle_S$ in the Hilbert space of the system can be expanded as

$$|\psi\rangle_S = \sum_{i=1}^{N} \psi_i |a_i\rangle_S \tag{5.5.1}$$

where the complex expansion-coefficients ψ_i are given by $\psi_i = {}_S\langle a_i|\psi\rangle_S$. The standard interpretation is that if a measurement is made of A, then (i) the result will necessarily be one of the eigenvalues of $\widehat{A}$; and (ii) the probability of getting a particular value a_i is $|\psi_i|^2$. However, prior to the measurement, the state $|\psi\rangle_S$ has no direct ontological interpretation vis-a-vis the observable A .

Such a view of $|\psi\rangle_S$ may be acceptable when applied to sub-atomic systems, but it seems problematic if the system concerned is Schrödinger's unfortunate cat or, even worse, if the components in (5.5.1) represent the different terms in the sum (5.4.14) of WKB solutions to the Wheeler-DeWitt equation. One might then prefer to interpret $|\psi\rangle_S$ as describing, for example, an ensemble of systems in which every element *possesses* a value for A, and the fraction having the value a_i is $|\psi_i|^2$; *i.e.*, an essentially classical probabilistic interpretation of the results of measuring A. However, such a situation is described quantum mechanically by the density matrix

$$\rho_{\text{mix}} = \sum_{i=1}^{N} |\psi_i|^2 \, |a_i\rangle_S \, {}_S\langle a_i| \tag{5.5.2}$$

[35] The number of linearly-independent eigenstates N can be finite or infinite, but for simplicity I have assumed that the spectrum of $\widehat{A}$ is discrete and non-degenerate.

whereas the density matrix associated with the pure state $|\psi\rangle_S$ is

$$\rho_\psi := |\psi\rangle_S\, {}_S\langle\psi| = \sum_{i=1}^{N}\sum_{j=1}^{N} \psi_i \psi_j^* \, |a_i\rangle_S\, {}_S\langle a_j| \tag{5.5.3}$$

which differs from (5.5.2) by off-diagonal terms. Of course, in the conventional interpretation of quantum theory, after the act of measurement the appropriate state *is* the mixed state $\rho_{\rm mix}$, and the transformation

$$|\psi\rangle_S\, {}_S\langle\psi| \rightarrow \rho_{\rm mix} \tag{5.5.4}$$

is what is meant by the 'reduction of the state vector'. Such a transformation can never arise from the effect of a unitary operator and hence, for example, cannot be described as the outcome of applying the Schrödinger equation to the combined system of object plus apparatus. The 'measurement problem' consists in reconciling this statement with the need to regard the constituents of an actual piece of equipment as being quantum mechanical.

The goal of decoherence is to show that there are many situations in which the replacement of the pure state $|\psi\rangle_S$ by the mixed state $\rho_{\rm mix}$ can be understood as a viable consequence of the theory itself *without* the need to invoke measurement acts as a primary concept. As emphasised by Zurek, no actual quantum system is really isolated: there is always some environment $\mathcal{E}$ to which it couples and which, presumably, can also be described quantum mechanically (Zurek 1981, Zurek 1982, Zurek 1983, Zurek 1986, Zurek 1991). Suppose the environment starts in a state $|\phi\rangle_{\mathcal{E}}$ and with a coupling between the system and environment such that the environment states become correlated with those of the system, *i.e.*, $|a_i\rangle_S|\phi\rangle_{\mathcal{E}}$ evolves to $|a_i\rangle_S|\phi_i\rangle_{\mathcal{E}}$ (thus the environment performs an 'ideal measurement' of A). Then, by the superposition principle, if the initial state of $\mathcal{S}$ is the vector $|\psi\rangle_S$ in (5.5.1), the state for the composite system $\mathcal{S}+\mathcal{E}$ will evolve as

$$\left(\sum_{i=1}^{N} \psi_i |a_i\rangle_S\right) |\phi\rangle_{\mathcal{E}} \rightarrow \sum_{i=1}^{N} \psi_i |a_i\rangle_S |\phi_i\rangle_{\mathcal{E}}. \tag{5.5.5}$$

This state is thoroughly entangled. However, if one is interested only in the properties of the system $\mathcal{S}$, the relevant state is the reduced density-matrix ρ_S obtained by summing over (*i.e.*, tracing out) the environment states. The result is

$$\rho_S = \sum_{i=1}^{N}\sum_{j=1}^{N} \psi_i \psi_j^* |a_i\rangle_S\, {}_{\mathcal{E}}\langle\phi_i|\phi_j\rangle_{\mathcal{E}}\, {}_{\mathcal{E}}\langle a_j| \tag{5.5.6}$$

which, if the environment states are approximately orthogonal (*i.e.*, ${}_{\mathcal{E}}\langle\phi_i|\phi_j\rangle_{\mathcal{E}} \simeq \delta_{ij}$), becomes

$$\rho_S \simeq \sum_{i=1}^{N} |\psi_i|^2 \, |a_i\rangle_S\, {}_S\langle a_i|. \tag{5.5.7}$$

Thus the system behaves 'as if' a state reduction has take place to the level of accuracy reflected in the $\simeq$ sign. This is what is meant by decoherence.

The mechanism works because the evolution of the traced-out density matrix satisfies a master equation rather than being induced by a unitary evolution (Caldeira & Leggett 1983, Joos & Zeh 1985). In effect, the environment 'continuously measures' the system, and hence gives rise to a continuous process of state reduction. This, it is claimed, is the reason why Schrödinger's cat is in fact always seen to be either dead or alive, never a superposition of the two. Numerical studies have shown that the reduction of real physical systems can take place very quickly. For example, the gas molecules surrounding a piece of equipment in the laboratory will serve very well, and even the background 3^0K radiation is sufficient to produce the desired effect in a very short time: a typical figure for a macroscopic object is $10^{-23}s$ (Joos & Zeh 1985, Zurek 1986, Unruh & Zurek 1989, Unruh 1991).

Note that the actual collection of states into which the initially pure state collapses (*i.e.*, whether they are eigenstates of this or that particular observable of $\mathcal{S}$) is determined by the coupling of the system to the environment. The essential requirement is that the operator concerned should commute with the interaction Hamiltonian describing the system-environment coupling (Zurek 1981).

5.5.2 Applications to Quantum Cosmology

For the purpose of this paper, the main interest in the above is the possibility of applying these techniques to the semiclassical solutions to the Wheeler-DeWitt equation discussed in the previous section. However, some modification of these ideas is clearly required if they are to be applied in the cosmological context: by definition, the universe in its entirety is the one system which has no environment. In practice, this is done by supposing that certain modes of the gravitational or matter fields can serve as an environment for the rest. For a brief, recent review see Kiefer (1992).

The idea that classical spacetime might 'emerge' in this way was discussed in Joos (1986) and Joos (1987). The first quantum-cosmology calculation seems to have been that of Kiefer (1987) who considered a scalar field coupled to gravity in a situation in which the inhomogeneous modes serve to decohere the homogeneous modes of both fields. Later papers by this author are Kiefer (1989*a*) and Kiefer (1989*b*); for a recent review see Kiefer (1992). Halliwell (1989) considered a homogeneous, DeSitter-space metric coupled to a scalar field. The idea was that some of the inhomogeneous scalar modes can act as the environment for the metric mode. Halliwell showed that decoherence occurred for a single $e^{iS[g]}$ solution and also that, in a Hartle-Hawking semiclassical solution of the form $e^{iS[g]} + e^{-iS[g]}$, the two terms $e^{iS[g]}$ and $e^{-iS[g]}$ decohered by the same mechanism; see also Morikawa (1989). Other relevant work is Mellor (1991), who discusses decoherence in the context of a Klein-Kaluza model, and

Padmanabhan (1989*a*) who studied the possibility of defining decoherence between different three-geometries in Riem(Σ).

The concept of decoherence is very interesting and could be of considerable importance in the context of quantum cosmology and the problem of time. However, several non-trivial problems arise that are peculiar to the use of this concept in quantum gravity and which need more study (see Kuchař (1992*b*) for a further critique). In particular:

1. There does not seem to be any general way of deciding how to separate the modes into those that are to be kept and those that are to serve as an environment.

2. In ordinary quantum theory, the transformation into the mixed state (5.5.6) is only valid so long as the environment modes are deliberately ignored: the 'true' density matrix is the one associated with the pure state (5.5.5) and still contains the off-diagonal interference terms. The transformation from pure to mixed is therefore only 'as if' and, although this may be more than adequate for normal purposes, it is not really clear what is going on in the case of quantum cosmology. There may be situations in which information is genuinely lost: for example down an event horizon, as in the suggestion by Hawking (1982) that the presence of a black hole can transform a pure state into a mixed state, and some of the models discussed in the literature involve a similar process using a cosmological event horizon (*e.g.*, Halliwell (1987)). However, most of the calculations that have been performed are not of this type, and their meaning remains unclear.

3. A rather serious technical problem arises when trying to adapt the ideas of decoherence to quantum gravity. The process of tracing-out modes requires a Hilbert space in which to take the traces. But, as we have seen, the Hilbert space problem for the Wheeler-DeWitt equation is still unsolved except, perhaps, in so far as the semi-classical Schrödinger equation (5.4.13) comes associated with a natural inner product. However, this equation arises only *after* the decision has been made to select a single $e^{iS[g]}$ solution, and so the corresponding inner product cannot be used to perform operator tracing in the calculation of a claimed decoherence effect. In practice, many people resort to the simple inner product (5.2.1) but, as we shall see in §6.1, this brings problems of its own and is hard to justify.

4. This problem is related to various opaque features that arise when decoherence is applied in discussions of the problem of time. In particular, there is a tendency to show (or try to show) that time itself is something that decoheres—in sharp contradistinction to normal quantum theory where decoherence is a *process* that happens *in* time. This is a natural consequence of the fact that time is not an external parameter in quantum gravity but rather something that must be constructed in some internal way. However, it is by no means clear if, or in what sense, time is represented by an operator in quantum gravity as is suggested by the idea that it decoheres. Indeed, this is one of the main distinctions between the internal Schrödinger interpretation

and those interpretations based on the Wheeler-DeWitt equation. This difficult concept of decohering time is deeply connected with the Hilbert space problem for the Wheeler-DeWitt equation, and is an area in which the consistent histories interpretation (to be discussed in §6.3) may have something to offer.

6 TIMELESS INTERPRETATIONS OF QUANTUM GRAVITY

6.1 The Naïve Schrödinger Interpretation

The general philosophy behind all 'timeless' interpretations of quantum gravity is the belief that it should be possible to construct a well-defined quantum formalism without the need to make a specific identification of time at any stage. Approaches of this type invariably employ some sort of 'internal clock' to measure the passage of time, but such a notion of time is understood to be purely phenomenological, and hence of no fundamental conceptual or technical significance. In particular, the choice of time plays no basic role in the construction of the theory. It is fully accepted that such a phenomenological time may only approximate the external time of Newtonian physics and that as a consequence a Schrödinger equation may arise at best as an approximate description of dynamical evolution. However, it is affirmed that the theory nonetheless admits a precise probabilistic interpretation with a well-defined Hilbert space structure.

The simplest example of such a scheme is the 'naïve [36] Schrödinger interpretation' whose central claim is that quantum gravity should be approached by quantising before constraining, and that the physically-correct inner product is (5.2.1)

$$\langle\Psi|\Phi\rangle := \int_{\mathrm{Riem}(\Sigma)} \mathcal{D}g\, \Psi^*[g]\, \Phi[g] \tag{6.1.1}$$

in which the measure $\mathcal{D}g$ is defined [37] on the space $\mathrm{Riem}(\Sigma)$ of Riemannian metrics on the three-manifold Σ.

At a first glance, the use of the scalar product (6.1.1) seems rather natural. After all, $L^2(\mathrm{Riem}(\Sigma), \mathcal{D}g)$ is the Hilbert space on which the canonical operators (5.1.14) and (5.1.15) are formally self-adjoint. Indeed, by starting with the canonical commutation relations (5.1.1–5.1.3) (or their affine generalisations (5.1.6–5.1.8)), the spectral theory associated with the abelian algebra generated by the commuting operators $\widehat{g}_{ab}(x)$ means the (rigorous version of the) scalar product (6.1.1) is bound to enter

[36] The appellation 'naïve' was coined by Unruh & Wald (1989).

[37] But recall my earlier caveats about the need to use distributional metrics, the problem of constructing a proper measure, *etc.*

the theory somewhere. The same Hilbert space is also used (sometimes implicitly) in discussions of whether or not the constraint operators $\widehat{\mathcal{H}}_\perp(x)$ and $\widehat{\mathcal{H}}_a(x)$ are self-adjoint.

The scalar product (6.1.1) defines a large class of square-integrable functions of $g_{ab}(x)$, but many of these are deemed to be unphysical. More precisely, the constraints (5.1.9–5.1.10)

$$\mathcal{H}_a(x;\widehat{g},\widehat{p})\Psi = 0 \tag{6.1.2}$$

$$\mathcal{H}_\perp(x;\widehat{g},\widehat{p})\Psi = 0 \tag{6.1.3}$$

serve to project out the 'physical' subspace of the Hilbert space $L^2(\mathrm{Riem}(\Sigma),\mathcal{D}g)$ of such functions. Of course, (6.1.3) reproduces the Wheeler-DeWitt equation (5.1.17). However, and unlike—for example—in the Klein-Gordon interpretation, the scalar product (6.1.1) is assumed to have a direct physical meaning with no specific reference to the Wheeler-DeWitt equation. This basic interpretation of $\Psi[g]$ is that the probability of 'finding' a hypersurface in $\mathcal{M}$ on which the three-metric g lies in the measurable subset B of Riem(Σ) is

$$\mathrm{Prob}(g \in B; \Psi) = \int_B \mathcal{D}g\, |\Psi[g]|^2. \tag{6.1.4}$$

This interpretation has often been used in studies of quantum cosmology, especially by Hawking and collaborators; for example Hartle & Hawking (1983), Hawking (1984*b*), Hawking (1984*a*), Hawking & Page (1986) and Hawking & Page (1988) (see also Castagnino (1988)). It has some attractive properties, not least of which is its simplicity and the fact that, unlike the Klein-Gordon pairing (5.2.12), (6.1.1) defines a genuine, positive-definite scalar product (modulo mathematical problems in constructing a proper measure theory). This interpretation also gives a clear 'wave-packet' picture of how a classical spacetime arises: one can say that the state $\Psi[g]$ is related to a specific Lorentzian spacetime γ if $\Psi[g]$ vanishes on almost every metric g except those that correspond to the restriction of γ to some spacelike hypersurface.

Notwithstanding these advantages, the naïve Schrödinger interpretation has some peculiar features that stem from the 'timeless' nature of the description. This is illustrated by the example above of how to recover a classical spacetime: it is the whole *spacetime* that is described by Ψ, not just the configuration of the physical variables on a single time slice. In general, a typical application of the naïve Schrödinger interpretation is to pose questions of the type 'What is the probability of finding this or that universe?' rather than questions dealing with this or that *evolution* of the same universe.

The issue becomes clearer if we think more carefully about what is intended by the statement that (6.1.4) is the probability of 'finding' g in some subset of Riem(Σ). By analogy with normal quantum theory, one would expect to talk about 'measuring' the three-metric, but it is hard to see what this means. As we have emphasised several times, measurements are usually made at a single value of a time parameter, and the results of time-ordered sequences of such measurements provide the dynamical evolution of the system. But such language is inappropriate here: the time parameter cannot be fixed since, in effect, it is *part* of the metric $g_{ab}(x)$. We might drop the use of measurement language and adopt a more realist stance by saying that (6.1.4) is the probability of g *being in* the subset B of $\mathrm{Riem}(\Sigma)$, although, to make sense, this probably needs to be augmented with some sort of many-worlds interpretation of the quantum theory.

However, the structure is still peculiar. This is partly because, as it stands, the interpretation given above could apply to *any* function $\Psi[g]$ of the $6 \times \infty^3$ variables $g_{ab}(x)$, $x \in \Sigma$, that are needed to specify a three-metric. The supermomentum constraints $\widehat{\mathcal{H}}_a(x)\Psi = 0$ remove $3 \times \infty^3$ variables [38], which leaves $3 \times \infty^3$. However, to specify a physical configuration of the gravitational field requires $2 \times \infty^3$ variables, and so the functions $\Psi[g]$ depend on an extra $1 \times \infty^3$ variables which, of course, correspond to an internal time function $\mathcal{T}(x, g]$. Thus the time variable is part of the configuration space $\mathrm{Riem}(\Sigma)$ and, in effect, is represented by an operator; that is why we get a timeless interpretation. For example, in the simple minisuperspace model discussed in §5.1.6, the wave function $\psi(\Omega, \phi)$ is interpreted as the probability amplitude for finding a matter configuration ϕ *and* a radius $a = \ln \Omega$. Note that the imposition of the super-Hamiltonian constraint $\widehat{\mathcal{H}}_\perp \Psi = 0$ does not remove this extra configuration variable (as it would if the constraint function $\mathcal{H}_\perp$ was linear, rather than quadratic, in $p^{ab}(x)$) but leads instead to the Wheeler-DeWitt equation on $\Psi[g]$.

The analogue in ordinary wave mechanics would be to interpret $|\psi(x,t)|^2$ as the probability density of 'finding the particle at point x *and* time to be t', in which t is regarded as an eigenvalue of some time operator $\widehat{T}$. The Schrödinger equation then seems to follow from imposing the constraint

$$(\widehat{p}_T + H(t, \widehat{x}, \widehat{p})\psi)(x,t) = 0 \tag{6.1.5}$$

on allowed state vectors with $\widehat{p}_T := i\hbar\partial/\partial t$. However, the conventional interpretation of such a state is that, for *fixed* t, $|\psi(x,t)|^2$ is the probability distribution in x. Thus the naïve Schrödinger interpretation of quantum gravity is based on an idea that involves a significant change in the quantum formalism. Note also that any solution to (6.1.5) will not be square-integrable in x and t, and hence, at best, it is possible to talk about *relative* probabilities only. This arises because the spectrum of the operator $\widehat{p}_T + \widehat{H}$ on $L^2(\mathbb{R}^2, dx\,dt)$ is *continuous*, and hence its eigenstates are not normalisable.

[38]The classical constraint functions $\mathcal{H}_a(x)$ generate the $\mathrm{Diff}(\Sigma)$ action on $\mathrm{Riem}(\Sigma)$ and fibre it into orbits. Then, modulo θ-vacuum effects, the constraints $\widehat{\mathcal{H}}_a(x)\Psi = 0$ can be interpreted as saying that the theory is really defined on the superspace $\mathrm{Riem}(\Sigma)/\mathrm{Diff}(\Sigma)$ of such orbits, in which case the inner product (6.1.1) must be replaced with an integral over $\mathrm{Riem}(\Sigma)/\mathrm{Diff}(\Sigma)$.

A notable property of this construction is that the time operator $\widehat{T}$ does not commute with the constraint operator $\widehat{p}_T + H(t, \widehat{x}, \widehat{p})$. In the quantum gravity case this is reflected in the fact that the projection operator onto a measurable subset B of Riem(Σ) does not commute with $\widehat{\mathcal{H}_\perp}(x)$ (since $[\widehat{g}_{ab}(x'), \widehat{\mathcal{H}_\perp}(x)] \neq 0$). In this sense, the naïve Schrödinger interpretation of probability is inconsistent with the Wheeler-DeWitt equation unless one studiously avoids asking about the state function *after* a hypersurface has been found with a given three-metric. What is at stake here is the crucial question of what is meant by an 'observable' $\widehat{A}$, and what role the concept plays in the construction of the physical Hilbert space. There is no problem if this means only that $[\widehat{A}, \widehat{\mathcal{H}_a}(x)] = 0$: one merely passes to the version of the formalism in which the states are defined on Riem(Σ)/Diff(Σ). The difficulty arises if it is required in addition that $\widehat{A}$ commutes with the super-Hamiltonian operators $\widehat{\mathcal{H}_\perp}(x)$. It does not help to impose the weaker condition that $[\widehat{A}, \widehat{\mathcal{H}_\perp}(x)] = 0$ only on physical states Ψ that satisfy $\widehat{\mathcal{H}_\perp}(x)\Psi = 0$: the operator $\widehat{g}_{ab}(x)$ maps such a state into one that is not annihilated by the super-Hamiltonian operator, and this is incompatible even with the weak condition.

Another problem is that, analogously to the operator $\widehat{p}_T + H(t, \widehat{x}, \widehat{p})$ in (6.1.5), the operators $\widehat{\mathcal{H}_\perp}(x)$, $x \in \Sigma$, defined on the Hilbert space $L^2(\text{Riem}(\Sigma), \mathcal{D}g)$ can be expected to have continuous spectra, and so the solutions to the Wheeler-DeWitt equation are not normalisable. Thus the inner product (6.1.1) does not induce an inner product on the physical states. This is a serious difficulty for the naïve Schrödinger interpretation and is one of the main attractions of the Klein-Gordon programme in which the aim is to define a scalar product only on *solutions* to the Wheeler-DeWitt equation, not on a general function of $g_{ab}(x)$.

6.2 The Conditional Probability Interpretation

6.2.1 The Main Ideas

The conditional probability interpretation is a development of the naïve Schrödinger approach that has been studied especially by Page and Wooters (Page & Wooters 1983, Hawking 1984*a*, Wooters 1984, Page 1986*b*, Page 1986*a*, Page 1989, Page 1991, Page & Hotke-Page 1992); see also Englert (1989), Deutsch (1990), Squires (1991) and Collins & Squires (1992). The Hilbert space is the one used before, *i.e.*, the space of all functionals $\Psi[g]$ that are square-integrable with respect to the inner product (6.1.1). However, the interpretation is different: $|\Psi[g]|^2$ is no longer regarded as the absolute probability density of finding a three-metric g but is instead thought of as the probability of finding the $2 \times \infty^3$ physical modes of g *conditional* on the remaining $1 \times \infty^3$ variables—the internal-time part of g—being equal to some specific function (I am assuming that the constraints $\widehat{\mathcal{H}_a}(x)\Psi = 0$ have already been solved). The claim or hope is that this does not require any specific split of g into physical parts and an internal time function; indeed, the interpretation is supposed to be correct for any such choice. Thus in the minisuperspace model in §5.1.6, the (suitably-normalised) wave-function $|\psi(\Omega,\phi)|^2$ can be regarded equally as the probability density in ϕ at fixed Ω (*i.e.*, regarding Ω as an internal metric-field time) or as the probability density in $\Omega = \ln a$ at fixed ϕ (*i.e.*, regarding ϕ as a matter field time).

The original work of Page and Wooters was not aimed at quantum cosmology alone but at the more general problem of the quantisation of any closed system. They argued, *pace* Bohr, that in the normal Schrödinger equation the time parameter t is an external parameter, and hence has no place in the quantum theory if the system is truly closed. Instead, time must be measured with a physical clock that is part of the system itself. Their interpretation of $|\psi(x,t)|^2$ is that it gives the probability distribution in x conditional on the value of the internal clock being t. The normal Schrödinger time-dependent equation is replaced by an eigenvalue equation

$$\widehat{H}_{\rm tot}\psi = E\psi \tag{6.2.1}$$

where $H_{\rm tot}$ is the Hamiltonian of the total system, which includes the physical clock and its interaction with the rest of the system. Whatever can be said about 'time development' has to be extracted from this equation. This is done by studying the dependence of conditional probabilities on the value of the internal-clock variable on which the probabilities are conditioned.

The relevance for quantum gravity of these ideas should be clear. The eigenvalue equation (6.2.1) becomes the super-Hamiltonian constraints (6.1.3), and the conditioning is on the internal time functional $\mathcal{T}(x,g]$ being equal to a specific function $\tau(x)$; *i.e.*, the internal clock is defined by a configuration of the gravitational field. Of course, if matter fields are present, they also can serve to define an internal time. Note that, as in the naïve Schrödinger interpretation, the internal clock is represented by a genuine operator on the Hilbert space of the total system.

6.2.2 Conditional Probabilities in Conventional Quantum Theory

To understand the novel features of this idea is it useful to recall briefly how the notion of conditional probability enters conventional quantum theory; this will also be helpful in our discussion in §6.3 of the consistent histories interpretation.

Let the (mixed) state of a quantum system at some time $t = 0$ be ρ_0. In the Schrödinger representation, the state ρ_t at time t is related to the $t = 0$ state by the unitary transformation

$$\rho_t = U(t)\,\rho_0\,U(t)^{-1} \tag{6.2.2}$$

where $U(t) := \exp(-it\widehat{H}/\hbar)$. Therefore, if a measurement of an observable A is made at time t_1, the probability that the result will lie in some subset α of the eigenvalue spectrum of the operator $\widehat{A}$ is

$$\begin{aligned}\text{Prob}(A \in \alpha, t_1; \rho_0) &= \text{tr}(P_\alpha^A\, \rho_{t_1}) \\ &= \text{tr}(P_\alpha^A(t_1)\, \rho_0)\end{aligned} \tag{6.2.3}$$

where $P_\alpha^A(t_1)$ is the Heisenberg-picture operator defined by

$$P_\alpha^A(t_1) := U(t_1)^{-1}\, P_\alpha^A\, U(t_1) \tag{6.2.4}$$

(with the reference time chosen to be $t = 0$) and P_α^A is the operator that projects onto the subset α; for example, if the spectrum of $\widehat{A}$ is a set $a_1, \ldots a_N$ of non-degenerate discrete eigenvalues, then

$$P_\alpha^A := \sum_{a_i \in \alpha} |a_i\rangle\langle a_i|. \tag{6.2.5}$$

If the measurement of A yields a result lying in α, any further predictions must be made using the density matrix

$$\rho_\alpha := \frac{P_\alpha^A(t_1)\, \rho_0\, P_\alpha^A(t_1)}{\text{tr}(P_\alpha^A(t_1)\, \rho_0)} \tag{6.2.6}$$

and the transformation

$$\rho_{t_1} \rightarrow \rho_\alpha = \frac{P_\alpha^A(t_1)\, \rho_0\, P_\alpha^A(t_1)}{\text{tr}(P_\alpha^A(t_1)\, \rho_0)} \tag{6.2.7}$$

is the analogue for density matrices of the familiar reduction of the state vector.

Now let the system evolve until time t_2 when a measurement of an observable B is made. According to the discussion above, the probability of finding B in a range β, given that (*i.e.*, conditional on) A was found to be in α at time t_1, is

$$\text{Prob}(B \in \beta, t_2 \,|\, A \in \alpha, t_1; \rho_0) = \text{tr}(P_\beta^B(t_2)\rho_\alpha) = \frac{\text{tr}(P_\beta^B(t_2)\, P_\alpha^A(t_1)\, \rho_0\, P_\alpha^A(t_1))}{\text{tr}(P_\alpha^A(t_1)\, \rho_0)}. \tag{6.2.8}$$

6.2.3 The Timeless Extension

The extension of the ideas above to the situation in which there is no external time parameter proceeds as follows. The physical observables in the theory are regarded as operators that commute with the total Hamiltonian H_{tot}. As a consequence, there is no difference between the Heisenberg and the Schrödinger pictures of time evolution. Indeed, there is no time evolution at all in the sense of a change with respect to any external parameter t; in particular, the density matrix ρ of the system satisfies $[H_{\text{tot}}, \rho] = 0$: a truly 'frozen' formalism. Nevertheless, it is assumed that much of the framework of conventional quantum mechanics is still applicable. In particular, it is deemed meaningful to talk about the conditional probability of finding B in the range β, given that A lies in α, and to assert that, if the state of the system is ρ, the value of this quantity is

$$\operatorname{Prob}(B \in \beta \,|\, A \in \alpha; \rho) = \frac{\operatorname{tr}(P^B_\beta\, P^A_\alpha\, \rho\, P^A_\alpha)}{\operatorname{tr}(P^A_\alpha\, \rho)}. \tag{6.2.9}$$

The extension to the quantum gravity situation is obvious and, again, there is no external time parameter.

The suggested form (6.2.9) should be compared carefully with the expression (6.2.8) of conventional quantum theory. There are no t-labels in (6.2.9) and therefore, in particular, no sense in which the quantities B and A are time-ordered (as they are in (6.2.8), with $t_2 > t_1 > t_0$). Furthermore, and unlike the case in conventional quantum theory, the expression (6.2.9) is not obtained via any process of state reduction. Rather it is simply *postulated* as one of the fundamental interpretative rules of the theory. Correspondingly, the concept of 'measurement' is only a secondary one: like time, it is not something that comes from 'outside' but is instead only a way of talking about a particular type of interaction between certain sub-elements of the closed system. As a consequence we are confronted almost inevitably with a many-worlds interpretation of the theory; indeed, supporters of the conditional probability interpretation of quantum gravity are almost always strong advocates of a post-Everett view of quantum mechanics.

Thus we have a timeless picture of quantum theory. This does not mean that the notion of time-evolution is devoid of any content, but the challenge is to recover it in some way from the conditional probability expression. This is done as follows. Let T be a quantity that we wish to use as an internal clock to measure the change in another quantity A. Then we study the probability $\operatorname{Prob}(A \in \alpha | T = \tau; \rho)$ and see how it varies with τ. This is the dynamical evolution in the theory.

The conditional probability interpretation is certainly attractive. It captures nicely the idea that the passage of time should be identified with correlations inside the system rather than reflecting changes with respect to an external parameter. In this respect it seems well-suited for application to the problem of time in quantum gravity, and it is certainly an improvement on the naïve Schrödinger interpretation. However, this new interpretation gives rise to various problems of its own, and these deserve careful consideration.

1. The central difficulty is that the probabilistic rule (6.2.9) constitutes a significant departure from conventional quantum theory, and the resulting structure may not be self-consistent. In particular, we need clear guidelines for deciding when a variable has the property that it is appropriate to condition on its values. For example, if applied to conventional quantum theory where there *is* an external time parameter t, the formalism makes sense only if T is a 'good' clock in the sense discussed in §2. If T is a bad clock, and hence can take on the same value τ at two different values of t, the probability of A conditional on $T = \tau$ is not well-defined. For a closed system,

this raises the general question of how we know whether or not a particular quantity affords a consistent choice for an internal time variable.

2. Another feature of the conditional probability interpretation is the peculiar lack of any sense of history. That is, there is no way of directly comparing things at different times. All statements are of the form 'the probability of A is this, when B is that' and, in that sense, always refer to the single 'now' at which the statement is made. Page defends the situation by citing the general philosophical position that statements about the past are really counter-factual claims that certain consistency conditions would be met if *present* records are examined. However, not everyone is convinced by this argument; in particular see the critical analysis in Kuchař (1992*b*) and the discussion between Page and Kuchař reported in Page & Hotke-Page (1992)

3. There is also a potential problem with the constraint $\widehat{H}_{\rm tot}\psi = E\psi$ or, in the gravitational case, the super-Hamiltonian constraint $\widehat{\mathcal{H}_\perp}(x)\Psi = 0$. As emphasised in Kuchař (1992*b*), if an internal clock is to function as such, it *cannot* commute with the constraint operator, and—in that sense—it is not a physical observable. In the gravitational context, this means that any derivation of (6.2.9) based on a state reduction with an internal time $\mathcal{T}(x, g]$ taking the value $\tau(x)$,

$$\rho \to P_\tau^{\mathcal{T}}\, \rho\, P_\tau^{\mathcal{T}} / \mathrm{tr}(P_\tau^{\mathcal{T}}\, \rho) \tag{6.2.10}$$

would be incompatible with the Wheeler-DeWitt equation for density matrices (which is $[\mathcal{H}_\perp(x), \rho] = 0$) since the reduced ρ violates this condition. In so far as (6.2.9) is simply *postulated*, this may not be a problem—indeed, expressions of this type are used frequently in the many-worlds interpretation of standard quantum theory without invoking the idea of a collapse—but, together with the absence of any time labels, it does show the extent to which the conditional probability interpretation deviates from conventional quantum ideas.

4. Finally, we still have the *spacetime* problem since in the form above the programme uses an internal time $\mathcal{T}(x, g]$ which, as has been mentioned several times, cannot lead to a local scalar function on the spacetime manifold $\mathcal{M}$. However, this problem might be addressed by applying the conditional probability interpretation to a system that includes the type of matter reference fluids discussed in §4.3.

6.3 The Consistent Histories Interpretation

6.3.1 Preamble

In some respects, the consistent histories approach to the problem of time can be regarded as a development of the conditional probability interpretation, with the particular advantage of enabling questions about the history of the system to be addressed directly. In particular, it takes account of the fact that decoherence is really a *process* that develops in time and that, for example, a system that has decohered could in principle recohere at some later time. The scheme culminates in a suggestion that gravity should be quantised with something like a functional integral over spacetime geometries.

In itself, this does not seem so novel: formal quantisation schemes of this type have been considered for many years and inevitably founder on intractable technical problems. Nor is it obvious how such an approach solves the problem of time, or thows any light on the related question of constructing the Hilbert space of states. For example, one way of defining a real-time functional integral is to start with the canonical Hilbert space quantum theory and then define the integral as the limit of a time-sliced approximation to some matrix element of the unitary evolution operator (*i.e.*, using the Trotter product formula). But this is no help in a situation in which the Hilbert space structure is one of the things we are trying to discover. The rival euclidean approach to quantum gravity does not help much either. For example, in the Hartle-Hawking ansatz the central object is the functional integral over all euclidean-signature metrics γ on a four-manifold $\mathcal{M}$ with a single three-boundary Σ

$$\Psi[g] := \int \mathcal{D}\gamma\, e^{-iS_E[\gamma]/\hbar}, \tag{6.3.1}$$

where S_E is the euclidean action, and where γ restricted to Σ is the given three-metric g (Hartle & Hawking 1983). The function $\Psi[g]$ can be shown formally to satisfy the Wheeler-Dewitt equation (Halliwell 1988, Halliwell 1991*b*), and this particular state is then regarded as the 'wave-function of the universe'. But this does not help with the problem of time since we are simply faced once more with the difficult question of how to interpret solutions of the Wheeler-DeWitt equation.

However, there is far more to the idea of consistent histories than simply a call to return to spacetime functional integrals. It stems from what is in fact a radical revision of the formalism of quantum theory in general. I must admit, when I first came across the idea I was not very enthusiastic. But my feelings have undergone a major change recently and I am inclined now [39] to rate it as one of the most significant developments in quantum theory during the last 25 years.

The consistent histories interpretation is a thorough-going post-Everett scheme in the sense that measurement is not a fundamental category; instead the theory itself prescribes when it is meaningful to say that a measurement has taken place. In particular, there is no external reduction of the state vector. In the original formalism, 'time' appears in the standard way as an external parameter. However, the relegation of measurement to an internal property gives rise to the hope that time may be treated likewise; indeed, for our present purposes, this is one of the main attractions of the approach.

---

[39] This enantiadromia was entirely the result of gentle pressure from Jim Hartle encouraging me to read the original papers and to think about them carefully. I most grateful to him for his efforts in this direction!

6.3.2 Consistent Histories in Conventional Quantum Theory

Let us start by summarising the idea in the context of non-gravitational quantum physics. The seminal papers are Griffiths (1984), Omnès (1988a, 1988b, 1988c, 1989, 1990), Gell-Mann (1987) and Gell-Mann &Hartle (1990a, 1990b, 1990c). Comprehensive recent reviews of both the gravitational and the non-gravitational case are Hartle (1991*a*) and Omnès (1992); see also Alberich (1990, 1991, 1992), Blencowe (1991), Dowker & Halliwell (1992) and Halliwell (1992*b*).

Consider first the description in conventional quantum theory of the process of making a series of measurements separated in time. Each measurement can be regarded as asking a set of questions—the answer to which is either yes or no—and each such question is represented by a hermitian projection operator; typically the projection onto a subset of the spectrum of some operator $\widehat{A}$ (so that the question is 'does the value of A lie in the given subset?'). Let Q denote the set of all questions pertaining to a particular measurement. Then if q is one such question, the associated projection operator will be written P^Q_q. We want the set of questions to be mutually exclusive (*i.e.*, the answer to at most one question is 'yes') and exhaustive (*i.e.*, the answer to at least one question is 'yes'), which means the projection operators must satisfy

$$P^Q_q \, P^Q_{q'} = \delta_{q\,q'} \, P^Q_q \tag{6.3.2}$$

and

$$\sum_{q \in Q} P^Q_q = 1. \tag{6.3.3}$$

Now consider making a series of measurements at times $t_1 < t_2 \ldots < t_N$ with corresponding sets $Q_1, Q_2, \ldots, Q_N$ of yes-no questions. The quantity of interest is the absolute probability $\text{Prob}(q_N \, t_N, q_{N-1} \, t_{N-1}, \ldots, q_1 \, t_1; \rho_0)$ of obtaining 'yes' to questions $q_1 \in Q_1$ at time t_1, $q_2 \in Q_2$ at time t_2, ..., $q_N \in Q_n$ at time t_N, given that the state at time $0 \leq t_1$ was the density matrix ρ_0. The discussion of conditional probabilities in §6.2 leading to (6.2.9) can be extended to show that

$$\text{Prob}(q_N \, t_N, q_{N-1} \, t_{N-1}, \ldots q_1 \, t_1; \rho_0) = \\ \text{tr}(P^{Q_N}_{q_N}(t_N) \, P^{Q_{N-1}}_{q_{N-1}}(t_{N-1}) \ldots P^{Q_1}_{q_1}(t_1) \, \rho_0 \, P^{Q_1}_{q_1}(t_1) \, \ldots P^{Q_{N-1}}_{q_{N-1}}(t_{N-1}) \, P^{Q_N}_{q_N}(t_N)) \tag{6.3.4}$$

where the projection operators are in the Heisenberg picture as defined by (6.2.2).

It must be emphasised that (6.3.4) is derived using conventional ideas of sequences of measurements and associated state-vector reductions like (6.2.7). However, the intention of the consistent histories interpretation of quantum theory is to sidestep this language completely by talking directly about the probability of a history; the idea of 'measurement' is then regarded as a secondary concept that can be described using the history language applied to the entire system (*i.e.*, including what used to be regarded as an observer). For this reason, following the example of John Bell, I shall talk about a hermitian operator $\widehat{A}$ representing a 'beable' rather than an 'observable'. Used in this way, a 'history' means any sequence of projection operators

$$P^{Q_N}_{q_N}(t_N)\, P^{Q_{N-1}}_{q_{N-1}}(t_{N-1}) \ldots P^{Q_1}_{q_1}(t_1) \tag{6.3.5}$$

satisfying the conditions (6.3.2–6.3.3). Note that this is a considerable generalisation of the notion of history as used in a standard path integral where it usually means a path in the configuration space of the system. A history of this particular type can regarded as a limit of a sequence of histories (6.3.5) of a special type in which the projection operators project onto vanishingly small regions of the configuration space, and the separation between time points tends to zero.

The desire to assign probabilities to histories is initially frustrated by the fact that this is precisely what *cannot* be done in conventional quantum theory. All that is possible there is to give a probability *amplitude* for a history, but then the passage to the probability itself introduces interference terms between different histories. This is seen most clearly in the Feynman path-integral approach. The amplitudes for paths $a(t)$, $b(t)$ in configuration space are $A[a] := e^{iS[a]/\hbar}$ and $A[b] := e^{iS[b]/\hbar}$ respectively, where $S[a]$ denotes the classical action evaluated on the path a. But then, generally speaking, $|\,A[a] + A[b]\,|^2 \neq |\,A[a]\,|^2 + |\,A[b]\,|^2$ because of the interference term $|\,A[a]\,A[b]\,|$. The classic example is the two-slit experiment.

The central idea in the consistent histories approach is that, although generic histories cannot be assigned probabilities, this may be possible for certain special families of histories: the so-called 'consistent" families. The key technical ingredient is the *decoherence functional* $D(h', h)$ which is a function of pairs of histories h', h associated to the same collection of questions $Q_1, Q_2, \ldots, Q_N$. If

$$h' := P^{Q_N}_{q'_N}(t_N)\, P^{Q_{N-1}}_{q'_{N-1}}(t_{N-1}) \ldots P^{Q_1}_{q'_1}(t_1) \tag{6.3.6}$$

and

$$h := P^{Q_N}_{q_N}(t_N)\, P^{Q_{N-1}}_{q_{N-1}}(t_{N-1}) \ldots P^{Q_1}_{q_1}(t_1) \tag{6.3.7}$$

then $D(h', h)$ is defined by

$$D(h', h) := \mathrm{tr}(P^{Q_N}_{q'_N}(t_N)\, P^{Q_{N-1}}_{q'_{N-1}}(t_{N-1}) \ldots P^{Q_1}_{q'_1}(t_1)\, \rho_0\, P^{Q_1}_{q_1}(t_1)\, \ldots P^{Q_{N-1}}_{q_{N-1}}(t_{N-1})\, P^{Q_N}_{q_N}(t_N)) \tag{6.3.8}$$

which provides a good measure of the size of the interference terms between the two histories. [40] The family of histories is said to be *consistent* if $D(h', h) = 0$ for all pairs h, h' for which $h \neq h'$. If this is so, we *assign* the probability to h given by (6.3.4). Thus, in this approach to quantum theory, the fundamental interpretative rule is

$$D(h', h) = \delta_{h'h}\, \mathrm{Prob}(q_N\, t_N, q_{N-1}\, t_{N-1}, \ldots q_1\, t_1; \rho_0) \tag{6.3.9}$$

[40] This is most easily seen by considering the special case where ρ is a pure state $|\psi\rangle\langle\psi|$.

where $\text{Prob}(q_N\, t_N, q_{N-1}\, t_{N-1}, \ldots q_1\, t_1; \rho_0)$ is *defined* by (6.3.4); but note again that this probability is assigned to the history *only* if the consistency conditions (6.3.9) are satisfied [41] for all histories in the family under consideration. It is straightforward to show that probabilities arrived at in this way obey all the basic rules of classical probability theory. It must be emphasised that the decoherence functional is *computed* using the mathematical techniques of standard quantum theory: it is only the probability *interpretation* that is new.

The main task is to find families of consistent histories. In practice, one may decide that exact consistency is not needed: it may be sufficient if (6.3.9) is *approximately* true [42], although the degree of approximation that is deemed appropriate will depend on the physical situation involved. However, even with this weakened requirement most families of histories will not satisfy the consistency condition. The most discriminating sets of projection operators $\{P_q^Q \mid q \in Q\}$ are those in which each operator P_q^Q projects onto a one-dimensional range. This would happen if $\{P_q^Q \mid q \in Q\}$ is the set of spectral projection operators for a complete set of commuting 'beables' with discrete eigenvalue spectra. Families of histories associated with collections $Q_1, Q_2, \ldots, Q_N$ of sets Q_i of questions of this type are least likely to be consistent. To gain consistency starting with such a family it will be necessary to *coarse-grain* the histories—another important concept in the general programme.

To coarse-grain a set Q of questions means to partition Q into subsets of less precise questions. If $\bar{Q}$ denotes the new set of questions, and if $\bar{q}$ is one of the partitions, then the projection operator corresponding to the new question $\bar{q}$ ('do any of the questions in the set $\{q \in \bar{q} \subset Q\}$ have the answer 'yes'?') is

$$P_{\bar{q}}^{\bar{Q}} = \sum_{q \in \bar{q}} P_q^Q, \tag{6.3.10}$$

and the associated decoherence functional is

$$D(\bar{h}', \bar{h}) = \sum_{h' \in \bar{h}'} \sum_{h \in \bar{h}} D(h', h). \tag{6.3.11}$$

The idea is that with an appropriate coarse-graining this new set of histories may be consistent. One extreme act of coarse-graining is to choose a single partition for one of the questions, at time t_j say. This results in the trivial question whose answer is always 'yes' and is represented by the unit operator; in effect that particular time t_j is removed from the sequence. Of course, there is a converse in which a family of histories can be 'fine-grained' by inserting a set of questions at a time intermediate between a consecutive pair in the original family.

The conventional interpretation of quantum physics can be recovered using the idea of a 'quasi-classical domain'. A quantum theory is said to have a quasi-classical

[41] In his original paper, Griffiths showed that it is sufficient if the *real* part of the off-diagonal parts of $D(h', h)$ vanish.

[42] This raises the intriguing notion of 'approximate probabilities'.

domain if there exists a consistent family of histories with the property that the values of certain, sufficiently coarse-grained 'beables' are correlated in time in a way that reproduces the equations of some piece of classical physics. Such variables could include the coarse-grained features of actual pieces of measuring equipment, with the histories involved describing, for example, the production of persistent records: a property that has frequently been seen as the signature of a successful 'measurement'.

It must be emphasised that consistency is a property of a complete *family* of histories. Many different such families may exist, giving different perspectives on the picture of reality portrayed by quantum theory. Properties like complementarity arise from the existence of families that are mutually incompatible. In a situation like this, a 'many-worlds' (or, 'many-histories') interpretation of quantum theory seems inevitable. But note that the 'many histories' involved come not only from the different histories associated with a fixed collection of questions $Q_1, Q_2, \ldots Q_N$; the collections of questions themselves are also variable—any collection leading to a consistent family of histories is admissible.

6.3.3 The Application to Quantum Gravity

The consistency condition (6.3.9) depends on the state ρ_0; in particular, this is true of the existence of quasi-classical domains. As emphasised by Gell-Mann and Hartle, this implies that the manifest existence in our current world of a quasi-classical domain depends ultimately on the initial state ρ_0 that existed shortly after the 'initial' big-bang. From this perspective, the classical features of our present-day world must be seen as a contingent property of the big-bang: they could have been otherwise. Indeed, for all we know, the quantum theory of our universe may admit other consistent families of histories with no quasi-classical domains at all; a valid concept in the context of a post-Everett interpretation of quantum theory.

Many discussions of this type can be carried out within the framework of a non-quantised background metric. However, problems arise when we come to quantum gravity itself. The very concept of a 'history' rests on the notion of a time parameter and, as we have seen, this is an elusive entity. Gell-Mann and Hartle address this problem by proposing an extension of the formalism in which the notion of 'history' becomes a primary one with no *a priori* reference to sequences of questions or beables ordered in any external time. The basic ingredients are:

1. families of mathematical objects called 'histories';

2. a notion of 'coarse-graining' whereby families of histories are partitioned into exclusive and exhaustive sub-families;

3. a 'decoherence functional' $D(h', h)$ defined on pairs of histories.

The decoherence functional must have the following properties:

- *Hermiticity*: $D(h', h) = D^*(h, h')$
- *Positivity*: $D(h, h) \geq 0$
- *Normalisation*: $\sum_{h',h} D(h', h) = 1$
- *The principle of superposition*: $D(\bar{h}', \bar{h}) := \sum_{h' \in \bar{h}'} \sum_{h \in \bar{h}} D(h', h)$ where $\bar{h}'$ and $\bar{h}$ are coarse-grained histories.

A particular family of histories is said to be *consistent* if $D(h', h) = 0$ unless $h' = h$, and then a probability $\mathrm{Prob}(h)$ is assigned to each member of such a family by the rule

$$D(h', h) = \delta_{h'h} \, \mathrm{Prob}(h). \tag{6.3.12}$$

As before, the strict equality might be replaced with an approximate equality where the approximation reflects the physical situation to which the formalism is applied.

These rules constitute the entire theory. In particular, there is no *prima facie* Hilbert space structure, although a 'phenomenological' one may 'emerge' in some domains of the theory. However, this absence of the standard mathematical formalism can cause problems when trying to implement the scheme. For example, finite or countably-infinite sums are used in (6.3.2–6.3.3) because the underlying Hilbert space is assumed to be separable. But it is not obvious that sums are sufficient in the absence of any such structure. Some of the collections of histories could well be non-countably infinite, which suggests that integrals are more appropriate, and this is likely to produce major technical problems. It also places in doubt the notion of a 'most-discriminative' set of histories from which all others can be obtained by coarse-graining. This happens already in the Hilbert space theory for an operator with a continuous spectrum, but the spectral theory for such operators enables one to avoid using integrals and to keep to the well-defined sums in (6.3.2–6.3.3). As we shall see, this problem is relevant to the application of these ideas in quantum gravity.

Several attempts have been made to apply this generalised formalism to general relativity. The first involves an extension of the conventional sum-over-histories formalism to situations in which, although there are paths in the configuration space, the theory is invariant under reparametrisations of these paths; in this sense the paths are a generalised form of a 'history'. Theories of this type were studied in some depth by Teitelboim (1982, 1983a, 1983b, 1983c, 1983d), but their development in the context of the consistent histories formalism has been mainly at the hands of Hartle who has emphasised the significance of the existence of many types of spacetime-oriented coarse-graining that have no analogue in a conventional Hamiltonian quantum theory (Hartle 1988*a*, Hartle 1991*b*); in particular, there may be no notion of a state being associated with a *spacelike* hypersurface of spacetime. Hartle (1988*b*) has also shown in a simple model how the notion of time, and conventional Hamiltonian quantum mechanics, can emerge from the formalism as a reading on a physical clock—the same general philosophy that underlies the conditional probability interpretation discussed earlier. These path-integral constructions are the subject of a careful critique in Kuchař (1992*b*).

More recently, Hartle (1991*a*) has proposed that the generalised consistent histories interpretation be extended to quantum gravity by defining a (most-discriminative) 'history' to be a Lorentzian metric γ on the spacetime manifold $\mathcal{M}$ plus a specification of the values of a set of spacetime fields ϕ. The decoherence functional is then defined as

$$D(h',h) := \int_{h'} \mathcal{D}\gamma'\,\mathcal{D}\phi' \int_{h} \mathcal{D}\gamma\,\mathcal{D}\phi\, e^{i(S[\gamma',\phi']-S[\gamma,\phi])/\hbar} \qquad (6.3.13)$$

where $S[\gamma,\phi]$ is the classical action, and where the integral is over the constituents of the coarse-grained histories h and h'. The motivation for this expression is that the analogous object in the path-integral version of normal Hamiltonian quantum mechanics is the correct choice to reproduce the conventional theory. Note that (6.3.13) apparently contains no reference to a state ρ. However, if desired, this can be thought of as boundary conditions that could be imposed on the spacetime geometries and matter field configurations appearing in the integrals. In particular, quantum cosmological considerations are coded into the behaviour of these fields near the big-bang region.

The development of the theory now proceeds as discussed earlier. Thus one seeks consistent families of such histories from which the probabilistic interpretation can be extracted. In particular, the problem of time reduces to studying the classical correlations between the various variables, including actual physical clocks, in a quasi-classical domain. Hence the view taken of 'time' is essentially the same as in the conditional probability interpretation, but the structural framework of the theory is better defined.

6.3.4 Problems With the Formalism

The consistent histories approach has many attractive features, but also some difficult problems and challenges that need to be taken seriously.

1. The expression (6.3.13) illustrates the problem mentioned earlier about the need to use integrals rather than sums. Functional integrals can provide valuable heuristic insights into the structure of a would-be quantum theory, but they are rarely well-defined mathematically. On the contrary, in the case of general relativity the theory is known to be perturbatively non-renormalisable, and hence the chances of making proper mathematical sense of (6.3.13) are not high. One might adopt a semi-classical approximation (*e.g.*, Kiefer (1991), Hartle (1991*a*)) but this is not terribly satisfactory given the lack of the proper theory that is supposedly being

approximated. In particular, in the absence of a non-perturbative evaluation, the functional integral (6.3.13) is at best a low-energy phenomenological description that must be cut-off at energies where the effects of the more basic theory (superstrings?) become significant. This may well be the best way of justifying the use of the WKB approach (via a saddle-point approximation to the functional integral), but it leaves unanswered the question of what happens at the Planck length and, in particular, the problem of time at that scale.

2. Even at a formal level, there is a problem attached to (6.3.13) that arises from the presumed Diff($\mathcal{M}$) invariance of the theory. This could be implemented by requiring the elements being integrated over to be Diff($\mathcal{M}$)-invariant equivalence classes of fields, but it is notoriously difficult to construct the Diff($\mathcal{M}$)-invariant measures needed to facilitate this process. The conventional, heuristic approach is to choose a gauge, define the gauge-fixed functional integrals in the standard way, and then to show that the theory is independent of the choice of gauge. However, in the case of gravity, fixing a gauge means making a choice of internal time *etc*, and then we must confront once more all the problems discussed in earlier sections. In other words, to construct the decoherence functional it may be necessary first to solve the problem of time, and so we are in danger of going round in circles. Thus further study is needed into the possibility of performing a functional integral like (6.3.13) *without* having to invoke a conventional Hilbert-space formalism. [43] Indeed, if something like (6.3.13) *could* be defined properly, it would be consistent with the general Gell-Mann-Hartle philosophy to expect the conventional, Hilbert-space structure to emerge only in some coarse-grained limit of the theory.

3. A major challenge is to find what type of coarse-graining is needed to produce a consistent family of histories using the spacetime fields γ and ϕ. At the very least, we need families that are consistent up to the approximations that may be inherent in pretending that (6.3.13) is a fundamental expression rather than a phenomenological reflection of a more basic theory.

4. This issue is connected to one of the major questions of quantum cosmology: 'What types of consistent families of histories give rise to a quasi-classical domain, and how is this related to the conditions in the early universe?' (*e.g.*, Gell-Mann & Hartle (1992)). A related issue is the extent to which the *only* relevant Planck-length era is that of the very early universe. More precisely, what physics would we find at the Planck scale if we could probe it here and now? The question is whether spacetime has some type of foam structure, and if so how this affects, or is reflected in, the consistent histories approach to quantum gravity.

Let me conclude by reaffirming my belief that the Gell-Mann-Hartle axioms constitute a significant generalisation of quantum theory. Their suggested implementation via the decoherence functional (6.3.13) represents a rather conservative approach to

[43] Of course, this does not rule out the possibility that the functional integral may be defined using *some* Hilbert space, but one that is not that of the conventional Hamiltonian formalism.

quantum gravity and runs into the difficulties mentioned above. But one can imagine more radical attempts involving, for example, some notion of generalised causal sets. The entire scheme certainly deserves very serious further study.

6.4 The Frozen Formalism: Evolving Constants of Motion

Rovelli has advocated recently an interesting approach to the problem of time that shares the central philosophy of the other 'timeless' schemes discussed earlier (Rovelli 1990, 1991a, 1991b, 1991c). Thus the main claim is that it is possible to construct a coherent quantum gravity scheme— including a probabilistic interpretation—without making any specific identification of time, which will rather emerge as a phenomenological concept associated with physical clocks and internal time variables.

Rovelli's starting point is his affirmation that, in the canonical version of classical general relativity, an *observable* is any functional $A[g,p]$ of the canonical variables $(g_{ab}(x), p^{cd}(x))$ whose Poisson bracket (computed using the basic relations (3.3.27–3.3.29)) with all the constraint functions vanishes:

$$\{A, \mathcal{H}_a(x)\} = 0 \quad (6.4.1)$$
$$\{A, \mathcal{H}_\perp(x)\} = 0. \quad (6.4.2)$$

Properly speaking, it is probably more correct to require these Poisson brackets to vanish only *weakly* (as in the right hand side of (3.3.40)), but this point is not addressed in the original papers and I shall not go into it here (it is not of any great significance).

Since the Hamiltonian (3.3.23) for the canonical theory is $H[N, \vec{N}] := \int_\Sigma d^3x\,(\vec{N}\mathcal{H}_\perp + N^a\mathcal{H}_a)$, these conditions imply that

$$\frac{dA}{dt}(g(t), p(t)) = 0. \quad (6.4.3)$$

Thus, as emphasised in §3.3.4, an observable is automatically a constant of motion with respect to evolution along the foliation associated with any choice of lapse function N and shift vector $\vec{N}$. This is the 'frozen formalism' of classical, canonical general relativity.

There are two different approaches to the construction of the quantum theory of this system. The first uses the group-theoretical scheme advocated inIsham (1984) with the aim of finding a self-adjoint operator representation of the classical Poisson-bracket algebra of all observables (or, perhaps, some selected subset of them) obeying (6.4.1–6.4.2). The feasibility of adopting such an approach lies in the observation that if $A[g,p]$ and $B[g,p]$ are a pair of functions which satisfy (6.4.1–6.4.2) then the Jacobi identity implies that $\{A, B\}$ also satisfies these conditions. Thus the set of all observables is closed under the Poisson bracket operation. If the resulting algebra is a genuine Lie algebra, a self-adjoint operator representation can be found by looking for unitary representations of the associated Lie group. Note that, by constructing the physical Hilbert space in this way one arrives at a probabilistic interpretation without making any specific identification of time.

Unfortunately, sets of observables that generate a true Lie algebra are rather rare, and the algebra seems more likely to be one in which the coefficient of the Poisson bracket of two generators is a non-trivial function of the canonical variables. It is difficult to find self-adjoint representations of algebras of this type because of awkward problems involving the ordering of the generators and their q-number coefficients. Certainly, no one has succeeded in constructing a proper quantum gravity scheme in this way, although this is partly due to the difficulty in finding *classical* functions that satisfy (6.4.1–6.4.2). An interesting model calculation is given in Rovelli (1990) (but note the criticism in Hájíček (1991) and the response of Rovelli (1991*c*)). For a comprehensive analysis of schemes of this type applied to finite-dimensional examples see Tate (1992).

An alternative approach is to start with the scheme employed in §5 which is based on an operator representation of the canonical commutation relations (5.1.1–5.1.3) (or their affine generalisation (5.1.6–5.1.8)) on some Hilbert space $\mathcal{H}$. The physical state space $\mathcal{H}_{\rm phys}$ is deemed to be all vectors in $\mathcal{H}$ that satisfy the operator constraints (5.1.9–5.1.10)

$$\mathcal{H}_a(x;\widehat{g},\widehat{p}]\Psi = 0 \tag{6.4.4}$$

$$\mathcal{H}_\perp(x;\widehat{g},\widehat{p}]\Psi = 0, \tag{6.4.5}$$

and a physical observable is then defined to be any operator $A[\widehat{g},\widehat{p}]$ that satisfies the operator analogue of (6.4.1–6.4.2)

$$[\widehat{A},\widehat{\mathcal{H}_\alpha}(x)] = 0 \tag{6.4.6}$$

$$[\widehat{A},\widehat{\mathcal{H}_\perp}(x)] = 0. \tag{6.4.7}$$

These equations are compatible with (6.4.4–6.4.5) in the sense that any operator satisfying them maps the physical subspace $\mathcal{H}_{\rm phys}$ into itself. A weaker version of (6.4.6–6.4.7) is to require the commutators to vanish only on the physical subspace.

The next step is to place a suitable scalar product on $\mathcal{H}_{\rm phys}$. As emphasised earlier, this cannot be done simply by regarding $\mathcal{H}_{\rm phys}$ as a subspace of the original Hilbert space $\mathcal{H}$: the continuous nature of the spectra of the constraint operators means that the vectors in $\mathcal{H}_{\rm phys}$ all have an infinite $\mathcal{H}$-norm. It is by no means clear how to set about finding the correct scalar product but presumably a minimal requirement is that the physical observables satisfying (6.4.6–6.4.7) should be self-adjoint in the new Hilbert space structure.

This issue raises the general question of how physical observables are actually to be constructed (this is also very relevant for the first approach). Rovelli claims that a particularly important class is formed by the so-called 'evolving constants of motion', which serve also to introduce some notion of dynamical evolution. The basic idea is best explained in a simple model with a single super-Hamiltonian constraint $H(q,p)$ defined on a finite-dimensional phase space $\mathcal{S}$. A classical physical observable is then defined to be any function $A(q,p)$ such that $\{A,H\}=0$ (or, perhaps, $\{A,H\}\approx 0$). The next step is to introduce some internal time function $\mathcal{T}(q,p)$ with the property that, for any $t\in\mathbb{R}$, the hypersurface

$$\mathcal{S}_t := \{(q,p)\in\mathcal{S}|\mathcal{T}(q,p)=t\} \tag{6.4.8}$$

intersects each dynamical trajectory (on the constraint surface) generated by H once and only once (of course, there may be global obstructions to finding such a function). Note that this requirement means that $\{\mathcal{T},H\}\neq 0$, and hence the internal time function $\mathcal{T}$ is *not* a physical observable in the sense above.

The key idea is to associate with each function F on $\mathcal{S}$ a one-parameter family of observables (*i.e.*, constants of motion) F_t, $t\in\mathbb{R}$, defined by the two conditions

$$\{F_t,H\} \;=\; 0 \tag{6.4.9}$$

$$F_t|_{\mathcal{S}_t} \;=\; F|_{\mathcal{S}_t} \tag{6.4.10}$$

i.e., the observable F_t is equal to F on the subspace $\mathcal{S}_t$ of the phase space $\mathcal{S}$. Dynamical evolution with respect to the internal time is then described by saying how the *family* of observables F_t, $t\in\mathbb{R}$, depends on t. This is therefore a classical analogue of the Heisenberg picture of time development in quantum theory.

A direct Poisson-bracket calculation shows that

$$\{\mathcal{T},H\}\frac{dF_t(q,p)}{dt}=\{F,H\}. \tag{6.4.11}$$

Note that if $\mathcal{T}$ is a 'perfect Hamiltonian clock' then, by definition, we have

$$H=p_{\mathcal{T}}+h \tag{6.4.12}$$

where the clock Hamiltonian is $p_{\mathcal{T}}$—the conjugate to the internal time function, so that $\{\mathcal{T},H\}=1$—and the Hamiltonian h describing the rest of the system is independent of $p_{\mathcal{T}}$. It follows that $\{F,h\}=\{F_t,h\}$, so that (6.4.11) becomes

$$\frac{dF_t(q,p)}{dt}=\{F_t,h\}, \tag{6.4.13}$$

which is the usual equation of motion. Thus (6.4.11) can be viewed as a *bona fide* generalisation of conventional mechanics to the situation where the only time variable is an internal one.

Rovelli's suggestion is that these evolving constants of motion should form the basis for a quantisation of the system. Thus, in the group-theoretical approach, the key algebra to be represented is the Poisson-bracket algebra generated by the classical quantities F_t, $t \in \mathbb{R}$. The main problem here will be to decide whether or not the set of all such objects forms a genuine Lie algebra. If it does, a unitary representation of the associated Lie group will yield the desired quantum observables. If—as seems more likely—it forms only a function algebra (*i.e.*, with q-number coefficients), it will be necessary to think again about how to find self-adjoint operator representations.

In the alternative, constraint-quantisation approach one needs operator equivalents of the defining equations (6.4.9–6.4.10). The hope is that the inner product on the physical states $\mathcal{H}_{\rm phys}$ can then be determined by the requirement that all operators $\widehat{F}_t$ are self-adjoint. A number of severe technical problems arise in this version of the programme and are articulated in Kuchař (1992*b*). For example:

- The operator form of (6.4.9) is ill-defined and ambiguous. Neither is it clear that, even if they could be defined properly, the conditions (6.4.9–6.4.10) are sufficient to yield a unique operator $\widehat{F}_t$ from a given operator $\widehat{F}$. This particular problem can be avoided by starting with the classical versions F_t, which *are* well-defined, and then trying to make them into operators. But severe operator-ordering problems will inevitably enter at this point and are likely to be intractable. This is because the classical object F_t can be obtained only by *solving* the classical equations of motion, and hence it is likely to be, at best, an implicit function of the starting function F.

- The global time problem means that no globally-defined internal time function exists. In this circumstance there is a good case for arguing that the associated operators $\widehat{F}_t$ should *not* be self-adjoint. This is the basis of the objection to Rovelli's procedure in Hájíček (1991).

- It seems most unlikely that a single Hilbert space can be used for all possible choices of an internal time function $\mathcal{T}$. Thus the multiple choice and Hilbert space problems appear once more.

These are real difficulties and need to be taken seriously. However, they are no worse than those that arise in any of the other approaches to the problem of time and Rovelli's scheme deserves serious attention, not least because it emphasises once again the importance of the still-debated question of what is to be regarded as an observable in a quantum theory of gravity.

7 CONCLUSIONS

We have discussed three main ways of approaching the central question of how time should be introduced into a quantum theory of gravity. In theories of type I, time is defined internally at a classical level: a procedure that is associated with the removal of all redundant variables before quantisation and which culminates in the production of a standard Schrödinger time-evolution equation for the physical modes of the gravitational and matter fields. This approach is relatively uncontroversial at a conceptual level but it runs into severe technical problems including obstructions to global existence, and local non-uniqueness. It also seems rather *ad hoc* and it is aesthetically unattractive.

In approaches of type II, all the canonical variables are quantised and the constraints are imposed at the quantum level *à la* Dirac as constraints on allowed state vectors. Unfortunately, there is no universally-agreed way of interpreting the ensuing Wheeler-DeWitt equation; certainly none of the ideas produced so far is satisfactory. However, it must be emphasised that there is no real justification for extending the Dirac approach to constraint generators that are *quadratic* functions of the momentum variables. Therefore, although it may be heretical to suggest it, the Wheeler-DeWitt equation—elegant though it be—may be completely the wrong way of formulating a quantum theory of gravity.

Approaches of type III differ from types I and II in ascribing to the concept of 'time' only a secondary, phenomenological status: a move that is inevitably associated with some change in the quantum formalism itself. Techniques of this sort are particularly well-suited for handling the deep philosophical issues that arise in quantum cosmology when quantum theory is applied to the universe as a whole. To my mind, the consistent-histories approach is the most far-reaching in its implications, but it needs further development, especially in the direction of finding a more adventurous definition of what is meant by a 'history' in the context of quantum gravity.

Let me emphasise once more that most of the problems of time in quantum gravity are *not* associated with the existence of ultraviolet divergences in the weak-field perturbative quantisation; in particular, many interpretative difficulties arise already in infinity-free, minisuperspace models. Therefore, I feel it is correct to say that the problems encountered in unravelling the concept of time in quantum gravity are grounded in a fundamental inconsistency between the basic conceptual frameworks of quantum theory and general relativity. In responding to this situation the main task is to decide whether 'time' should preserve the basic role it plays in classical general relativity—something that is most naturally achieved by incorporating it into the quantum formalism by the application of a quantization algorithm to the classical theory—or if it is a concept that should emerge phenomenologicaly from a theoretical framework based on something very different from 'quantising' classical general relativity.

If the former is true, which suggests a type I approach to the problem, the best bet could be some 'natural' choice of internal time dictated by the technical requirements of mathematical consistency in a quantisation scheme; for example the programme currently being pursued by Abhay Ashtekar and collaborators.

If the latter is true, two key questions arise: (i) what is this new framework?, and (ii) how, if at all, does it relate to the existing approaches to quantum gravity, especially the semi-classical scheme? In particular, how does the framework yield conventional quantum theory and our normal ideas of space and time in their appropriate domains?

The most widely-studied scheme of this sort is superstring theory but, in its current manifestation, this is not well-suited for addressing these basic questions. The idea of strings moving in a spacetime already presupposes a great deal about the structure of space and time; and the quantisation techniques employed presuppose most of structure of standard quantum theory, particularly at a conceptual level. It may well be that a new, non-perturbative approach to superstring theory will involve a radical reappraisal of the ideas of space, time and quantum theory; but this remains a task for the future. Perhaps the answer is to find a superstring version of Ashtekar's formalism (or an Ashtekarisation of superstring theory), and with the conceptual aspects of quantum theory being handled by a consistent-histories formalism. A nice challenge for the next few years!

Acknowledgements

I would like to reiterate my remarks in the preamble concerning my great indebtedness to Karel Kuchař for sharing his ideas with me. I have also enjoyed recent fruitful discussions and correspondence on the problem of time with Julian Barbour, Jim Hartle and Ranjeet Tate. Finally, I would like to thank the organisers of the Advanced Study Institute for their kindness and friendship during the course of a very pleasant meeting.

References

Albrecht, A. (1990), 'Identifying decohering paths in closed quantum systems'. preprint.

Albrecht, A. (1991), 'Investigating decoherence in a simple system'. preprint.

Albrecht, A. (1992), Two perspectives on a decohering spin, *in* J. Halliwell, J. Perez-Mercader & W. Zurek, eds, 'Physical Origins of Time Asymmetry', Cambridge University Press, Cambridge.

Alvarez, E. (1989), 'Quantum gravity: an introduction to some recent results', *Rev. Mod. Phys.* **61**, 561–604.

Arnowitt, R., Deser, S. & Misner, C. (1959*a*), 'Dynamical structure and definition of energy in general relativity', *Phys. Rev.* **116**, 1322–1330.

Arnowitt, R., Deser, S. & Misner, C. (1959*b*), 'Quantum theory of gravitation: General formalism and linearized theory', *Phys. Rev.* **113**, 745–750.

Arnowitt, R., Deser, S. & Misner, C. (1960*a*), 'Canonical variables for general relativity', *Phys. Rev.* **117**, 1595–1602.

Arnowitt, R., Deser, S. & Misner, C. (1960*b*), 'Consistency of the canonical reduction of general relativity', *J. Math. Phys.* **1**, 434–439.

Arnowitt, R., Deser, S. & Misner, C. (1960*c*), 'Energy and the criteria for radiation in general relativity', *Phys. Rev.* **118**, 1100–1104.

Arnowitt, R., Deser, S. & Misner, C. (1960*d*), 'Finite self-energy of classical point particles', *Phys. Rev. Lett.* **4**, 375–377.

Arnowitt, R., Deser, S. & Misner, C. (1961*a*), 'Coordinate invariance and energy expressions in general relativity', *Phys. Rev.* **122**, 997–1006.

Arnowitt, R., Deser, S. & Misner, C. (1961*b*), 'Wave zone in general relativity', *Phys. Rev.* **121**, 1556–1566.

Arnowitt, R., Deser, S. & Misner, C. (1962), The dynamics of general relativity, *in* L. Witten, ed., 'Gravitation: An Introduction to Current Research', Wiley, New York, pp. 227–265.

Ashtekar, A. (1986), 'New variables for classical and quantum gravity', *Phys. Rev. Lett.* **57**, 2244–2247.

Ashtekar, A. (1987), 'New Hamiltonian formulation of general relativity', *Phys. Rev.* **D36**, 1587–1602.

Ashtekar, A. (1991), *Lectures on Non-Perturbative Canonical Gravity*, World Scientific Press, Singapore.

Ashtekar, A. & Stachel, J., eds (1991), *Conceptual Problems of Quantum Gravity*, Birkhäuser, Boston.

Baierlein, R., Sharp, D. & Wheeler, J. (1962), 'Three-dimensional geometry as carrier of information about time', *Phys. Rev.* **126**, 1864–1865.

Banks, T. (1985), 'TCP, quantum gravity, the cosmological constant and all that', *Nucl. Phys.* **B249**, 332–360.

Barbour, J. (1990), 'Time, gauge fields, and the Schrödinger equation in quantum cosmology'. Preprint.

Barbour, J. & Smolin, L. (1988), 'Can quantum mechanics be sensibly applied to the universe as a whole?'. Yale University preprint.

Barvinsky, A. (1991), 'Unitarity approach to quantum cosmology'. University of Alberta preprint.

Bergmann, P. & Komar, A. (1972), 'The coordinate group of symmetries of general relativity', *Int. J. Mod. Phys.* **5**, 15–28.

Blencowe, M. (1991), 'The consistent histories interpretation of quantum fields in curved spacetime', *Ann. Phys. (NY)* **211**, 87–111.

Blyth, W. & Isham, C. (1975), 'Quantisation of a Friedman universe filled with a scalar field', *Phys. Rev.* **D11**, 768–778.

Bohr, N. & Rosenfeld, L. (1933), 'Zur frage der messbarkeit der elektromagnetischen feldgrossen', *Kgl. Danek Vidensk. Selsk. Math.-fys. Medd.* **12**, 8.

Bohr, N. & Rosenfeld, L. (1978), On the question of the measurability of electromagnetic field quantities (English translation), *in* R. Cohen & J. Stachel, eds, 'Selected Papers by Léon Rosenfeld', Reidel, Dordrecht.

Brout, R. (1987), 'On the concept of time and the origin of the cosmological temperature', *Found. Phys.* **17**, 603–619.

Brout, R. & Venturi, G. (1989), 'Time in semiclassical gravity', *Phys. Rev.* **D39**, 2436–2439.

Brout, R., Horowitz, G. & Wiel, D. (1987), 'On the onset of time and temperature in cosmology', *Phys. Lett.* **B192**, 318–322.

Brown, J. & York, J. (1989), 'Jacobi's action and the recovery of time in general relativity', *Phys. Rev.* **D40**, 3312–3318.

Caldeira, A. & Leggett, A. (1983), 'Path integral approach to quantum Brownian motion', *Physica* **A121**, 587–616.

Carlip, S. (1990), 'Observables, gauge invariance, and time in 2 + 1 dimensional quantum gravity', *Phys. Rev.* **D42**, 2647–2654.

Carlip, S. (1991), Time in 2 + 1 dimensional quantum gravity, *in* 'Proceedings of the Banff Conference on Gravitation, August 1990'.

Castagnino, M. (1988), 'Probabilistic time in quantum gravity', *Phys. Rev.* **D39**, 2216–2228.

Coleman, S. (1988), 'Why is there something rather than nothing: A theory of the cosmological constant', *Nucl. Phys.* **B310**, 643–668.

Collins, P. & Squires, E. (1992), 'Time in a quantum universe'. In press, Foundations of Physics.

Deutsch, D. (1990), 'A measurement process in a stationary quantum system'. Oxford University preprint.

DeWitt, B. (1962), The quantization of geometry, *in* L. Witten, ed., 'Gravitation: An Introduction to Current Research', Wiley, New York.

DeWitt, B. (1965), *Dynamical Theory of Groups and Fields*, Wiley, New York.

DeWitt, B. (1967*a*), 'Quantum theory of gravity. I. The canonical theory', *Phys. Rev.* **160**, 1113–1148.

DeWitt, B. (1967*b*), 'Quantum theory of gravity. II. The manifestly covariant theory', *Phys. Rev.* **160**, 1195–1238.

DeWitt, B. (1967*c*), 'Quantum theory of gravity. III. Applications of the covariant theory', *Phys. Rev.* **160**, 1239–1256.

Dirac, P. (1958*a*), 'Generalized Hamiltonian dynamics', *Proc. Royal Soc. of London* **A246**, 326–332.

Dirac, P. (1958*b*), 'The theory of gravitation in Hamiltonian form', *Proc. Royal Soc. of London* **A246**, 333–343.

Dirac, P. (1965), *Lectures on Quantum Mechanics*, Academic Press, New York.

Dowker, H. & Halliwell, J. (1992), 'The quantum mechanics of history: The decoherence functional in quantum mechanics', *Phys. Rev.*

Duff, M. (1981), Inconsistency of quantum field theory in a curved spacetime, *in* C. Isham, R. Penrose & D. Sciama, eds, 'Quantum Gravity 2: A Second Oxford Symposium', Clarendon Press, Oxford, pp. 81–105.

Earman, J. & Norton, J. (1987), 'What price spacetime substantialism?', *Brit. Jour. Phil. Science* **38**, 515–525.

Englert, F. (1989), 'Quantum physics without time', *Phys. Lett.* **B228**, 111–114.

Eppley, K. & Hannah, E. (1977), 'The necessity of quantizing the gravitational field', *Found. Phys.* **7**, 51-68.

Fischer, A. & Marsden, J. (1979), The initial value problem and the dynamical formulation of general relativity, *in* S. Hawking & W. Israel, eds, 'General Relativity: An Einstein Centenary Survey', Cambridge University Press, Cambridge, pp. 138–211.

Fredenhagen, K. & Haag, R. (1987), 'Generally covariant quantum field theory and scaling limits', *Comm. Math. Phys* **108**, 91–115.

Friedman, J. & Higuchi, A. (1989), 'Symmetry and internal time on the superspace of asymptotically flat geometries', *Phys. Rev.* **D41**, 2479–2486.

Fulling, S. (1990), 'When is stability in the eye of the beholder? Comments on a singular initial value problem for a nonlinear differential equation arising in semiclassical cosmology', *Phys. Rev.* **D42**, 4248–4250.

Gell-Mann, M. (1987), 'Superstring theory—closing talk at the 2nd Nobel Symposium on Particle Physics', *Physica Scripta* **T15**, 202–209.

Gell-Mann, M. & Hartle, J. (1990*a*), Alternative decohering histories in quantum mechanics, *in* K. Phua & Y. Yamaguchi, eds, 'Proceedings of the 25th International Conference on High Energy Physics, Singapore, August, 2–8, 1990', World Scientific, Singapore.

Gell-Mann, M. & Hartle, J. (1990*b*), Quantum mechanics in the light of quantum cosmology, *in* S. Kobayashi, H. Ezawa, Y. Murayama & S. Nomura, eds, 'Proceedings of the Third International Symposium on the Foundations of Quantum Mechanics in the Light of New Technology', Physical Society of Japan, Tokyo, pp. 321–343.

Gell-Mann, M. & Hartle, J. (1990*c*), Quantum mechanics in the light of quantum cosmology, *in* W. Zurek, ed., 'Complexity, Entropy and the Physics of Information, SFI Studies in the Science of Complexity, Vol. VIII', Addison-Wesley, Reading, pp. 425–458.

Gell-Mann, M. & Hartle, J. (1992), 'Classical equations for quantum systems'. UCSB preprint UCSBTH-91-15.

Gerlach, U. (1969), 'Derivation of the ten Einstein field equations from the semiclassical approximation to quantum geometrodynamics', *Phys. Rev.* **117**, 1929–1941.

Giddings, S. & Strominger, A. (1988), 'Baby universes, third quantization and the cosmological constant', *Nucl. Phys.* **B231**, 481–508.

Greensite, J. (1990), 'Time and probability in quantum cosmology', *Nucl. Phys.* **B342**, 409–429.

Greensite, J. (1991*a*), 'Ehrenfests' principle in quantum-gravity', *Nucl. Phys.* **B351**, 749–766.

Greensite, J. (1991*b*), 'A model of quantum gravitational collapse', *Int. J. Mod. Phys.* **A6**, 2693–2706.

Griffiths, R. (1984), 'Consistent histories and the interpretation of quantum mechanics', *J. Stat. Phys.* **36**, 219–272.

Groenwold, H. (1946), 'On the principles of elementary quantum mechanics', *Physica* **12**, 405–460.

Hajicek, P. & Kuchař, K. (1990*a*), 'Constraint quantization of parametrized relativistic gauge systems in curved spacetimes', *Phys. Rev.* **D41**, 1091–1104.

Hajicek, P. & Kuchař, K. (1990*b*), 'Transversal affine connection and quantisation of constrained systems', *J. Math. Phys.* **31**, 1723–1732.

Hájíček, P. (1986), 'Origin of nonunitarity in quantum gravity', *Phys. Rev.* **D34**, 1040–1048.

Hájíček, P. (1988), 'Reducibility of parametrised systems', *Phys. Rev.* **D38**, 3639–3647.

Hájíček, P. (1989), 'Topology of parametrised systems', *J. Math. Phys.* **20**, 2488–2497.

Hájíček, P. (1990*a*), 'Dirac quantisation of systems with quadratic constraints', *Class. Quan. Grav.* **7**, 871–886.

Hájíček, P. (1990*b*), 'Topology of quadratic super-Hamiltonians', *Class. Quan. Grav.* **7**, 861–870.

Hájíček, P. (1991), 'Comment on 'Time in quantum gravity: An hypothesis'', *Phys. Rev.* **D44**, 1337–1338.

Halliwell, J. (1987), 'Correlations in the wave function of the universe', *Phys. Rev.* **D36**, 3626–3640.

Halliwell, J. (1988), 'Derivation of the Wheeler-DeWitt equation from a path integral for minisuperspace models', *Phys. Rev.* **D38**, 2468–2481.

Halliwell, J. (1989), 'Decoherence in quantum cosmology', *Phys. Rev.* **D39**, 2912–2923.

Halliwell, J. (1990), 'A bibliography of papers on quantum cosmology', *Int. J. Mod. Phys.* **A5**, 2473–2494.

Halliwell, J. (1991*a*), Introductory lectures on quantum cosmology, *in* S. Coleman, J. Hartle, T. Piran & S. Weinberg, eds, 'Proceedings of the Jerusalem Winter School on Quantum Cosmology and Baby Universes', World Scientific, Singapore.

Halliwell, J. (1991*b*), The Wheeler-DeWitt equation and the path integral in minisuperspace quantum cosmology, *in* A. Ashtekar & J. Stachel, eds, 'Conceptual Problems of Quantum Gravity', Birkhäuser, Boston, pp. 75–115.

Halliwell, J. (1992*a*), The interpretation of quantum cosmology models, *in* 'Proceedings of the 13th International Conference on General Relativity and Gravitation, Cordoba, Argentina'.

Halliwell, J. (1992*b*), 'Smeared Wigner functions and quantum-mechanical histories', *Phys. Rev.*

Halliwell, J. & Hawking, S. (1985), 'Origin of structure in the universe', *Phys. Rev.* **D31**, 1777–1791.

Halliwell, J., Perez-Mercander, J. & Zurek, W., eds (1992), *Physical Origins of Time Asymmetry*, Cambridge University Press, Cambridge.

Hartle, J. (1986), Prediction and observation in quantum cosmology, *in* B. Carter & J. Hartle, eds, 'Gravitation and Astrophysics, Cargese, 1986', Plenum, New York.

Hartle, J. (1988*a*), 'Quantum kinematics of spacetime. I. Nonrelativistic theory', *Phys. Rev.* **D37**, 2818–2832.

Hartle, J. (1988*b*), 'Quantum kinematics of spacetime. II. A model quantum cosmology with real clocks', *Phys. Rev.* **D38**, 2985–2999.

Hartle, J. (1991*a*), The quantum mechanics of cosmology, *in* S. Coleman, P. Hartle, T. Piran & S. Weinberg, eds, 'Quantum Cosmology and Baby Universes', World Scientific, Singapore.

Hartle, J. (1991*b*), 'Spacetime grainings in nonrelativistic quantum mechanics', *Phys. Rev.* **D44**, 3173–3195.

Hartle, J. & Hawking, S. (1983), 'Wave function of the universe', *Phys. Rev.* **D28**, 2960–2975.

Hawking, S. (1982), 'The unpredictability of quantum gravity', *Comm. Math. Phys* **87**, 395–416.

Hawking, S. (1984*a*), Lectures in quantum cosmology, *in* B. DeWitt & R. Stora, eds, 'Relativity, Groups and Topology II', North-Holland, Amsterdam, pp. 333–379.

Hawking, S. (1984*b*), 'The quantum state of the universe', *Nucl. Phys.* **B239**, 257–276.

Hawking, S. & Ellis, G. (1973), *The Large Scale Structure of Space-Time*, Cambridge University Press, Cambridge.

Hawking, S. & Page, D. (1986), 'Operator ordering and the flatness of the universe', *Nucl. Phys.* **B264**, 185–196.

Hawking, S. & Page, D. (1988), 'How probable is inflation?', *Nucl. Phys.* **B298**, 789–809.

Henneaux, M. & Teitelboim, C. (1989), 'The cosmological constant and general covariance', *Phys. Lett.* **B222**, 195–199.

Higgs, P. (1958), 'Integration of secondary constraints in quantized general relativity', *Phys. Rev. Lett.* **1**, 373–375.

Hojman, S., Kuchař, K. & Teitelboim, C. (1976), 'Geometrodynamics regained', *Ann. Phys. (NY)* **96**, 88–135.

Horowitz, G. (1980), 'Semiclassical relativity: The weak-field limit', *Phys. Rev.* **D21**, 1445–1461.

Horowitz, G. (1981), Is flat space-time unstable?, *in* C. Isham, R. Penrose & D. Sciama, eds, 'Quantum Gravity 2: A Second Oxford Symposium', Clarendon Press, Oxford, pp. 106–130.

Horowitz, G. & Wald, R. (1978), 'Dynamics of Einstein's equations modified by a higher-derivative term', *Phys. Rev.* **D17**, 414–416.

Horowitz, G. & Wald, R. (1980), 'Quantum stress energy in nearly conformally flat spacetimes', *Phys. Rev.* **D21**, 1462–1465.

Isenberg, J. & Marsden, J. (1976), 'The York map is a canonical transformation', *Ann. Phys. (NY)* **96**, 88–135.

Isham, C. (1984), Topological and global aspects of quantum theory, *in* B. DeWitt & R. Stora, eds, 'Relativity, Groups and Topology II', North-Holland, Amsterdam, pp. 1062–1290.

Isham, C. (1985), Aspects of quantum gravity, *in* A. Davies & D. Sutherland, eds, 'Superstrings and Supergravity: Proceedings of the 28th Scottish Universities Summer School in Physics, 1985', SUSSP Publications, Edinburgh, pp. 1–94.

Isham, C. (1987), Quantum gravity—grg11 review talk, *in* M. MacCallum, ed., 'General Relativity and Gravitation: Proceedings of the 11th International Conference on General Relativity and Gravitation', Cambridge University Press, Cambridge.

Isham, C. (1992), Conceptual and geometrical problems in quantum gravity, *in* H. Mitter & H. Gausterer, eds, 'Recent Aspects of Quantum Fields', Springer-Verlag, Berlin, pp. 123–230.

Isham, C. & Kakas, A. (1984*a*), 'A group theoretical approach to the canonical quantisation of gravity: I. Construction of the canonical group', *Class. Quan. Grav.* **1**, 621–632.

Isham, C. & Kakas, A. (1984*b*), 'A group theoretical approach to the canonical quantisation of gravity: II. Unitary representations of the canonical group', *Class. Quan. Grav.* **1**, 633–650.

Isham, C. & Kuchař, K. (1985*a*), 'Representations of spacetime diffeomorphisms. I. Canonical parametrised spacetime theories', *Ann. Phys. (NY)* **164**, 288–315.

Isham, C. & Kuchař, K. (1985*b*), 'Representations of spacetime diffeomorphisms. II. Canonical geometrodynamics', *Ann. Phys. (NY)* **164**, 316–333.

Isham, C. & Kuchař, K. (1994), *Quantum Gravity and the Problem of Time*. In preparation.

Jackiw, R. (1992*a*), Gauge theories for gravity on a line, *in* 'Proceedings of NATO Advanced Study Institute 'Recent Problems in Mathematical Physics, Salamanca, Spain, 1992'', Kluwer, The Netherlands.

Jackiw, R. (1992*b*), Update on planar gravity (≈ physics of infinite cosmic strings), *in* H. Sato, ed., 'Proceedings of the Sixth Marcel Grossman Meeting on General Relativity', World Scientific, Singapore.

Joos, E. (1986), 'Why do we observe a classical spacetime?', *Phys. Lett.* **A116**, 6–8.

Joos, E. (1987), Quantum theory and the emergence of a classical world, *in* D. Greenberger, ed., 'New Techniques and Ideas in Quantum Measurement', New York Academic of Sciences, New York.

Joos, E. & Zeh, H. (1985), 'The emergence of classical properties through interaction with the environment', *Zeitschrift für Physik* **B59**, 223–243.

Kaup, D. & Vitello, A. (1974), 'Solvable quantum cosmological models and the importance of quantizing in a special canonical frame', *Phys. Rev.* **D9**, 1648–1655.

Kibble, T. (1981), Is a semi-classical theory of gravity viable?, *in* C. Isham, R. Penrose & D. Sciama, eds, 'Quantum Gravity 2: A Second Oxford Symposium', Clarendon Press, Oxford, pp. 63–80.

Kiefer, C. (1987), 'Continuous measurement of mini-superspace variables by higher multipoles', *Class. Quan. Grav.* **4**, 1369–1382.

Kiefer, C. (1989*a*), 'Continuous measurement of intrinsic time by fermions', *Class. Quan. Grav.* **6**, 561–566.

Kiefer, C. (1989*b*), 'Quantum gravity and Brownian motion', *Phys. Lett.* **A139**, 201–203.

Kiefer, C. (1991), 'Interpretation of the decoherence functional in quantum cosmology', *Class. Quan. Grav.* **8**, 379–392.

Kiefer, C. (1992), Decoherence in quantum cosmology, *in* 'Proceedings of the Tenth Seminar on Relativistic Astrophysics and Gravitation, Postdam, 1991', World Scientific, Singapore.

Kiefer, C. & Singh, T. (1991), 'Quantum gravitational corrections to the functional Schrödinger equation', *Phys. Rev.* **D44**, 1067–1076.

Klauder, J. (1970), Soluble models of guantum gravitation, *in* M. Carmeli, S. Flicker & Witten, eds, 'Relativity', Plenum, New York.

Komar, A. (1979*a*), 'Consistent factoring ordering of general-relativistic constraints', *Phys. Rev.* **D19**, 830–833.

Komar, A. (1979*b*), 'Constraints, hermiticity, and correspondence', *Phys. Rev.* **D19**, 2908–2912.

Kuchař, K. (1970), 'Ground state functional of the linearized gravitational field', *J. Math. Phys.* **11**, 3322–3344.

Kuchař, K. (1971), 'Canonical quantization of cylindrical gravitational waves', *Phys. Rev.* **D4**, 955–985.

Kuchař, K. (1972), 'A bubble-time canonical formalism for geometrodynamics', *J. Math. Phys.* **13**, 768–781.

Kuchař, K. (1973), Canonical quantization of gravity, *in* 'Relativity, Astrophysics and Cosmology', Reidel, Dordrecht, pp. 237–288.

Kuchař, K. (1974), 'Geometrodynamics regained: A Lagrangian approach', *J. Math. Phys.* **15**, 708–715.

Kuchař, K. (1976*a*), 'Dynamics of tensor fields in hyperspace III.', *J. Math. Phys.* **17**, 801–820.

Kuchař, K. (1976*b*), 'Geometry of hyperspace I.', *J. Math. Phys.* **17**, 777–791.

Kuchař, K. (1976*c*), 'Kinematics of tensor fields in hyperspace II.', *J. Math. Phys.* **17**, 792–800.

Kuchař, K. (1977), 'Geometrodynamics with tensor sources IV', *J. Math. Phys.* **18**, 1589–1597.

Kuchař, K. (1981*a*), Canonical methods of quantisation, *in* C. Isham, R. Penrose & D. Sciama, eds, 'Quantum Gravity 2: A Second Oxford Symposium', Clarendon Press, Oxford, pp. 329–374.

Kuchař, K. (1981*b*), 'General relativity: Dynamics without symmetry', *J. Math. Phys.* **22**, 2640–2654.

Kuchař, K. (1982), 'Conditional symmetries in parametrized field theories', *J. Math. Phys.* **25**, 1647–1661.

Kuchař, K. (1986*a*), 'Canonical geometrodynamics and general covariance', *Found. Phys.* **16**, 193–208.

Kuchař, K. (1986*b*), 'Covariant factor ordering for gauge systems', *Phys. Rev.* **D34**, 3044–3057.

Kuchař, K. (1986*c*), 'Hamiltonian dynamics of gauge systems', *Phys. Rev.* **D34**, 3031–3043.

Kuchař, K. (1987), 'Covariant factor ordering of constraints may be ambiguous', *Phys. Rev.* **D35**, 596–599.

Kuchař, K. (1991*a*), 'Does an unspecified cosmological constant solve the problem of time in quantum gravity?', *Phys. Rev.* **D43**, 3332–3344.

Kuchař, K. (1991*b*), The problem of time in canonical quantization, *in* A. Ashtekar & J. Stachel, eds, 'Conceptual Problems of Quantum Gravity', Birkhäuser, Boston, pp. 141–171.

Kuchař, K. (1992*a*), 'Extrinsic curvature as a reference fluid in canonical gravity', *Phys. Rev.* **D45**, 4443–4457.

Kuchař, K. (1992*b*), Time and interpretations of quantum gravity, *in* 'Proceedings of the 4th Canadian Conference on General Relativity and Relativistic Astrophysics', World Scientific, Singapore.

Kuchař, K. & Ryan, M. (1989), 'Is minisuperspace quantization valid? Taub in Mixmaster', *Phys. Rev.* **D40**, 3982–3996.

Kuchař, K. & Torre, C. (1989), 'World sheet diffeomorphisms and the cylindrical string', *J. Math. Phys.* **30**, 1769–1793.

Kuchař, K. & Torre, C. (1991*a*), 'Gaussian reference fluid and the interpretation of geometrodynamics', *Phys. Rev.* **D43**, 419–441.

Kuchař, K. & Torre, C. (1991*b*), 'Harmonic gauge in canonical gravity', *Phys. Rev.* **D44**, 3116–3123.

Kuchař, K. & Torre, C. (1991*c*), Strings as poor relatives of general relativity, *in* A. Ashtekar & J. Stachel, eds, 'Conceptual Problems of Quantum Gravity', Birkhäuser, Boston, pp. 326–348.

Lapchinski, V. & Rubakov, V. (1979), 'Canonical quantization of gravity and quantum field theory in curved spacetime', *Acta. Phys. Pol.* **B10**, 1041.

Lee, J. & Wald, R. (1990), 'Local symmetries and constraints', *J. Math. Phys.* **31**, 725–743.

Linden, N. & Perry, M. (1991), 'Path integrals and unitarity in quantum cosmology', *Nucl. Phys.* **B357**, 289–307.

Manojlović & Miković (1992), 'Gauge fixing and independent canonical variables in the ashtekar formulation of general relativity'.

McGuigan, M. (1988), 'Third quantization and the Wheeler-DeWitt equation', *Phys. Rev.* **D38**, 3031–3051.

McGuigan, M. (1989), 'Universe creation from the third quantized vacuum', *Phys. Rev.* **D39**, 2229–2233.

Mellor, F. (1991), 'Decoherence in quantum Kaluza-Klein theories', *Nucl. Phys.* **B353**, 291–301.

Misner, C. (1957), 'Feynman quantization of general relativity', *Rev. Mod. Phys.* **29**, 497–509.

Misner, C. (1992), Minisuperspace, *in* J. Klauder, ed., 'Magic without Magic: John Archibald Wheeler, A Collection of Essays in Honor of his 60th Birthday', Freeman, San Francisco.

Møller, C. (1962), The energy-momentum complex in general relativity and related problems, *in* A. Lichnerowicz & M. Tonnelat, eds, 'Les Théories Relativistes de la Gravitation', CNRS, Paris.

Morikawa, M. (1989), 'Evolution of the cosmic density matrix', *Phys. Rev.* **D40**, 4023–4027.

Newman, E. & Rovelli, C. (1992), 'Generalized lines of force as the gauge invariant degrees of freedom for general relativity and Yang-Mills theory'. University of Pittsburgh preprint.

Omnès, R. (1988*a*), 'Logical reformulation of quantum mechanics. I. Foundations', *J. Stat. Phys.* **53**, 893–932.

Omnès, R. (1988*b*), 'Logical reformulation of quantum mechanics. II. Interferences and the Einstein-Podolsky-Rosen experiment', *J. Stat. Phys.* **53**, 933–955.

Omnès, R. (1988*c*), 'Logical reformulation of quantum mechanics. III. Classical limit and irreversibility', *J. Stat. Phys.* **53**, 957–975.

Omnès, R. (1989), 'Logical reformulation of quantum mechanics. III. Projectors in semiclassical physics', *J. Stat. Phys.* **57**, 357–382.

Omnès, R. (1990), 'From Hilbert space to common sense: A synthesis of recent progress in the interpretation of quantum mechanics', *Ann. Phys. (NY)* **201**, 354–447.

Omnès, R. (1992), 'Consistent interpretations of quantum mechanics', *Rev. Mod. Phys.* **64**, 339–382.

Padmanabhan, T. (1989*a*), 'Decoherence in the density matrix describing quantum three-geometries and the emergence of classical spacetime', *Phys. Rev.* **D39**, 2924–2932.

Padmanabhan, T. (1989*b*), 'Semiclassical approximations to gravity and the issue of back-reaction', *Class. Quan. Grav.* **6**, 533–555.

Padmanabhan, T. (1990), 'A definition for time in quantum cosmology', *Pramana Jour. Phys.* **35**, L199–L204.

Page, D. (1986*a*), 'Density matrix of the universe', *Phys. Rev.* **D34**, 2267–2271.

Page, D. (1986*b*), Hawking's wave function for the universe, *in* R. Penrose & C. Isham, eds, 'Quantum Concepts in Space and Time', Clarendon Press, Oxford, pp. 274–285.

Page, D. (1989), 'Time as an inaccessible observable'. ITP preprint NSF-ITP-89-18.

Page, D. (1991), Intepreting the density matrix of the universe, *in* A. Ashtekar & J. Stachel, eds, 'Conceptual Problems of Quantum Gravity', Birkhäuser, Boston, pp. 116–121.

Page, D. & Geilker, C. (1981), 'Indirect evidence for quantum gravity', *Phys. Rev. Lett.* **47**, 979–982.

Page, D. & Hotke-Page, C. (1992), Clock time and entropy, *in* J. Halliwell, J. Perez-Mercander & W. Zurek, eds, 'Physical Origins of Time Asymmetry', Cambridge University Press, Cambridge.

Page, D. & Wooters, W. (1983), 'Evolution without evolution: Dynamics described by stationary observables', *Phys. Rev.* **D27**, 2885–2892.

Pilati, M. (1982), 'Strong coupling quantum gravity I: solution in a particular gauge', *Phys. Rev.* **D26**, 2645–2663.

Pilati, M. (1983), 'Strong coupling quantum gravity I: solution without gauge fixing', *Phys. Rev.* **D28**, 729–744.

Randjbar-Daemi, S., Kay, B. & Kibble, T. (1980), 'Renormalization of semi-classical field theories', *Phys. Rev. Lett.* **91B**, 417–420.

Rosenfeld, L. (1963), 'On quantization of fields', *Nucl. Phys.* **40**, 353–356.

Rovelli, C. (1990), 'Quantum mechanics without time: A model', *Phys. Rev.* **D42**, 2638–2646.

Rovelli, C. (1991*a*), 'Ashtekar formulation of general relativity and loop-space non-perturbative quantum gravity: A report.'. Pittsburgh University preprint.

Rovelli, C. (1991*b*), Is there incompatibility between the ways time is treated in general relativity and in standard quantum theory?, *in* A. Ashtekar & J. Stachel, eds, 'Conceptual Problems of Quantum Gravity', Birkhäuser, Boston, pp. 126–140.

Rovelli, C. (1991*c*), 'Quantum evolving constants. Reply to 'Comments on time in quantum gravity: An hypothesis'', *Phys. Rev.* **D44**, 1339–1341.

Rovelli, C. (1991*d*), 'Statistical mechanics of gravity and the problem of time'. University of Pittsburgh preprint.

Rovelli, C. (1991*e*), 'Time in quantum gravity: An hypothesis', *Phys. Rev.* **D43**, 442–456.

Rovelli, C. & Smolin, L. (1990), 'Loop space representation of quantum general relativity', *Nucl. Phys.* **B331**, 80–152.

Salisbury, D. & Sundermeyer, K. (1983), 'Realization in phase space of general coordinate transformations', *Phys. Rev.* **D27**, 740–756.

Simon, J. (1991), 'The stability of flat space, semiclassical gravity and higher derivatives', *Phys. Rev.* **D43**, 3308–3316.

Singer, I. (1978), 'Some remarks on the Gribov ambiguity', *Comm. Math. Phys* **60**, 7–12.

Singh, T. & Padmanabhan, T. (1989), 'Notes on semiclassical gravity', *Ann. Phys. (NY)* **196**, 296–344.

Smarr, L. & York, J. (1978), 'Kinematical conditions in the construction of space-time', *Phys. Rev.* **D17**, 2529–2551.

Smolin, L. (1992), 'Recent developments in nonperturbative quantum gravity'. Syracuse University Preprint.

Squires, E. (1991), 'The dynamical role of time in quantum cosmology', *Phys. Lett.* **A155**, 357–360.

Stachel, J. (1989), Einstein's search for general covariance, 1912-1915, *in* D. Howard & J. Stachel, eds, 'Einstein and the History of General Relativity: Vol I', Birkhäuser, Boston, pp. 63–100.

Stone, C. & Kuchař, K. (1992), 'Representation of spacetime diffeomorphisms in canonical geometrodynamics under harmonic coordinate conditions', *Class. Quan. Grav.* **9**, 757–776.

Suen, W. (1989*a*), 'Minkowski spacetime is unstable in semi-classical gravity', *Phys. Rev. Lett.* **62**, 2217–2226.

Suen, W. (1989*b*), 'Stability of the semiclassical Einstein equation', *Phys. Rev.* **D40**, 315–326.

Tate, R. (1992), 'An algebraic approach to the quantization of constrained systems: Finite dimensional examples'. PhD. Dissertation.

Teitelboim, C. (1982), 'Quantum mechanics of the gravitational field', *Phys. Rev.* **D25**, 3159–3179.

Teitelboim, C. (1983*a*), 'Causality versus gauge invariance in quantum gravity and supergravity', *Phys. Rev. Lett.* **50**, 705–708.

Teitelboim, C. (1983*b*), 'Proper time gauge in the quantum theory of gravitation', *Phys. Rev.* **D28**, 297–309.

Teitelboim, C. (1983*c*), 'Quantum mechanics of the gravitational field', *Phys. Rev.* **D25**, 3159–3179.

Teitelboim, C. (1983*d*), 'Quantum mechanics of the gravitational field in asymptotically flat space', *Phys. Rev.* **D28**, 310–316.

Teitelboim, C. (1992), *in* 'Proceedings of NATO Advanced Study Institute 'Recent Problems in Mathematical Physics, Salamanca, Spain, 1992'', Kluwer, The Netherlands.

Torre, C. (1989), 'Hamiltonian formulation of induced gravity in two dimensions', *Phys. Rev.* **D40**, 2588–2597.

Torre, C. (1991), 'A complete set of observables for cylindrically symmetric gravitational fields', *Class. Quan. Grav.* **8**, 1895–1911.

Torre, C. (1992), 'Is general relativity an "already parametrised" theory?'.

Unruh, W. (1988), Time and quantum gravity, *in* M. Markov, V. Berezin & V. Frolov, eds, 'Proceedings of the Fourth Seminar on Quantum Gravity', World Scientific, Singapore, pp. 252–268.

Unruh, W. (1989), 'Unimodular theory of canonical quantum gravity', *Phys. Rev.* **D40**, 1048–1052.

Unruh, W. (1991), Loss of quantum coherence for a damped oscillator, *in* A. Ashtekar & J. Stachel, eds, 'Conceptual Problems of Quantum Gravity', Birkhäuser, Boston, pp. 67–74.

Unruh, W. & Wald, R. (1989), 'Time and the interpretation of quantum gravity', *Phys. Rev.* **D40**, 2598–2614.

Unruh, W. & Zurek, W. (1989), 'Reduction of the wave-packet in quantum Brownian motion', *Phys. Rev.* **D40**, 1071–1094.

Valentini, A. (1992), 'Non-local hidden-variables and quantum gravity'. SISSA preprint 105/92/A.

Van Hove, L. (1951), 'On the problem of the relations between the unitary transformations of quantum mechanics and the canonical transformations of classical mechanics', *Acad. Roy. Belg.* **37**, 610–620.

Vilenkin, A. (1988), 'Quantum cosmology and the initial state of the universe', *Phys. Rev.* **D39**, 888–897.

Vilenkin, A. (1989), 'Interpretation of the wave function of the universe', *Phys. Rev.* **D39**, 1116–1122.

Vink, J. (1992), 'Quantum potential interpretation of the wavefunction of the universe', *Nucl. Phys.* **B369**, 707–728.

Wheeler, J. (1962), Neutrinos, gravitation and geometry, *in* 'Topics of Modern Physics. Vol 1', Academic Press, New York, pp. 1–130.

Wheeler, J. (1964), Geometrodynamics and the issue of the final state, *in* C. DeWitt & B. DeWitt, eds, 'Relativity, Groups and Topology', Gordon and Breach, New York and London, pp. 316–520.

Wheeler, J. (1968), Superspace and the nature of quantum geometrodynamics, *in* C. DeWitt & J. Wheeler, eds, 'Batelle Rencontres: 1967 Lectures in Mathematics and Physics', Benjamin, New York, pp. 242–307.

Wooters, W. (1984), 'Time replaced by quantum corrections', *Int. J. Theor. Phys.* **23**, 701–711.

York, J. (1972*a*), 'Mapping onto solutions of the gravitational initial value problem', *J. Math. Phys.* **13**, 125–130.

York, J. (1972*b*), 'Role of conformal three-geometry in the dynamics of gravitation', *Phys. Rev. Lett.* **28**, 1082–1085.

York, J. (1979), Kinematics and dynamics of general relativity, *in* L. Smarr, ed., 'Sources of Gravitational Radiation', Cambridge University Press, Cambridge, pp. 83–126.

Zeh, H. (1986), 'Emergence of classical time from a universal wave function', *Phys. Lett.* **A116**, 9–12.

Zeh, H. (1988), 'Time in quantum gravity', *Phys. Lett.* **A126**, 311–317.

Zurek, W. (1981), 'Pointer basis of quantum apparatus: Into what mixture does the wave packet collapse?', *Phys. Rev.* **D24**, 1516–1525.

Zurek, W. (1982), 'Environment-induced superselection rules', *Phys. Rev.* **D26**, 1862–1880.

Zurek, W. (1983), Information transer in quantum measurements:Irreversibility and amplification, *in* P. Meystre & M. Scully, eds, 'Quantum Optics, Experimental Gravity, and Measurement Theory', Plenum, New York, pp. 87–116.

Zurek, W. (1986), Reduction of the wave-packet: How long does it take?, *in* G. Moore & M. Scully, eds, 'Frontiers of Nonequilibrium Statistical Physics', Plenum, New York, pp. 145–149.

Zurek, W. (1991), Quantum measurements and the environment-induced transition from quantum to classical, *in* A. Ashtekar & J. Stachel, eds, 'Conceptual Problems of Quantum Gravity', Birkhäuser, Boston, pp. 43–66.

HIGHER SYMMETRIES IN LOWER-DIMENSIONAL MODELS *

R. Jackiw

Center for Theoretical Physics
Laboratory for Nuclear Science
and Department of Physics
Massachusetts Institute of Technology
Cambridge, Massachusetts 02139 U.S.A.

CONTENTS

I. Introduction

Symmetries of physical theories aid in model building by limiting available options, illuminate the structure of the model and assist in finding solutions. The adjective "higher" refers to unusual and unexpected symmetries that may be operative, for

* This work is supported in part by funds provided by the U. S. Department of Energy (D.O.E.) under contract #DE-AC02-76ER03069.

L. A. Ibort and M. A. Rodríguez (eds.), Integrable Systems, Quantum Groups, and Quantum Field Therapy 289–316.

example the infinite conformal symmetry found in various two-dimensionsal theories. Here we discuss symmetries of several (2 + 1)-dimensional and (1 + 1)-dimensional field theories. The former arise in planar physics and also as effective descriptions of four-dimensional dynamics in the eikonal limit. The latter are scalar-tensor gravity theories, which have been the focus of research for some time, most recently as toy models for quantum black holes.

II. Finite and Infinite Symmetries in (2 + 1)-Dimensional Field Theory

2.1. Overview

We shall discuss finite- and infinite-dimensional conformal symmetries of *field theories with non-relativistic kinematics*. Such field theories also describe the second quantization of *non-relativistic particle mechanics*. Particle mechanics, with its second order in time dynamics, has the structure of a *relativistic field theory* in one time and zero space dimensions, and a relativistic field theory in any dimension can enjoy conformal symmetry. Thus there are family relationships between the conformal symmetries of non-relativistic field theory, non-relativistic particle mechanics and relativistic field theory, and our first task is to describe these interrelations.

A conformal transformation in $(D+1)$-dimensional relativistic field theory changes the *independent* variables, *viz.* the space-time coordinates x^μ of the fields (fields are *dependent* variables), and infinitesimally reads

$$\delta_f x^\mu = -f^\mu(x) \tag{2.1.1}$$

where f^μ is a *conformal Killing vector*, *i.e.* f^μ satisfies the *conformal Killing equation.*

$$\partial_\mu f_\nu + \partial_\nu f_\mu = \frac{2}{D+1} g_{\mu\nu} \partial_\alpha f^\alpha \tag{2.1.2}$$

Here $g_{\mu\nu}$ is the Minkowski metric tensor with signature $(1, -1, -1, \ldots)$ and D is the spatial dimensionality.

As is well-known, Eq. (1.2) has the *finite* number of $\frac{1}{2}(D+2)(D+3)$ solutions for $D > 1$, and conformal transformations form an $SO(2, D+1)$ group. The solutions to (1.2) comprise

$$\begin{aligned}
&D+1 \text{ space-time translations} && f^\mu(x) = a^\mu \quad , \quad a^\mu \text{ constant} && (2.1.3a)\\
&\tfrac{1}{2}D(D+1) \text{ space-time rotations} && f^\mu(x) = \omega^\mu{}_\nu x^\nu \quad , \quad \omega_{\mu\nu} = -\omega_{\nu\mu} && (2.1.3b)\\
&\text{a single scale transformation} && f^\mu(x) = a x^\mu \quad , \quad a \text{ constant} && (2.1.3c)\\
&D+1 \text{ special conformal transformations} && f^\mu(x) = 2c\cdot x\, x^\mu - c^\mu x^2 \quad , \quad c^\mu \text{ constant} && (2.1.3d)
\end{aligned}$$

The finite versions of these are, respectively,

$$x^\mu \to x^\mu + a^\mu \tag{2.1.4a}$$
$$x^\mu \to \Lambda^\mu{}_\nu x^\mu \ , \qquad \Lambda^\mu{}_\alpha \Lambda^\nu{}_\beta g_{\mu\nu} = g_{\alpha\beta} \tag{2.1.4b}$$
$$x^\mu \to e^a x^\mu \tag{2.1.4c}$$
$$x^\mu \to \frac{x^\mu - c^\mu x^2}{1 - 2c \cdot x + c^2 x^2} \tag{2.1.4d}$$

The last, the finite special conformal transformation, can also be seen as an inversion, $x^\mu \to x^\mu / x^2$, followed by a translation and another inversion, *i.e.* a translation in the inverted coordinate.

At $D = 1$ there exists an *infinite* number of solutions to (1.2) corresponding to arbitrary redefinition of $x^\pm = \frac{1}{\sqrt{2}}(x^0 \pm x^1)$ and forming an infinite parameter group. Infinitesimally we have

$$\delta_f x^\pm = -f^\pm(x^\pm) \ , \qquad f^\pm \text{ arbitrary} \tag{2.1.5}$$

while the finite version reads

$$x^\pm \to X^\pm(x^\pm) \ , \qquad X^\pm \text{ arbitrary} \tag{2.1.6}$$

A linear conformal transformation on a space-time multiplet of Lorentz covariant relativistic fields φ, *i.e.* on the dependent variables, can be taken as

$$\delta_f \varphi = f^\alpha \partial_\alpha \varphi + \partial_\alpha f_\beta \left(\frac{\Delta}{D+1} g^{\alpha\beta} + \frac{1}{2} \Sigma^{\alpha\beta} \right) \varphi \tag{2.1.7}$$

Here $\Sigma^{\alpha\beta}$ is the spin-matrix, acting on the space-time components of φ and Δ is a constant, called the scale-dimension of φ. When the Lagrange density for φ possesses a conventional relativistic kinetic term — quadratic in derivatives for Bose fields, linear for Fermi fields — the kinetic action is invariant against conformal transformations (1.7) provided

$$\Delta = \frac{D-1}{2} \qquad \text{bosons} \tag{2.1.8a}$$
$$\Delta = \frac{D}{2} \qquad \text{fermions} \tag{2.1.8b}$$

[These values for Δ correspond to the dimensionality of a field in units of inverse length when $\hbar$ and c are scaled to unity.] Also the Bose field monomial

$$\mathcal{L}_I = \varphi^{2\left(\frac{D+1}{D-1}\right)} \tag{2.1.9}$$

leads to an invariant action $\int d^{D+1}x \mathcal{L}_I$.

At $D = 1$, Bose fields become dimensionless, see (1.8a), and the conformally invariant monomial (1.9) cannot be formed. Nevertheless, there exists a non-trivial conformally invariant theory — the completely integrable Liouville theory,

$$\mathcal{L}_{\text{Liouville}} = \frac{1}{2}\partial_\mu\varphi\partial^\mu\varphi - \frac{\mu^2}{\beta^2}e^{\beta\varphi} \tag{2.1.10}$$

whose action is invariant provided the single-component scalar field φ is transformed according to an inhomogenous generalization of (1.7),

$$\delta_f\varphi = f^\alpha\partial_\alpha\varphi + \frac{1}{\beta}\partial_\alpha f^\alpha \tag{2.1.11a}$$

or equivalently

$$\delta_f e^{\beta\varphi} = \partial_\alpha\left(f^\alpha e^{\beta\varphi}\right) \tag{2.1.11b}$$

The hallmark of a conformally invariant theory is that an energy-momentum tensor $T^{\mu\nu}$ can be constructed, which is conserved (as a consequence of translation invariance) symmetric (as a consequence of Lorentz invariance) and traceless (as a consequence of conformal invariance). Thus conformal invariance in a relativistic field may be summarized by a relation between the energy density $\mathcal{E} = T^{00}$ and the trace of the spatial stress tensor T^{ij}.

$$\mathcal{E} = \sum_{i=1}^{D} T^{ii} \tag{2.1.12}$$

The currents j_f^μ that are conserved as a consequence of conformal invariance are then constructed in the Bessel–Hagen form from a projection of $T^{\mu\nu}$ on the conformal Killing vector,

$$j_f^\mu = T^{\mu\nu}f_\nu \tag{2.1.13a}$$

and the constants of motion read

$$C_f = \int d^D r\,(\mathcal{E}f^0 - \mathcal{P}\cdot\mathbf{f}) \tag{2.1.13b}$$

where $\mathcal{P}$ is the momentum density.

$$\mathcal{P}^i = T^{0i} \tag{2.1.14}$$

[Frequently it is necessary to "improve" the energy-momentum tensor obtained by Noether's theorem or by general relativistic considerations.]

The kinetic Lagrangian for non-relativistic motion of point-particles in d-dimensional space is quadratic in derivatives with respect to time, which is the single independent variable. Hence it has the structure of a $(0+1)$-dimensional relativistic "field theory," where the "field" is particle position $\mathbf{r}(t)$, now a dependent variable, while the d-spatial dimensions form an "internal" space and $\mathbf{r}$ is a vector in this space.

"Conformal" transformations degenerate into reparametrization of the single independent variable, *i.e.* time. The previous discussion can be taken at $D = 0$, but the conformal Killing equation (1.2) becomes vacuous. Nevertheless, one easily shows that (1.3) and (1.4) [with (1.3b) and (1.4b) absent] are invariances of the kinetic term provided the dependent variable $\mathbf{r}$ transforms according to

$$\delta\mathbf{r} = f\dot{\mathbf{r}} - \frac{1}{2}\dot{f}\mathbf{r} \tag{2.1.15}$$

when the independent variable t changes by

$$\delta t = -f(t) \tag{2.1.16a}$$

$$f(t) = a, at, at^2 \tag{2.1.16b}$$

These comprise the $D = 0$ restriction of (1.7) and (1.8a) with $\Delta = -1/2$, and form an $SO(2,1)$ group of transformations. It is seen that $\mathbf{r}$ has scale dimension of $-1/2$, *i.e.* it scales as $\sqrt{t}$ — this is a consequence of having scaled $\hbar$ to unity and with non-relativistic kinematics one can take m to be dimensionless. Further from (1.9) at $D = 0$ one sees that the r^{-2} potential also gives an invariant action since $\dot{r}^2$ and $1/r^2$, the two terms comprising a Lagrangian or Hamiltonian, scale in the same way. When

$$H = \frac{1}{2}m\dot{r}^2 + \frac{\lambda}{r^2} \quad . \tag{2.1.17}$$

the three constants of motion

$$C_f = Hf - \frac{m}{4}(\mathbf{r}\cdot\dot{\mathbf{r}} + \dot{\mathbf{r}}\cdot\mathbf{r})\dot{f} + \frac{m}{4}r^2\ddot{f} \tag{2.1.18}$$

also generate the transformation (1.15) when the canonical momentum $\mathbf{p}$, conjugate to $\mathbf{r}$, is taken to be $m\dot{\mathbf{r}}$, and their algebra realizes the $SO(2,1)$ Lie algebra. One can view (1.18) as the one-time, zero-space analog of the Bessel–Hagen expression (1.13).

In specific spatial dimensions d, other interactions in addition to the $1/r^2$ potential preserve conformal invariance. Examples for $d = 3$ and 2 are, respectively, interaction with a Dirac magnetic monopole and a point vortex. For these (1.17) and (1.18) retain the same form, but the relation between canonical momentum and velocity is modified by the presence of a vector potential

$$\mathbf{p} = m\dot{\mathbf{r}} + \mathbf{A} \tag{2.1.19}$$

where at $d = 3$, $\mathbf{A}$ is the Dirac vector potential that gives rise to a monopole magnetic field of strength g_m

$$\mathbf{A} = \mathbf{A}_D \ , \qquad \mathbf{B} = \nabla \times \mathbf{A}_D = \frac{g_m \mathbf{r}}{r^3} \tag{2.1.20}$$

while in planar physics at $d = 2$, $\mathbf{A}$ is the vortex potential,

$$\mathbf{A} = \frac{\Phi}{2\pi}\nabla\theta \ , \qquad \mathbf{B} = \nabla \times \mathbf{A} = \Phi\delta^2(\mathbf{r}) \tag{2.1.21}$$

with $\tan\theta = y/x$ and Φ being the flux of the vortex. [In the above we use the amusing distributional formula $\nabla \times \nabla\theta = 2\pi\,\delta^2(\mathbf{r})$.]

Furthermore, at $d = 2$, the δ-function potential also scales as r^{-2}, hence it also appears to be conformally invariant. In the following, we shall consider the two-dimensional particle model, with a vortex (1.21) and δ-function interactions; *i.e.* the Hamiltonian is

$$H = \frac{1}{2}m\dot{r}^2 - g\,\delta^2(\mathbf{r}) = \frac{1}{2m}(\mathbf{p} - \mathbf{A})^2 - g\,\delta^2(\mathbf{r}) \tag{2.1.22}$$

Non-relativistic particle quantum mechanics may be second quantized, and in this way one is led to a non-relativistic quantum field theory, with field-theoretic symmetries that encode the above $SO(2,1)$ particle symmetries, but now realized in an action on the dependent field variable ψ, which is a function of the independent variables t, and $\mathbf{r}$, where $\mathbf{r}$ is a two-dimensional vector.

Here we shall explain these symmetries of non-relativistic field theory, at $d = 2$ — planar physics. Also we shall show how the $SO(2,1)$ group can expand to an infinite-dimensional group of conformal reparametrizations of the two-dimensional spatial plane.

Contexts, wherein recently there is encountered a non-relativistic, planar field theory, are the following two:

(i) *Second quantized, non-relativistic particles with Abelian or non-Abelian charge, interacting with a gauge field whose kinetic dynamics is provided by the Chern–Simons action.* The field theoretic action in the Abelian case, with gauge potentials eliminated in terms of matter variables, is

$$I = \int dt\, d^2r \left\{ i\psi^*\partial_t\psi - \frac{1}{2m}\left|\mathbf{D}\psi\right|^2 + \frac{g}{2}\rho^2 \right\} \tag{2.1.23}$$

where the covariant derivative $\mathbf{D}$ involves a gauge potential $\mathbf{A}$,

$$\mathbf{D} = \nabla - i\mathbf{A} \tag{2.1.24}$$

which is determined by the matter density $\rho = |\psi|^2$

$$\mathbf{A}(t,\mathbf{r}) = \nabla \times \frac{1}{2\pi\kappa}\int d^2r' \ln|\mathbf{r} - \mathbf{r}'|\,\rho(t,\mathbf{r}') \tag{2.1.25}$$

so that the Chern–Simons Gauss law is satisfied.

$$B = \nabla \times \mathbf{A} = -\frac{1}{\kappa}\rho \tag{2.1.26}$$

Here $1/\kappa$ measures the interaction strength and without loss of generality we may take it to be non-negative. [In the plane, the cross product of two vectors defines a scalar, and cross multiplication with a single vector results again in a vector; in components: $s = \epsilon^{ij}v^i_{(1)}v^j_{(2)}$; $v^i_{(1)} = \epsilon^{ij}v^j_{(2)}$.] Also there is present in (1.23) a quartic self-interaction of strength g, which is the second-quantized description of a two-body δ-function interaction. Thus (1.23) provides the second quantization of (1.22); the action may also be presented as

$$I = \int dt\, d^2r\, \{i\psi^* \partial_t \psi - \mathcal{E}\} \tag{2.1.27}$$

where the energy density is given by the formula[1]

$$\mathcal{E} = \frac{1}{2m} |\mathbf{D}\psi|^2 - \frac{g}{2}\rho^2 \tag{2.1.28}$$

(ii) *Effective action for gravity or Abelian vector gauge theories in the eikonal (large-s, fixed-t) limit.* It is found that in the eikonal regime, the conventional action [Einstein–Hilbert for gravity, Maxwell for gauge theory] can be written as a total derivative on a two-dimensional space-time plane imbedded in four-dimensional space-time. By integrating the total derivative onto a curve (parametrized by τ) forming the boundary of that two-dimensional plane, the action [without sources] becomes

$$I_{\text{eikonal}} = \frac{1}{2} \int d\tau\, d^2r \left(\partial_i \Omega^+ \partial_i \dot{\Omega}^- - \partial_i \Omega^- \partial_i \dot{\Omega}^+ \right) \tag{2.1.29}$$

where the overdot denotes differentiation with respect to τ, while $\mathbf{r}$ and ∂_i, $i = 1, 2$, refer to the remaining two spatial directions, and $\Omega^\pm$ are the surviving field (gravitational, vector) degrees of freedom.[2] Upon defining

$$\psi = \frac{1}{\sqrt{2}} (\partial_x + i\partial_y)(\Omega^+ - i\Omega^-) \tag{2.1.30}$$

(1.29) may be rewritten, apart from total derivative contributions, as

$$I_{\text{eikonal}} = \int d\tau\, d^2r\, i\psi^* \partial_\tau \psi \tag{2.1.31}$$

Precisely the same form as (1.27) is revealed, except now the energy density vanishes.

2.2. Symmetries

The field theoretic Lagrangian

$$L = \int d^2r\, i\, \psi^* \partial_t \psi - H \tag{2.2.1a}$$

$$H = \int d^2r\, \mathcal{E} \tag{2.2.1b}$$

represents both the non-relativistic Chern–Simons model (1.27), (1.28) and the eikonal limits of relativistic theory (1.29), (1.31), where in the latter case $\mathcal{E}$ vanishes. So we use (2.1) as the basis for our discussion of symmetries in both cases, keeping in mind that the vanishing of $\mathcal{E}$ for the latter, renders much of the analysis vacuous, but as we shall see, not without relevance to the former. Throughout we shall solely deal with the Abelian Chern–Simons theory, though as far as symmetry properties are concerned, the non-Abelian model behaves similarly. Also discussion is confined to classical symmetries of the field theory, viewed as a classical non-linear system. Anomalies in the symmetries due to quantum effects will only be mentioned in the Conclusion.

The energy density (1.28) of the Chern–Simons model

$$\mathcal{E} = \frac{1}{2m}\left|\mathbf{D}\psi\right|^2 - \frac{g}{2}\rho^2 \tag{2.2.2a}$$

is identically equal to

$$\mathcal{E} = \frac{1}{2m}\left|D\psi\right|^2 - \frac{1}{2}\left(\boldsymbol{\nabla}\times\mathbf{j} + \frac{1}{m}B\rho\right) - \frac{g}{2}\rho^2 \tag{2.2.2b}$$

where the current $\mathbf{j}$ is

$$\mathbf{j} = \frac{1}{m}\Im\psi^*\mathbf{D}\psi \tag{2.2.3}$$

and D is the holomorphic, gauge covariant derivative.

$$D \equiv D_x - iD_y \tag{2.2.4}$$

The curl of $\mathbf{j}$ will not contribute to a variational derivation of the equations of motion, nor will it contribute to the integrated total energy, provided the current is sufficiently well-behaved at the edge of space [which lies at infinity]. With the assumption of requisite regularity for $\mathbf{j}$ and the use of the Chern–Simons Gauss law (1.26), the energy/Hamiltonian may be presented as

$$E = H = \int d^2\mathbf{r}\mathcal{H} \ , \qquad \mathcal{H} = \frac{1}{2m}\left|D\psi\right|^2 - \frac{1}{2}\left(g - \frac{1}{m\kappa}\right)\rho^2 \tag{2.2.5}$$

This is the form of the Hamiltonian density that we shall scrutinize in regards to the symmetries of the model.

The symmetries are of two kinds: a) symmetries of the action, *i.e.* transformations which leave the action invariant and lead to constants of motion by Noether's theorem — these are well-known and include the obvious Galileo transformations, and the $SO(2,1)$ time-reparametrization conformal symmetries specific to the planar model with which we are here concerned; and b) symmetries of the critical points of the action, *i.e.* transformations which leave selected equations of motion invariant, map solutions into solutions, but do not give rise to constants of motion because they do not leave invariant the action away from its critical points. These symmetries of the Chern–Simons model have not been previously studied systematically, though their occurrence in static solutions at $g = 1/m\kappa$ had been noted: they comprise conformal reparametrization symmetries of the two-dimensional plane.[3,4]

2.2.1 *Finite-Dimensional Symmetry Group of the Action*

As befits any respectable field theory, our model is invariant against time translation, space translation and rotation, as well as against Galileo boosts because dynamics is non-relativistic. The last invariance is perhaps unexpected in the presence of gauge fields, which conventionally are invariant against the *Lorentz* boosts of *special relativity* (indeed this led to the invention of special relativity!). What distinguishes the present situation is that the gauge dynamics are of the Chern–Simons variety, and the Chern–Simons term, being topological, is invariant against *all* space-time transformation, while the non-relativistic matter system is only Galileo invariant. The transformation laws on the fields are familiar. For the first three,

$$\text{time translations} \qquad \begin{aligned} t' &= t + a \\ \mathbf{r}' &= \mathbf{r} \end{aligned} \tag{2.2.6}$$

$$\text{space translation} \qquad \begin{aligned} t' &= t \\ \mathbf{r}' &= \mathbf{r} + \mathbf{a} \end{aligned} \tag{2.2.7}$$

$$\text{space rotation} \qquad \begin{aligned} t' &= t \\ r'^i &= R^{ij}(\omega) r^j \end{aligned} \tag{2.2.8}$$

$[R^{ij}(\omega)$ is the rotation matrix through angle $\omega]$ the field transforms as a scalar.

$$\psi'(t', \mathbf{r}') = \psi(t, \mathbf{r}) \tag{2.2.9}$$

The Galileo transform

$$\text{Galileo boost} \qquad \begin{aligned} t' &= t \ , \\ \mathbf{r}' &= \mathbf{r} + \mathbf{v}t \end{aligned} \tag{2.2.10}$$

requires a 1-cocycle in the field transformation law.

$$\psi'(t', \mathbf{r}') = e^{im\mathbf{v}\cdot\left(\mathbf{r}+\frac{1}{2}\mathbf{v}t\right)} \psi(t, \mathbf{r}) \tag{2.2.11}$$

Additionally, our system is invariant against conformal reparametrizations of time. These include three $SO(2,1)$ transformations of time: translation (2.6) and (2.9),

$$\text{time dilation} \qquad \begin{aligned} t' &= at \\ \mathbf{r}' &= \sqrt{a}\,\mathbf{r} \end{aligned} \tag{2.2.12}$$

for which the field transformation law acquires a weight factor,

$$\psi'(t', \mathbf{r}') = \frac{1}{\sqrt{a}} \psi(t, \mathbf{r}) \tag{2.2.13}$$

and translation of inverse time,

$$\begin{array}{ll}\text{conformal time} & \dfrac{1}{t'} = \dfrac{1}{t} + a \\ \text{transformation} & \mathbf{r}' = \dfrac{1}{1+at}\mathbf{r}\end{array} \tag{2.2.14}$$

where the field transformation law has both a weight factor and 1-cocycle.

$$\psi'(t', \mathbf{r}') = (1+at)\, e^{\frac{-ima r^2}{2(1+at)}}\, \psi(t, \mathbf{r}) \tag{2.2.15}$$

One can check that owing to the weight factors, which are square roots of the Jacobian, the density ρ transforms with the Jacobian, J,

$$\rho'(t', \mathbf{r}') = J\rho(t, \mathbf{r}) \tag{2.2.16}$$

$$J \equiv \det\left\{\frac{\partial r^i}{\partial r'^j}\right\} \tag{2.2.17}$$

and the vector potential, defined by (1.25), transforms covariantly.

$$A'^i(t', \mathbf{r}') = A^j(t, \mathbf{r})\frac{\partial r^j}{\partial r'^i} \tag{2.2.18}$$

The action is invariant, and the conserved generators can be obtained from Noether's theorem.

Alternatively one records the formula for the energy momentum tensor components:

$$\text{energy density} \qquad T^{00} \equiv \mathcal{E} = \frac{1}{2m}|\mathbf{D}\psi|^2 - \frac{g}{2}\rho^2 \tag{2.2.19}$$

$$\text{momentum density} \qquad \boldsymbol{\mathcal{P}} = m\mathbf{j} = \Im\psi^*\mathbf{D}\psi \tag{2.2.20}$$

These satisfy continuity equations with energy flux $\mathbf{T}$,

$$\text{energy flux} \qquad \mathbf{T} = -\frac{1}{2}\left((D_t\psi)^*\,\mathbf{D}\psi + (\mathbf{D}\psi)^*\,D_t\psi\right) \tag{2.2.21}$$

and momentum flux — the stress tensor T^{ij}.

$$\begin{aligned}\text{momentum flux} \qquad T^{ij} &= \frac{1}{2}\left((D_i\psi)^*(D_j\psi) + (D_j\psi)^*(D_i\psi) - \delta^{ij}(D_\kappa\psi)^*(D_\kappa\psi)\right) \\ &+ \frac{1}{4}\left(\delta^{ij}\nabla^2 - 2\partial_i\partial_j\right)\rho + \delta^{ij}\mathcal{E}\end{aligned} \tag{2.2.22}$$

Here $D_t = \partial_t + iA^0$, where A^0 solves the Chern–Simons equation that supplements (1.26).

$$A^0(t, \mathbf{r}) = -\frac{1}{2\pi\kappa}\int d^2r'\, \epsilon^{ij}\frac{(r^i - r'^i)}{|\mathbf{r}-\mathbf{r}'|^2} j^j(t, \mathbf{r}') \tag{2.2.23}$$

The continuity equations read

$$\partial_t \mathcal{E} + \nabla \cdot \mathbf{T} = 0 \tag{2.2.24}$$

$$\partial_t \mathcal{P}^i + \partial_j T^{ij} = 0 \tag{2.2.25}$$

Note that energy flux $\mathbf{T}$ does not equal momentum density $\boldsymbol{\mathcal{P}}$, since our theory is not Lorentz invariant. But it is rotationally invariant; that is why the stress-tensor is symmetric in its spatial indices. Also T^{ij} satisfies

$$2\mathcal{E} = \sum_{i=1}^{2} T^{ii} \tag{2.2.26}$$

and this reflects the $SO(2,1)$ invariance, being the non-relativistic analog of (1.12).

Of course, the theory is also phase invariant; this produces one more continuity equation.

$$\partial_t \rho + \nabla \cdot \mathbf{j} = 0 \tag{2.2.27}$$

The proportionality of the matter flux current $\mathbf{j}$ to the momentum density (2.20) is a consequence of Galileo invariance.

The constants of motion are now constructed from moments of the energy momentum tensor and ρ. They are, respectively

energy $$E = \int d^2r\, \mathcal{E} \tag{2.2.28}$$

momentum $$\mathbf{P} = \int d^2r\, \boldsymbol{\mathcal{P}} \tag{2.2.29}$$

angular momentum $$M = \int d^2r\, \mathbf{r} \times \boldsymbol{\mathcal{P}} \tag{2.2.30}$$

Galileo boost $$\mathbf{B} = t\mathbf{P} - m\int d^2r\, \mathbf{r}\rho \tag{2.2.31}$$

dilation $$D = tE - \frac{1}{2}\int d^2r\, \mathbf{r} \cdot \boldsymbol{\mathcal{P}} \tag{2.2.32}$$

special conformal $$K = -t^2E + 2tD + \frac{m}{2}\int dr^2\, r^2\rho \tag{2.2.33}$$

matter number $$N = \int d^2r\, \rho \tag{2.2.34}$$

It is straightforward to verify that all these are conserved, as a consequence of the continuity equations and the special properties (symmetry and trace) of the stress tensor. Note that collectively the constants may be written analogously to (1.18) and to the Bessel–Hagen expression (1.13), as

$$C_f = \int d^2r\, \mathcal{E} f_1 - \int d^2r\, \boldsymbol{\mathcal{P}} f_2 + \int d^2r\, \rho f_3 \tag{2.2.35}$$

for suitable f_i.

The above transformations may be generalized in the following interesting manner. One may consider an *arbitrary* reparametrization of time $t \to T(t)$ [rather than the specific forms (2.6), (2.12), (2.14)]. Also one may shift $\mathbf{r}$ by a vector $\mathbf{r}_0(t)$ with *arbitrary* time dependence [rather than constant (2.7) or linear (2.10) in time]. Finally, one may rotate $\mathbf{r}$ as in (2.8), but with a *time-dependent* angle $\omega(t)$. Of course, these transformations are no longer symmetry operations of our theory, but they map our model onto another, closely related one: it is found that the above transformations introduce interactions with external fields. Specifically, after these transformations are carried out on the fields according to the rule

$$\psi'(t', \mathbf{r}') = \frac{1}{\sqrt{\dot{T}(t)}} e^{i\gamma(t,\mathbf{r})} \psi(t, \mathbf{r}) \tag{2.2.36}$$

where the transformed coordinates are given by

$$t' = T(t) \quad , \qquad r'^i = \sqrt{\dot{T}(t)}\, R^{ij}(\omega(t)) \left(r^j + r_0^j(t)\right) \tag{2.2.37}$$

and the 1-cocycle γ takes the form

$$\begin{aligned}\gamma(t, \mathbf{r}) &= \frac{m}{4} \left(\mathbf{r} + \mathbf{r}_0(t)\right)^2 \frac{\ddot{T}(t)}{\dot{T}(t)} + m r^i \left(\dot{r}_0^i(t) - \dot{\omega}(t) \epsilon^{ij} r_0^j(t)\right) + \tilde{\gamma}(t) \\ \dot{\tilde{\gamma}}(t) &= \frac{m}{2} r_0^2(t) \sqrt{\dot{T}(t)} \frac{d^2}{dt^2} \frac{1}{\sqrt{\dot{T}(t)}} + \frac{m}{2} \left(\dot{r}_0^i(t) - \dot{\omega}(t) \epsilon^{ij} r_0^j(t)\right)^2\end{aligned} \tag{2.2.38}$$

one finds that in the transformed system there arise external electric and magnetic fields, determined by the parameters of the transformations [$T(t)$, $\omega(t)$, $\mathbf{r}_0(t)$], hence the fields are time-dependent but constant in space; additionally there is an external harmonic force field, with time-varying frequency.[5] [With specific time-dependence, the parameters can conspire to produce static electric and magnetic fields as well as time-independent harmonic forces; also one can suppress selectively any of the external effects.]

Note that the symmetry transformations (2.6) – (2.15) also follow the more general rules (2.36) – (2.38). One may understand the presence of a one-cocycle in (2.36) by recognizing that when the kinetic term of a *particle* Lagrangian is transformed according to (2.37), external electromagnetic and harmonic force fields are again generated, and also the Lagrangian changes by a total time derivative which is $\frac{d}{dt}\gamma(t, \mathbf{r}(t))$.

Higher symmetries are widely studied these days in field theory, but it seems that rarely do they provide specific dynamical information about a model — rather they give an elegant frame for describing solutions and other properties.

As an exception that proves the rule, we now show that the conformal symmetries allow deriving the following useful result about the highly non-linear dynamics of our Chern–Simons theory: all static solutions carry zero energy.[6] This follows immediately from (2.32) and/or (2.33): the left sides are time independent, and so are the last terms in the right sides for time independent ρ and $\mathcal{P} = m\mathbf{j}$, *i.e.* for static solutions. Thus $H = E$, and also but less importantly D, must vanish. Similarly, from (2.31) one sees that $\boldsymbol{P}$ must vanish, but this is not surprising — we expect static solutions to carry no momentum. Note however: angular momentum need not vanish for static configurations; it can be constructed from the current $m\mathbf{j} = \boldsymbol{\mathcal{P}}$, which in the static case must be divergence-free, according to (2.27).

Since E can be given by (2.5) (provided there is sufficient regularity so that the integral $\int d^2r\, \boldsymbol{\nabla} \times \mathbf{j}$ vanishes) we see that static solutions can exist only for $g \geq 1/m\kappa$. Especially interesting is the borderline case $g = 1/m\kappa$, where the integrand is non-negative and therefore must vanish on static solutions. In this way the $SO(2,1)$ conformal symmetry demands that *all* static solutions (at $g = 1/m\kappa$) satisfy

$$D\psi = 0 \tag{2.2.39}$$

Together with the Chern–Simons constraint (1.26) this implies that ρ satisfies the Liouville equation,

$$\nabla^2 \ln \rho = -\frac{2}{\kappa}\rho \tag{2.2.40}$$

which can be integrated explicitly in terms of two arbitrary functions, which are further specified by the physical requirements that one may wish to impose on static solutions.[1,3]

[In the non-Abelian case, the analogous equations, with ψ in the same adjoint representation as the gauge fields, realize a two-dimensional reduction of four-dimensional self-dual gauge field equations in a space with signature $(+ + - -)$ and lead to many integrable systems, principally the Toda system.[4]]

As is well-known and was remarked in the Overview , the Liouville equation is invariant against conformal redefinition of the two-dimensional plane. In Euclidean space this involves the complex variable $x + iy = z$ transforming into an arbitrary function of z, but not of z^*: Our next task is to understand the properties of the action (1.23) (at $g = 1/m\kappa$) that are responsible for this infinite symmetry. In fact its stationary points are conformally invariant.

But before turning to this topic, we point out that the above described transformations can be used to generate interesting new solutions from the explicitly determined static solutions. First by Galileo [(2.10), (2.11)] or conformal [(2.14), (2.15)] boosting of static solutions, one obtains time dependent solutions to the Chern–Simons model. Moreover, by performing transformations with time-dependent parameters [(2.36), (2.37)], one finds time-dependent solutions to the Chern–Simons model with external, appropriately constructed electric and magnetic fields as well as an external harmonic force field.[7]

2.2.2. *Infinite-Dimensional Symmetry Group of Stationary Points of the Action*

The dilation transformation (2.12) and (2.13) rescales the spatial coordinate $\mathbf{r}$. Here we inquire about the response of the action (2.1a), (2.5) to a conformal redefinition of spatial coordinates,

$$\mathbf{r}' = \mathbf{r}'(\mathbf{r}) \tag{2.2.41a}$$

where

$$x' + iy' \equiv z' = z'(z) \tag{2.2.41b}$$

and time is unchanged, $t' = t$.

Generalizing (2.13) and (2.36), we posit a field transformation law with a weight,

$$\psi'(\mathbf{r}') = \frac{\partial z^*}{\partial z'^*}\psi(\mathbf{r}) \tag{2.2.42}$$

which apart from a phase is the square root of the Jacobian, as in (2.13) and (2.15), while the choice of phase is dictated by the Hamiltonian (2.5), see below. [Since time is not transformed, we suppress the time argument.] This has the consequence that the density transforms with the Jacobian as in (2.16).

$$\rho'(\mathbf{r}') = J\rho(\mathbf{r}) \tag{2.2.43}$$

$$J = \det\left\{\frac{\partial r^i}{\partial r'^j}\right\} = \left|\frac{\partial z}{\partial z'}\right|^2 \tag{2.2.44}$$

For infinitesimal $\delta z = -f(z)$, this transformation law coincides with (1.11b), taken in Euclidean space and $e^{\beta\varphi}$ identified with ρ. It further follows that the gauge potential transforms covariantly.

$$A'^i(\mathbf{r}') = A^j(\mathbf{r})\frac{\partial r^j}{\partial r'^i} \tag{2.2.45}$$

This is most easily proven by first noting that $\mathbf{A}$, when given by Eq. (1.25), is transverse and satisfies $\boldsymbol{\nabla}_{\mathbf{r}} \times \mathbf{A}(\mathbf{r}) = -\frac{1}{\kappa}\rho(\mathbf{r})$, and then verifying that $\mathbf{A}'(\mathbf{r}')$ in (2.45) also is transverse and satisfies $\boldsymbol{\nabla}_{\mathbf{r}'} \times \mathbf{A}'(\mathbf{r}') = -\frac{1}{\kappa}J\rho(\mathbf{r}) = -\frac{1}{\kappa}\rho'(\mathbf{r}')$. [In carrying out the differentiations it is useful to pass the complex variables.] It follows that $D\psi \equiv (\partial_x - i\partial_y - iA^x + A^y)\psi$ transforms with the Jacobian, because the holomorphic derivative does not act on $\partial z^*/\partial z'^*$.

$$D_{\mathbf{r}'}\psi'(\mathbf{r}') = JD_{\mathbf{r}}\psi(\mathbf{r}) \tag{2.2.46}$$

So finally we can state the transformation law for the Lagrange density.

$$\begin{aligned}\mathcal{L} &= i\psi^*\partial_t\psi - \frac{1}{2m}|D\psi|^2 + \frac{1}{2}\left(g - \frac{1}{m\kappa}\right)\rho^2 \\ &= i\psi^*\partial_t\psi - \mathcal{H}\end{aligned} \tag{2.2.47}$$

Evidently it is true that

$$\mathcal{L}'(\mathbf{r}') = Ji\psi^*(\mathbf{r})\partial_t\psi(\mathbf{r}) - J^2\mathcal{H}(\mathbf{r}) \tag{2.2.48}$$

and the Lagrangian transforms as

$$L' = \int d^2r'\, \mathcal{L}'(\mathbf{r}') = \int d^2ri\psi^*(\mathbf{r})\partial_t\psi(\mathbf{r}) - \int d^2\mathbf{r}\, J\mathcal{H}(\mathbf{r}) \tag{2.2.49}$$

One factor of the Jacobian disappears when changing variables in the spatial integration, and the symplectic form $\int d^2r\, i\psi^*\partial_t\psi$ is invariant. But the Hamiltonian density $\mathcal{H}$ remains with one factor J, hence the total Lagrangian is not in general invariant, and neither is the action, the time integral of L — because t is not changed in the present transformation rules [in contrast to (2.12) and (2.14)]. [It does not appear possible to find a transformation of time that would restore invariance.]

However, for static solutions we know that $E = \int d^2r\, \mathcal{H}$ vanishes. If this vanishing is due to the local vanishing of $\mathcal{H}$, as is true at $g = 1/m\kappa$, then the static critical points of the action are invariant. This then shows that static solutions with zero $\mathcal{H}$ will be mapped into each other by spatial conformal transformations — the dilation (2.12) expands to an infinite symmetry group on the solutions, but there are no new constants of motion.

[Since in the non-Abelian generalization, with matter in the adjoint representation, the corresponding static Chern--Simons equations are dimensional reductions of self-dual Yang–Mills equations in four dimensions,[4] the *finite-dimensional* conformal invariance of the latter[8] is seen to survive the dimensional reduction, and in two dimensions expands to the *infinite-dimensional* conformal group.]

On the other hand, in the effective field theories for the eikonal regime (1.29), (1.31), where there is no Hamiltonian to begin with, the transformations (2.41), (2.42) *are* symmetries of the action, and also τ may be arbitrarily reparametrized. Note that owing to the derivative relation (1.30) between $\Omega^\pm$ and ψ: $\psi = \sqrt{2}\frac{\partial}{\partial z^*}\,(\Omega^+ - i\Omega^-)$, the transformation law for $\Omega^\pm$ is without the weight factor,

$$\Omega'^{\pm}(\mathbf{r}') = \Omega^{\pm}(\mathbf{r}) \tag{2.2.50}$$

which arises for ψ, as in (2.42), when the derivative is taken.

3. Conclusion

The rigid scale invariance of the action for non-relativistic $(2+1)$-dimensional field theory with quartic self-interaction and coupling to a Chern–Simons gauge field, expands at the static critical points of the action to the infinite conformal group on the plane. The scale symmetry allows establishing the important result that static solutions carry zero energy, and the infinite conformal symmetry "explains" why the static system is completely integrable. The kinetic action of effective eikonal field theories also possesses the infinite symmetry.

The Chern–Simons model at $g = 1/m\kappa$ is the bosonic partner of an $N = 2$ supersymmetric theory with fermions and the invariance of the extended action against the supersymmetric generalization of the bosonic symmetries (2.6) – (2.15) has been established.[9] While the invariances of the static critical points in the supersymmetric action have not been explicitly checked, they too presumably enjoy an infinite conformal symmetry, because the supersymmetric static equations retain the form of the bosonic equations.

In our considerations, the possibility of quantum symmetry breaking anomalies has been ignored. It is known that the quartic self-interaction, which as we have seen is formally scale invariant, suffers from quantum scale anomalies.[10] This is particularly clear in the first quantized framework, where the two-dimensional δ-function potential, while scaling classically as r^{-2}, does not give rise to energy-independent phase shifts, as is required by scale invariance and is explicitly realized by the scale invariant $1/r^2$ potential. There is a quantum scale anomaly — the simplest example of the anomaly phenomenon.[11] On the other hand, anomalies in the theory with *both* quartic self-coupling and Chern–Simons interaction have thus far not been assessed; in fact there is some indication of anomaly cancellation, even without supersymmetry.[12] Further research on this question would be interesting.[13]

References

1. For a discussion of non-relativistic Chern–Simons theory and its relation through second quantization to the particle mechanics of (1.22), see *e.g.* R. Jackiw and S.-Y. Pi, *Phys. Rev. D* **42**, 3500 (1990).

2. For gravity: H. Verlinde and E. Verlinde, *Nucl. Phys.* **B371**, 246 (1992); for Maxwell theory: R. Jackiw, D. Kabat and M. Ortiz, *Phys. Lett. B* **277**, 148 (1992).

3. For Abelian Chern–Simons interactions: R. Jackiw and S.-Y. Pi, *Phys. Rev. Lett.* **64**, 2969 (1990); (C) **66**, 2682 (1992) and Ref. [1].

4. For non-Abelian Chern–Simons interactions: B. Grossman, *Phys. Rev. Lett.* **65**, 3230 (1990); G. Dunne, R. Jackiw, S.-Y. Pi and C. Trugenberger, *Phys. Rev. D* **43**, 1332 (1991); G. Dunne, *Commun. Math. Phys.* (in press).

5. S. Takagi, *Prog. Theor. Phys.* **84**, 1019 (1990), **85**, 463, 723 (1991), **86**, 783 (1991).

6. D. Freedman and A. Newell (unpublished).

7. Z. Ezawa, M. Hotta and Z. Iwazaki, *Phys. Rev. Lett.* **67**, 441 (1991); *Phys. Rev. D* **44**, 452 (1991); R. Jackiw and S.-Y. Pi, *Phys. Rev. Lett.* **67**, 415 (1991) and *Phys. Rev. D* **44**, 2524 (1991).

8. R. Jackiw and C. Rebbi, *Phys. Rev. D* **14**, 517 (1977).

9. M. Leblanc, G. Lozano and H. Min, *Ann. Phys.* (NY) (in press).

10. O. Bergman, MIT preprint CTP#2045 (1991).

11. R. Jackiw, in *M. A. B. Bég Memorial Volume*, A. Ali and P. Hoodbhoy, eds. (World Scientific, Singapore, 1991).

12. G. Lozano, *Phys. Lett. B* (in press).

13. O. Bergman, in preparation.

III. Gauge Symmetries in Lineal Gravity Theories

3.1. Overview

We study lower-dimensional gravity both for pedagogical reasons — one expects that the dimensional reduction effects sufficient simplification to permit thorough analysis, while still retaining useful content to inform the physical (3+1)-dimensional problem — and also, if one is lucky, there are practical applications — *e.g.* idealized cosmic strings are described by (2 + 1)-dimensional gravity, while the still lower-dimensional models are used in statistical mechanics.

The drastic dimensional reduction to (1 + 1) dimensions — gravity on a line, *i.e.*, *lineal* gravity — is not devoid of interest, provided dynamical equations are not based on the Einstein tensor $G_{\mu\nu} = R_{\mu\nu} - \frac{1}{2} g_{\mu\nu} R$, which vanishes identically in two dimensions.

In a proposal of several years ago,[3] it was suggested that gravity equations be based on the Riemann scalar R, the simplest entity that encodes in two dimensions all local geometric information about space-time. Moreover, in an action formulation it is necessary to introduce an additional scalar field, which acts as a Lagrange multiplier that enforces the equation of motion for R. Thus we are dealing with scalar-tensor theories, or — to use the contemporary string nomenclature — "dilaton" gravities.

Since the initial proposal, various models have been studied. Here I shall describe two that are selected by their group theoretical properties: they can be formulated as gauge theories based on groups relevant to space-time: de Sitter or anti-de Sitter (in (1 + 1)-dimensions both groups are $SO(2,1)$, although the geometries are different) and Poincaré. The first of these is the one proposed originally;[1] it is governed by the action

$$I_1 = \int d^2x \sqrt{-g}\, \eta (R - \Lambda) \tag{3.1.1}$$

The second is "string-inspired" and has been recently studied for purposes of modeling (on a line!) black hole physics;[2] its action is

$$\bar{I}_2 = \int d^2x \sqrt{-\bar{g}}\, e^{-2\varphi} \left(\bar{R} - 4\bar{g}^{\mu\nu} \partial_\mu \varphi \partial_\nu \varphi - \Lambda \right) \tag{3.1.$\bar{2}$}$$

(Notation: time and space carry the metric tensor $g_{\mu\nu}$ with signature $(1, -1)$. The two-vector $x^\mu = (t, x)$ will be frequently presented in light-cone components $x^\pm \equiv \frac{1}{\sqrt{2}}(t \pm x)$. Tangent space components are labeled by Latin letters $a, b, \ldots$, and the

Minkowski metric tensor $h_{ab} = \text{diag}(1,-1)$ raises/lowers these indices. Also we use the anti-symmetric tensor ϵ^{ab}, $\epsilon^{01} = 1$.)

In (1.1), R is the scalar curvature built from $g_{\mu\nu}$, η is a world scalar Lagrange multiplier related to the dilaton, while Λ is a cosmological constant. In ($1.\bar{2}$) we temporarily use an over-bar to denote a differently scaled metric tensor $\bar{g}_{\mu\nu}$ from which $\bar{R}$ is constructed, while φ is the dilaton. Formula ($1.\bar{2}$) arises naturally from string theory, restricted to a two-dimensional target space, with the anti-symmetric tensor field identically vanishing. In the string context, matter is taken to couple to $\bar{g}_{\mu\nu}$; for our purposes in the absence of matter it is convenient to redefine variables by $\bar{g}_{\mu\nu} = e^{2\varphi} g_{\mu\nu}$, $\eta = e^{-2\varphi}$. Then ($1.\bar{2}$) becomes

$$I_2 = \int d^2x \, \sqrt{-g} \, (\eta R - \Lambda) \tag{3.1.2}$$

but it is to be remembered that because of the redefinition, the "physical" metric tensor is $g_{\mu\nu}/(-2\eta)$. Note that (1.2) is invariant against shifting η by a constant, because $\sqrt{-g}\,R$ is a total derivative.

It is seen that the two models (1.1) and (1.2) differ in the placement of the Lagrange multiplier with the cosmological term: in (1.1) η multiplies Λ, in (1.2) the η factor is absent from Λ. Of course in the limit $\Lambda = 0$, the difference disappears.

We now describe the interesting gauge group structure of (1.1) and (1.2) which we name *(anti) de Sitter gravity* and *extended Poincaré gravity*, respectively.

3.2. (Anti) de Sitter Gravity

The equations of motion that follow from varying η and $g_{\mu\nu}$ in (1.1) are

$$R = \Lambda \tag{3.2.1}$$

$$\left(\mathcal{D}_\mu \mathcal{D}_\nu - g_{\mu\nu}\mathcal{D}^2\right)\eta + \frac{\Lambda}{2} g_{\mu\nu}\eta = 0 \tag{3.2.2a}$$

The second equation, with $\mathcal{D}_\mu$ the space-time covariant derivative, can be decomposed into traceless and trace parts.

$$\left(\mathcal{D}_\mu \mathcal{D}_\nu - \frac{1}{2} g_{\mu\nu}\mathcal{D}^2\right)\eta = 0 \tag{3.2.2b}$$

$$\left(\mathcal{D}^2 - \Lambda\right)\eta = 0 \tag{3.2.2c}$$

The above geometric dynamics may be presented in a gauge theoretical fashion.[3] To this end one uses the (anti) de Sitter group with Lorentz generator J and translation generators P_a satisfying the $SO(2,1)$ algebra (for $\Lambda \neq 0$).

$$[P_a, J] = \epsilon_a{}^b P_b \ , \qquad [P_a, P_b] = -\frac{\Lambda}{2}\epsilon_{ab} J \tag{3.2.3}$$

The gauge connection one-form is introduced $A = A_\mu \, dx^\mu$ and expanded in terms of the generators,

$$A = e^a P_a + \omega J \tag{3.2.4}$$

where e^a_μ is the *Zweibein* and ω_μ is the spin-connection. The curvature two-form

$$F = dA + A^2 \tag{3.2.5}$$

becomes

$$F = f^a P_a + fJ = (De)^a P_a + \left(d\omega - \frac{\Lambda}{4} e^a \epsilon_{ab} e^b \right) J \tag{3.2.6}$$

$$(De)^a \equiv de^a + \epsilon^a{}_b \omega e^b \tag{3.2.7}$$

It is seen that $d\omega$ is proportional to the scalar curvature density and f^a is the torsion density, each expressed in terms of e^a and ω, which at this stage are independent variables.

The Lagrange density

$$\begin{aligned} \mathcal{L}_1' &= \sum_{A=0}^{2} \eta_A F^A = \eta_a (De)^a + \eta_2 \left(d\omega - \frac{\Lambda}{4} e^a \epsilon_{ab} e^b \right) \\ F^A &= (f^a, f) \quad , \qquad \eta_A = (\eta_a, \eta_2) \end{aligned} \tag{3.2.8}$$

is gauge invariant: the three field strengths F^A transform covariantly according to the three-dimensional adjoint representation, while the Lagrangian multiplier triplet η_A transforms by the coadjoint representation.

The equation obtained from (2.8) by varying η_a gives the condition of vanishing torsion, and allows evaluating the spin connection in terms of the *Zweibein*.

$$\omega = e^a \left(h_{ab} \epsilon^{\mu\nu} \partial_\mu e^b_\nu \right) / \det e \tag{3.2.9}$$

The equation which follows upon variation of η_2 regains (2.1) once (2.9) is used. Variation of e^a and ω produces equations for the Lagrange multipliers η_a and η_2, respectively, the latter of course coinciding with η in the geometric formulations (1.1), (2.1) and (2.2).

$$d\eta_a + \epsilon_a{}^b \omega \eta_b - \frac{\Lambda}{2} \epsilon_{ab} \eta_2 e^b = 0 \tag{3.2.10a}$$

$$d\eta_2 + \eta_a \epsilon^a{}_b e^b = 0 \tag{3.2.10b}$$

Upon taking a space-time covariant derivative of (2.10b) and using (2.10a) to eliminate η_a, we recover (2.2). Finally we see that when ω is eliminated from $\mathcal{L}_1'$ with the help of (2.9), so that the torsion (2.7) vanishes, what remains is the Lagrange density of (1.1), expressed in terms of *Zweibeine*.

Thus the geometric formulation of this gravity theory is contained within the (anti) de Sitter group theoretical framework for solutions with $\det e \neq 0$, but see below.

Explicit classical solutions to the equations are easy to find. Working within the geometric framework, we use coordinate invariance to choose a conformally flat metric tensor.

$$g_{\mu\nu} = h_{\mu\nu} \exp 2\sigma \tag{3.2.11}$$

Then (2.1) becomes the Liouville equation,

$$\Box 2\sigma = \Lambda \exp 2\sigma \tag{3.2.12}$$

Its general solution depends on two arbitrary functions of the two light-cone variables, $F(x^+)$, $G(x^-)$,

$$\exp 2\sigma = \frac{F'(x^+)G'(x^-)}{\left(1 - \frac{\Lambda}{4}FG\right)^2} \tag{3.2.13}$$

whose derivatives fulfill the consistency condition $F'G' > 0$. But the residual coordinate invariance within the conformal gauge allows choosing $F(x^+) = x^+, G(x^-) = x^-$, hence

$$\exp 2\sigma = \frac{1}{\left(1 - \frac{\Lambda}{8}x^2\right)^2} \tag{3.2.14}$$

In conformal gauge, (2.2b) reduces to

$$\partial_\mu V_\nu + \partial_\nu V_\mu - h_{\mu\nu} h^{\alpha\beta} \partial_\alpha V_\beta = 0 \tag{3.2.15}$$

where V_μ is defined by

$$V_\mu \exp 2\sigma = \partial_\mu \eta \tag{3.2.16}$$

Equation (2.15) is just the (flat-space) conformal Killing equation with solutions in terms of arbitrary functions of a single light-cone variable.

$$V_- = V_-(x^+) \ , \qquad V_+ = V_+(x^-) \tag{3.2.17}$$

Finally the remaining equation (2.2c) together with (2.16) restricts these functions, so that the solution for η takes the form

$$\eta = \frac{\alpha_a x^a + \alpha_2 \left(1 + \frac{\Lambda}{8}x^2\right)}{1 - \frac{\Lambda}{8}x^2} \tag{3.2.18}$$

where α_a is a constant two-vector and α_2 is a constant scalar.

The *Zweibein* and spin connection of the gauge theoretical formulation are given by related formulas. The former, the "square root" of the metric tensor, becomes (apart from an arbitrary Lorentz transformation on the tangent-space indices)

$$e^a_\mu = \delta^a_\mu \exp\sigma = \frac{1}{1-\frac{\Lambda}{8}x^2}\delta^a_\mu \tag{3.2.19}$$

while the latter is

$$\omega_\mu = -h_{\mu\alpha}\epsilon^{\alpha\beta}\partial_\beta\sigma \tag{3.2.20}$$

The Lagrange multiplier η_2 coincides with η, while Eq. (2.10) for η_a is solved by

$$\eta_a \exp\sigma = \epsilon_a{}^\mu \partial_\mu \eta \tag{3.2.21}$$

Of course the general solution is an arbitrary coordinate transformation of the above.

Finally we observe that the gauge theoretical formulation allows an alternative group theoretical presentation of solutions. The field equations following from (2.8), upon respective variation of η_A and A, are

$$F = 0 \tag{3.2.22}$$

$$dH + [A, H] = 0 \tag{3.2.23}$$

A, F and $H = \eta_a h^{ab} P_b + \frac{2}{\Lambda}\eta_2 J$ belong to the $SO(2,1)$ algebra (the factor $2/\Lambda$ is a consequence of the group metric). Equation (2.22) implies that A is a pure gauge given by an arbitrary element U of the $SO(2,1)$ group,

$$A = U^{-1}dU \tag{3.3.24}$$

while the Lagrange multiplier is then determined by (2.23) to be

$$H = U^{-1}\Phi U \tag{3.2.25}$$

where Φ is a constant element in the algebra. The explicit group and algebra elements that correspond to the above solution, Eqs. (2.18) – (2.21), are

$$U = \exp\left(\frac{i\pi}{\sqrt{1-\frac{\Lambda}{8}x^2}}\left(-\frac{1}{2}x^a\epsilon_a{}^b P_b + J\right)\right) \tag{3.2.26}$$

and

$$\Phi = \frac{2}{\Lambda}\alpha_a\epsilon^{ab}P_b - \alpha_2 J \tag{3.2.27}$$

U is unique up to a constant gauge transformation.

Within the gauge theoretical framework, an even simpler solution to (2.22) and (2.23) is available: $A = 0$, $H = \Phi$, which makes no sense geometrically, because not only $\det e$, but both the connections e^a and ω vanish! But in fact use can be made of such solutions: when presented with a geometrically singular configuration, perform any gauge transformation producing non-singular connections, for example with the group element U above. So we see that the group theoretical framework, even in its $\det e = 0$ sector, contains adequate information for encoding the gravity theory.

3.3. Extended Poincaré Gravity

Equations of motion of the string-inspired gravitational theory (1.2) are, from varying η

$$R = 0 \tag{3.3.1}$$

and from varying $g_{\mu\nu}$

$$\left(\mathcal{D}_\mu \mathcal{D}_\nu - g_{\mu\nu}\mathcal{D}^2\right)\eta + \frac{\Lambda}{2} g_{\mu\nu} = 0 \tag{3.3.2a}$$

which is equivalent to

$$\mathcal{D}_\mu \mathcal{D}_\nu \eta = \frac{\Lambda}{2} g_{\mu\nu} \tag{3.3.2b}$$

Note that (3.2a) differs from (2.2a) by the absence of η in the last term.

To give a gauge theoretical formulation,[4] we make use of the *centrally extended* Poincaré group, whose algebra is

$$[P_a, J] = \epsilon_a{}^b P_b \ , \qquad [P_a, P_b] = \epsilon_{ab} I \tag{3.3.3}$$

where the central element I commutes with P_a and J. Consequently the connection A and curvature F now become

$$A = e^a P_a + \omega J + aI \tag{3.3.4}$$

$$\begin{aligned} F &= dA + A^2 = f^a P_a + fJ + gI \\ &= (De)^a P_a + d\omega J + \left(da + \frac{1}{2} e^a \epsilon_{ab} e^b\right) I \end{aligned} \tag{3.3.5}$$

Here a and g are the additional connection and curvature associated with the central element in the algebra.

This magnetic-like extension of the Poincaré group may be viewed as an unconventional contraction of the de Sitter group: The ordinary Poincaré algebra [Eq. (3.3) without the central element] is the $\Lambda \to 0$ contraction of the $SO(2,1)$ algebra (2.3). However, owing to the well-known ambiguity of two-dimensional angular momentum, in (2.3) one may replace J by $J - 2I/\Lambda$ before taking the $\Lambda \to 0$ limit, which then leaves (3.3).

The extension reflects a 2-cocycle in the composition law for representatives of the Poincaré group. If the group acts on coordinates x^a by

$$x^a \longrightarrow \bar{x}^a = \mathcal{M}^a{}_b x^b + q^a \tag{3.3.6a}$$

where $\mathcal{M}$ is a Lorentz transformation

$$\mathcal{M}^a{}_b = \delta^a{}_b \cosh\alpha + \epsilon^a{}_b \sinh\alpha \tag{3.3.6b}$$

and q^a is a translation, the composition law for these is

$$\mathcal{M}_{(12)} = \mathcal{M}_1\mathcal{M}_2 \tag{3.3.7a}$$
$$q_{(12)} = q_1 + \mathcal{M}_1 q_2 \tag{3.3.7b}$$

However, the composition law for a representation $G(\mathcal{M}, q)$ containing the extension (3.3) in its algebra acquires a 2-cocycle.

$$G(\mathcal{M}_1, q_1)\, G(\mathcal{M}_2, q_2) = \exp\left\{\frac{i}{2} q_1^a \epsilon_{ab} (\mathcal{M}_1 q_2)^b\right\} G(\mathcal{M}_1\mathcal{M}_2, q_1 + \mathcal{M}_1 q_2) \tag{3.3.8}$$

(I is represented by $i \equiv \sqrt{-1}$.)

A gauge transformation, generated by the gauge function Θ,

$$\Theta = \theta^a P_a + \alpha J + \beta I \tag{3.3.9}$$

produces the following transformations on the connections.

$$\begin{aligned} e^a \to \bar{e}^a &= \left(\mathcal{M}^{-1}\right)^a{}_b \left(e^b + \epsilon^b{}_c \theta^c \omega + d\theta^b\right) \\ \omega \to \bar{\omega} &= \omega + d\alpha \\ a \to \bar{a} &= a - \theta^a \epsilon_{ab} e^b - \frac{1}{2}\theta^2 \omega + d\beta + \frac{1}{2} d\theta^a \epsilon_{ab} \theta^b \end{aligned} \tag{3.3.10}$$

The multiplet of curvatures $F^A = (f^a, f, g)$ transforms by the adjoint 4×4 representation of the extended group,

$$\begin{aligned} f^a \to \bar{f}^a &= \left(\mathcal{M}^{-1}\right)^a{}_b \left(f^b + \epsilon^b{}_c \theta^c f\right) \\ f \to \bar{f} &= f \\ g \to \bar{g} &= g - \theta^a \epsilon_{ab} f^b - \frac{1}{2}\theta^2 f \end{aligned} \tag{3.3.11}$$

or

$$\begin{aligned} F^A \to \bar{F}^A &= \sum_{B=0}^{3} \left(U^{-1}\right)^A{}_B F^B \\ U &= \begin{pmatrix} \mathcal{M}^a{}_b & -\epsilon^a{}_c \theta^c & 0 \\ 0 & 1 & 0 \\ \theta^c \epsilon_{cd} \mathcal{M}^d{}_b & -\theta^2/2 & 1 \end{pmatrix} \end{aligned} \tag{3.3.12}$$

The upper left 3×3 block in U comprises the adjoint representation of the conventional Poincaré group with q^a of (3.6) identified with $-\epsilon^a{}_c \theta^c$, while the fourth row and column arise from the extension. Note that in the above realization of the gauge action on F, the extension is not visible: I is represented by $\mathbf{O}$. On the other hand, an additional connection and curvature (a, g) are present.

In this representation, the extended algebra possesses a non-singular Killing metric, which is unavailable without the extension.

$$h_{AB} = \begin{pmatrix} h_{ab} & 0 & 0 \\ 0 & 0 & -1 \\ 0 & -1 & 0 \end{pmatrix} \tag{3.3.13}$$

It is true that ${}^{T}UhU = h$; this allows raising and lowering the indices (A, B).

An invariant Lagrange density is now constructed with an extended multiplet of Lagrange multipliers η_A,

$$\begin{aligned} \mathcal{L}_2' &= \sum_{A=0}^{3} \eta_A F^A = \eta_a (De)^a + \eta_2 d\omega + \eta_3 \left(da + \frac{1}{2} e^a \epsilon_{ab} e^b \right) \\ F^A &= (f^a, f, g) \quad , \qquad \eta_A = (\eta_a, \eta_2, \eta_3) \end{aligned} \tag{3.3.14}$$

which obey the coadjoint transformation law,

$$\eta_A \to \bar{\eta}_A = \sum_{B=0}^{3} \eta_B U^B{}_A \tag{3.3.15}$$

or in components

$$\begin{aligned} \eta_a &\to \bar{\eta}_a = (\eta_b - \eta_3 \epsilon_{bc} \theta^c) \mathcal{M}^b{}_a \\ \eta_2 &\to \bar{\eta}_2 = \eta_2 - \eta_a \epsilon^a{}_b \theta^b - \frac{1}{2} \eta_3 \theta^2 \\ \eta_3 &\to \bar{\eta}_3 = \eta_3 \end{aligned} \tag{3.3.16}$$

Using the invariant metric (3.13), other group invariants may be constructed.

$$\mathcal{F}^2 = \sum_{A,B=0}^{3} {}^*F^A h_{AB} F^B \tag{3.3.17}$$

$$M = -\frac{2}{\Lambda} \sum_{A,B=0}^{3} \eta_A h^{AB} \eta_B \tag{3.3.18}$$

where ${}^*F^A$ is the 0-form $\frac{1}{2}\epsilon^{\mu\nu} F^A_{\mu\nu}$, dual to the 2-form F^A.

We recognize in (3.14) the torsion $(De)^a$ and curvature $d\omega$ densities, which vanish as a consequence of varying η_a and η_2, respectively. Thus Eq. (3.1) is regained. The Lagrange multiplier η in (1.2) corresponds to η_2 in the present formulas and the equation for it, obtained by varying ω, is as in the (anti) de Sitter model, (2.10b),

$$d\eta_2 + \eta_a \epsilon^a{}_b e^b = 0 \tag{3.3.19a}$$

while the equation for η_a, obtained by varying e^a, differs from (2.10a),

$$d\eta_a + \epsilon_a{}^b \omega \eta_b + \eta_3 \epsilon_{ab} e^b = 0 \tag{3.3.19b}$$

We need a value for η_3 to close the system (3.19). The equation for that multiplier is obtained by varying a,

$$d\eta_3 = 0 \tag{3.3.19c}$$

and a constant, cosmological solution

$$\eta_3 = -\frac{\Lambda}{2} \tag{3.3.19d}$$

renders (3.19b) similar to (2.10a),

$$d\eta_a + \epsilon_a{}^b \omega \eta_b - \frac{\Lambda}{2} \epsilon_{ab} e^b = 0 \tag{3.3.19e}$$

except that there is no factor of η_2 in the last, cosmological term of (3.19e). This of course has the consequence that when (3.19a) and (3.19e) are combined as before, the second order equation that emerges for $\eta = \eta_2$ reproduces (3.2).

The remaining equation of the gauge theoretical formulation, obtained by varying η_3

$$da = -\frac{1}{2} e^a \epsilon_{ab} e^b \tag{3.3.20}$$

and allowing evaluation of a, has no counterpart in the geometric formulation. Equation (3.20) can always be locally integrated because the right side is a two-form, hence closed in two dimensions. However in general, there will be singularities in a, since upon integrating (3.20) over a two-space, the right side gives the total "volume," which could be a well-defined non-vanishing quantity, while the left side always integrates to zero if the manifold is closed and bounded, and a is non-singular. It is seen that the "volume" coincides with the flux of a "magnetic" field arising from a.

Note that upon eliminating ω in $\mathcal{L}_2'$ with the zero-torsion equation $(De)^a = 0$ and evaluating η_3 at $-\Lambda/2$, $\mathcal{L}_2'$ coincides with the Lagrange density in (1.2), now expressed in terms of *Zweibeine*, apart from the total derivative $-\Lambda/2\, da$. This does not affect the equations of motion, but it does contribute to the action a topological quantity: $-\frac{\Lambda}{2}\int da$. Thus the cosmological constant enters as a coefficient of a topological term — the "magnetic" flux — in an analogy to the vacuum angle in gauge theories.

Thus here again, the group theoretical formulation reproduces the geometric one, for solutions with $\det e \neq 0$, but again see below. However, the former is more flexible: Eq. (3.19c) is satisfied with vanishing η_3; this corresponds to a vanishing cosmological constant. Thus the gauge theory built on the *extended* Poincaré group possesses as a solution a *non-extended* system. It is interesting therefore that here the cosmological term is an integration constant, and not inserted *a priori* into the theory.

Finding explicit solutions is straightforward. In the geometric formulation, (3.1) is solved by a flat metric tensor.

$$g_{\mu\nu} = h_{\mu\nu} \tag{3.3.21}$$

Then (3.2) immediately gives

$$-2\eta = M - \frac{\Lambda}{2}(x - x_0)^2 \tag{3.3.22}$$

with M and x_0 being integration constants, the former reflecting the η-translation invariance mentioned earlier.

Interest in the model[2] derives precisely from the above "black-hole" solution [in terms of the "physical" metric $g_{\mu\nu}/(-2\eta)$] with mass M, located at x_0. An arbitrary coordinate transformation of this configuration produces the general solution.

The gauge theoretical counterparts of the above are a flat *Zweibein* (apart from a constant tangent-space Lorentz transformation)

$$e^a_\mu = \delta^a_\mu \tag{3.3.23}$$

and a vanishing spin connection.

$$\omega = 0 \tag{3.3.24}$$

Taking in (3.19c) the cosmological solution for η_3, allows solving (3.19e) for η_a

$$\eta_a = \frac{\Lambda}{2}\epsilon_{a\mu}\left(x^\mu - x_0^\mu\right) \tag{3.3.25}$$

and from (3.19a) $\eta_2 = \eta$ is recovered to be as in (3.22). Finally (3.20) is solved for a.

$$a_\mu = \frac{1}{2}\epsilon_{\mu\nu}x^\nu \tag{3.3.26}$$

with a pure gauge contribution $\partial_\mu\chi$ left arbitrary. The potential in (3.26) corresponds to a constant "magnetic" field, as is appropriate with our "magnetic-like" extension of translations.

Note the two invariants defined in (3.17) and (3.18): $\mathcal{F}^2$ vanishes since F^A does, while M is recognized as the "black hole" mass.

The gauge theoretical solution may of course also be presented in a group theoretical fashion, since the equations are of the same form as in (2.22) and (2.23), with all quantities belonging to the *extended* algebra and group. The explicit formulas, corresponding to the "black hole" solution, Eqs. (3.21) - (3.26), are as follows. The group element U that leads to the pure gauge connection $A = U^{-1}dU$ is

$$U = \exp x^a P_a \tag{3.3.27}$$

up to a constant gauge transformation. The constant algebra element Φ that gives $H = \eta_a h^{ab} P_b - \eta_3 J - \eta_2 I = U^{-1}\Phi U$ is [placement of η_2 and η_3 dictated by the group metric (3.13), *viz.* $\eta^A = h^{AB}\eta_B$]

$$\Phi = \frac{\Lambda}{2} x_0^a \epsilon_a{}^b P_b + \frac{\Lambda}{2} J + \left(\frac{M}{2} - \frac{\Lambda}{4} x_0^2\right) I \tag{3.3.28}$$

As in the (anti) de Sitter model, we see that after a further gauge transformation we pass to the geometrically singular configuration $A = 0$, $H = \Phi$. This gives an especially succinct account of the relevant geometric information : Φ encodes the integration constants, which characterize the intrinsic geometry: the cosmological constant Λ, the "black hole" mass M and location x_0. A geometry is built with these characteristics once a gauge transformation is performed, say with the above U, to obtain non-singular connections.

3.4. CONCLUSION

The two models here considered are special: their geometric dynamics access a gauge theoretical formulation. The extended Poincaré model exhibits the intriguing possibility of a cosmological term that is an integration constant, as are the "black hole" mass M and location x_0; all three are encoded in the Lagrange multipliers of the theory.

Both models can also be obtained by dimensional reduction from $(2+1)$ dimensions: To obtain (anti) de Sitter gravity in its geometric formulation one begins[1] with the Einstein theory/Hilbert action (with cosmological term), suppresses dependence on the third dimension, sets $g_{\mu 2}$ to zero for $\mu = 0, 1$ and g_{22} to η^2; for the gauge theoretical formulation one starts with the *Dreibein*-spin connection form of the theory, which also is equivalent to a Chern–Simons, $\mathcal{O}(2,2)$ or $\mathcal{O}(3,1)$ model.[5] Extended Poincaré gravity can be similarly constructed, but the higher-dimensional theory has to be suitably extended by an Abelian ideal. Indeed it is found that *both* the (anti) de Sitter and extended Poincaré $(1+1)$ dimensional theories arise as *different* dimensional reductions of a *single*, extended $(2+1)$-dimensional gravity.[6] This and another interesting topic — the coupling of matter consistently with the gauge principle[7] — are beyond the scope of our review. In yet a further investigation one could study non-topological theories in which invariants (3.17) and/or (3.18) are added to the Lagrange density (3.14).

In conclusion, we note that dynamics determined by a group has been familiar in physics since the invention of Yang–Mills theory. However, the examples described here offer a new possibility: in the Lie algebra that determines a gauge theory one can allow an extension. This gives rise to richer dynamics within the same group theoretical structure, and in the gravity model studied above produces the cosmological constant as a (boundary) value for a dynamical variable, rather than a parameter in the Lagrangian.[8]

References

1. C. Teitelboim, *Phys. Lett.* **126B**, 41 (1983) and in *Quantum Theory of Gravity*, S. Christensen, ed. (Adam Hilger, Bristol, 1984); R. Jackiw, in *Quantum Theory of Gravity*, S. Christensen, ed. (Adam Hilger, Bristol, 1984) and *Nucl. Phys.* **B252**, 343 (1985).

2. H. Verlinde, in *The Sixth Marcel Grossman Meeting on General Relativity*, H. Sato, ed. (World Scientific, Singapore, 1992); C. Callan, S. Giddings, A. Harvey and A. Strominger, *Phys. Rev. D* **45**, 1005 (1992). Quantization of the model is discussed in these papers.

3. T. Fukuyama and K. Kamimura, *Phys. Lett.* **160B**, 259 (1985); K. Isler and C. Trugenberger, *Phys. Rev. Lett.* **63**, 834 (1989); A. Chamseddine and D. Wyler, *Phys. Lett. B* **228**, 75 (1989). Quantization of the model is discussed in these papers.

4. D. Cangemi and R. Jackiw, *Phys. Rev. Lett.* **69**, 233 (1992). My lecture was prepared in collaboration with D. Cangemi, particularly in finding explicit solutions; this I gratefully acknowledge.

5. A. Achúcarro and P. Townsend, *Phys. Lett. B* **180**, 89 (1986); E. Witten, *Nucl. Phys.* **B311**, 46 (1988/89).

6. D. Cangemi, in preparation.

7. G. Grignani and G. Nardelli, in preparation.

8. We note that the cosmological constant has appeared in this guise previously: A. Aurilia, H. Nicolai and P. Townsend, *Nucl. Phys.* **B176**, 509 (1980); S. Hawking, in *Shelter Island II*, R. Jackiw, N. Khuri, S. Weinberg and E. Witten, eds. (The MIT Press, Cambridge, MA, 1985). M. Henneaux and C. Teitelboim, *Phys. Lett.* **143B**, 415 (1984), **B222**, 195 (1989). In these works dynamical generation of the cosmological constant is an *option* that is taken when building a model. In contrast, our gauge theoretical formulation *requires* the mechanism, which arises naturally owing to the extension in the algebra.

Three Lectures on Supersymmetry and Extended Objects

P K Townsend
DAMTP, University of Cambridge
Silver Street, Cambridge, U.K

Contents

1: Extended objects revisited

A few years ago I gave some lectures at the Trieste spring school [1] on supermembranes and, to a lesser extent, supersymmetric p-dimensional objects, or p-branes for short. The emphasis there was on the 11-dimensional supermembrane [2], and the main issue was the nature of its quantum spectrum. As explained in [1], there was good evidence that the supermembrane, unlike the bosonic membrane, contains zero energy states, which it was hoped would correspond to the massless particles of $d = 11$ supergravity, but the non-uniqueness of the configuration of a collapsed object for $p > 1$ suggested that the spectrum might be continuous, which would preclude a particle-like interpretation. Shortly afterwards, de Wit, Lüscher and Nicolai [3] showed, for the regularized version of the $d = 11$ supermembrane reviewed in [1], that the spectrum *is* continuous, from zero. Although it has been questioned whether this result will continue to hold for the full, unregularized, supermembrane, or for extended objects of higher dimension, where the mathematics of [3] may fail, convincing physical arguments in favour of this scepticism have not been found.

However, as was also explained in [1], the theory of super p-branes is also of interest in the context of topological defects in supersymmetric field theories, some examples of which are provided by instantons of Euclidean d-dimensional field theories interpreted as p-dimensional topological defects of the $(d+p+1)$-dimensional Lorentzian field theory (provided that $(d+p+1)$ does not exceed the maximum dimension for which this field theory has a supersymmetric extension). One such example is the familiar Yang-Mills (YM) instanton, which becomes a fivebrane of supersymmetric $d = 10$ YM theory [4]. An attempt to extend this to a solution of the $d = 10$ supergravity/YM theory that serves as a low-energy approximation to the heterotic string was made by Strominger [5], but what he found was really a solution of the Chapline-Manton theory, as I shall explain. A number of related solutions have been found, at least one of which is known to be an *exact* solution of *classical* heterotic string theory [6,7], but it is unclear whether any of them can survive as approximate solutions of the quantum theory.

As an introduction I shall begin with a discussion of bosonic extended objects, as in [1] (and using the same notation as far as possible) but where there is an overlap I shall make use of alternative methods, which will hopefully clarify points skipped over in [1]; I shall

L. A. Ibort and M. A. Rodríguez (eds.), Integrable Systems, Quantum Groups, and Quantum Field Therapy 317–345.

also take the opportunity to correct a few mis-statements. In lecture 2 I shall review the connection between supersymmetry and solitons, generalizing the original observations of Witten and Olive [8]. The most intriguing aspect of fivebranes, taken up in lecture 3, is the electric-magnetic-type duality between strings and fivebranes in $d = 10$, as first pointed out by Nepomechie [9] in the context of 'elementary' magnetic sources. The implications for the effective low-energy supergravity theory of string and/or fivebrane theory have been pursued by Duff and Lu [10,11,12] to the extent that the issue of quantum consistency deserves to be reassessed. One barrier to progress is the fact that the heterotic fivebrane, as a *six-dimensional worldvolume field theory*, has yet to be constructed; its field content must differ from the standard super-fivebrane action [13] by the inclusion of fields required for coupling to a YM background (as for the heterotic string) but it is not known how this can be done in a way that is consistent with all required symmetries.

A convenient starting point is the action for a p-dimensional extended object in d-dimensional spacetime $\mathcal{M}$ with Minkowski metric η. Let W be the $(p+1)$-dimensional worldvolume, with coordinates $\{\xi^i; i = 0, 1, \ldots, p\}$, and let $\{X^\mu; \mu = 0, 1, \ldots, d-1\}$ be coordinates for $\mathcal{M}$. The embedding of W in $\mathcal{M}$ is specified by a map $f : W \to \mathcal{M}$ such that $\xi^i \mapsto X^\mu(\xi)$, by virtue of which η induces a metric M on W with components

$$M_{ij} = \partial_i X^\mu \partial_j X^\nu \eta_{\mu\nu} . \tag{1.1}$$

We take as our action

$$S = -T \int d^{p+1}\xi \sqrt{-\det M} , \tag{1.2}$$

where T is the p-volume tension. As in [1], we shall suppose that the object is closed, i.e. has no boundary, so the only boundary conditions on the integral that need to be specified are at the initial and final times. For $p = 0$ this action is that of a relativistic point particle of mass $m = T$, while for $p = 1$ it is the Nambu-Goto action for a (closed, bosonic) relativistic string. We are using units in which c, the speed of light, is unity. An interesting point [14] here is that if the factors of c are reinstated, the non-relativistic limit $c \to \infty$ is straightforward only for $p = 0$. In that case the Lagrangian in the gauge $X^0 = ct$ is $L = mc^2 + (1/2)mv^2 + O(c^{-2})$ so the $c \to \infty$ limit can be taken after discarding the constant rest-mass energy. For the string, the leading term in the expansion of the Lagrangian in inverse powers of c is Tlc^2, where $l = \oint d\sigma \sqrt{|\mathbf{X}'|^2}$ is the length of the string (σ being the string's coordinate and the prime indicating a derivative with respect to it); this is again the rest-mass energy but now it cannot simply be discarded because it is *not* constant. The relativistic string is intrinsically relativistic in a way that the relativistic particle is not because the dynamics of even a free string does not separately conserve rest-mass energy. Similar remarks of course apply for $p > 1$.

As explained in [1], an action equivalent to (1.2) (in the sense that the classical Euler-Lagrange equations are equivalent) is

$$S = -\frac{1}{2}T \int d^{p+1}\xi \sqrt{-\det\gamma}\left[\gamma^{ij}M_{ij} - (p-1)\right] , \tag{1.3}$$

where γ^{ij} are the components of the inverse of an *independent* worldvolume metric γ. This action can be interpreted as a set of scalar fields $X^\mu(\xi)$ of a $(p+1)$-dimensional field theory coupled to 'gravity', and with a cosmological constant proportional to $(p-1)$. For $p=0$ we may set $T=m$ and $\gamma_{00}=-m^2V^2$ to arrive at the action

$$S=\int dt\left[\frac{1}{2V}\dot{X}^\mu\dot{X}^\nu\eta_{\mu\nu}+m^2V\right] \tag{1.4}$$

for a relativistic particle of mass m. Taking $m\to 0$ we get the action of a massless particle for which the 'cosmological' term vanishes. It might at first appear that there is no analogous limit for $p>0$ because for a string (1.3) reduces to the well-known action

$$S=-\frac{1}{2}T\int d^2\xi\,\sqrt{-\det\gamma}\,\gamma^{ij}\partial_iX\cdot\partial_jX\ , \tag{1.5}$$

for which there is already no 'cosmological term', while for $p>1$ this term is present but cannot be consistently omitted. However, an alternative form of the action, equivalent (classically) to (1.2), but more closely analogous to the particle action (1.4), is

$$S=\int d^{p+1}\xi\,\frac{1}{2}\left[\frac{1}{V}\det M-T^2V\right]\ , \tag{1.6}$$

where V is an independent worldvolume *scalar density*. We may now take the $T\to 0$ limit for any p to arrive at the action of the *null* p-brane [15]

$$S=\int d^{p+1}\xi\,\frac{1}{2V}\det M\ . \tag{1.7}$$

This action has the feature that it is *spacetime* conformal invariant. Specifically, if k^μ is a conformal Killing vector ($k_{(\mu;\nu)}=(1/d)g_{\mu\nu}k^\lambda{}_{;\lambda}$) then S is invariant under the infinitesimal transformations

$$\delta X^\mu=k^\mu \qquad \delta V=-2\frac{(p+1)}{d}Vk^\mu{}_{;\mu}\ . \tag{1.8}$$

The physics of null objects is clearer from the phase-space form of the action. Make the worldvolume space/time split $\xi^i\to(t,\boldsymbol{\sigma})$, where $\boldsymbol{\sigma}=(\sigma^I,\ I=1,\ldots,p)$ are the p-brane's coordinates. Then an action equivalent to (1.7) is

$$S=\int dt\oint d^p\sigma\,[\dot{X}^\mu p_\mu-\ell p^2-s^Ip\cdot\partial_IX]\ , \tag{1.9}$$

where $\ell(t,\boldsymbol{\sigma})$ and $s^a(t,\boldsymbol{\sigma})$ are simultaneously gauge fields for, respectively, the time and space reparametrization invariances and Lagrange multipliers for the constraints (cf. general relativity). The equivalence can be shown by eliminating p_μ and s^a by their equations of motion and making the identification $V=\ell\det M_{ab}$. Note in particular the constraint $p^2(\boldsymbol{\sigma})=0$ which shows that *all points of a null extended object move at the speed of light.* This is not very sensible physically but the null p-brane is a useful starting point for the discussion of some mathematical issues.

The spacetime conformal invariance of the null p-brane action should not be confused with the well-known *worldsheet* Weyl invariance of the string action (1.5). Weyl invariance is often supposed to be an exclusive property of string theory but consider the following action for a p-dimensional extended object [16]:

$$S = -T\int d^{p+1}\xi\,\sqrt{-\det\gamma}\left[\frac{\gamma^{ij}M_{ij}}{(p+1)}\right]^{\frac{p+1}{2}} . \tag{1.10}$$

This action is Weyl invariant for *any* value of p. It is (classically) equivalent to our original action (1.2) because this is what it reduces to on elimination of the metric γ by means of its (algebraic) field equation (for $p=0$ this step is unnecessary) and reduces to (1.5) for $p=1$. The action (1.10) is *not* spacetime conformal invariant since, as a consequence of Weyl invariance, a rescaling of the metric now has no effect.

One interpretation of the action (1.2) is as an effective action at low energy, i.e. long wavelength, for a p-dimensional topological defect of a d-dimensional field theory (note that d is the dimension of spacetime whereas p is number of *spatial* dimensions of the defect). To fix ideas I shall concentrate in this lecture on domain walls in d=4 scalar field theories. This has the virtue of potential physical relevance, as well as simplicity. As our first example we shall consider a complex scalar field A with the action

$$S = \int d^4x\,\left[\partial\bar{A}\cdot\partial A - |U'(A)|^2\right] , \tag{1.11}$$

where U is a holomorphic function of A and U' is its derivative. This choice of potential is motivated by supersymmetry, in which context U is called the superpotential. We shall suppose that U has at least one critical point so that the minimum of the potential is at zero. If there is at least one other critical point then there will be at least two possible vacua. If A has one vacuum value as, say, $x\to-\infty$ and another one as $x\to+\infty$ then there must be a domain wall between. Let us look for a domain wall solution to the A equations of motion for A independent of y and z, assuming periodicity in the y and z directions in order to achieve finite total energy. The energy per unit area $\mathcal{E}$ can then be written as [17]

$$\begin{aligned}\mathcal{E} &= \int dx\,\left[|\dot{A}|^2 + |\partial_x A|^2 + |U'|^2\right]\\ &= \int dx\,\left[|\dot{A}|^2 + |\partial_x A - e^{i\alpha}\overline{U'(A)}\,|^2\right] \;+\; \mathcal{R}e(e^{-i\alpha}\mathcal{T}) ,\end{aligned} \tag{1.12}$$

where

$$\mathcal{T} = 2\big|U(A)\big|_{x=-\infty}^{x=\infty} \tag{1.13}$$

is a complex *topological* charge (depending only on the boundary conditions at $x=\pm\infty$). By choosing $\alpha = \arg\mathcal{T}$ we deduce the energy density bound

$$\mathcal{E} \geq |\mathcal{T}| , \tag{1.14}$$

with equality for a static solution satisfying the *first-order* equation

$$\partial_x A = e^{i\alpha}\overline{U'(A)} . \tag{1.15}$$

Multiplying by U' and integrating over x we find that $\mathcal{T} = -e^{i\alpha}2\int_{-\infty}^{\infty} dx\mathcal{L}(x)$, where $\mathcal{L}(x) = |U'(A(x))|^2$ is the Lagrangian density for the solution $A(x)$ of (1.15), so that α is indeed the argument of $\mathcal{T}$.

We now wish to consider [18] what might be the effective action governing the dynamics of fluctuations of this domain wall away from the static configuration assumed above. Let $X^\mu(\xi)$ be the locus in spacetime of the wall's worldvolume, W. The cartesian coordinates of a point p near W can be written as

$$x^\mu = X^\mu(\xi) + xn^\mu(\xi) , \tag{1.16}$$

where x is the perpendicular distance of p from W and n^μ are the components of an outward normal, n, to W passing through p and the point on W with coordinates ξ^i, as shown in Fig. 1.1.

Fig. 1.1: Coordinates of a point in $\mathcal{M}$ near W.

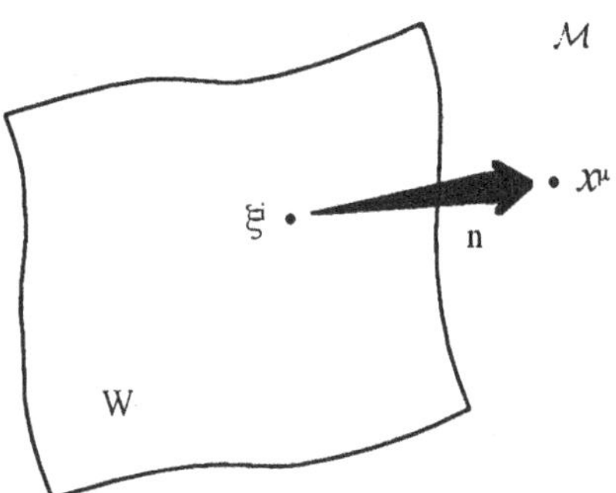

We must suppose that x is less than the radii of curvature of W in order to ensure the uniqueness of n. The metric of $\mathcal{M}$ at p is

$$ds^2 = \eta_{\mu\nu}\, dx^\mu dx^\nu . \tag{1.17}$$

Now (1.16) provides a change of coordinates from x^μ to (ξ^i, x). In terms of these new coordinates the metric is

$$ds^2 = \left[M_{ij} - 2x\Omega_{ij} + O(x^2)\right] d\xi^i d\xi^j + (dx)^2 , \tag{1.18}$$

where

$$\Omega_{ij} = n \cdot X,_{ij} \tag{1.19}$$

are the components of the extrinsic curvature Ω, or second fundamental form, of W as a hypersurface in $\mathcal{M}$. Thus, if R^{-1} is the lowest eigenvalue of Ω then

$$ds^2 = M_{ij}d\xi^i d\xi^j + (dx)^2 + O(x/R) , \tag{1.20}$$

and the volume element of $\mathcal{M}$ is therefore

$$\sqrt{-\det M}\,(1 + O(x/R))d^3\xi dx . \tag{1.21}$$

In these new coordinates the locus of W is $x = 0$. Neglecting the domain wall's curvature we may assume that it is produced by the static scalar field configuration $A(x)$. Then, to order (x/R), $\mathcal{L}(A)$ is a function only of x and so

$$\begin{aligned} S &= \Big[\int dx \mathcal{L}(x)\Big] \int d^3\xi \sqrt{-\det M}\big(1 + O(x/R)\big) \\ &= -|T| \int d^3\xi \sqrt{-\det M}\big(1 + O(x/R)\big) \ . \end{aligned} \tag{1.22}$$

Neglecting the O(x/R) terms we arrive at (1.2) with $T = |T|$. Note that the terms neglected can be expected to include not only corrections proportional to the extrinsic curvature but also interaction terms between the domain wall and the scalar field A. It is a curious feature of extended object solutions of *relativistic* field theories that, to leading order, the dynamics of the object decouples from that of the fields from which it is composed. The difficulty with the non-relativistic limit mentioned previously is presumably a reflection of the fact that such a decoupling does not happen for extended objects in non-relativistic field theories (e.g. vortices in superfluids). This is reminiscent of the fact that the discovery of the relativistic invariance of Maxwell's equations made the ether hypothesis unnecessary.

Some interesting new issues are raised by another type of scalar field theory with domain walls [19]. The Lagrangian density has the form

$$\mathcal{L} = \frac{1}{2}(\partial\phi^I \cdot \partial\phi^J - m^2 k^I(\phi)k^J(\phi))g_{IJ}(\phi) \ . \tag{1.23}$$

The scalar fields constitute a map $\phi : \mathcal{M} \rightarrow M$, from spacetime to an n-dimensional target space M with coordinates $\{\phi^I,\ I = 1, 2 \ldots, n\}$ and metric g_{IJ}. M is assumed to have a Killing vector k, and k^I are its components. A particularly interesting case, again motivated by supersymmetry, is (M, g) *hyper-Kähler* and k *triholomorphic*. These terms require some explanation. A hyper-Kähler manifold has the following properties:

(i) A quaternionic structure. That is, a triplet of complex structures $\mathbf{J}(\phi)$ obeying the algebra of the quaternions,

$$(J^a)_I{}^K (J^b)_K{}^J = -\delta^{ab}\delta_I{}^J + \varepsilon^{abc}(J^c)_I{}^J \tag{1.24}$$

(ii) A metric g that is Hermitian with respect to all three complex structures. That is, the second rank tensors

$$\mathbf{\Omega}_{IJ} = (\mathbf{J})_I{}^K g_{KJ} \tag{1.25}$$

are *antisymmetric*. They can therefore be taken to be the components of a triplet of two-forms $\mathbf{\Omega}$, called the *Kähler two-forms.*

(iii) The Kähler two forms are closed. That is, $d\mathbf{\Omega} = 0$.

A triholomorphic Killing vector k is one for which $\mathcal{L}_k\mathbf{\Omega} = 0$, where $\mathcal{L}_k$ is the Lie derivative with respect to k. Since $\mathcal{L}_k = di_k + i_k d$, where i_k indicates the interior product with k (i.e. contraction) we have $d(i_k\mathbf{\Omega}) = 0$, i.e. $i_k\mathbf{\Omega}$ is a triplet of closed one-forms on M. Its pullback $f^*(i_k\mathbf{\Omega})$ is a triplet of closed one-forms on $\mathcal{M}$. Again assuming independence of the fields on y and z, we identify

$$\mathbf{Q} = \int_{-\infty}^{\infty} dx\, \phi'^I k^J \mathbf{\Omega}_{IJ} \tag{1.26}$$

as a *topological* 3-vector charge, where the prime now indicates differentiation with respect to x. There is also, of course, the Noether charge

$$Q_0 = \int_{-\infty}^{\infty} dx\, \dot{\phi}^I k^J g_{IJ} , \tag{1.27}$$

associated with invariance of (1.23) under $\delta\phi^I \propto k^I(\phi)$.

As in the previous example we now write the energy density as

$$\begin{aligned} E &= \int_{-\infty}^{\infty} dx\, \frac{1}{2} g_{IJ}(\dot{\phi}^I\dot{\phi}^J + \phi'^I\phi'^J + m^2 k^I k^J) \\ &= \int_{-\infty}^{\infty} dx \Big\{ \frac{1}{2} g_{IJ}(\phi'^I - m(\mathbf{n}\cdot\mathbf{J})^I{}_K k^K)(\phi'^J - m(\mathbf{n}\cdot\mathbf{J})^J{}_L k^L) \\ &\qquad + \frac{1}{2} g_{IJ}(\dot{\phi}^I - m n_0 k^I)(\dot{\phi}^J - m n_0 k^J)\Big\} \\ &\quad + m(n_0 Q_0 + \mathbf{n}\cdot\mathbf{Q}) , \end{aligned} \tag{1.28}$$

where $(n_0, \mathbf{n})$ is a *unit* Euclidean 4-vector, i.e. $n_0^2 + \mathbf{n}\cdot\mathbf{n} = 1$. By choosing $n = (n_0, \mathbf{n})$ parallel to $Q = (Q_0, \mathbf{Q})$ we obtain from (1.28) the energy bound

$$E \geq m|Q| = m\sqrt{Q_0^2 + \mathbf{Q}\cdot\mathbf{Q}} , \tag{1.29}$$

which is saturated by solutions of the first-order equations

$$\dot{\phi}^I = m n_0 k^I \qquad \phi'^I = m(\mathbf{n}\cdot\mathbf{J})^I{}_J k^J . \tag{1.30}$$

Conversely, any solution of eqs. (1.30) has the property that the 4-vector Q is *parallel* to the 4-vector n. This can be shown by substitution of eqs. (1.30) into (1.26) and (1.27) and use of the quaternion algebra satisfied by the three complex structures.

Consider now the special case for which (M, g) is the four-dimensional Eguchi-Hanson manifold. There exist coordinates $(\phi^0, \boldsymbol{\phi})$ for which the metric takes the form [20]

$$ds^2 = \Phi^{-1}(d\phi^0 + \boldsymbol{\omega}\cdot d\boldsymbol{\phi})^2 + \Phi d\boldsymbol{\phi}\cdot d\boldsymbol{\phi} , \tag{1.31}$$

where

$$\Phi = \frac{1}{2}\Big[\frac{1}{|\boldsymbol{\phi} - \boldsymbol{\phi}_0|} + \frac{1}{|\boldsymbol{\phi} + \boldsymbol{\phi}_0|}\Big] \tag{1.32}$$

and $\boldsymbol{\omega}$ is a solution to $\boldsymbol{\nabla}\times\boldsymbol{\omega} = \pm\boldsymbol{\nabla}\Phi$ (in the usual notation of Euclidean 3-vector calculus). The three closed Kahler 2-forms are

$$\boldsymbol{\Omega} = (d\phi^0 + \boldsymbol{\omega}\cdot d\boldsymbol{\phi})d\boldsymbol{\phi} - \Phi d\boldsymbol{\phi}\times d\boldsymbol{\phi} , \tag{1.33}$$

where the wedge product of forms is understood. The metric (1.31) is singular at $\boldsymbol{\phi} = \pm\boldsymbol{\phi}_0$ but this is only a coordinate singularity if we make the identification $\phi^0 \sim \phi^0 + 2\pi$. The tri-holomorphic Killing vector is $\partial/\partial\phi^0$, except at the points $\boldsymbol{\phi} = \pm\boldsymbol{\phi}_0$ where ϕ^0 is not defined.

Clearly, k vanishes at these points. The scalar field potential therefore has two isolated minima at $\boldsymbol{\phi} = \pm\boldsymbol{\phi}_0$ and we expect there to exist kink-type solutions interpolating between them with topological charge $\mathbf{Q} = 2\boldsymbol{\phi}_0$. To find them we look for solutions to eqs. (1.30), which now read

$$\dot{\phi}^0 = mn_0 \qquad \dot{\boldsymbol{\phi}} = 0 \qquad \phi'^0 = 0 \qquad \boldsymbol{\phi}' = m\Phi^{-1}\mathbf{n}\ . \tag{1.34}$$

For $\mathbf{n}$ parallel to $\boldsymbol{\phi}_0$ they have the finite energy solutions

$$\phi^0 = \varphi + mn_0 t \qquad \boldsymbol{\phi} = \boldsymbol{\phi}_0 \tanh\left(m|\mathbf{n}|(x - x_0)\right)\ , \tag{1.35}$$

where x_0 and φ are constants. The novel features here are:

(i) In addition to the static solutions there are also stable *time-dependent* solutions with Noether charge $Q_0 = (n_0/|\mathbf{n}|)|\mathbf{Q}|$, and hence energy

$$E = \frac{m}{\sqrt{1 - n_0^2}}|\mathbf{Q}|\ . \tag{1.36}$$

(ii) In addition to depending on a position x_0 in space, the kink solution (1.35) also depends on the additional parameter φ.

The latter point has implications for the effective action because in addition to its motion in space the domain wall's motion in the 'internal' space parametrized by φ must be taken into account. It is not difficult to see how this should be done. Although we started with a $d = 4$ field theory, it can be viewed as a $d = 5$ theory with a spacetime of the form $\mathcal{M} \times S^1$ and a prescribed dependence of the fields on the extra coordinate. Specifically, starting from the Lagrangian (1.23) but *without a potential*, and taking x^5 to be the S^1 coordinate, the Lagrangian with the potential is found on imposing

$$\frac{\partial}{\partial x^5}\phi^I = mk^I\ , \tag{1.37}$$

which is consistent with (1.35) if $\varphi = m^{-1}x^5$. Motion in the internal S^1 is therefore the same as motion in the extra dimension. A time-dependent kink solution can be viewed as static solution that has been boosted in the extra dimension; the square root factor in (1.36) is just the usual relativistic factor in the expression for the energy of a moving particle. The effective action of the domain walls in this model is therefore presumably

$$S = -T\int d^3\xi\, \sqrt{-\det(M_{ij} + m^2\partial_i\varphi\partial_j\varphi)}\ . \tag{1.38}$$

I leave the verification of this as an exercise for the student.

2. Solitons and Supersymmetry

A soliton is a solution of a classical field theory with a finite localized energy density, so that at scales large compared to its size it appears to be a point particle. In the original meaning of the word 'soliton' there was also a condition that the particles scatter elastically with an S-matrix determined entirely by phase shifts. This is possible only in dimension 2 but may be good approximation for low-energy scattering in higher dimensions. In any case we shall here use the term in its looser sense of 'particle-like object'. Note that here we avoid the terminology 'extended' particle (which indicates that it is not actually point-like but only apparently so at large wavelengths) because of possible confusion with extended objects.

In the previous lecture we saw how, for a particular field theory, a lower bound on the energy could be deduced by expressing it as a sum of squares. Not all field theories with solitons have this property, but those that do have the simplifying feature that the soliton solutions can be found by solving associated *first-order* equations. The possibility of expressing the energy as a sum of squares is also a feature of supersymmetric field theories. In fact, any purely bosonic field theories with soliton solutions saturating a Bogomolnyi-type bound is *supersymmetrizable* in the sense that there exists a supersymmetric theory of which it is the purely bosonic sector. In this supersymmetric theory the first-order equations that are solved by the soliton field configuration can alternatively be deduced from the requirement that some of the supersymmetry of the vacuum solution be preserved. This connection is well-illustrated by the supersymmetric extension of the first model considered in the previous lecture, i.e. the Wess-Zumino model. Since we are at present interested in solitons rather than domain walls, we immediately dimensionally reduce to $d = 2$ by requiring that no field depend on y or z. It is then convenient to introduce the notation

$$\partial_{\ddagger} = \partial_t + \partial_x \qquad \partial_{=} = \partial_t - \partial_x \; . \tag{2.1}$$

These derivatives are Lorentz covariant; they scale with weight +2 or −2, respectively, under Lorentz transformations. The scale weight, or Lorentz 'charge', is indicated by a suffix with that number of plus or minus signs. The fermion partners of the complex scalar field A, which are for $d = 4$ the four real components of a Majorana spinor, may here be taken to be two *complex* anticommuting chiral spinor fields ψ_+ and ψ_- (the chirality being given by the sign of the Lorentz charge). The $d = 2$ action is

$$\begin{aligned} S = \int dt dx \Big[& \partial_{=} A \partial_{\ddagger} \bar{A} - |U'(A)|^2 \\ & + i\bar{\psi}_+ \partial_{=} \psi_+ + i\bar{\psi}_- \partial_{\ddagger} \psi_- - iU''(A)\psi_+\psi_- - i\overline{U''(A)}\bar{\psi}_+\bar{\psi}_- \Big] \; . \end{aligned} \tag{2.2}$$

The supersymmetry transformations are

$$\begin{aligned} \delta A &= i\epsilon_+ \psi_- + i\epsilon_- \psi_+ \\ \delta\psi_+ &= -\partial_{\ddagger} A \bar{\epsilon}_- - \overline{U'(A)}\epsilon_+ \\ \delta\psi_- &= -\partial_{=} A \bar{\epsilon}_+ + \overline{U'(A)}\epsilon_- \; . \end{aligned} \tag{2.3}$$

Note that we have one complex spinor parameter of one chirality and another of the other chirality, which means that we have an action with (2,2) supersymmetry, as expected from its $d = 4$ origin. The corresponding (real) Noether charges may be parametrized by two (real) phases β and γ as follows:

$$
\begin{aligned}
Q_+(\beta) &= \frac{1}{\sqrt{2}}\left\{e^{-\frac{i}{2}\beta}\int_{-\infty}^{\infty} dx\left[\partial_{\ddagger} A\bar{\psi}_+ - U'(A)\psi_-\right] \; + \; \text{c.c.}\right\} \\
Q_-(\gamma) &= \frac{1}{\sqrt{2}}\left\{e^{-\frac{i}{2}\gamma}\int_{-\infty}^{\infty} dx\left[\partial_{=} A\bar{\psi}_- + U'(A)\psi_+\right] \; + \; \text{c.c.}\right\} .
\end{aligned}
\tag{2.4}
$$

From (2.2) we see that $\dot{A} = \partial L/\partial\dot{A} := \overline{\Pi_A}$, where Π_A is the variable conjugate to A. Making the replacement $\dot{A} \to \overline{\Pi_A}$ and using the (anti)commutation relations

$$
[\Pi_A\,, A] = -i \qquad \{\psi_+, \bar{\psi}_+\} = \{\psi_-, \bar{\psi}_-\} = 1 \,, \tag{2.5}
$$

from canonical quantization of (2.2) (setting $\hbar = 1$), we find that

$$
\begin{aligned}
\left(Q_+(\beta)\right)^2 = H + P \qquad\qquad & \left(Q_-(\gamma)\right)^2 = H - P \\
\{Q_+(\beta), Q_-(\gamma)\} = 2\mathcal{R}e&\left(e^{-\frac{i}{2}(\beta+\gamma)}\mathcal{T}\right) ,
\end{aligned}
\tag{2.6}
$$

where H is the Hamiltonian, P the total momentum and $\mathcal{T}$ the topological charge introduced in the previous lecture, which now appears in the (2,2) supersymmetry algebra as a *central charge*. Now consider the particular (hermitian) supersymmetry charge

$$
S = \frac{1}{\sqrt{2}}\left(Q_+(\alpha) + Q_-(\alpha)\right) , \tag{2.7}
$$

where $\alpha = \arg \mathcal{T}$. This charge satisfies

$$
S^2 = H - T \,, \tag{2.8}
$$

where $T = |\mathcal{T}|$, as before. As the left hand side is a positive definite operator we deduce that

$$
H \geq T \,. \tag{2.9}
$$

If $|Sol\rangle$ is the eigenstate of H representing the static soliton, and if we identify $\mathcal{E}$ as its eigenvalue, then we recover from (2.9) the energy bound of (1.6). Moreover, as we saw previously, this bound is *saturated* by the static soliton solution so

$$
S^2|Sol\rangle = 0 \;\Rightarrow\; \langle Sol|S^2|Sol\rangle = 0 \;\Rightarrow\; \left\|S|Sol\rangle\right\|^2 = 0 \,, \tag{2.10}
$$

the last step following from the Hermiticity of S. Assuming a positive definite Hilbert space we conclude that

$$
S|Sol\rangle = 0 \,. \tag{2.11}
$$

It is not difficult to show that there are two such hermitian charges annihilating $|Sol\rangle$ so precisely half of the supersymmetry is preserved by the soliton. This can also deduced by an analysis of the transformation laws (2.3) in the soliton background. In this background, $\psi_+ = \psi_- = 0$ so δA vanishes. Since $\partial_{\ddagger} A = -\partial_{=} A = \partial_x A = e^{i\alpha}\overline{U'(A)}$, for the soliton background, we see from (2.3) that $\delta\psi_+ = \delta\psi_- = 0$ provided that $\epsilon_+ + e^{i\alpha}\bar{\epsilon}_- = 0$. This is one complex condition on two complex parameters, so precisely *half the supersymmetry is broken by the soliton, and half preserved.* This of course remains true if we now re-interpret the $d = 2$ soliton as a string in $d = 3$ or a membrane in $d = 4$.

The above analysis was introduced in [7] in the context of a (1,1) supersymmetric model for which the topological charge is real. The possibility of a complex central charge in the (2,2) model introduces some interesting new features [21,16,22] when we consider whether the collision of two solitons of masses $M_1 = |\mathcal{T}_1|$ and $M_2 = |\mathcal{T}_2|$ can produce a third soliton of mass $M_3 = |\mathcal{T}_3|$. By topological charge conservation, $\mathcal{T}_3 = \mathcal{T}_1 + \mathcal{T}_2$, so

$$M_3 = |\mathcal{T}_1 + \mathcal{T}_2| \leq M_1 + M_2 \ , \tag{2.12}$$

with equality only if the phases of $\mathcal{T}_1$ and $\mathcal{T}_2$ are equal. If the phases differ then $M_3 < M_1 + M_2$ and the third soliton is stable in the sense that energy is required to disassociate it. Given the existence of this third soliton the stability of the first two requires that $M_1 < M_2 + M_3$ and $M_2 < M_3 + M_1$, i.e that *the phases of all three solitons differ.* This is not possible for real charges, for which at least two have the same sign and hence the same phase, but for complex charges we may have a situation in which any two of three types of soliton can fuse to form the third. The simplest model for which this occurs has the superpotential

$$U(A) = A^4 - 4A \ . \tag{2.13}$$

This has three critical points; at the cube roots of unity. To each pair of critical points at A_a and A_b is associated the topological charge

$$\mathcal{T}_{ab} = 2\big[U(A_b) - U(A_a)\big] = -6(A_b - A_a) \ . \tag{2.14}$$

There are therefore three *possible* topological charges, which form an equilateral triangle, as shown in Fig. 2.1.

Fig. 2.1: Topological charges for the superpotential $U = A^4 - 4A$.

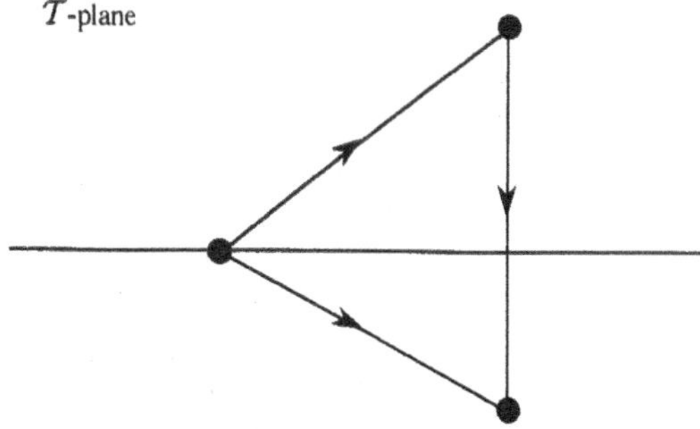

The three vertices of this triangle represent the three different vacua. For this example symmetry considerations ensure that each vertex is indeed connected to each of the other two by a soliton solution. Hence any two solitons could *in principle* fuse to form a third of lower energy, radiating quanta of the scalar field A in the process. However, it happens that the model with superpotential (2.13) belongs to a class of quantum field theory for which there is no radiation produced by the collision of solitons and for which the S-matrix for soliton scattering can be found *exactly* [21]. It is crucial for this integrability that the topological charge triangle be equilateral. For a generic fourth-order superpotential the triangle formed by the three possible topological charges will not be equilateral, but in this case there is no guarantee that there is a soliton solution corresponding to every leg of the triangle. Further analysis [17] shows that for some ranges of the parameters defining the superpotential there are only two distinct solitons, so that two of the three vacua are not directly connected to each other, but for other ranges of these parameters, which of course include the special case of (2.13), all three solitons exist. In the latter case the process of soliton fusion described above is possible and since the model is integrable only for the special case of (2.13) there will generically be a non-zero probability for soliton fusion to occur. In the context of the original $d = 4$ Wess-Zumino model this process of soliton fusion corrsponds to the fusion of two domain walls to form a third one of lower tension. This produces a system of *intersecting* walls [17] (although it seems that when gravity is included this is no longer possible [23]). This is more easily visualized for strings in $d = 3$ as shown in Fig. 2.2:

Fig. 2.2: Fusing of two strings (in $d = 3$) to produce intersections.

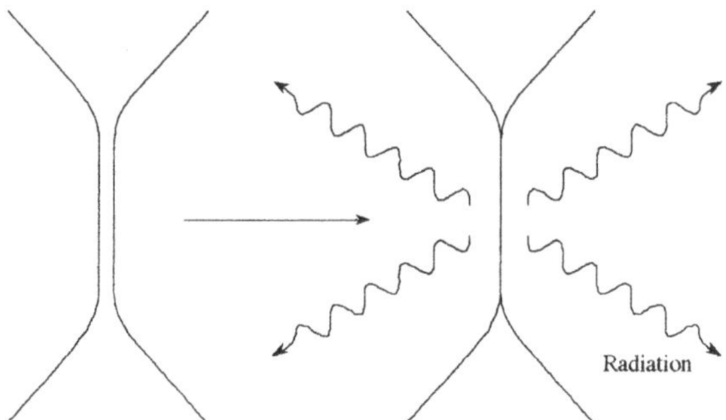

Similar remarks apply to the second model considered in the previous lecture, i.e the hyper-Kähler sigma model with a potential given by the length squared of a triholomorphic Killing vector, although there are several differences. Firstly, the maximal supersymmetry is now (4,4). Secondly, the Noether charge Q_0 *also* appears as a central charge in the supersymmetry algebra, which is not surprising given that this charge can be considered, as we saw, to be the momentum in an extra dimension.

An example of a soliton in a $d > 2$ field theory is provided by the vortex of the $d = 3$ Abelian-Higgs model. The maximally supersymmetric extension has an $N = 4$ supersymmetry. It can be obtained by dimensional reduction from $d = 6$, where the vortex solution has the interpretation of a threebrane [24], but we shall remain here in $d = 3$. The bosonic Lagrangian density of the maximally supersymmetric Abelian-Higgs model is

$$\mathcal{L} = (D_\mu\phi_1)\overline{(D^\mu\phi_1)} + (D_\mu\phi_2)\overline{(D^\mu\phi_2)} - \frac{1}{4}F_{\mu\nu}F^{\mu\nu} - V(\phi_1,\phi_2) \,, \tag{2.15}$$

where ϕ_1 and ϕ_2 are two complex scalar fields of charge e, i.e. $D_\mu\phi = \partial_\mu + ieA_\mu\phi$, and A_μ is an Abelian gauge potential with field strength $F_{\mu\nu}$. The potential V is given by

$$V = \frac{1}{2}e^2\Big[(|\phi_1|^2 - \lambda^2)^2 + |\phi_2|^2(|\phi_2|^2 + \lambda^2)\Big] , \tag{2.16}$$

in terms of another (real) parameter λ, which we may assume to be positive. The potential is minimized when

$$\phi_2 = 0 \qquad \phi_1 = \lambda e^{i\theta} , \tag{2.17}$$

for arbitrary phase $\theta(x)$.

The Hamiltonian associated with L is

$$\begin{aligned} H = \int d^2x \Big\{ \sum_r |\pi^r|^2 + \frac{1}{2}\mathbf{E}\cdot\mathbf{E} + \sum_r |(\boldsymbol{\nabla} + ie\mathbf{A})\phi_r|^2 + \frac{1}{2}B^2 + V \\ + A_0\Big[\boldsymbol{\nabla}\cdot\mathbf{E} - ie\sum_r(\pi^r\phi_r - \bar{\pi}^r\bar{\phi}_r)\Big]\Big\} , \end{aligned} \tag{2.18}$$

where $B = F_{12}$ is the magnetic field, $\mathbf{E}$ (the electric field) is the variable conjugate to $\mathbf{A}$, and π^r are the variables conjugate to ϕ_r, $r = 1, 2$. For configurations with $\mathbf{E} = 0$, $\pi^r = 0$ and $\phi_2 = 0$ the Gauss law constraint is satisfied and the energy is

$$\begin{aligned} \mathcal{E} &= \int d^2x \{\mathbf{D}\phi_1 \cdot \overline{\mathbf{D}\phi_1} + \frac{1}{2}B^2 + \frac{1}{2}e^2(|\phi_1|^2 - \lambda^2)^2\} \\ &= \int d^2x \Big\{|(D_1 \pm iD_2)\phi_1|^2 + \frac{1}{2}\big[B \mp e(|\phi_1|^2 - \lambda^2)\big]^2\Big\} + T , \end{aligned} \tag{2.19}$$

where (D_1, D_2) are the components of $\mathbf{D}\phi = \big(\boldsymbol{\nabla} + ie\mathbf{A}\big)\phi$, and

$$T = \pm e \oint d\mathbf{l} \cdot \left[(|\phi_1|^2 - \lambda^2)\mathbf{A} + i\phi_1\boldsymbol{\nabla}\bar{\phi}_1\right] \tag{2.20}$$

is a topological charge since the line integral is taken over the circle at spatial infinity. Assuming that ϕ_1 tends to one of its vacuum values parametrized by the angular variable θ as $|x| \to \infty$, we find that $T = \pm e\lambda^2 \oint d\mathbf{l} \cdot \boldsymbol{\nabla}\theta$. We therefore deduce [25] the energy bound

$$\mathcal{E} \geq (2\pi)e\lambda^2|\nu| , \tag{2.21}$$

where ν is an integer. This bound is saturated by solutions of the first-order equations

$$(D_1 + \text{sgn}(\nu)D_2)\phi_1 = 0 \qquad B = \text{sgn}(\nu)e(|\phi_1|^2 - \lambda^2) . \tag{2.22}$$

Vortex solutions of (2.22) exist for any integer ν. It can be shown that the topological charge $T = (e\lambda^2)\nu$ appears in the $d = 3$, $N = 4$, supersymmetry algebra as a central charge; if Q^a_α, $a = 1, \ldots, 4$ are the four real two-component spinor charges, then

$$\{Q^a_\alpha, Q^b_\beta\} = \delta^{ab}(\Gamma^\mu C)_{\alpha\beta}P_\mu + T\Omega^{ab}C_{\alpha\beta} \tag{2.23}$$

where C is the (antisymmetric) charge-conjugation matrix, and Ω^{ab} are the components of a real antisymmetric matrix (the details for a different model, but one with the same algebra, may be found in [26]). For this model there is a single real topological charge but the consequences are otherwise the same as before, i.e. the algebra (2.23) implies the bound (2.21) and the vortex solutions, which saturate the bound, break half the supersymmetry.

We now turn to the question of the effective action for solitons in supersymmetric field theories. We shall concentrate here on the $d = 3$ vortices since the $d = 2$ case has some non-generic features. One approach, reviewed in [1], starts from the observation that since a soliton breaks translation invariance, its low-energy dynamics should be governed by an associated Goldstone variable $\mathbf{X}(t)$. This can be promoted to the Lorentz-vector $X^\mu(t)$ by requiring worldline parametrization invariance of the action because the unphysical X^0 variable can then be 'gauged away'. The usual particle action is then found to be the lowest dimension one with the required symmetry properties. These ideas can be put on a firmer foundation by means of the theory of non-linear realizations of spacetime symmetries. We shall not pursue this direction here because the details are rather involved, especially in the supersymmetric case [27,28,29]; instead we shall rely on educated guesswork. A first guess at the effective action in the supersymmetric case might be

$$S = -m \int dt\, \sqrt{\omega \cdot \omega}\,, \tag{2.24}$$

where

$$\omega^\mu = \dot{X}^\mu - i\bar{\theta}_a \Gamma^\mu \dot{\theta}_a \tag{2.25}$$

since we expect fermionic Goldstone variables θ_a, and ω^μ is invariant under the $N = 4$ supersymmetry transformations

$$\delta X^\mu = i\bar{\epsilon}_a \Gamma^\mu \theta_a \qquad \delta\theta_a = \epsilon_a\,. \tag{2.26}$$

This cannot be right because the supersymmetry Noether charges of the action (2.24) obey the usual supersymmetry algebra, without a central charge. In fact, *any* Lagrangian invariant under (2.26) will have this property. In order to circumvent it we need a Lagrangian that is *not* invariant, but since the *action* must remain invariant the Lagrangian is limited to change by a total derivative. A term of this type which also has the desired effect of modifying the algebra of charges is called a Wess-Zumino (WZ) term (no connection with the Wess-Zumino model, other than the authors). It happens that there is a unique WZ term of the same dimension as (2.24). By its inclusion we arrive at the massive superparticle action

$$S = -m \int dt\, \sqrt{-\omega \cdot \omega} \;+\; iT\Omega^{ab} \int dt \bar{\theta}_a \dot{\theta}_b\,. \tag{2.27}$$

It can be shown that the supersymmetry charges of this action satisfy the algebra (2.23), as required. There remains one point to check. We know that when the bound $m \geq T$ is saturated only half of the supersymmetry is broken so only half the components of θ_a are needed as Goldstone variables. Remarkably, this is taken into account by the superparticle action as a result of a fermionic gauge invariance, or 'kappa symmetry', which appears precisely when $m = T$ (as explained in [1] for general p).

Once we have understood the principles that underlie the construction of the effective action we can run the previous arguments in reverse. That is, we can restrict the possibilities for the existence of solitons in supersymmetric field theories by the requirement that there must exist an effective action for them. This restriction is disappointingly weak; for example, the massive superparticle action exists in *any* spacetime dimension (but possibly requiring extended supersymmetry). However, once we take into account the fact that a soliton solution of a d-dimensional field theory can generally be used to represent a p-dimensional object in a higher-dimensional field theory for some $p \geq 1$ we may also require the existence of an effective action for these objects. This turns out to be very restrictive. By arguments similar to those used above, and as explained in [1], the effective action analogous to (2.27) for p-dimensional extended objects in a d-dimensional spacetime is the super p-brane action

$$S = -T \int d^{p+1}\xi \sqrt{-\det(\omega_i^\mu \omega_j^\nu \eta_{\mu\nu})} + TS_{WZ} , \tag{2.28}$$

where

$$\omega_i^\mu := \partial_i X^\mu - i\bar{\theta}\Gamma^\mu \partial_i \theta \tag{2.29}$$

generalizes (2.25). I refer to [1] for more details. Here it will suffice to recall that the WZ term can be constructed only for specific values of (p, d) [13], as summarized in the 'Brane-Scan' of Fig.2.3. Note the four sequences labelled **R**,**C**,**H**,**O** with common co-dimension $d - p - 1 = 1, 2, 4, 8$, respectively. A feature of the Brane-Scan not shown on the figure is that the number N of supersymmetries (counting each "minimal" spinor as one) is restricted to $N > 1$ for $p > 1$. The maximal number of supersymmetries within each of the **R**,**C**,**H**, and **O** sequences, *counting separately each real spinor component*, is therefore 4, 8, 16 and 32 respectively.

The particular field theories that we have been studying provide explicit examples of the **R** and **C** sequences. A candidate for the **H** sequence is the four-dimensional Euclidean Yang-Mills (YM) instanton, viewed as a soliton in $d = 5$. In agreement with the brane-scan, its maximally supersymmetric extension is obtainable by dimensional reduction from $d = 10$, where it can be interpreted as a fivebrane with a YM core [4]. This example has an unsatisfactory feature however in that the $d = 5$ instantonic soliton has an arbitrary size and perturbations away from an exact static solution will cause it to spread out indefinitely or shrink to a singularity. This can be overcome by considering instead the $d = 4$ YM/Higgs monopole, which has a definite size determined by the vacuum value of the Higgs field. The existence of the $d = 4$ monopole (in the BPS limit, where the YM/Higgs system is supersymmetrizable) does not contradict the Brane-Scan because, like the Q-kink solution considered in lecture 1, its effective action depends on an additional S^1 variable which can be interpreted as an extra dimension. The $d = 4$ monopole may therefore be regarded as a soliton in a $d = 5$ spacetime of the form $\mathcal{M}_4 \times S^1$.

A candidate for a $d = 9$ soliton of the **O** sequence is the 'octonionic instanton' of eight dimensional Euclidean SO(8) YM theory [30], which can be viewed as a string in $d = 10$ [31], but this has rather different properties. For instance, it breaks 15/16 of the supersymmetry, rather than half, and it cannot account for the eleven-dimensional supermembrane.

Fig.2.3: The Brane Scan

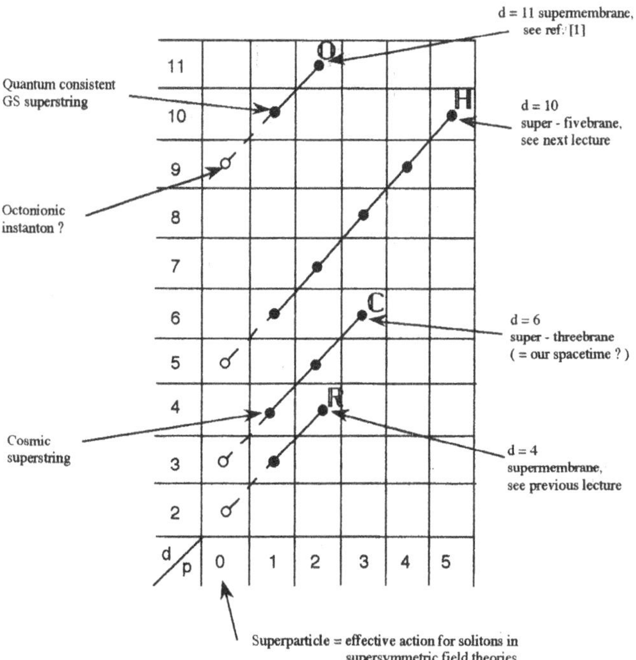

An alternative way to understand the restrictions on (p, d) imposed by the Brane-Scan is as a consequence of (i)*worldvolume supersymmetry*, and (ii) *the absence of worldvolume fields of spin* $> 1/2$. The worldvolume supersymmetry is at first surprising because only *spacetime* supersymmetry is built into the construction described above; the worldvolume supersymmetry emerges only after fixing the gauge invariances. I refer to [18] for details of how this happens for the $d = 4$ supermembrane, for which the worldvolume superspace form of the action has recently been found [28]. On second thoughts the worldvolume supersymmetry should not be too surprising because the fact that a soliton breaks only half the supersymmetry implies that the gauge-fixed action must be invariant under some 'linearly-realized' fermionic symmetry. By 'linear' I mean that the transformations do not contain field-independent terms so that, in particular, the fermion field variations vanish when the fermion fields do. This is sufficient for us to be able to invoke the Haag-Lopusanski-Sohnius theorem to the effect that any fermionic symmetry must be supersymmetry. The requirement that worldvolume fields have spins $\leq 1/2$ follows directly from the assumption that all of them are Goldstone fields resulting from the spontaneous breakdown, *at the locus of the object's worldvolume*, of (super)symmetries, but this assumption is not always justifiable since worldvolume fields may also appear for topological reasons unrelated to symmetry, as we shall see in the following lecture. If worldvolume vectors or antisymmetric tensors are permitted then various new possibilities arise for $d = 10$. In fact, fivebrane [6] and threebrane [32,33] solutions of type II $d = 10$ supergravity theories have been found that would not be allowed by the brane-scan. Although their full effective actions have not yet been constructed the field content is known and does include a worldvolume vector or antisymmetric tensor. There are no known examples for which there are higher worldvolume spins, but neither do I know of a theorem forbidding them; the absence of any example with worldvolume spin two is the major obstacle to an interpretation of our (3+1)-dimensional spacetime as the worldvolume of a threebrane embedded in a higher-dimensional spacetime.

A feature of the vortex and other solitons that is absent in the simpler scalar field models is the existence of *multi-vortex* solutions, corresponding to $\nu > 1$. The phase of ϕ_1 increases by $2\pi\nu$ as the circle at spatial infinity is circumscribed once, which means that ϕ_1 must have ν zeros in the interior. The space of vortex solutions of charge ν is therefore the space of polynomials of order ν in one complex variable [34]. The coefficients of this polynomial are the moduli, i.e. coordinates, of the solution space. The Gauss-law constraint in (2.18) can in principle be solved for an infinite set of unconstrained momenta P_A conjugate to a set of gauge-invariant variables Q^A, the coordinates of the quotient space $\mathcal{Q}$ of all field configurations $(\phi_r, \mathbf{A})$ modulo gauge transformations. The kinetic term in the Hamiltonian will then take the form $T = g^{AB}P_AP_B$, which defines a metric on the space $\mathcal{Q}$. This induces a metric on the ν complex dimensional subspace of ν-vortex solutions. The effective action for (non-relativistic) multi-vortex solutions is therefore a sigma-model with the moduli space as the target space and the low-energy scattering of vortices is given by geodesic motion on this space [35]. This constitutes an interesting generalization of the ideas discussed above for one-soliton solutions (although it seems unlikely that it can be made relativistic).

The importance of supersymmetry in this context is that the effective action is then a *supersymmetric* sigma model with half the supersymmetry of the original field theory, the half that is preserved when the Bogomoln'yi bound is saturated (this has recently been demonstrated directly [36] for multi-solitons in the model of [26], following methods used in [37] for instantons). This fact has interesting consequences because extended supersymmetry imposes strong constraints on the target space metric. For the vortex solutions, for example, we started with a maximum of 8 supersymmetries (counting each component as one) so we expect an effective sigma model with 4 supersymmetries. But this number (corresponding to (2,2) supersymmetry in $d = 2$) requires that the moduli space metric be Kähler, as indeed it is [34]. Similarly the maximally supersymmetric model for monopoles is the $N = 4$ YM theory which have a total of 16 supersymmetries. The effective sigma-model action will therefore have 8 supersymmetries (corresponding to (4,4) supersymmetry in $d = 2$) and this implies [38] that the metric on the multi-monopole space must be hyper-Kähler, which is again known to be true [39].

3. String/fivebrane duality

There is a natural generalization of the super p-brane action that describes the dynamics of supersymmetric extended objects in the presence of background supergravity fields, as briefly explained in [1]. In particular, if there is a $(p+1)$-form gauge potential in the supergravity multiplet then it couples to the p-brane via a generalization of the Lorentz coupling of an electrically charged particle to the electromagnetic gauge potential, and this is all that we shall need here. It is customary in string theory to denote by B the (two-form) potential in the $d=10$, $N=1$, supergravity multiplet but for generality, and because the heterotic string theory case requires further modifications, let us here denote the $(p+1)$-form gauge potential by A. Thus we consider

$$A = \frac{1}{(p+1)!} dX^{\mu_1} \cdots dX^{\mu_{p+1}} A_{\mu_1 \ldots \mu_{p+1}} \tag{3.1}$$

coupled to a p-brane via the worldvolume integral $q_e \int_W f^*(A)$, where q_e is an 'electric' charge density (i.e. charge per unit p-volume). This can be rewritten as the *spacetime* integral

$$\frac{1}{(p+1)!} \int d^d x \sqrt{-g}\, J_e^{\mu_1 \ldots \mu_{p+1}}(x) A_{\mu_1 \ldots \mu_{p+1}}(x) \ , \tag{3.2}$$

where the $(p+1)$th rank antisymmetric 'electric' current tensor J_e is given by

$$\sqrt{-g}(J_e)^{\mu_1 \ldots \mu_{p+1}}(x) = q_e \int_{t_i}^{t_f} dt \int d^p\sigma\, \delta^d(x - X(t,\sigma)) \varepsilon^{i_1 \ldots i_{p+1}} \partial_{i_1} X^{\mu_1} \cdots \partial_{i_{p+1}} X^{\mu_{p+1}} \ , \tag{3.3}$$

and g is the determinant of the spacetime metric; this current satisfies the conservation condition $\partial_\nu(\sqrt{-g} J_e^{\nu\mu_1 \ldots \mu_p}) \equiv 0$ except at the initial and final times t_i and t_f.

If we take the spacetime action for A to be

$$S = -\frac{1}{2(p+2)!} \int d^d x \sqrt{-g} F_{\mu_1 \ldots \mu_{p+2}} F^{\mu_1 \ldots \mu_{p+2}} \ , \tag{3.4}$$

where $F_{\mu_1 \ldots \mu_{p+2}} = (p+2)\partial_{[\mu_1} A_{\mu_2 \ldots \mu_{p+2}]}$ are the components of $F = dA$, then variation of the combined action with respect to A yields

$$\frac{1}{\sqrt{-g}} \partial_\mu (\sqrt{-g}\, F^{\mu\nu_1 \ldots \nu_{p+1}}) = (J_e)^{\nu_1 \ldots \nu_{p+1}} \ , \tag{3.5}$$

which we can rewrite as

$$\star d \star F = J_e \ , \tag{3.6}$$

where J_e is the differential form with components $(J_e)_{\mu_1 \ldots \mu_{p+1}}$, and the star indicates the Hodge dual. This equation and the Bianchi identity $dF \equiv 0$ generalize to p-branes the electrodynamics of point particles [40]. The total charge density q_e is given by the integral

$$q_e = \int_{\Sigma_{d-p-1}} \star J_e \ , \tag{3.7}$$

where Σ_{d-p-1} is a $(d-p-1)$-dimensional spacelike subspace of $\mathcal{M}$ which intersects W once, as shown in Fig. 1.2 (but with p dimensions supressed):

Fig 3.1: Spacelike subspace intersects W on p-brane.

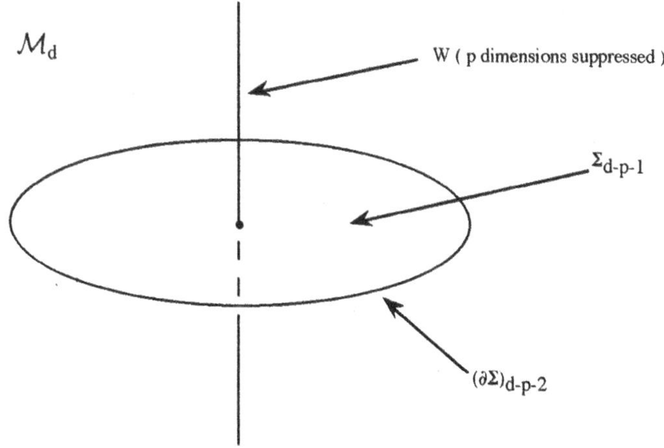

The formula (3.7) is an identity for the singular source assumed above (generalizing a point particle) but is valid generally. Using (3.6), q_e can be expressed as the surface integral

$$q_e = \int_{(\partial\Sigma)_{d-p-2}} \star F \,, \tag{3.8}$$

in direct analogy with Gauss' law in electrodynamics. As in electrodynamics, there is a generalization that allows for magnetic sources as well as electric ones [9,40]. The equations for F in the presence of both types of source are

$$\star d \star F = J_e \qquad \star\, dF = J_m \,, \tag{3.9}$$

where J_m is a $(d-p-3)$-form 'magnetic' current. In the absence of the electric source we could solve the first of equations (3.9) for a $(d-p-3)$-form, *magnetic dual*, potential $\tilde{A}$ by setting

$$\star F := G = d\tilde{A} \,. \tag{3.10}$$

This equation then becomes the Bianchi identity $dG \equiv 0$ while the second equation of (3.9) now reads $\star d \star G = J_m$. By comparison with (3.6) it can be seen that this is an equation for the potential $\tilde{A}$ in the presence of a magnetic source of dimension

$$\tilde{p} = d - p - 4 \,. \tag{3.11}$$

The total magnetic charge density (i.e. charge per unit $\tilde{p}$-volume) is given by the formula analogous to (3.8),

$$q_m = \int_{\Sigma_{p+3}} \star J_m = \int_{(\partial\Sigma)_{p+2}} \star G \,. \tag{3.12}$$

Equivalently,

$$q_m = \int_{(\partial\Sigma)_{p+2}} F \ . \tag{3.13}$$

As for monopoles in electrodynamics, quantum consistency requires the Dirac-like quantization condition [9,40,41]

$$\frac{q_e q_m}{2\pi\hbar} = \text{integer} \ . \tag{3.14}$$

The formula (3.11) confirms that the magnetic objects dual to electric particles in four-dimensional electrodynamics are particle-like, but we also learn that *the magnetic dual of a string in ten dimensions is a fivebrane.* The usual conformal gauge worldsheet action for the bosonic string, with tension $T = 1/2\pi\alpha'$, in a metric and antisymmetric tensor background is the sigma-model action

$$S = \frac{1}{2\pi\alpha'}\int d^2\xi \, \frac{1}{2}\left[\delta^{ij}\partial_i X^\mu \partial_j X^\nu G_{\mu\nu} + \epsilon^{ij}\partial_i X^\mu \partial_j X^\nu B_{\mu\nu}\right] . \tag{3.15}$$

This shows that in order to apply the quantization condition (3.19) to string theory we should set (e.g.) $q_e = 1$ and $\hbar = 2\pi\alpha'$. The magnetic charge of a fivebrane is therefore quantized in integer units of

$$Q \equiv \frac{1}{4\pi^2\alpha'}\int_{(\partial\Sigma)_3} H \ , \tag{3.16}$$

which need not vanish because H need not be *globally* exact.

In all the discussion so far we have supposed the magnetic source to be a higher-dimensional analogue of the *singular* Dirac magnetic monopole. Is there a string theory with a *non-singular magnetic fivebrane*, analogous to the 'tHooft-Polyakov monopole? As a first approach [5] to this question, we might take as our starting point the leading order effective ($d = 10$) supergravity/YM theory for the heterotic string, which can be found to by a one-loop sigma-model calculation [42] followed by a 'by hand' supersymmetrization. Instead, I shall begin with what might be described as a zeroth order approach in which the starting point is 'ordinary' $d = 10$ supergravity/YM theory, i.e. the Chapline-Manton theory [43], with gauge group $SO(32)$ (for simplicity I ignore the $E_8 \times E_8$ possibility). I shall make use of the formulation of this theory (but not all of the conventions) given in [44] (which also contains results on the supersymmetrization of the string-induced corrections, which we shall subsequently have to confront). After rescaling some of the fields, the bosonic part of the action in units for which the gravitational coupling constant is unity is

$$S = \int d^{10}x\, \sqrt{-g}e^{-2\phi}\left[R + 4(\partial\phi)^2 - \frac{1}{2.3!}H^{MNP}H_{MNP} - \frac{1}{8}\alpha'\mathrm{tr}(F^{MN}F_{MN})\right] , \tag{3.17}$$

where R is the scalar curvature of the $d = 10$ spacetime metric g_{MN} in spacetime coordinates $\{x^M, M = 0, 1\ldots 9\}$, ϕ is the dilaton, F_{MN} is the field strength tensor of an $SO(32)$ YM potential and *tr* indicates a trace in the *vector* representation. The third rank antisymmetric tensor H_{MNP} defines the 'modified' field strength three-form

$$H = dB + \frac{1}{4}\alpha' K_3 \ , \tag{3.18}$$

where K_3 is the (Chern-Simons) three-form potential for the four-form

$$dK_3 = \mathrm{tr}(F \wedge F) \ . \tag{3.19}$$

The constant α' will later be identified as the inverse of 2π times the string tension, but in the present context it is just the inverse square root of the YM coupling constant. Note that H now satisfies the 'anomalous' Bianchi identity

$$dH = \frac{1}{4}\alpha' \mathrm{tr}(F \wedge F) \ . \tag{3.20}$$

The fermion fields of the theory are the gravitino ψ_M and gaugino χ, which are chiral, and the dilatino ϕ, which is antichiral. We shall need their supersymmetry transformation laws, which are

$$\begin{aligned}
\delta\psi_M &= (\partial_M - \frac{1}{4}(\omega_+)_{MAB}\Gamma^{AB})\epsilon \\
\delta\lambda &= -\frac{1}{4}(\Gamma^M \partial_M \phi - \frac{\sqrt{2}}{12}\Gamma^{MNP} H_{MNP})\epsilon \\
\delta\chi &= \frac{1}{4}(\Gamma^{MN} F_{MN})\epsilon \ .
\end{aligned} \tag{3.21}$$

where $\epsilon(x)$ is a $d = 10$ chiral spinor parameter, $\{\Gamma^A\}$ are the $d = 10$ Dirac matrices, and

$$(\omega_\pm)_{MAB} = \omega_{MAB} \pm \frac{1}{2} E_M{}^C H_{CAB} \tag{3.22}$$

is a Lorentz connection with torsion ($E_M{}^A$ is the zehnbein and ω_{MAB} the standard torsion-free connection).

Rather than attempt to solve the second order equations that follow from the action (3.17) we reduce the problem to solving a set of first-order equations, as in previous examples, by seeking field configurations that partially preserve the supersymmetry of the 'vacuum', by which we mean here the trivial configuration for which ϕ is constant, g_{MN} flat and all other fields vanish. In can be shown, as before, that such configurations automatically solve the full field equations. For a purely bosonic configuration to partially preserve supersymmetry there must exist at least one non-vanishing solution for the spinor parameter ϵ of the equations obtained by requiring the supersymmetry variations of the fermion fields to vanish. We shall seek solutions of these conditions for which the ten-dimensional spacetime is a direct product $M_6 \times T_4$ of six-dimensional Minkowski spacetime, M_6, to be identified as the worldsheet of an infinite static fivebrane, and a four-dimensional 'transverse' space, T_4. Let $\{x^\mu;\ \mu = 0, 1, \ldots, 5\}$ be the worldvolume coordinates and $\{y^m;\ m = 1, 2, 3, 4\}$ the coordinates of the transverse space, which we shall assume to be conformally flat. Thus, by assumption,

$$g_{\mu\nu} = \eta_{\mu\nu} \qquad g_{mn} = e^{2f(y)}\delta_{mn} \ , \tag{3.23}$$

where e^{2f} is the conformal factor. We further assume the vanishing of all other fields with worldvolume indices, and that the remaining fields depend only on the transverse coordinates. The problem is thereby reduced to solving four-dimensional Euclidean equations for the function $f(y)$, the dilaton $\phi(y)$, the antisymmetric tensor $B_{mn}(y)$ and the YM potential

$A_m(y)$. Moreover, we shall restrict A_m to take values in an SO(3) subgroup of SO(32) such that the 496-dimensional adjoint of $SO(32)$ has the $SO(3)\times SO(29)$ decomposition

$$(\mathbf{3},\mathbf{1})\oplus(\mathbf{1},\mathbf{406})\oplus(\mathbf{3},\mathbf{29})\ . \tag{3.24}$$

With these restrictions the fermion variations of (3.22) reduce to

$$\begin{aligned}
\delta\psi_\mu &= \partial_\mu\epsilon \\
\delta\psi_m &= \partial_m\epsilon - \frac{1}{8}e^{-2f}\gamma_{pq}\big(H_{mpq} + 2e^{2f}\varepsilon_{mpqn}\partial_n f\,\gamma_5\big)\epsilon \\
\delta\lambda &= \frac{e^{-3f}}{24\sqrt{2}}\gamma_{mpq}\big(H_{mpq} + \sqrt{2}e^{2f}\varepsilon_{mpqn}\partial_n\phi\,\gamma_5\big)\epsilon \\
\delta\chi &= -\frac{1}{8}e^{-2f}\gamma_{mn}\big(F_{mn} - \frac{1}{2}\varepsilon_{mnpq}F_{pq}\,\gamma_5\big)\epsilon\ ,
\end{aligned} \tag{3.25}$$

where γ_5 is the product of the four (flat space) transverse $d=10$ Dirac matrices, and satisfies $\gamma_5^2=1$.

Let $\epsilon_\pm$ be an eigenspinor of γ_5 with eigenvalue ±1. Then, by choosing $\epsilon=\epsilon_\pm$ we find that $\delta\psi_M=0$ has a solution for constant ϵ provided that $H_{mpq}=\mp 2e^{2f}\varepsilon_{mnpq}\partial_n f$. Given this, $\delta\lambda=0$ requires

$$f=\frac{1}{\sqrt{2}}\phi\ . \tag{3.26}$$

Let us choose $\epsilon=\epsilon_+$. Then we have

$$g_{mn}=e^{\sqrt{2}\phi}\delta_{mn}\qquad H_{mnp}=-\varepsilon_{mnpq}\partial_q\big(e^{\sqrt{2}\phi}\big)\ . \tag{3.27}$$

The remaining equation $\delta\chi=0$ is solved by any Euclidean instanton solution for the $SO(3)$-vector potential $A^a_m(y)$ $(a=1,2,3)$. Let us choose $\epsilon=\epsilon_+$ and the one instanton solution of size ρ,

$$A^a_m(y)=\frac{2}{(r^2+\rho^2)}\eta^a{}_{mn}y^n\ , \tag{3.28}$$

where r is the radial distance (in the Euclidean metric) from the YM core of the fivebrane and $\eta^a{}_{mn}$ is 't Hooft's 3-vector valued self-dual tensor [45].

It remains to check that H as given in (3.27) satisfies the Bianchi identity (3.20). Let us first pause to consider some general implications of this identity. Integrating it over the four-dimensional transverse space, and using the definition of Q in (3.16), we find that

$$Q=\frac{1}{16\pi^2}\int_{T_4}\mathrm{tr}(F\wedge F) \tag{3.29}$$

which is just the instanton number, and therefore an integer (this is the correct normalization of the instanton number for the trace in the vector representation of $SO(3)$; the often seen normalization of one over $8\pi^2$ is for the fundamental representation of $SU(2)$). This should make us happy because we saw earlier that Q had to be an integer because of a Dirac-like quantization condition in string theory, but before we are overcome by euphoria we should recall that we have yet to establish a relation between the constant α' that appears in the action (3.17) (and in the definition of H) and the constant α' that appears in the string action (3.15). The connection between them is established by comparison of (3.17) with the low-energy effective action obtained by requiring conformal invariance of the two-dimensional sigma-model defined by the string action. A two-loop calculation is needed to fix the coefficient of F^2 in this action but a one-loop sigma-model anomaly calculation [46] suffices to fix the coefficient of the anomalous term in the Bianchi identity (3.20). The result of the latter calculation is precisely (3.20), so the two *a priori* different constants α' are actually the same. Moreover, since we chose a solution with instanton number one we find that

$$Q = 1 . \tag{3.30}$$

(I therefore disagree with the $Q = 8$ value given in the literature.) Now we substitute our result for H_{mnp} into the Bianchi identity (3.20). This gives the equation

$$\Box(e^{\sqrt{2}\phi}) = -\frac{1}{16}\alpha' \varepsilon_{mnpq} \mathrm{tr}(F_{mn}F_{pq}) . \tag{3.31}$$

This is the four-dimensional version of Poisson's equation with a non-singular source of size ρ centered at the origin. Clearly there is a solution and, if we impose the boundary condition that $\phi \to 0$ as $r \to \infty$, its asymptotic form is

$$e^{\sqrt{2}\phi} \sim 1 + \frac{\alpha' Q}{r^2} . \tag{3.32}$$

Thus, there is a non-singular fivebrane solution of the Chapline-Manton theory that breaks half the supersymmetry. By taking the $\rho \to 0$ limit one can find an explicit 'elementary fivebrane' solution [11] for which the asymptotic result (3.32) is exact. This corresponds to a singular source at $r = 0$. The metric $g_{mn} = e^{\sqrt{2}\phi}\eta_{mn}$ is also singular in this limit, with the result that the singularity is removed to the 'end' of an infinite wormhole throat with cross-section S^3. This might be considered acceptable in supergravity but in string theory e^{ϕ} is the effective string coupling constant and this diverges as $r \to 0$ if the asymptotic form (3.32) is exact. Consequently, the elementary fivebrane solution cannot be expected to survive quantum corrections.

To extend the non-singular solution of Chapline-Manton theory to string theory we must, in particular, further modifify the Bianchi identity (3.20) to

$$dH = X_4 \equiv \frac{1}{4}\alpha' \left[\mathrm{tr}(F \wedge F) - \mathrm{tr}(R_- \wedge R_-)\right] , \tag{3.33}$$

as required by the Green-Schwarz anomaly cancellation [47]. Observe that the curvature two-form occurring in this formula is the one for the connection ω_-. For the solution of the Chapline-Manton theory found above, R_- is a self-dual two-form. This can be seen as follows: Firstly, it can be verified that ω_+ is self-dual on the $SO(4)$ group indices, which implies that R_+ is too. Secondly, $R_{abcd}(\omega_+) = R_{cdab}(\omega_-)$ so that R_- is self-dual on the form indices. As a confirmation of this note that the supersymmetry transformation of the supercovariant gravitino curvature ψ_{ab} (in a purely bosonic background) is [44]

$$\delta_\epsilon \psi_{ab} = -\frac{1}{4}\gamma_{mn} R_{mnab}(\omega_-)\epsilon . \tag{3.34}$$

which indeed vanishes for $\epsilon = \epsilon_+$ if R_- is a self-dual two-form. Clearly, the additional R_--dependent term in (3.33) leads to a modification of the previous fivebrane solution because the source for the dilaton equation is now different (a difference that was not taken into account in [5]). Leaving aside, for the moment, the issue of higher-derivative corrections in R_-, we are now faced with a problem of self-consistency. Given R_- we can solve Poisson's equation for ϕ, but it is precisely this solution that determines R_-. One self-consistent solution is found [6] by identifying ω_+ with the YM connection (this is possible because the self-duality of ω_+ in its $SO(4)$ indices means that it actually takes values in $SO(3)$). In this case $\Box(e^{\sqrt{2}\phi}) = 0$ and, as for the elementary fivebrane, the asymptotic solution (3.32) is exact. This, 'symmetric', solution has the additional merit that all higher-order corrections vanish, so it is an exact solution of (classical) heterotic string theory. However, it also has the defect that it is unlikely to survive as an approximate solution of the quantum theory.

The exactness of the 'symmetric' fivebrane solution can be deduced from the fact that the action (in conformal gauge) for a string in this background is a sigma model with (4,4) supersymmetry, and it is known that such theories are conformally invariant to all orders of perturbation theory [48]. All other fivebrane solutions must correspond to sigma models with at least (4,0) supersymmetry, because this is the sigma-model equivalent of the condition of half-breaking of supersymmetry that we used to find these solutions [6]. In particular, the solution envisaged by Strominger must be in this class so further progress on this front can be made by investigation of the conditions for the finiteness (strictly speaking, conformal invariance) of $(4,0)$ sigma models. There exist arguments that purport to prove the finiteness of all (4,0) sigma models [49], but they could be vitiated by chiral (worldsheet) anomalies. Also, there was initially some puzzlement as to how the finiteness of $(4,0)$ sigma models could be compatible with corrections to the spacetime supersymmetry transformations, but it has recently been shown [50], up to three-loop order, that *finite* local counterterms are needed at each order to maintain (4,0) supersymmetry (which lead to modifications of the spacetime supersymmetry transformations), and that these are precisely the counterterms required for conformal invariance. This would appear to be strong evidence that there does exist a 'non-symmetric' solution of heterotic string theory generalizing the fivebrane solution of the Chapline-Manton theory.

We now turn to the question of the effective worldvolume action for fivebranes. The effective action for the elementary fivebrane is expected to be that of the standard $d = 10$ super-fivebrane of the brane-scan. The evidence for this is that the linearized limit of the gauge-fixed super fivebrane is the free six-dimensional hypermultiplet, for which the field content is four scalars and one complex spinor (the supersymmetry of this action is to be expected from the fact that the fivebrane configuration preserves precisely half of the original $d = 10$ supersymmetry). This is known to be the physical field content of the effective action for the 'elementary' fivebrane [7]. It is believed, however, that this theory is anomalous, and if this is the case then additional worldvolume fields will be needed. Strominger's fivebrane solution does have additional 'heterotic' worldvolume fields because the Atiyah-Singer index theorem implies that the gluino equation has zero-modes, one (for a one-instanton solution) for each gaugino triplet. From (3.24) we see that there are a total of 30 gluino triplets so there will be a total of 30 gluino zero modes. One of these, the $SO(29)$ singlet is the zero mode associated with the partial breaking of supersymmetry which produces the fermion content of the, non-heterotic, super-fivebrane. This means that there are 29 additional zero modes, the coefficients of which will appear as additional 'heterotic' worldvolume fermions. Supersymmetry implies additional bosonic variables and the final result of the zero mode analysis is that there are an *additional 29 'heterotic hypermultiplets'* [7]. We don't know whether this ensures anomaly freedom but at least there is a chance. Part of the difficulty in answering this question is that the fully relativistic and spacetime supersymmetric *heterotic fivebrane* action that includes these additional fields has not yet been found.

The fact that the fivebrane is the magnetic dual of the string suggests that it should be possible to reformulate string theory as a fivebrane theory. A start on this program can be made by considering what the effects of such a string/fivebrane duality should be at the level of the effective supergravity theory [10]. It has been appreciated for a long time that there is a dual formulation of the supergravity/YM action in which the two-form potential B is replaced by a six-form potential $\tilde{B}$ [51,52]. This dual form may be found from (3.17) by introducing $\tilde{B}$ as a Lagrange multiplier for the Bianchi identity of H. This amounts, after an integration by parts, to the addition of the term

$$\int G \wedge \left[H - \frac{1}{4}\alpha' K_3 \right] , \tag{3.35}$$

where $G = d\tilde{B}$ and K_3 is the Chern-Simons (CS) three-form potential for $X_4 = dK_3$ (although, in the Chapline-Manton action the Lorentz CS contribution is absent). The three-form H may now be treated as an independent auxiliary field which can be eliminated by its field equation $H = e^{2\phi} \star G$. This results in a G^2 term appearing with a factor $e^{2\phi}$ rather than $e^{-2\phi}$, spoiling the factorization of the exponential of ϕ. To remedy this we rescale the metric: $g_{MN} \to e^{2\phi/3} g_{MN}$. This leads to the action (note the absence of a $(\partial\phi)^2$ term)

$$\begin{aligned} S = \int d^{10}x \sqrt{-g} \Big\{ e^{\frac{2}{3}\phi} [R - \frac{1}{2.7!} G^{MNPQRST} G_{MNPQRST}] - \alpha' \mathrm{tr} F^{MN} F_{MN} \Big\} \\ + \alpha' \int \tilde{B} \wedge X_4 \, . \end{aligned} \tag{3.36}$$

Although we have now arranged for a common factor of $e^{(2/3)\phi}$ in the pure supergravity sector, this has been at the expense of uniformity with the YM sector. However, at the one *string*-loop level we should include F^4 terms in the action. After rescaling the metric these appear precisely with a factor of $e^{(2/3)\phi}$. Taking this into account we may rewrite the Lagrangian as

$$e^{\frac{2}{3}\phi} L_o + L_1 + \dots , \tag{3.37}$$

where L_0 has the form $(R + G^2 + F^4)$. Duff and Lu now interpret $e^{-(2/3)\phi}$ as a *fivebrane-loop* counting parameter [10]. From this point of view $e^{(2/3)\phi} L_0$ is the classical Lagrangian and L_1 a fivebrane one-loop correction. Evidence for this interpretation is that $e^{(2/3)\phi} L_0$, now to be considered as the field-theory limit of a hypothetical fundamental fivebrane theory, admits *string* solutions [12], as one would expect from string/fivebrane duality. Duff and Lu further argue that the *complete* action as a double expansion in two-dimensional sigma-model and string loops can be re-interpreted as a similar expansion in six-dimensional sigma-model and fivebrane loops.

At the string one-loop level we should also take into account the effects of $d = 10$ chiral anomalies. Recall [53] that these are encoded in a 12-form X_{12} which, provided the gauge group is $SO(32)$ or $E_8 \times E_8$, factorizes as $X_{12} = X_4 \wedge X_8$ where X_4 is the four-form of (3.19) and

$$X_8 \equiv dK_7 = \Big[\mathrm{tr}(F \wedge F \wedge F \wedge F) - \frac{1}{8}\mathrm{tr}(F \wedge F)\mathrm{tr}(R \wedge R) + \frac{1}{32}\mathrm{tr}(R \wedge R)\mathrm{tr}(R \wedge R) + \frac{1}{8}\mathrm{tr}(R \wedge R \wedge R \wedge R)\Big] . \tag{3.38}$$

Because of the anomalous Bianchi identity for H the two-form B acquires an anomalous YM and Lorentz transformation. The non-invariance of the one string-loop effective action is then cancelled [47] by the anomalous variation of certain local terms, in particular the term $c \int B \wedge X_8$, where c is a calculable constant (but one which I have not calculated for the conventions used here). This is taken into account in the dual theory by modifying G to $G = d\tilde{B} - cK_7$, so that G now has the anomalous Bianchi identity

$$dG = -cX_8 , \tag{3.39}$$

while, as we see from (3.34), the effect of the anomalous Bianchi identity for H is to produce the term $\int \tilde{B} \wedge X_4$ in the action. This term is no longer YM and Lorentz invariant, however, because of the anomalous transformation of $\tilde{B}$ required by (3.37), and this fact is responsible for anomaly cancellation in the dual theory [51].

These facts, and a number of others that I haven't mentioned, support the conjecture that string theory has a dual formulation as a fivebrane theory. Here I wish to make a further observation in this connection. The fivebranes discussed here are all of infinite extent. It is this assumption that allows the possibility of finding *static* solutions. This greatly simplifies the analysis but it is a physical idealization unless the five spatial dimensions of the worldvolume are periodically identified. This means that the natural framework for the discussion of static fivebranes solutions is a toroidally compactified spacetime where space is at least a five-torus and the fivebrane is wrapped around five compact spatial dimensions. In this case the fivebranes appear as *particles* in the lower dimension. Consider a toroidal compactification to $d = 4$ on a six-torus. In addition to the fivebranes wrapped around the homology 5-cycles of the six-torus, which are now to be viewed as magnetic monopoles, we also have the strings wrapped around the homology 1-cycles. String fivebrane duality would require the spectra of these two objects to be the same. They are, by *Poincaré duality*!

As mentioned previously, the anomalous Bianchi identity for the two-form H can be derived from worldsheet considerations [46]. Specifically, the worldsheet chiral fermions, which include 32 'heterotic' fermions that couple to the background $SO(32)$ YM potentials, lead to anomalous one-loop diagrams with two insertions of either the background spin connection ω_- or the YM potential. Because the two vertices are associated with *background* gauge potentials that are functions of the worldsheet fields X^μ, rather than with independent 'fundamental' gauge potentials, the anomaly is called a 'sigma-model anomaly'. Its physical interpretation differs from that of the usual non-abelian gauge anomaly (in two dimensions), but its form is the same. In particular, it is characterized by a four form. One of the miracles of string theory is that this four form is the same as the four form X_4 appearing in the factorization of the spacetime anomaly twelve-form X_{12}, as a result of which the anomalous transformation of the two-form potential B required by the GS anomaly cancellation mechanism is exactly what is required for cancellation of the worldsheet sigma-model anomaly. String/fivebrane duality would suggest that the anomalous transformation of the dual six-form potential $\tilde{B}$, required by the Bianchi identity (3.39), should be precisely what is required for the cancellation of *six-dimensional sigma-model anomalies.* In order for this to be true, the eight-form characterizing these anomalies must be the eight form X_8 of (3.38).

A partial check of this can be made on the basis of the little information we already have about the heterotic fivebrane. Each of the heterotic hypermultiplets of the heterotic fivebrane contains an $SU(2)$-Majorana *chiral* spinor λ^r $(r = 1, 2)$ [54]. The YM fields in the $SO(3) \times SO(29)$ subgroup of $SO(32)$ couple to these fermions via the Lagrangian density

$$\mathcal{L} = \frac{i}{2}\bar{\lambda}^i_I \Gamma^\mu \lambda^j_J \left(\varepsilon_{ij} A^{IJ}_\mu + \delta^{IJ} A_{\mu\, ij}\right) . \tag{3.40}$$

The chirality of the six-dimensional spinors means that these vertices will produce a sigma-model anomaly from the square diagram of Fig. 3.2 in which the external lines at the four vertices are background YM potentials (there are more anomalous diagrams but they need not be computed as their contribution is determined by consistency conditions, as for the usual non-abelian chiral anomalies).

Fig. 3.2: Worldvolume sigma-model anomaly

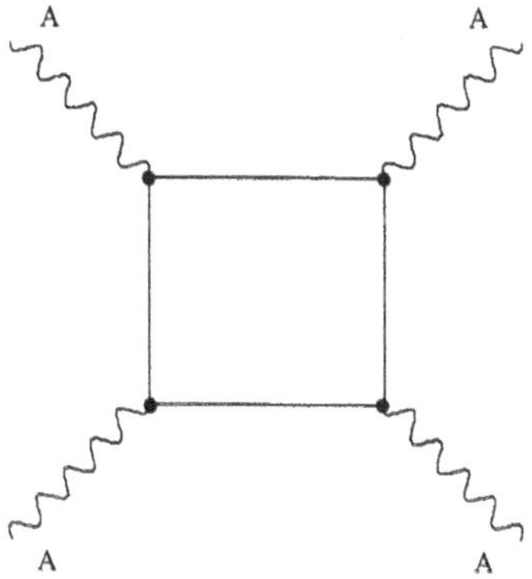

The anomaly from this diagram is encoded in the eight form $\mathrm{tr}(F^4)$. Fortunately for the string/fivebrane duality conjecture, this is precisely the purely YM part of X_8. The agreement is non-trivial since X_8 might have contained a $[\mathrm{tr}(F^2)]^2$ term, but in fact does not. Thus we recover from fivebrane worldvolume considerations part of the anomalous YM transformation of $\tilde{B}$. Further progress will require the construction of the complete heterotic fivebrane action. I leave this as an exercise for the student.

Acknowledgements: I am grateful to Edward Abraham, Eric Bergshoeff and J. Gomis for discussions on the contents of these lectures, and to Lee London for producing the figures.

References

[1] P.K. Townsend, *Three lectures on supermembranes*, in *Superstrings '88*, eds. M. Green, M. Grisaru, R. Iengo, E. Sezgin and A. Strominger, (World Scientific 1989).

[2] E. Bergshoeff, E. Sezgin and P.K. Townsend, Phys. Lett. **189B** (1987) 75; Ann. Phys. (N.Y.) **185** (1988) 350.

[3] B. de Wit, M. Lüscher and H. Nicolai, Nucl. Phys. **B320** (1989) 135.

[4] P.K. Townsend, Phys. Lett. **202B** (1988) 53.

[5] A. Strominger, Nucl. Phys. **B343** (1990) 167.

[6] C. Callan, J. Harvey and A. Strominger, Nucl. Phys. **B359** (1991) 611.

[7] C. Callan, J. Harvey and A. Strominger, Nucl. Phys. **B367** (1991) 60.

[8] E. Witten and D. Olive, Phys. Lett. **78B** (1978) 97.

[9] R. Nepomechie, Phys. Rev. **D31** (1985) 1921

[10] M.J. Duff and J.X. Lu, Nucl. Phys. **B354** (1991) 129; Nucl. Phys. **B357** (1991) 534.

[11] M.J. Duff and J.X. Lu, Nucl. Phys. **B354** (1991) 141

[12] M.J. Duff and J.X. Lu, Phys. Rev. Lett, **66** (1991) 1402.

[13] A. Achúcarro, J. Evans, D. Wiltshire and P.K. Townsend, Phys. Lett.**198B** (1987) 441.

[14] J. Gomis, *unpublished*; C. Yastremiz, Class. and Quantum Grav. **9** (1992) 2395.

[15] A. Karlhede and U. Lindström, Class. Quantum Grav. **3** (1986) L73.

[16] M.S. Alves and J. Barcelos-Neto, Europhys. Lett. **7** (1988) 395; *erratum*, **8** (1989) 90; U. Lindström and M. Roček, Phys. Lett. **218B** (1989) 207.

[17] E.R.C. Abraham and P.K Townsend, Nucl. Phys. **B351** (1991) 313.

[18] A. Achúcarro, J. Gauntlett, K. Itoh and P.K. Townsend, Nucl. Phys. **B314** (1989) 129.

[19] E.R.C. Abraham and P.K Townsend, Phys. Lett. **291B** (1992) 85; *More on Q-kinks: a (1+1)-dimensional analogue of dyons*, Phys. Lett. **B** *in press.*

[20] G.W. Gibbons and P.J. Ruback, Commun. Math. Phys. **115** (1988) 267.

[21] P. Fendley, W. Lerche, S.D. Mathur and N.P. Warner, Phys. Lett. **243B** (1990) 257.

[22] M. Cvetic, F. Quevedo and S.-J. Rey, Phys. Rev. Lett. **63** (1991) 1836.

[23] M. Cvetic, S. Griffies and S.-J. Rey, Nucl. Phys. **B381** (1992) 301; M. Cvetic and S. Griffies, Phys. Lett. **285B** (1992) 27.

[24] J. Hughes, J. Liu and J. Polchinski, Phys. Lett. **180**B (1986) 370.

[25] E.B. Bogomol'nyi, Sov. J. Nucl. Phys. **24** (1976) 449.

[26] P. Ruback, Commun. Math. Phys. **116** (1988), 645.

[27] J.P. Gauntlett, K. Itoh and P.K. Townsend, Phys. Lett. **238B** (1990) 65.

[28] E.A. Ivanov and A.A. Kapustnikov, Phys. Lett. **252B** (1990) 212.

[29] J.P. Gauntlett, Phys. Lett. **272B** (1991) 25.

[30] D.B. Fairlie and J. Nuyts, J.Phys. **A17** (1984) 2867; S. Fubini and H. Nicolai, Phys. Lett. **155B** (1985) 369.

[31] J. Harvey and A. Strominger, Phys. Rev. Lett. **66** (1991) 549.

[32] G.T. Horowitz and A. Strominger, Nucl. Phys. **B360** (1991) 197.

[33] M.J. Duff and J.X. Lu, Phys. Lett. **B273** (1991) 409.

[34] T. Samols, Commun. Math. Phys. **145** (1992) 149.

[35] N.S. Manton, Phys. Lett. **B110** (1982) 54.

[36] J.P. Gauntlett, *Low energy dynamics of supersymmetric solitons*, Nucl. Phys. **B**, *in press.*

[37] J. Harvey and A. Strominger, *String theory and the Donaldson Polynomial* preprint EFI-91-30.

[38] L. Alvarez-Gaumé and D.Z. Freedman, Commun. Math. Phys. **80** (1981) 443.

[39] M. Atiyah and N. Hitchin, *The geometry and Dynamics of Monopoles* (Princeton University Press 1988).

[40] C. Teitelboim, Phys. Lett. **B167** (1986) 63; *ibid* 69.

[41] R. Rohm and E. Witten, Ann. Phys. (N.Y.) **170** (1986) 454.

[42] C.G. Callan, D. Friedan, E. Martinec and M.J. Perry, Nucl. Phys. **B262** (1985) 593.

[43] G.F. Chapline and N.S. Manton, Phys. Lett. **120B** (1983) 105.

[44] E.A. Bergshoeff and M. de Roo, Nucl. Phys. **B238** (1989) 439.

[45] R. Rajaraman, *Solitons and instantons* (North Holland 1982).

[46] C.M. Hull and E. Witten, Phys. Lett. **160B** (1985) 398; C.M. Hull and P.K. Townsend, Phys. Lett. **178B** (1986) 187.

[47] M.B. Green and J.H. Schwarz, Phys. Lett. 149B (1984) 117.

[48] L. Alvarez-Gaumé and D.Z. Freedman, Commun. Math. Phys. **80** (1981) 443.

[49] P.S. Howe and G. Papadopoulos, Nucl. Phys. **B289** (1987) 264; Class. Quantum Grav. **5** (1987) 501; E. Sokatchev and K.S. Stelle, Class. Quantum. Grav. **4** (1987) 501; C. Becchi and O. Piguet, Nucl. Phys. **B347** (1990) 596.

[50] G. Papadopoulos and P.S. Howe, Nucl. Phys. **B381** (1992) 360.

[51] S.J. Gates and H. Nishino, Phys. Lett. **157B** (1985) 157.

[52] E. Bergshoeff and M. de Roo, Phys. Lett. **247B** (1990) 530; Phys. Lett. **249B** (1990) 27.

[53] M.B. Green, J,H. Schwarz and E. Witten, *Superstring Theory: Vol 2*, (CUP 1987).

[54] P.S. Howe, G. Sierra and P.K. Townsend, Nucl. Phys. **B221** (1983) 331.

THE GEOMETRIC PHASE IN QUANTUM PHYSICS*

A. BOHM

The Physics Department
The University of Texas
Austin, Texas 78712-1081, USA

CONTENTS

---

* Lectures at the NATO-ASI on Recent Problems in Mathematical Physics, Salamanca, Spain, June, 1992.

L. A. Ibort and M. A. Rodríguez (eds.), Integrable Systems, Quantum Groups, and Quantum Field Therapy 347-415.

I. INTRODUCTION

The geometrical phase has been ignored in quantum physics for half a century. It had not been forgotten, but it had been shown by V. Fock in 1928[1)] that the extra phase factor which occurs for time-dependent Hamiltonians can be chosen to unity. Though Fock's proof was limited to non-cyclic evolution, this phase choice was generally used until around 1980 when, Mead et al. and Berry reconsidered cyclic evolutions.

Cyclic evolutions can occur if one considers a quantum physical system in a periodically changing environment. This environment can be classical (magnetic moment in a precessing external magnetic field) or quantal (a fast moving electron in the slowly changing quantal environment given by the collective motion of the molecule as a whole). The importance of the quantum geometric phase and the gauge potential connected with it was discovered in 1978 in molecular physics by Mead et al[2)] based upon some earlier observations by Herzberg and Longuet-Higgins.[4)] The motion of the molecule is naturally divided into two "parts", the fast motion of the electrons and the slow collective vibrations and rotations of the molecule as a whole.[3)] In the past this problem was treated in the following way: One first investigated the motion of the fast variables considering the slow variables as *fixed* parameters. After the dynamics of the fast variables has been solved for all values of these fixed parameters, one turns to the dynamics of the slow variables. In this treatment the division into the slow and the fast moving parts is trivial. These are the ideas that underly the molecular Born-Oppenheimer approximation.[5] If one does not consider the nuclear coordinates R as fixed parameters but as quantum observables whose (eigen)-values change (e.g. rotate) in time, then the gauge potential emerges naturally from the Born-Oppenheimer method.[3)]

Conceptually simpler is the discussion of Hamiltonians which depend upon slowly changing parameters. This was done in 1983 in a beautiful paper by M.V. Berry[6)] for any quantum system in a slowly changing ("adiabatic") classical environment. Independently he arrived at the same ("Berry") gauge potentials and ("Berry") phase that were derived from the Born-Oppenheimer method by Mead.[7)]

It is easy to show in general that exact adiabatic cyclic evolutions do not exist;[8)] the Berry phase (and the Berry potential) is only an approximation of the geometric phase of real cyclic evolutions. This quantum geometric phase was introduced by Anandan and Aharonov in 1987.[9)]

The geometric phase has observable consequences in many problems of physics and chemistry.[10)] This is one reason why it has become an important subject. But the geometric phase is also one of the beautiful examples of what Wigner called "the unreasonable effectiveness of mathematics in natural sciences." Immediately after Berry introduced his phase Simon[11)] noticed that it is the holonomy of fibre bundle and that the gauge potential is the connection of this bundle. And it was this relation to the beautiful mathematics of fibre bundles which caused the geometric phase to become a fashion in mathematical physics. When fibre bundles were created and when the mathematics of the universal classifying bundles was developed, one had no idea that this could have anything to do with a quantum mechanical phase factor which could be measured in an interference experiment. The Stiefel connection of the universal classifying bundle was the "natural" mathematical object to define. It is incredible that this mathematical entity is exactly the vector potential whose integral over a closed path of states gives the Anandan-Aharonov[9] phase with the Berry[7] potential as its limiting case and that this connection is related to the vector potential which would be discovered by Mead[2] in the molecular structure.

In addition to its usefulness for physical applications and its attractiveness for the mathematical formulation, the geometric phase also teaches us something new about the meaning of understanding in science.

Understanding means the separation of complicated physical systems into simpler subsystems, the parts. In classical physics, separation into parts usually means reduction to the simpler objects, the constituents. In quantum mechanics the "parts" are described by subspaces of the Hilbert space[12] though these subspaces do not have to represent constituents (they may, e.g., represent collective vibrations or rotations as in the Born-Oppenheimer method).[5] The dissection of complicated quantum systems results in the trivial direct product of the states for the subsystems and the combinaton of two parts is given by the direct product of the Hilbert spaces for the parts (if the parts are (identical) particles the many body Hilbert space is the (symmetrized or antisymmetrized) tensor product of the one-particle Hilbert spaces).

In contrast to this quantum mechanical hypothesis, the geometrical phase — if it is non-trivial — shows that a part of the complicated physical system is not what one trivially expects. The fast motion effects the slow moving part and modifies its dynamics such that the many body Hilbert space is not always the tensor product of the parts.

In this review we shall discuss only quantum systems in a *classical* time-dependent environment. Quantum systems in a changing quantal environment, that appear in the dissection of complicated quantum systems into their parts are discussed in a number of publications.[2] [3] With a few exceptions[13] they are restricted to molecular physics.[14]

Our presentation in sections 2 and 3 does not make use of differential geometric notions, like fibre bundles, but it uses some of its nomenclature. Section 2 is restricted to the adiabatic approximation and discusses the general abelian case. In section 3 we discuss the standard example in detail but in greater generality than usual. We start with the adiabatic approximation in section 3.2 and then introduce the geometric phase for non-adiabatic change of the environment (Anandan-Aharonov phase). *In section 4 we discuss general cyclic (non-adiabatic) evolution. We introduce the mathematics of fibre bundles and use some of its results to de-

---

* Sections 2 and 3 are based on Chapter XXII of the 3rd edition of the book in

scribe the relation between the adiabatic Berry phase and the geometric phase for general cyclic evolution of a pure state. In this review we restrict ourselves to the abelian, $U(1)$ phase. These results generalize in a straightforward way[8] to states with $\mathcal{N}$-fold degeneracy (using a $U(\mathcal{N})$ phase[15]). Some features of these concepts also extend to cyclic evolutions of states described by density operators.[16]

References

1. V. Fock. *Z. Phys.* **49**, 323 (1928).
2. C. Mead and D. Truhlar, *J. Chem. Phys.* **70**, 2284 (1979); C.A. Mead, *Chem. Phys.* **49**, 23 and 33 (1980).
3. J. Moody, A. Shapere, F. Wilczek, *Phys. Rev. Lett.* **56**, 893 (1986); R. Jackiw, *Int. J. Mod. Phys.* A**3**, 285 (1988); A. Bohm, Lectures on *Symmetry in Science III*, 1988; p. 85, B. Gruber and F. Iachello (editors) Plenum Press; B. Zygelman, Phys. Rev. Lett. **64**, 256 (1990).
4. G. Herzberg and H.C. Longuet-Higgins, *Discuss, Faraday Soc.* **35**, 77 (1963); H.C. Longuet-Higgins, *Proc. Roy. Soc. London Ser.* A**344**, 147 (1975).
5. M. Born and J. Oppenheimer, *Ann. Phys.* **84**, 457 (1927).
6. M.V. Berry, Proc. Roy. Soc. London Ser. A**392**, 45 (1984).
7. For earlier "anticipations" of the Berry phase, see M. Berry, *Physics Today*, December, 1990, p. 34.
8. A. Bohm, *The Berry Connection as a Limit of the Stiefel Connection*, Proceedings of the XIX International Group Theory Colloquium, Salamanca (1992).
9. Y. Aharonov and J. Anandan, Phys. Rev. Lett. **58**, 1593 (1987); J. Anandan and Y. Aharonov, *Phys. Rev.* D**38**, 1863 (1988).
10. A. Shapere, F. Wilczek, *Geometric Phases in Physics* (Review and Reprint volume), World Scientific (1989); J. Zwanziger, M. Koenig, A. Pines, Berry's Phase, *Annual Rev. Phys. Chem.*, **41**, 601 (1990); J.W. Zwanziger, S.P. Rucher, G.C. Chingas, Phys. Rev. A**43**, 3232 (1991).
11. This phase was shown to be the holonomy of a fibre bundle with Berry connection; B. Simon, *Phys. Rev. Lett.* **51**, 2167 (1983).
12. More precisely, of the space of physical states Φ which is a subspace of the Hilbert space, A. Bohm, M. Gadella, *Dirac Kets, Garnow Vectors and Gelfand Triplets*, Springer-Verlag, Berlin (1989).
13. A. Bulgac, Phys. Rev. C**41**, 2333, (1990); R.S. Nikam, P. Ring, Y. Sun, E.R. Marshalek, Phys. Lett. **235**B, 215, (1990). M. Rho, International Workshop on Baryons as Skyrme Solitons, Siegen, September 1992.
14. There exist a few reviews and expositions of this subject: C. Alden Mead, *The Geometric Phase in Molecular Systems. Reviews Modern Physics* **64**, 51

reference 14, which contains details of the calculations.

(1992); Chapter XXIII, Third Edition of A. Bohm, *Quantum Mechanics: Foundations and Applications* to appear in Springer-Verlag, New York, 1993/94.
15. F. Wilczek, A. Zee, Phys. Rev. Lett. **52**, 2111 (1984).
16. A. Uhlmann, Parallel Transport of Phases. Lecture Notes of Physics, Vol. **379**, 55, Springer-Verlag (1991).

2. QUANTAL PHASE FACTORS FOR ADIABATIC CHANGES

The observables of a quantum system which is not isolated from its environment are described by operators that depend upon parameters $R = (x, y \ldots)$. We assume that the parameter space is a differentiable manifold and the observables, in particular the Hamiltonian $h(R)$, are nice (continuous, infinitely differentiable) and single-valued functions on the parameter space M. These parameters R (which we also write as $\mathbf{R}$ if we want to suggest that it is a three-vector) describe the classical environment in which the quantum physical system is immersed. If the physical system is in a quantal environment then the parameters are (generalized) eigenvalues of the observables of the quantal environment (like *e.g.* the position operator of the internuclear axis of a diatomic molecule).

An example of a quantum physical system whose Hamiltonian depends upon environmental parameters is a (quantum) magnetic moment $\mathbf{m}$ in a rotating (classical) magnetic field $\mathbf{B}$ of constant magnitude $B \equiv |\mathbf{B}|$. The parameter dependent Hamiltonian is given according to (IX.3.1)[1)] by

$$H = H_0 - \mathbf{m} \cdot \mathbf{B} = H_0 - Bg\frac{e}{2mc}\hat{\mathbf{R}}\cdot\mathbf{J} = H_0 + b\hat{\mathbf{R}}\cdot\mathbf{J} \tag{2.1}$$

Here H_0 is the Hamiltonian without the magnetic field, which we shall ignore. The quantum system's magnetic moment is given in terms of its angular momentum operator $\mathbf{J}$ according to (IX.3.18)[1)] by

$$\mathbf{m} = +\frac{e}{2mc}g\mathbf{J}, \tag{2.2}$$

$\hat{R}(t) = \mathbf{B}/B$ is the unit vector pointing in the direction of the magnetic field, and $b = -Bg(\frac{e}{2mc})$ is a constant. The parameter dependent term of the Hamiltonian is $h(\hat{\mathbf{R}}) = b\hat{\mathbf{R}} \cdot \mathbf{J}$ and the parameter space of the environment is the unit sphere $\{\hat{\mathbf{R}}; |\hat{\mathbf{R}}| = 1\}$. We can use as the parameters the polar angles $\hat{\mathbf{R}} = (\theta, \phi)$ of the unit vectors $\hat{\mathbf{R}}$.

The evolution of states of the quantum system in the external environment is described by the Schrödinger equation

$$i\frac{d\psi(t)}{dt} = h(R)\psi(t) \tag{2.3a}$$

(if the state is pure and described by a vector $\psi(t)$, and for general states described by the statistical operator $W(t)$ the evolution is given by the Schrödinger-von Neumann equation

$$i\frac{dW(t)}{dt} = [h(R), W(t)] \tag{2.3b}$$

This is the generalization of a basic postulate of quantum mechanics for conservative systems.[2)] The evolution of states of the non-conservative system is determined by the dynamical equation (2.3) and by the environmental process, i.e., by the way in which the environmental parameters $\mathbf{R}$ change. For a given physical situation, the way in which the parameters change, i.e., the path in the environment's parameter space, must be specified. In specifying the environmental process it is usually convenient to give the environmental parameters a time parameterization $\mathbf{R}(t) = (x(t), y(t), \cdots)$. One then obtains from $h(\mathbf{R})$ a time-dependent Hamiltonian $h(t) = h(\mathbf{R}(t))$. In our example (2.1) the environmental processes are described by how the direction $\hat{\mathbf{R}}$ of the magnetic field changes. The direction of the magnetic field may change periodically; e.g., it may perform rotations in the 1-2 plane with angular velocity $\omega, \hat{\mathbf{R}}(t) = \mathbf{e}_1 \cos\omega t - \mathbf{e}_2 \sin\omega t$, or it may run through other closed paths $\mathbf{C}$ in the parameter space. In this case the Hamiltonian $h(t) = h(\hat{\mathbf{R}}(t))$ returns to its original form as time progresses from $t = 0$ to the period $t = T = 2\pi/\omega$.

We postulate that the space of physical states does not only contain the solutions of (2.3) for one given fixed value of the parameter $\mathbf{R}$ or for one given environmental process $t \to \mathbf{R}(t)$ but for all values $\mathbf{R} \in$ parameter space M. This means that there is one space of physical states $\mathcal{H}$ for all values of $\mathbf{R}$. For any given value of $\mathbf{R}$, one may choose an orthonormal basis of eigenvectors $|n; \mathbf{R}\rangle$ of the parameter dependent Hamiltonian $h(\mathbf{R})$,*

$$\begin{aligned} h(\mathbf{R})|n; \mathbf{R}\rangle &= E_n(\mathbf{R})|n; \mathbf{R}\rangle, \\ \langle m; \mathbf{R}|n; \mathbf{R}\rangle &= \delta_{m,n}, \end{aligned} \tag{2.4}$$

and write the Hamiltonian according to its spectral resolution (I.4.10d):[1)]

$$h(\mathbf{R}) = \sum_n E_n(\mathbf{R})|n; \mathbf{R}\rangle\langle n; \mathbf{R}|. \tag{2.5}$$

Given any environmental process along with a time parameterization $\mathbf{R}(t)$ one obtains from the parameter dependent Hamiltonian $h(\mathbf{R})$ in (2.5) a time-dependent Hamiltonian $h(t) = h(\mathbf{R}(t))$ along with its spectral resolution (of I.4.10d):

$$h(\mathbf{R}(t)) = \sum_n E_n(\mathbf{R}(t))|n; \mathbf{R}(t)\rangle\langle n; \mathbf{R}(t)|. \tag{2.6}$$

The projection operators

$$\Lambda_n(\mathbf{R}(t)) \equiv |n;\ \mathbf{R}(t)\rangle\langle n;\ \mathbf{R}(t)|$$

change in general with time.

---

* We assume here, for simplicity, that the spectrum of $h(\mathbf{R})$ is discrete. If in addition to $h(\mathbf{R})$ other observables $A_i(\mathbf{R}),\ i = 1, \cdots, N-1$ are needed to form a c.s.c.o. for any given value of $\mathbf{R}$, then their eigenvalues $a_i(\mathbf{R})$ will also be needed to label the basis vectors $|n; \mathbf{R}\rangle = |n, a_1, \cdots, a_{N-1}; \mathbf{R}\rangle$.

We have assumed that the observables are single-valued as functions of $\mathbf{R}$ over the whole parameter space of the environment. Single-valuedness of the observables means that if the same value of $\mathbf{R}$ occurs more than once (i.e., at different times) during a process, then the observables are the same at each occurence. In particular, if the environmental process is closed, i.e., if the environmental parameters $\mathbf{R}(t)$ traverse a closed path $\mathbf{C}$ and return, after some period T, to their original values,

$$\mathbf{C}: \ \mathbf{R}(0) \to \mathbf{R}(t) \to \mathbf{R}(T) = \mathbf{R}(0), \tag{2.7}$$

then the Hamiltonian and also its eigenvalues and projection operators, which are uniquely defined by (2.5), are the same at $\mathbf{R}(T)$ as they are at $\mathbf{R}(0)$:

$$h(\mathbf{R}(T)) = h(\mathbf{R}(0)), \tag{2.8}$$
$$E_n(\mathbf{R}(T)) = E_n(\mathbf{R}(0)), \tag{2.9}$$
$$|n; \mathbf{R}(T)\rangle\langle n; \mathbf{R}(T)| = |n; \mathbf{R}(0)\rangle\langle n; \mathbf{R}(0)|. \tag{2.10}$$

Though the observables are single-valued functions of $\mathbf{R}$ the basis vectors $|n; \mathbf{R}\rangle$ themselves will in general not be single-valued functions of $\mathbf{R}$ over the whole parameter space. Usually it is necessary to use different parameterization over different patches of the parameter space. For the example $h(\mathbf{R}) = b\mathbf{R}(\theta, \varphi) \cdot \mathbf{J}$ this means that the vectors in, $\mathbf{R}(\theta, \varphi)\rangle$ are different functions of the polar angles (θ, φ) for different patches of the unit sphere, as explained in detail in section 3, (cf. equations (3.11) and (3.19)). Thus in general (2.10) will not imply

$$|n; \mathbf{R}(T)\rangle = |n; \mathbf{R}(0)\rangle \qquad \text{for} \quad \mathbf{R}(T) = \mathbf{R}(0) \tag{2.11}$$

but only

$$|n; \mathbf{R}(T)\rangle = e^{i\zeta_n} |n; R(0)\rangle \qquad \text{for} \quad \mathbf{R}(T) = R(0) \tag{2.11a}$$

where $e^{i\zeta_n}$ is a phase factor, because the $|n; \mathbf{R}\rangle$ are determined by (2.4) only up to a phase factor. We can define a new system of eigenvector $|n; \mathbf{R}\rangle$ by making phase transformations (also called gauge transformations as shall be explained below),

$$|n; \mathbf{R}\rangle \quad \to \quad |n; \mathbf{R}\rangle' = e^{i\zeta_n(\mathbf{R})}|n; \mathbf{R}\rangle, \tag{2.12}$$

where the $\zeta_n(\mathbf{R})$ are arbitrary real phase angles. Any such basis constitutes just as valid a basis of eigenvectors of the Hamiltonian as the $|n; \mathbf{R}\rangle$. We will restrict ourselves to transformations for which the phase factors $e^{i\zeta_n(\mathbf{R}(t))}$ are single-valued functions. Only these are called gauge transformations.

In general, if we go from one patch $O_1, \subset M$ of the parameter space into a neighboring patch $O_2 \subset M$ with a different parameterization, then eigenvectors of $h(R)$ in the overlap region $R \in O_1 \bigcap O_2$ are related by the gauge transformation (2.12). We will assume however that the closed path $\mathbf{C}$ of (2.7) can always be placed into one single patch $O \subset M$ and the basis vectors can be chosen to be single-valued functions. We shall then choose for the basis vectors single-valued functions over parameter space so that (2.11) is fulfilled. The gauge transformation (2.12) will transform single-valued basis vectors into vectors which are also single valued.

So far we considered the observables (the $|n;\mathbf{R}(t)\rangle$ are also observables), which change in time due to the change of the environmental parameters. We now want to consider the states $W(t)$, or the pure states $|\psi(t)\rangle\langle\psi(t)|$ (which can also be described by the state vector $\psi(t)$). The states evolve in time due to the Schrödinger or von Neumann equation (2.3).

The time evolution can also be described by an operator $U^\dagger(t)$ (which is unitary due to the hermiticity of $h(\mathbf{R}(t)) = h(t)$)

$$\psi(t) = U^\dagger(t)\psi(0), \tag{2.13}$$

which fulfills as a consequence of (2.3) the integral equation

$$U^\dagger(t) = I - i\int_0^t h(t')U^\dagger(t')dt'. \tag{2.14}$$

An expression for $U^\dagger(t)$ in terms of $h(t)$ is then given by successively substituting the right-hand side of (2.14) for the $U^\dagger(t)$ that appears in the integrand:

$$\begin{aligned}U^\dagger(t) &= I + \frac{1}{i}\int_0^t dt' h(t') + \left(\frac{1}{i}\right)^2\int_0^t dt'\int_0^{t'} dt'' h(t')h(t'')U^\dagger(t'')\\ &= I + \frac{1}{i}\int_0^t dt' h(t') + \left(\frac{1}{i}\right)^2\int_0^t dt'\int_0^{t'} dt'' h(t')h(t'') + \cdots\\ &\quad + \left(\frac{1}{i}\right)^n\int_0^t dt_1\int_0^{t_1} dt_2\cdots\int_0^{t_{n-1}} dt_n h(t_1)h(t_2)\cdots h(t_n) + \cdots.\end{aligned} \tag{2.15}$$

Note that the operators appearing in the integrands have decreasing time arguments reading from left to right, $t \geq t' \geq t'' \geq 0$; $t \geq t_1 \geq t_2 \geq \ldots \geq t_n \geq 0$, and that their order is important since they do not in general commute:

$$[h(t), h(t')] \neq 0 \qquad \text{for } t' \neq t. \tag{2.16}$$

For the case that $h(t)$ commutes at different times,

$$[h(t), h(t')] = 0 \qquad \text{for all } t, t', \tag{2.17}$$

the order of the operators in (2.15) is not important and it can be shown (Problem 2.1)*

$$U^\dagger(t) = \sum_{n=0}^{\infty}\frac{1}{n!}\left(\frac{1}{i}\right)^n\int_0^t dt_1\ldots\int_0^t dt_n h(t_1)\ldots h(t_n) = e^{\frac{1}{i}\int_0^t dt' h(t')}. \tag{2.18}$$

* *Problem 2.1.* Show that (2.18) is obtained from (2.15) if (2.17) holds (cf. e.g. reference 3).

According to (II.4.45a),[1)] (2.17) means that the projection operators $\Lambda_n(\mathbf{R}(t)) = |n;\mathbf{R}(t)\rangle\langle n;\mathbf{R}(t)|$ commute:

$$[\Lambda_n(\mathbf{R}(t)), \Lambda_{n'}(\mathbf{R}(t'))] = 0 \qquad \text{for all } n,\ n' \text{ and } t,\ t'. \tag{2.19}$$

Together with the assumed continuity of the $\Lambda_n(\mathbf{R}(t))$ this means that the projection operators and eigenspaces of $h(t)$ are in fact time-independent with only the eigenvalues $E_n(t)$ depending upon time. The projection operators may then be written as $\Lambda_n(\mathbf{R}(t)) = \Lambda_n = |n\rangle\langle n|$ with time-independent vectors $|n\rangle$. The resulting spectral resolution (2.6) of $h(t)$ may then be used in (2.18) to give the following spectral resolution of $U^\dagger(t)$:

$$U^\dagger(t) = \sum_n e^{\frac{1}{i}\int_0^t dt' E_n(t')}|n\rangle\langle n|. \tag{2.20}$$

(If the eigenvalues are also independent of time, i.e., if the Hamiltonian is time-independent, $h(t) = h$, then (2.19) goes into the standard expression for conservative systems (XII.1.16)[1)] $U^\dagger(t) = e^{\frac{1}{i}th}$). But for general time-dependent Hamiltonians, if (2.16) holds, (2.14) cannot be integrated. (Formally one obtains the time-ordered product of the expression on the right-hand side of (2.18)[3)] but this is not of much practical value).

We now solve (2.3) under various approximations. We use the initial condition that at $t = 0$ the state is an eigenstate of $h(R(0))$ with energy $E_n(R(0))$ which means

$$|\psi(0)\rangle\langle\psi(0)| = |n;R(0)\rangle\langle n;R(0)| \qquad \text{or} \qquad \psi(0) = |n;R(0)\rangle \tag{2.21}$$

The second form of this initial condition follows after fixing an arbitrary phase factor to unity. First we consider R = "fixed" parameter (This is the assumption that has been used for the old Born-Oppenheimer approximation in molecular physics). Then the solution of (2.3) with (2.21) is

$$\psi(t) = e^{-iE_n(R)t}|n;R\rangle = e^{-iE_n(R)t}\psi(0) \tag{2.22}$$

Essentially the same result we obtain if $R(t)$ changes, but only in such a way that (2.17) is fulfilled. Then (2.20) with the initial condition (2.21) gives immediately:

$$\psi(t) = e^{-i\int_0^t dt' E_n(R(t'))}|n;R(0)\rangle = e^{-i\int_0^t dt' E_n(R(t'))}\psi(0) \tag{2.23}$$

We now consider the less drastic assumption, called the adiabatic approximation.[4] For a Hamiltonian $h(\mathbf{R}(t))$ whose parameter change in time the interaction with the environment can cause the physical system to jump from the n-th eigenstate at $t = 0$ into all other eigenstates $|m,\mathbf{R}(t)\rangle\langle m,\mathbf{R}(t)|, m \neq n$ at a later time t. A very particular situation arises if this is not the case and if the state remains an eigenstate of $h(\mathbf{R}(t))$ at all time t with the same energy quantum number n. This means that $|\psi(t)\rangle\langle\psi(t)|$ changes in such a way that at all times t

$$|\psi(t)\rangle\,\langle\psi(t)| \overset{\substack{\text{adiabatic}\\ \downarrow}}{=} |n;\mathbf{R}(t)\rangle\,\langle n;\mathbf{R}(t)| = \Lambda_n(\mathbf{R}(t)) \tag{2.24}$$

This time-development is called adiabatic time-development. It represents an approximation not a limiting case. States which do not change in time, i.e. for which

$$|\psi(t)\rangle\langle\psi(t)| = |\psi(0)\rangle\langle\psi(0)| \qquad \text{for all } t, \tag{2.25}$$

are called stationary states. Clearly the solutions (2.22) and (2.23) describe stationary states:

$$|\psi(t)\rangle\langle\psi(t)| = |n;R(0)\rangle\langle n;R(0)| \tag{2.26}$$

One can show that stationary states of conservative systems are always energy eigenstates, cf. (XII.1.41)(XII.1.44).[1] For general time-dependent Hamiltonians this is not the case. The assumption of adiabaticity (2.24) is a generalization of the assumption that the state is stationary (2.25). Because $\Lambda_n(R(t))$ changes in time, the adiabatic state $W(t) = |\psi(t)\rangle\langle\psi(t)|$ of (2.24) changes in time, whereas the stationary state (2.25) does not change. The state (2.24) will always be the n-th eigenstate of $h(R(t))$ but its eigenvalue $E_n(R(t))$ may change in time; and even if that happens to be not the case than $|n;R(t)\rangle\langle n;R(t)|$ can still change. The path $t \to |\psi(t)\rangle\langle\psi(t)|$ of (2.24) describes a curve in the set of projection operators $|\psi(t)\rangle\langle\psi(t)|$ whereas the path of (2.25) consists of one point.

The set of one-dimensional projection operators $W(t) = |\psi(t)\rangle\langle\psi(t)|$ of the Hilbert space $\mathcal{H}$ is denoted by $\mathcal{P}(\mathcal{H})$. It is the projective space of (pure) physical states. The adiabaticity assumption (2.24) then means that if $R(t)$ traverses through a path in the parameter space M then the pure state $|\psi(t)\rangle\langle\psi(t)|$ traverses through a path of Hamiltonian eigenstates $t \to \Lambda_n(\mathbf{R}(t))$ in $\mathcal{P}(\mathcal{H})$. In particular if $\mathbf{R}(t)$ traverses through the closed path $\mathbf{C}$ of (2.7) in the parameter space M then because of (2.10) the pure Hamiltonian eigenstate traverses through a closed path

$$\mathcal{C} : \Lambda_n(\mathbf{R}(0)) = W(0) \to W(t) \to W(T) \overset{\substack{\text{adiabatic}\\ \downarrow}}{=} \Lambda_n(R(T)) = \Lambda_n(R(0)) \tag{2.27}$$

If the eigenprojectors traverse a closed path $\mathcal{C}$ in $\mathcal{P}(\mathcal{H})$ then the eigenvectors $|n;\mathbf{R}(t)\rangle$ traverse a closed path in $\mathcal{H}$

$$|n;R(0)\rangle \to |n;R(t)\rangle \to |n;R(T)\rangle = |n;R(0)\rangle\ , \tag{2.28}$$

because these basis vectors had been chosen to be single-valued functions in the patch that contained the closed curve $\mathbf{C}$ of (2.7).

That $W(t) = |\psi(t)\rangle\langle\psi(t)|$ traverses a closed path in the state space $\mathcal{P}(\mathcal{H})$ does not necessarily mean that the normalized state vector $\psi(t)$, which fulfills the Schrödinger equation (2.3) also traverse a closed path in $\mathcal{H}$. In general the path

$$C \quad : \quad t \to \psi(t) \qquad 0 \leq t \leq T \quad ; \quad (\psi(t), \psi(t)) = 1 \tag{2.29}$$

is not closed in $\mathcal{H}$ but fulfills

$$\psi(T) = e^{-i\alpha_\psi}\psi(0), \tag{2.30}$$

For the case of a time-independent Hamiltonian (2.22) or for (2.23), the phase factor is:

$$e^{-i\alpha_\psi} = e^{-iE_n T} \qquad \text{or} \qquad e^{-i\alpha_\psi} = e^{-i\int_0^t dt'\, E_n(t')} \tag{2.31}$$

which is called the *dynamical phase factor.*

We will instantly show that in case of a time-dependent Hamiltonian $h(\mathbf{R}(t))$ there is in general an additional phase factor which is called *geometric phase* for Berry phase (We shall in general use the term Berry phase for the geometrical phase obtained in the adiabatic approximation (2.24) only). But before we derive the geometric phase we want to discuss the meaning of the adiabatic approximation.

The dynamical equation (2.3) and the adiabaticity assumption (2.24) are two separate conditions on the state $|\psi(t)\rangle\langle\psi(t)|$ and may, therefore, not be compatible with each other. This is indeed the case, which can immediately be shown by inserting (2.24) into (2.3b):

$$i\frac{dW(t)}{dt} = [h(R(t)), W(t)] = [h(R(t)), \Lambda_n(R(t))] = 0 \tag{2.32}$$

which means that $W(t)$ does not change in time

$$W(t) \equiv |\psi(t)\rangle\langle\psi(t)| = |\psi(0)\rangle\langle\psi(0)| \equiv W(0) \tag{2.33}$$

This means that any adiabatic evolution (2.24) of a state (which obeys (2.3)) must be a stationary evolution (2.25) and cannot be a non-trivial cyclic evolution

$$W(0) \to W(t) \to W(T) = W(0) \tag{2.34}$$

for which $W(t)$ changes in time.

It is, of course, not clear that for a given Hamiltonian $h(\mathbf{R}(t))$ there exist at all a cyclic solution (2.34) of the dynamical equation (2.3). But it is plausible that for a periodic Hamiltonian, i.e. a time-dependent Hamiltonian fulfilling (2.8), there exist some (a countable set of) cyclic solutions fulfilling (2.34). (In section 3 we shall demonstrate this statement by an example). However, as we have seen from (2.32), these cyclic solutions cannot be Hamiltonian eigenstates $\Lambda_n(\mathbf{R}(t))$.

Exact adiabatic cyclic evolutions do not exist. The adiabatic equality $\stackrel{\text{adiabatic}}{\doteq}$ in (2.24) can only be an approximation, not a limiting case (except for $T \to \infty$ when the time-development becomes stationary and the evolution a trivial cyclic evolution (2.25)). Nevertheless we can use the approximation (2.24) and determine the adiabatic geometrical phase or Berry phase.

To do this we expand $\psi(t)$ with respect to the basis system $|n; \mathbf{R}(t)\rangle$:

$$\psi(t) = \sum_m c_m(t)|m; \mathbf{R}(t)\rangle = \sum_m a_m(t) e^{\frac{1}{i}\int_0^t E_m(t')dt'}|m; \mathbf{R}(t)\rangle. \tag{2.35}$$

In the second equality we have separated the dynamical phase of the expansion coefficients $c_m(t)$. For the special case (2.24) of the adiabatic approximation the expansion (2.35) becomes

$$\psi(t) \stackrel{\text{adiabatic}}{\doteq} c_n(t)|n, \mathbf{R}(t)\rangle = e^{\frac{1}{i}\int_0^t E_n(t')dt'} a_n(t)|n, \mathbf{R}(t)\rangle \tag{2.36}$$

where because of (2.21)

$$a_n(0) = 1 \tag{2.21'}$$

This is not an equality but an approximate equality which we indicate by the qualification "adiabatic" at the equality sign. Inserting (2.36) into the Schrödinger equation (2.3) and using (2.4) we obtain after a little calculation

$$\frac{d}{dt} a_n(t) \equiv \dot{a}_n(t) = -a_n \langle n; \mathbf{R}(t)| \left(\frac{d}{dt}|n; \mathbf{R}(t)\rangle\right). \tag{2.37}$$

This can be integrated:

$$\int_{a_n(0)}^{a_n(t)} \frac{da_n}{a_n} = -\int_0^t \langle n; \mathbf{R}(t)| \frac{d}{dt}|n; \mathbf{R}(t)\rangle dt \tag{2.38}$$

and leads with (2.21') to:

$$a_n(t) = e^{i\int_{\mathbf{R}(0)}^{\mathbf{R}(t)} i\langle n;\mathbf{R}(t)| \frac{\partial}{\partial \mathbf{R}^i}|n;\mathbf{R}(t)\rangle d R^i} \equiv e^{i\gamma_n(t)}. \tag{2.39}$$

The coefficients $c_n(t)$ and $a_n(t)$ in (2.36) are phase factors (their absolute value is one), because $\psi(t)$ is obtained from $\psi(0)$ of (2.21) by a unitary time-development (2.13) and is therefore a normalized vector. This justifies the definition of $\gamma_n(t)$ as a real phase angle by (2.39). Note that $\gamma_n(t)$ is only defined up to a multiple of 2π.

The phase angle $\gamma_n(t)$ is defined in terms of an integral over a vector valued function

$$\mathbf{A}^n(\mathbf{R}) \equiv i\langle n; \mathbf{R}|\nabla|n; \mathbf{R}\rangle \tag{2.40}$$

which is called the Mead-Berry vector potential or the Mead-Berry connection.[5] It is defined as the scalar product of the eigenvectors of $h(t)$ and its derivative with respect to the parameter $\mathbf{R}, \nabla|n;\mathbf{R}\rangle$. For the definition (2.40) of $\mathbf{A}(\mathbf{R})$ one uses single valued basis vectors only. As a single-valued $|n;\mathbf{R}\rangle$ can in general not be found on the whole parameter space but only on open subsets (patches) the same is true for the $A(\mathbf{R})$. * We will discuss this problem in detail for the example (2.1) in section 3. Here we will assume that the curve $\mathbf{C}$ lies in one of these patches and that the $|n;\mathbf{R}\rangle$ are single-valued. The integral

$$\begin{aligned}\gamma_n(t) &= \int_0^t i\langle n;\mathbf{R}(t)|\frac{d}{dt}|n;\mathbf{R}(t)\rangle dt = \int_{R(0)}^{R(t)} i\langle n;\mathbf{R}(t)|\frac{\partial}{\partial R^i}|n;\mathbf{R}(t)\rangle dR^i \\ &= \int_{R(0)}^{R(t)} i\langle n;\mathbf{R}(t)|d|n;\mathbf{R}(t)\rangle = \int_{R(0)}^{R(t)} A_i^n(\mathbf{R})dR^i \equiv \int A^n\end{aligned} \tag{2.41}$$

is independent of the parameterization as expressed by the third term of (2.41) as line integral of a differential along the curve $t \to \mathbf{R}(t)$.

For the sake of simplicity we have so far pretended that the environmental parameter is a three-vector $\mathbf{R}$, then a differential one-form is defined by

$$A^n \equiv i\langle n;R^i|\frac{\partial}{\partial R^i}|n;R^i\rangle dR^i = A_i^n dR^i = i\langle n;R^i|d|n;R^i\rangle, \tag{2.42}$$

where A_i^n are the components of the three-vector (2.40). In general the parameter space is (assumed to be) a differentiable manifold and $R^i; i = 1, 2, \dots f$; are its arbitrary (local) coordinates. The differentials dR^i are covariant vectors and form a basis dual to the basis $\frac{\partial}{\partial R^i}$ of contravariant vectors and $A^n \equiv i\langle n;R|d|n;R\rangle$ is a differential one-form independent of the parameterization.[6)] The expressions for the phase angle $\gamma_n(t)$ thus apply also to an f-dimensional parameter space in which case the Berry vector potential A_i^n is an f-dimensional contravariant vector.

Returning to our calculation of $\psi(t)$ we obtain by inserting (2.39) into (2.36) for the state vector in the adiabatic approximation with initial condition (2.21):

$$\psi(t) \overset{\text{adiabatic}}{\underset{\downarrow}{=}} e^{\frac{1}{i}\int^t E_n(t')dt'} e^{i\gamma_n(t)}|n;\mathbf{R}(t)\rangle \tag{2.43}$$

In addition to the dynamical phase factor (2.31) one obtains thus the phase factor $e^{i\gamma_n(t)}$ given in terms of the eigenvectors of $h(R(t))$ by (2.41).

According to (2.12) the $|n;R(t)\rangle$ are only determined up to a phase factor. Thus, the additional phase factor in (2.43) can be transformed away by a phase transformation. From the requirement that the (eigen)vectors $\psi(t)$ have the least

* It is, therefore, also called a local representative of the Mead-Berry connection.

possible oscillatory behavior, $\|\frac{d\psi(t)}{dt}\| = min$, Fock[7] derived in 1928 the condition $\langle n;t|\frac{d}{dt}|n;t\rangle = 0$ which — according to (2.41) — leads to the phase choice of unity for this extra phase factor. Though Fock did not consider cyclic time change as in (2.7) or (2.27), (2.34) his phase choice was universally accepted, and the extra phase factor was ignored for half a century. In the following we shall first show how γ_n can be eliminated and then discuss that it can in general not be eliminated for cyclic processes.

From the definition (2.40) it follows that under a gauge tranformation (2.12) the Berry vector potential transforms like:

$$\begin{aligned}\mathbf{A}^n(R) \to \mathbf{A}'^n(R) &= i\,\langle n;R|'(\nabla|n,R\rangle') \\ &= i\,\langle n,R|e^{-i\zeta_n(R)}(\nabla\, e^{i\zeta_n(R)}|n;R\rangle) \\ &= i\,\langle n;R|\nabla|n;R\rangle + i\,e^{-i\zeta_n(R)}(\nabla\, e^{i\zeta_n(R)})\end{aligned}$$

or

$$\mathbf{A}^n(R) \longrightarrow \mathbf{A}'^n(R) = \mathbf{A}^n(R) - \nabla\zeta_n(R). \tag{2.44}$$

Thus

$$\gamma_n(t) = \int_{R(0)}^{R(t)} \mathbf{A}^n(R)d\mathbf{R} \to \int_{R(0)}^{R(t)} \mathbf{A}'^n(R)d\mathbf{R} = \gamma_n(t) - \zeta_n(R(t)) + \zeta_n(R(0)) = \gamma'_n(t). \tag{2.45}$$

We shall now do the calculations that led to (2.43) using the $|n;R\rangle'$ in place of the $|n;R\rangle$. Then we obtain (2.43) with the primed quantities on the right hand side. Then using (2.12) we obtain for the primed quantities:

$$e^{i\gamma'_n(t)}|n,R(t)\rangle' = e^{i\gamma'_n(t)}e^{i\zeta_n(R(t))}|n;R(t)\rangle.$$

If $\zeta_n(R(t))$ is an arbitrary single-valued function modulo 2π we can choose it such that the phase factor $e^{i\gamma'_n(t)}e^{i\zeta_n(R(t))}$ becomes unity and we obtain in place of (2.43):

$$\psi(t) = e^{-i\int_0^t E_n(t')dt'}|n,R(t)\rangle\ . \tag{2.46}$$

This also fulfills the initial condition (2.21). Since $|n,R\rangle'$ is as valid a basis system (fulfilling (2.4)) as is $|n,R\rangle$, we can use it and thus describe the time-developement of the state vector by (2.46) with the dynamical phase factor only. This is Fock's result.[7]

The above arguments made use of the fact that $\zeta_n(R(t))$ was arbitrary. If after some period T the environmental parameters return to their original value as for the closed path $\mathbf{C}$ (2.7), then one cannot choose $\zeta_n(R(T))$ freely to remove $\gamma_n(T)$.

Since $e^{i\zeta_n(R)}$ is a single-valued function of R we must have for $R(T) = R(0)$

$$e^{i\zeta_n(R(T))} = e^{i\zeta_n(R(0))} \qquad \text{or} \qquad \zeta_n(R(T)) = \zeta_n(R(0)) + 2\pi\cdot\text{integer}. \tag{2.47}$$

Therefore according to (2.45)

$$\begin{aligned}\gamma_n(T) \to \gamma'_n(T) \equiv \oint_{\mathbf{C}} \mathbf{A}'^n(R)d\mathbf{R} &= \gamma_n(T) - 2\pi \cdot \text{integer} \\ &= \oint_{\mathbf{C}} \mathbf{A}^n(R)d\mathbf{R} - 2\pi \cdot \text{integer}\end{aligned} \tag{2.48}$$

Thus $\gamma_n(T)$– which is only defined modulo 2π– is an invariant of the gauge transformation (2.12) and can therefore not be removed. We therefore have

$$\psi(T) = e^{-i\int_0^T dtE_n(t)+i\gamma_n(T)}|n; R(T)\rangle \tag{2.49}$$

with the $\gamma_n(T)$ given by the loop integral over the closed path $\mathbf{C}$ of (2.7):

$$\gamma_n(\mathbf{C}) \equiv \gamma_n(T) = \oint_{\mathbf{C}} i\langle n; R| \frac{\partial}{\partial R^i}|n; R\rangle dR^i = \oint_{\mathbf{C}} \mathbf{A}^n(R)d\mathbf{R} \text{ modulo } 2\pi\,. \tag{2.50}$$

If we insert the initial condition (2.21) and (2.11) into (2.49) we obtain

$$\psi(T) = e^{-i\int_0^T dtE_n(t)}e^{i\gamma_n(\mathbf{C})}\psi(0). \tag{2.51}$$

The phase angle $\gamma_n(T)$ is called the Berry phase angle, $e^{i\gamma_n(T)}$ is called the Berry phase factor.

We have shown that for a closed path the extra phase factor cannot be transformed away. This does not mean that $\gamma_n(C)$ could not be zero. Indeed this will turn out to be the case for many Hamiltonians $h(R)$. In this case the vector potential $A^n(R)$ need not be zero but will be trivial, which means given by

$$\mathbf{A}^n(R) = \nabla\zeta(R) = -ie^{-i\zeta(R)}\nabla e^{i\zeta(R)}, \tag{2.52}$$

where $\zeta(R)$ is a well-defined function of R. Then, according to (2.44), it can be transformed to zero by the gauge transformation with $e^{-i\zeta(R)}$. The cases that we are interested in are those for which the Berry phase is different from zero. The Hamiltonian (2.1) is such a case, as we shall discuss in detail in the next section.

From (2.44) we see that the Berry vector potential transforms under the phase transformation (2.12) like the vector potential of electrodynamics transforms under a gauge transformation. The set of phase factors $e^{i\zeta_n(R)}$ form the group $U(1)$, they are continuous, differentiable, single-valued functions of the parameters R. We therefore have here a gauge theory with gauge group $U(1)$ and gauge potential $\mathbf{A}^n(R)$. This is the reason for which we call the phase transformation (2.12) a $U(1)$-gauge transformation. Whereas $\mathbf{A}^n(R)$ is not an invariant with respect to a gauge transformation but transforms like (2.44), the Berry phase is gauge invariant. Precisely, the phase factor $e^{i\gamma_n(T)}$ is $U(1)$ gauge invariant and the phase angle $\gamma_n(T)$

modulo 2π is $U(1)$ gauge invariant (it is only defined modulo 2π). If the parameter space is three-dimensional and the parameter $R = \mathbf{R}$ is the three-dimensional coordinate vector then we have complete analogy with the electromagnetic theory. However, the physical meaning of these quantities is different, because the gauge potential (2.40) is defined in terms of the eigenvectors of the Hamiltonian and has nothing to do with electromagnetism. For an f-dimensional parameter space we have again a $U(1)$-gauge theory, but the gauge transformations and gauge potentials now depend upon an f-dimensional parameter R and the $A_i(R)$; $i = 1, 2, \ldots f$, consist of f components. It is called a $U(1)$ gauge theory over an f-dimensional base (= parameter) space.

In analogy to electrodynamics we can define a gauge field

$$F_{ij}^n = \frac{\partial}{\partial R^i} A_j^n - \frac{\partial}{\partial R^j} A_i^n \quad i, j = 1, 2, \ldots f_i; \tag{2.53a}$$

or (if $\mathbf{R}$ is a 3-vector),

$$\mathbf{F}^n = \nabla \wedge \mathbf{A}^n. \tag{2.53b}$$

The corresponding 2-form is:

$$F^n = dA^n = \frac{\partial A_j^n}{\partial R^i} dR^i \wedge dR^j = \frac{1}{2} F_{ij}^n dR^i \wedge dR^j. \tag{2.53c}$$

This gauge field F^n with A^n given by (2.42) is also called Berry curvature. From (2.44) it follows immediately that the gauge field - like in every abelian gauge theory - is gauge invariant:

$$F^n \to F'^n = \nabla \wedge \mathbf{A}^n = \mathbf{F}^n \tag{2.54}$$

In formula (2.50) Stokes' theorem may be used to convert the line integral over the closed path $\mathbf{C}$ in the environments parameter space into a surface integal over *any* surface S enclosed by the path $\mathbf{C}$. For 3-dimensional $\mathbf{R}$ this is:

$$\gamma_n(T) = \oint_{\mathbf{C}} \mathbf{A}^n(R) d\mathbf{R} = \iint_S d\mathbf{S}(\nabla \wedge \mathbf{A}^n) \mod 2\pi, \tag{2.55}$$

and for an f–dimensional parameter space

$$\gamma_n(T) = \oint_{\mathbf{C}} i \langle n; R| \frac{\partial}{\partial R^i} |n; R\rangle \, dR^i = \oint_{\mathbf{C}} A^n = \iint_S dA^n = \iint_S F^n \mod 2\pi \,. \tag{2.55'}$$

S is the surface that subtends $\mathbf{C}$ and $d\mathbf{S}$ denotes the surface element in parameter space. The direction of $d\mathbf{S}$ is normal to the surface S and the line element $d\mathbf{R}$ of $\mathbf{C}$ is traversed in the right-hand sense with respect to the normal.

Using (2.42) in (2.55′) we can express $\gamma_n(T)$ also in the form

$$\gamma_n(T) = \iint\limits_S F^n = \int d(i\langle n; R|d|n; R\rangle) \quad \text{mod } 2\pi, \tag{2.56}$$

or for the 3-dimensional case

$$\gamma_n(T) = \iint\limits_S d\mathbf{S} \cdot \mathbf{F} = \iint d\mathbf{S}\, i\, \nabla \wedge (\langle n; \mathbf{R}|\nabla|n; \mathbf{R}\rangle)\,. \tag{2.56′}$$

We have added mod 2π in these equations because $\gamma_n(T)$ is defined only up to an integer times 2π. In these integrals the Berry curvature is:

$$\begin{aligned} F^n &= i\,(d\langle n; R|) \wedge d|n; R\rangle = -\text{ Im } d\,\langle n; R|d|n; R\rangle \\ &= -\text{Im}\left(\frac{\partial}{\partial R^i}\,\langle n; R|\right)\,\frac{\partial}{\partial R^j}\,|n, R\rangle\, dR^i \wedge dR^j\,, \end{aligned} \tag{2.57}$$

or for the three-dimensional case

$$\mathbf{F}^n = -Im(\nabla\langle n; \mathbf{R}|) \wedge \nabla|n; \mathbf{R}\rangle. \tag{2.57′}$$

Equation (2.56′) shows that the Berry phase $\gamma_n(T)$ is analogous to the magnetic flux of the electromagnetic theory.

To see how non-trivial phases arise when two energy levels become degenerate for some values of the parameter, we express the Berry curvature in terms of the eigenvalues $E_n(R)$. We continue the calculation of F in (2.57). Inserting the complete basis system

$$1 = \sum_m |m; R\rangle\,\langle m; R| \tag{2.58}$$

we obtain

$$\begin{aligned} F^n &= -\text{Im}\sum_m (\nabla\langle n; R|)|m; R\rangle \wedge \langle m; R|\nabla|n; R\rangle \\ &= -\text{Im}\overline{\langle n; R|\nabla|n; R\rangle} \wedge \langle n; R|\nabla|n; R\rangle \\ &\quad - \text{Im} \sum_{m\neq n} \overline{\langle m; R|\nabla|n, R\rangle} \wedge \langle m; R|\nabla|n; R\rangle. \end{aligned}$$

The first term is zero because $\langle n; R|\nabla|n; R\rangle$ is purely imaginary. In the second term we use

$$\langle m; R|\nabla|n; R\rangle = \langle m; R|(\nabla h(R))\,|n, R\rangle\, \frac{1}{E_n(R) - E_m(R)} \quad \text{for } n \neq m \tag{2.59}$$

which can be easily proven (problem 2.2). * Then we obtain

$$\mathbf{F}^n = -\mathrm{Im} \sum_{m \neq n} \frac{\langle n; R| \left[\nabla h(R)\right] |m, R\rangle \wedge \langle m; R| \left[\nabla h(R)\right] |n; R\rangle}{\left[E_n(R) - E_m(R)\right]^2} , \tag{2.60}$$

and the Berry phase angle is given in tems of this quantity:

$$\gamma_n(\mathbf{C}) = \iint_S d\mathbf{S}\, \mathbf{F}^n (\text{modulo } 2\pi) , \tag{2.61}$$

where S is any surface in the parameter space which subtends the closed curve $\mathbf{C}$ (we write this also as $\mathbf{C} = \partial S$ and call ∂S the boundary of S).

The form (2.60) does not contain any phase factor of the basis vectors $|m; R\rangle$. Whereas (2.40) is only defined for single-valued $|n, R\rangle$ and can therefore be defined only on a patch where single-valued $|n; R\rangle$ exist, the formula (2.60) can also be used for an area where single-valued $|n; R\rangle$ do not exist.

The formula (2.60) also shows that the singularities of F^n occur at those values of $R = R_0$ where the eigenvalues are degenerate $E_n(R_0) = E_m(R_0)$.

We conclude this part with the following remark: The Berry phase (2.50) has been calculated with single-valued basis vectors $|n; R\rangle$ fulfilling (2.11). If we use instead vectors fulfilling (2.11a) (which may have their origin in non-single-valued phase transformations $e^{i\zeta_n(R(T))} = e^{i\zeta_n} e^{i\zeta_n(R(0))}$:

$$|\widetilde{n;\ R(T)}\rangle = e^{i\zeta_n(R(T))} |n; R(T)\rangle = e^{i\zeta_n} |n; R(0)\rangle)$$

then there will be again an extra phase factor for $\psi(T)$ which is now given by

$$\psi(T) = e^{-i \oint_0^T dt' E_n(t')} e^{i\tilde{\gamma}_n(C)} e^{i\zeta_n} \psi(0)$$

where $\tilde{\gamma}_n(T)$ is the phase calculated from (2.50) with the $|n; R\rangle$ replaced by the $|\widetilde{n; R}\rangle$. This phase angle; calculated with the non-single-valued basis vectors is not the Berry phase angle; the Berry phase angle is now given by $\tilde{\gamma}_n + \zeta_n$. The Berry phase is thus only defined by (2.50) if the basis vectors are single valued.

From the above formulas we see that $\gamma_n(\mathbf{C})$ is independent of how the closed loop $\mathbf{C}$ is traversed (provided it is such that all our conditions are fufilled, which means for the case under consideration, that the loop is traversed slowly enough for the adiabatic approximation (2.24) to hold). The Berry phase angle $\gamma_n(T)$ thus does not depend upon the dynamics of the quantum system or upon the details of the environmental changes. It only depends upon the path $\mathbf{C}$ (with the provision of adiabaticity) and is thus a quantity determined by the geometry of the parameter space. It is therefore called the adiabatic geometric phase. Later we will also consider non-adiabatic cyclic evolutions and shall reserve the name geometric phase for the non-adiabatic generalization of the Berry phase $\gamma_n(\mathbf{C})$.

---

* *Problem 2.2.* Show that for the eigenvectors $|n; \mathbf{R}\rangle$ defined by (2.4) the relation (2.59) holds.

References

1. All equation numbers of this kind refer to A. Bohm, *Quantum Mechanics: Foundations and Applications*, Second Edition, Springer-Verlag, New York, 1986, The roman number gives the chapter, the other number the equation.
2. A. Bohm, ibid, Chapter XII.
3. See e.g. C. Itzykson, J.B. Zuber, *Quantum Field Theory*, Section 4-1-4, McGraw-Hill publishers (1980).
4. M. Born and V. Fock, *Z. Phys.* **51**, 165 (1928).
5. See reference 2 and reference 6 of section 1.
6. C.J. Isham, *Modern Differential Geometry for Physicists*, World Scientific Publishers (1989).
7. Reference 1, section 1.

3. A SPINNING QUANTUM SYSTEM IN AN EXTERNAL MAGNETIC FIELD

3.1. The Parameterization of the Basis Vectors

In this lecture we will discuss in detail a quantum particle with magnetic moment $\mathbf{m} = \mu_B g \mathbf{J}$ in an external magnetic field, $\mathbf{B}(t) = B\mathbf{R}(t)$ whose direction $\mathbf{R}(t)$ is changing periodically. In particular we will consider the case in which the magnetic field direction precesses around the 3-direction. If the direction rotates slowly ("adiabatically") this system provides an application of the general ideas presented in section 2. The Schrödinger equation for a magnetic moment in a precessing magnetic field has been solved exactly.[1)] Therefore we need not restrict ourselves in this example to the adiabatic approximation and will obtain the non-adiabatic (Aharonov-Anandan) geometric phase.[2)] With the help of this example we will then introduce in section 3.4 the non-adiabatic geometric phase for a general cyclic evolution.[3)]

The Hamiltonian is given according to (2.1) by:

$$h(\hat{\mathbf{R}}(t)) = -Bg\frac{e}{2mc}\hat{\mathbf{R}}(t)\cdot\mathbf{J} = b\hat{\mathbf{R}}(t)\cdot\mathbf{J} \tag{3.1}$$

where $\mathbf{J}$ is the angular momentum operator of the quantum physical system, $\hat{\mathbf{R}}(t)$ is the parameter that describes the changing environment, and $b = -Bg\frac{e}{2mc}$ is a constant. ($\frac{e\hbar}{2mc} = \mu_B$ is the Bohr magneton and g is the Landé factor of the quantum particle. b is a frequency and $b/(-g) \equiv \omega_L$ is called the Larmor frequency).

The parameter space of the environment (the set of all $\hat{\mathbf{R}}$) is the unit sphere in two-dimensions, $\mathcal{S}^2$. We parameterize $\mathcal{S}^2$ by the polar angles (θ, φ) according to

$$\hat{\mathbf{R}} = \quad \hat{\mathbf{R}}(\theta,\varphi) = \begin{pmatrix} \sin\theta\cos\varphi \\ \sin\theta\sin\varphi \\ \cos\theta \end{pmatrix}, \qquad \begin{matrix} 0 \le \theta \le \pi \\ 0 \le \varphi < 2\pi \end{matrix}. \tag{3.2}$$

This parameterization associates unique values of the pair (θ, φ) to each unit vector $\hat{\mathbf{R}}$ except for the unit vector

$$\mathbf{e}_3 \equiv \begin{pmatrix} 0 \\ 0 \\ 1 \end{pmatrix} \tag{3.3}$$

of the north pole and the unit vector $-\mathbf{e}_3$ of the south pole. $\mathbf{e}_3$ and $-\mathbf{e}_3$ are given, respectively, by $\theta = 0$ and $\theta = \pi$ for all values of φ in the range $0 \le \varphi < 2\pi$; the value of φ is not determined when $\hat{\mathbf{R}} = \pm\mathbf{e}_3$.

The special case in which the magnetic field precesses uniformly about the 3-axis, is described by

$$\mathbf{B}(t) = B(\sin\theta\cos\omega t, \sin\theta\sin\omega t, \cos\theta) = B\hat{R}(\theta, \omega t) \tag{3.4}$$

with B, θ and $\omega = \varphi/t$ being constants.

The precession of the magnetic field is shown in Fig. 3.1. For $t = 0$ the magnetic field lies in the 1-3 plane

$$\mathbf{B}(0) = B(\sin\theta, 0, \cos\theta) = (B_{12}, 0, B_3) = B\mathbf{R}(\theta, 0)$$

Figure 3.1 also shows the angles (θ, φ) and a new angle $\tilde{\theta}$ which we will define later.

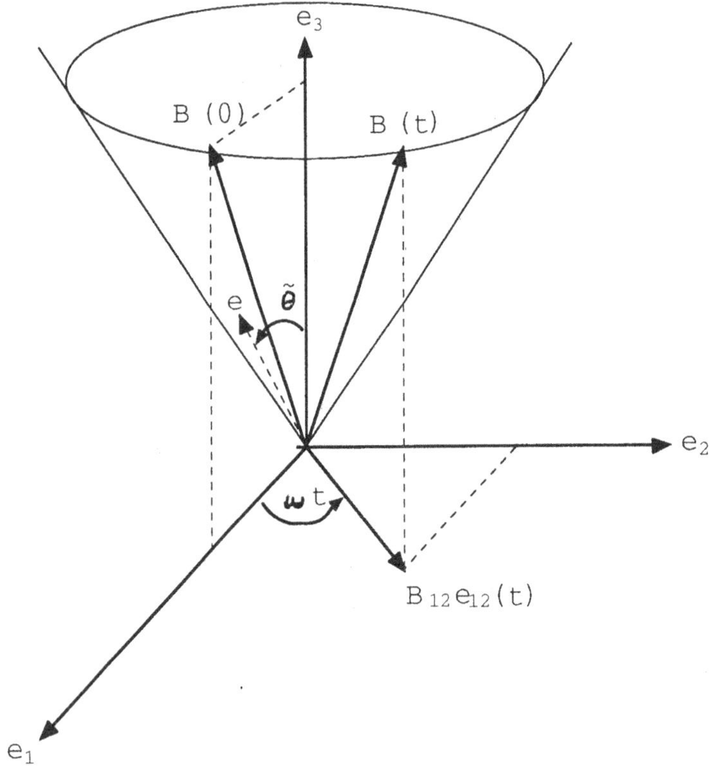

Figure 3.1. A quantal magnetic moment in an external magnetic field precesses uniformly around a cone of semiangle θ with e_3. (θ, φ) are the polar angles for the rotation of the external magnetic field. $(\tilde{\theta}, \varphi)$ are the "polar angles" for the evolution of the state of the magnetic moment.

The eigenvectors (2.4) for this example are given by

$$h(R)|k;R(\theta(t),\varphi(t))\rangle = b\hat{\mathbf{R}}\cdot\mathbf{J}|k;R\rangle = bk|k;R\rangle \tag{3.5}$$

They are eigenvectors of the operator $\hat{\mathbf{R}}(t)\cdot\mathbf{J}$. The place of the energy quantum number n in (2.4) is taken here by k which is the quantum number for the component of angular momentum along the changing direction of the external magnetic field.

The eigenvalue $E_k(R(t)) = bk$ of the time-dependent Hamiltonian $h(R(t))$ is for this example a constant, but the eigenvectors $|n;R(t)\rangle$ and the eigenprojectors $\Lambda_k(R(t)) = |k;R\rangle\langle k;R|$ change in time. The eigenvalue k is a constant but its physical interpretation changes in time, it is the value of the observable $\mathbf{R}(t)\cdot\mathbf{J}$ where the direction $\mathbf{R}(t)$ changes with respect to the laboratory frame (which is the observable given by the direction e_3).

If the vectors $|k;\ R\rangle$ are also eigenvectors of $\mathbf{J}^2$, then the possible values of k are given by

$$k=-j, \qquad \text{or} \quad -j+1,\ldots, \qquad \text{or} \quad +j; \qquad \text{where} \quad j = \text{integer or half integer} \tag{3.6}$$

and $j(j+1)$ is the eigenvalue of the angular momentum $\mathbf{J}^2$. We assume here that j and any other additional quantum numbers, which may be connected with the eigenvalues of the rotationally invariant part H_0 of the total Hamiltonian $H^{total} = H_0 + h(\hat{\mathbf{R}})$, have fixed values and we suppress them throughout the current discussion as labels of the vectors $|k;R\rangle$.

But it may also be that the $|k;\hat{\mathbf{R}}\rangle$ are not eigenvectors of $\mathbf{J}^2$ and thus do not have a definite value of j associated to them. This kind of vectors will be needed in the example of a diatomic molecule where k is the eigenvalue of a quantum mechanical observable $\mathbf{X}\cdot\mathbf{J}$. The quantum mechanical observable $\mathbf{X}$ is the position operator (direction) of the internuclear axis of the molecule, which rotates in space. This observable describes the slowly changing quantal environment for the electron. The electron is the fast quantum system that follows instantaneously the motion of the molecule as a whole. The electronic angular momentum is not a good quantum number, because one does not have spherical symmetry but only axial symmetry about the internuclear axis $\mathbf{X}$. So the total Hamiltonian commutes with $\mathbf{X}\cdot\mathbf{J}$ but not with $\mathbf{J}^2$ ($\mathbf{J}$ being in this case the electronic angular momentum or electronic spin) and the molecular states are not $\mathbf{J}^2$ eigenstates.

In this case the possible values, that k can take, are not given by (3.6) and $2k$ need not have *a priori* an integer value. We shall, however, see below that also in this case one can prove that k must be an integer or half integer. Thus k can be interpreted as the component of angular momentum along the changing internuclear axis.

The vectors of (3.5) are parameterized by the unit vector $\hat{\mathbf{R}}$, or by the polar angles (θ,φ). They can be obtained by applying (θ,φ)-dependent rotations to an eigenvector $|k;\hat{\mathbf{e}}_3\rangle$ of the component of angular momentum in the direction of the north pole, $J_3 = \mathbf{e}_3 \cdot \mathbf{J}$.

There are many (θ,φ)-dependent rotations $\mathcal{R}(\theta,\varphi)\in$SO(3) which when applied to $\mathbf{e}_3$ give the arbitrary unit vector $\hat{\mathbf{R}}(\theta,\varphi)$. We choose the product of rotation*

$$\begin{aligned}\mathcal{R}(\theta,\varphi)\mathbf{e}_3 &= \mathcal{R}_3(\varphi)\mathcal{R}_2(\theta)\mathcal{R}_3(-\varphi)\mathbf{e}_3 = \mathcal{R}_3(\varphi)\mathcal{R}_2(\theta)\mathbf{e}_3 = \mathcal{R}_3(\varphi)\hat{\mathbf{R}}(\theta,0)\\ &= \hat{\mathbf{R}}(\theta,\varphi)\end{aligned} \tag{3.7}$$

where $\mathcal{R}_3(-\varphi)$ does nothing to the unit vector $\mathbf{e}_3$ and where $\mathcal{R}_2(\theta)$ produces the unit vector $\hat{\mathbf{R}}(\theta,0)$ which lies in the 1-3 plane at an angle θ with respect to the $\mathbf{e}_3$-axis. The rotation $\mathcal{R}_3(-\varphi)$ has been included in the definition of $\mathcal{R}(\theta,\varphi)$ in order for the rotation $\mathcal{R}(0,\varphi)$ to be independent of φ.

Rotations, like any other continuous transformations, are represented in the space of quantum physical states by unitary operators (representing the group SU(2)). The unitary operators that represent the rotations $\mathcal{R}_3(\varphi)$ and $\mathcal{R}_2(\theta)$ are given by

$$U_3(\varphi) = e^{-i\varphi J_3} \equiv I + \frac{\varphi}{i}J_3 + \frac{1}{2!}\left(\frac{\varphi}{i}J_3\right)^2 + \cdots . \tag{3.8}$$

$$U_2(\theta) = e^{-i\theta J_2} \tag{3.9}$$

and an analogous expression holds for rotations about the $\mathbf{e}_1$-axis. The product of two or more rotations is represented by the product of the corresponding operators. The rotation $\mathcal{R}(\theta,\varphi)$ of (3.7) is thus represented by the operator

$$U(\theta,\varphi) = U_3(\varphi)U_2(\theta)U_3(-\varphi) = e^{-i\varphi J_3}e^{-i\theta J_2}e^{i\varphi J_3}. \tag{3.10}$$

We now choose a fixed normalized eigenvector $|k;\mathbf{e}_3\rangle$ of $\mathbf{e}_3\cdot\mathbf{J} = \hat{\mathbf{R}}(0,0)\cdot\mathbf{J}$ and apply to it the unitary operator $U(\theta,\varphi)$. The resulting vector

$$|k;\theta\,\varphi\rangle \equiv U(\theta,\varphi)|k;\hat{\mathbf{e}}_3\rangle = e^{-i\varphi J_3}e^{-i\theta J_2}e^{i\varphi J_3}|k;\hat{\mathbf{e}}_3\rangle, \tag{3.11}$$

---

* *Problem 3.1.* Show that the rotation $\mathcal{R}(\theta,\varphi)$ defined by (3.7) transforms the vector e_3 represented by (3.3) into the vector $\hat{\mathbf{R}}(\theta,\varphi)$ represented by (3.2). Use the standard representation matrices for $\mathcal{R}_3(\varphi)$ and $\mathcal{R}_2(\theta)$:

$$\mathcal{R}_3(\varphi) = \begin{pmatrix}\cos\varphi & -\sin\varphi & 0\\ \sin\varphi & \cos\varphi & 0\\ 0 & 0 & 1\end{pmatrix} \qquad \mathcal{R}_2(\theta) = \begin{pmatrix}\cos\theta & 0 & \sin\theta\\ 0 & 1 & 0\\ -\sin\theta & 0 & \cos\theta\end{pmatrix}$$

(Note the change in sign for the angle θ).

Problem 3.2. Find a rotation $\mathcal{R}'(\theta,\varphi)$ which has the following properties

i. It transforms $\mathbf{e}_3$ into the vector $\hat{\mathbf{R}}(\theta,\varphi)$.

ii. The rotation $\mathcal{R}'(\theta=\pi,\varphi)$ is independent of φ.

is an eigenvector of the operator $\hat{\mathbf{R}}(\theta,\varphi)\cdot\mathbf{J}$ with eigenvalue k:

$$\hat{\mathbf{R}}(\theta,\varphi)\cdot\mathbf{J}\ |k;\theta\,,\varphi\rangle = k\ |k;\theta\,,\varphi\rangle\,. \tag{3.12}$$

To prove (3.12) one has to use the following transformation property of the angular momentum operators J_i which follow from their commutation relations:[4)]

$$e^{-i\theta J_2}J_3e^{i\theta J_2} = J_3\cos\theta + J_1\sin\theta, \tag{3.13}$$
$$e^{-i\varphi J_3}J_1e^{i\varphi J_3} = J_1\cos\varphi + J_2\sin\varphi. \tag{3.14}$$
$$e^{-i\varphi J_3}J_2e^{i\varphi J_3} = J_1(-\sin\varphi) + J_2\cos\varphi \tag{3.15}$$

The vector (3.11) is a (continuous, differentiable) vector-valued function of (θ,φ). This vector-valued function gives a unique vector for all $\hat{\mathbf{R}}$ except for the south pole $\hat{\mathbf{R}} = -\mathbf{e}_3$. At the south pole, $\theta=\pi$, (3.13) becomes

$$e^{-i\pi J_2}J_3e^{i\pi J_2} = -J_3 \tag{3.16}$$

which implies

$$e^{-i\pi J_2}e^{-i\varphi J_3}e^{i\pi J_2} = e^{i\varphi J_3}\,. \tag{3.17}$$

and may be used to obtain

$$\begin{aligned} |k;\pi\ \varphi\rangle &= e^{-i\varphi J_3}e^{-i\pi J_2}e^{i\varphi J_3}\,|k;\hat{\mathbf{e}}_3\rangle \\ &= e^{-i\pi J_2}e^{2i\varphi J_3}\,|k;\hat{\mathbf{e}}_3\rangle \\ &= e^{-i\pi J_2}e^{2ik\varphi}\,|k;\hat{\mathbf{e}}_3\rangle\,. \end{aligned} \tag{3.18}$$

This shows that at the south pole different vectors of the unit ray are obtained as φ varies in the range $0\le\varphi<2\pi$. φ also varies in the range $0\le\varphi<2\pi$ at the north pole $\mathbf{e}_3 = \hat{\mathbf{R}}(0,\varphi)$ but $|k;\theta,\ \varphi\rangle$ is single-valued at the north pole because $|k;0\varphi\rangle$ does not depend upon φ as a result of the inclusion of the rotation $\mathcal{R}_3(-\varphi)$ in the definition of $\mathcal{R}(\theta,\varphi)$. $|k;\theta,\ \varphi\rangle$ is thus a single-valued vector function everywhere on $\mathcal{S}^2$ except on the south pole.

A vector-valued function which is well-defined at the south pole but not at the north pole is obtained by a gauge transformation (2.12):

$$|k;\theta,\varphi\rangle' = e^{-i2k\varphi}|k;\theta,\varphi\rangle = e^{-i\varphi J_3}e^{-i\theta J_2}e^{-i\varphi J_3}|k;0,0\rangle. \tag{3.19}$$

The new vector $|k;\theta,\varphi\rangle'$ differs from $|k;\theta,\varphi\rangle$ by the gauge transformation $e^{i\zeta(\theta,\varphi)} = e^{-i2k\varphi}$. At the south pole $|k;\theta,\varphi\rangle'$ evaluates to a single vector

$$|k;\pi,\varphi\rangle' = e^{-i\pi J_2}|k;\hat{\mathbf{e}}_3\rangle \tag{3.20}$$

but at the north pole it evaluates to many vectors

$$|k;0,\varphi\rangle' = e^{-2ik\varphi}|k;\hat{\mathbf{e}}_3\rangle \tag{3.21}$$

It therefore can be used everywhere on $\mathcal{S}^2$ except on the north pole. Either vector (3.11) or (3.19) can be used in the overlap region of the two open patches which consists of $\mathcal{S}^2$ with both poles excluded.

The vector $|k;\pi,\varphi\rangle'$ is an eigenvector of J_3 with eigenvalue $-k$ and an eigenvector of $\hat{\mathbf{R}}(\pi,\varphi)\cdot\mathbf{J} = -\mathbf{e}_3\cdot\mathbf{J} = -J_3$ with eigenvalue k (Problem 3.3). *

We thus see that two different parametrizations of $|k;\hat{\mathbf{R}}\rangle$ are needed. Since $|k;\theta,\varphi\rangle$ and $|k;\theta,\varphi\rangle'$ differ only by the phase factor $e^{-i2k\varphi}$ the projection operators and subspaces are, however, the same:

$$|k;\theta,\varphi\rangle\langle k;\theta,\varphi| = |k;\theta,\varphi\rangle'\ '\langle k;\theta,\varphi|. \tag{3.22}$$

3.2 Berry Connection and Berry Phase for Adiabatic Evolutions — Magnetic Monopole Potentials

We now calculate the gauge potential (Mead-Berry connection) for the adiabatic evolution of the Hamiltonian (3.1). The easiest is to calculate it as the one-form

$$A^{k'k}(R^i) = A_i^{k'k}dR^i = i\langle k';R^j|\frac{\partial}{\partial R^i}|k;R^j\rangle dR^i\,. \tag{3.23}$$

As the vectors (3.11) and (3.19) are parameterized by the angles θ,φ we will use for R^i the polar coordinates: $\mathbf{R} = (r\sin\theta\cos\varphi, r\sin\theta\sin\varphi, r\cos\theta)$.

We are mainly interested in the diagonal matrix elements $A^{kk}\equiv A^k$ but include in our derivation also the off-diagonal matrix elements without much additional effort.

If dR^i are the cartesian components, then $A_i^{k'k}$ are the cartesian components of the vector

$$\begin{aligned}
\mathbf{A}^{k'k}(\theta,\varphi) &= i\langle k';\theta\varphi|\nabla_R|k;\theta\varphi\rangle \\
&= \langle k';\hat{\mathbf{e}}_3|iU^\dagger(\theta,\varphi)\left(\hat{\mathbf{e}}_r\frac{\partial}{\partial r} + \hat{\mathbf{e}}_\theta\frac{1}{r}\frac{\partial}{\partial\theta} + \hat{\mathbf{e}}_\varphi\frac{1}{\sin\theta\partial\varphi}\right)U(\theta,\varphi)|k;\hat{\mathbf{e}}_3\rangle \\
&= \hat{\mathbf{e}}_r\hat{A}_r^{k'k}(\theta,\varphi) + \hat{\mathbf{e}}_\theta\hat{A}_\theta^{k'k}(\theta,\varphi) + \hat{\mathbf{e}}_\varphi\hat{A}_\varphi^{k'k}(\theta,\varphi)\,.
\end{aligned} \tag{3.24}$$

$\hat{\mathbf{e}}_r, \hat{\mathbf{e}}_\theta, \hat{\mathbf{e}}_\varphi$ are the unit vectors along the $r-,\theta-$ and $\varphi-$direction respectively and the standard form for the vector ∇ has been used. $\hat{A}_r, \hat{A}_\theta, \hat{A}_\varphi$ are the spherical orthonormal components of $\mathbf{A}$. As $|k;\theta,\varphi\rangle = U(\theta,\varphi)|k;\hat{e}_3\rangle$ is independent or r

$$\hat{A}_r^{k'k} = 0 \tag{3.25}$$

---

* *Problem 3.3.* Show that the vector (3.20) is an eigenvector of J_3 with eigenvalue $(-k)$ and an eigenvector of $h(R(\theta=\pi,\varphi))$ fulfilling (3.5).

The connection 1-form therefore is

$$\begin{aligned} A^{k'k}(\theta,\varphi) &= \hat{A}_\theta r d\theta + \hat{A}_\varphi r\sin\theta d\varphi \\ &= i\langle k';\theta,\varphi|\ \frac{\partial}{\partial\theta}\ |k;\theta,\varphi\rangle d\theta + i\langle k';\theta,\varphi|\ \frac{\partial}{\partial\varphi}\ |k;\ \theta,\varphi\rangle d\varphi \\ &\equiv A_\theta^{k'k} d\theta + A_\varphi^{k'k} d\varphi \end{aligned} \tag{3.26}$$

where $A_\theta^{k'k}$, $A_\varphi^{k'k}$ are the spherical covariant components of $\mathbf{A}$.

It is easiest to calculate the covariant components and then — if one wants to — obtain the spherical orthonormal components from comparison of the first and third row of (3.26).

We first use the vectors (3.11). In this gauge we obtain

$$A_\theta^{k'k} = i\langle k';\theta,\varphi|\frac{\partial}{\partial\theta}|k;\theta,\varphi\rangle = \langle k'\hat{\mathbf{e}}_3|iU^\dagger(\theta,\varphi)\frac{\partial}{\partial\theta}U(\theta,\varphi)\,|k;\hat{\mathbf{e}}_3\rangle = r\hat{A}_\theta \tag{3.27}$$

and

$$A_\varphi^{k'k} = i\langle k'\theta,\varphi|\frac{\partial}{\partial\varphi}|k;\theta,\varphi\rangle = \langle k';\hat{\mathbf{e}}_3|iU^\dagger(\theta,\varphi)\frac{\partial}{\partial\varphi}U(\theta,\varphi)\,|k;\hat{\mathbf{e}}_3\rangle = r\sin\theta\hat{A}_\varphi \tag{3.28}$$

We calculate:

$$\begin{aligned} iU^\dagger(\theta,\varphi)\,\frac{\partial U(\theta,\varphi)}{\partial\theta} &= ie^{-i\varphi J_3}e^{i\theta J_2}e^{i\varphi J_3}\,\frac{\partial}{\partial\theta}\,e^{-i\varphi J_3}e^{-i\theta J_2}e^{i\varphi J_3} \\ &= e^{-i\varphi J_3}J_2 e^{i\varphi J_3} \\ &= \ (J_2\cos\varphi - J_1\sin\varphi) \end{aligned} \tag{3.29}$$

and

$$\begin{aligned} iU^\dagger(\theta,\varphi)\,\frac{\partial U(\theta,\varphi)}{\partial\varphi} &= \left[U^\dagger(\theta,\varphi)J_3U(\theta,\varphi) - U^\dagger(\theta,\varphi)U(\theta,\varphi)J_3\right] \\ &= \left[-(J_1\cos\varphi + J_2\sin\varphi)\sin\theta + J_3(\cos\theta - 1)\right]. \end{aligned} \tag{3.30}$$

Inserting this into (3.27), (3.28) we obtain:

$$A_\theta^{k'k}(\theta,\varphi) = \langle k';\hat{\mathbf{e}}_3|(J_2\cos\varphi - J_1\sin\varphi)|k;\hat{\mathbf{e}}_3\rangle \tag{3.31}$$

$$A_\varphi^{k'k}(\theta,\varphi) = \langle k';\hat{\mathbf{e}}_3|\left[-(J_1\cos\varphi + J_2\sin\varphi)\sin\theta + J_3(\cos\theta - 1)\right]|k;\hat{\mathbf{e}}_3\rangle\,. \tag{3.32}$$

For the diagonal matrix elements, J_1 and J_2 do not contribute so that we obtain

$$A_\theta^k(\theta,\varphi) = 0, \tag{3.33}$$

$$A_\varphi^k(\theta,\varphi) = \langle k;\hat{\mathbf{e}}_3|J_3(\cos\theta - 1)|k;\hat{\mathbf{e}}_3\rangle = -k(1-\cos\theta), \quad \theta \neq \pi\,. \tag{3.34}$$

Thus the vector potential $\mathbf{A}$ has only a component in the φ-direction which is given by:

$$\hat{\mathbf{A}}^k(\theta,\varphi) = \hat{\mathbf{e}}_\varphi\,\frac{k(\cos\theta - 1)}{r\sin\theta}, \quad \theta \neq \pi\,. \tag{3.35}$$

We now calculate the connection using the vector-functions $|n;\hat{\mathbf{R}}\rangle'$ of (3.19) which are single-valued everywhere except the north pole. We obtain in a similar way as above

$$A_\theta^{k'k}(\theta,\varphi) = \langle k';\hat{\mathbf{e}}_3|(J_2\cos\varphi - J_1\sin\varphi)|k;\hat{\mathbf{e}}_3\rangle', \tag{3.36}$$

$$A_\varphi^{k'k}(\theta,\varphi) = \langle k';\hat{\mathbf{e}}_3|\,[-(J_1\cos\varphi + J_2\sin\varphi)\sin\theta + J_3(\cos\theta-1) + 2\,k]\,|k;\hat{\mathbf{e}}_3\rangle' \tag{3.37}$$

The diagonal matrix elements are:

$$A_\theta^{\prime k}(\theta,\varphi)=0, \tag{3.38}$$

$$A_\varphi^{\prime k}(\theta,\varphi)=\langle k;\hat{\mathbf{e}}_3|\,[J_3(\cos\theta-1)+2k]\,|k;\hat{\mathbf{e}}_3\rangle'= k(\cos\theta+1), \quad \theta\neq 0 \tag{3.39}$$

and the vector potential is given by:

$$\mathbf{A}^{\prime k}(\theta,\varphi) = \hat{\mathbf{e}}_\varphi\,\frac{k(\cos\theta+1)}{r\sin\theta} \qquad \theta\neq 0 \tag{3.40}$$

According to the general theory of section 2 we expect from (2.44) and $\zeta_k = -2k\varphi$:

$$A^{\prime k} - A^k = -d\zeta_k(\theta,\varphi) = d(2k\varphi) = 2kd\varphi \tag{3.41}$$

or for the vectors

$$\mathbf{A}^{\prime k} - \mathbf{A}^k = -\nabla\zeta_k = 2k\frac{1}{r\sin\theta}\,\mathbf{e}_\varphi\,. \tag{3.42}$$

Comparing (3.39) with (3.34) we have indeed $A_\varphi^{\prime k} - A_\varphi^k = 2k$ in agreement with (3.41) and (3.42).

We now calculate the Berry curvature. According to (2.53c):

$$F^k = dA^k = \frac{\partial A_\theta^k}{\partial\varphi}\,d\varphi\wedge d\theta + \frac{\partial A_\varphi^k}{\partial\theta}\,d\theta\wedge d\varphi \tag{3.43}$$

where we have already made use of (3.25). Using (3.33) and (3.34) we obtain

$$F^k = -k\sin\theta\;d\theta\wedge d\varphi \equiv F_{\theta\varphi}^k\;d\theta\wedge d\varphi\,. \tag{3.44}$$

Thus except for the component

$$F_{\theta\varphi}^k = -F_{\varphi\theta}^k = -k\sin\theta \tag{3.45}$$

all spherical covariant components of F_{ij} are zero.

We now use (2.55) to calculate the Berry phase angle for a closed path $\mathbf{C}$ on the unit sphere $\mathcal{S}^2$:

$$\begin{aligned}\gamma_k(\mathbf{C}) &= \oint_{\mathbf{C}} A^k = \oint_{\mathbf{C}} i\langle k;\theta(t),\varphi(t)|d|k;\theta(t),\varphi(t)\rangle \\ &= \int_S F^k = -k\int_S \sin\theta d\theta\wedge d\varphi = -k\int_S d\Omega \qquad \text{mod } 2\pi\end{aligned} \tag{3.46}$$

where S is the surface (any surface) which spans the closed curve $\mathbf{C}$ (i.e. for which $\mathbf{C}$ is the boundary, $\mathbf{C}=\partial S$) and where $d\Omega$ is the solid angle element (area element of the unit sphere). We will denote by $\Omega(\mathbf{C})$ the solid angle that $\mathbf{C}$ subtends:

$$\Omega(\mathbf{C}) = \int_S \sin\theta d\theta \wedge d\varphi\,. \tag{3.47}$$

Then (3.46) can be written

$$\gamma_k(\mathbf{C}) = -k\Omega(\mathbf{C}) \qquad \bmod 2\pi\,. \tag{3.48}$$

According to (3.46) S is the surface that lies above $\mathbf{C}$ when one traverses the curve $\mathbf{C}$ (the direction in which $\mathbf{C}$ is traversed points along the fingers and the normal of S, i.e. the direction of $d\theta \wedge d\varphi$ points along the thumb of the right hand; cf. Fig. 3.2). Now let us consider the surface S' below $\mathbf{C}$. Then again

$$\gamma_k(\mathbf{C}) = \int_{S'} F^k = -k\int_{S'} \sin\theta d\theta \wedge d\varphi = -k\left(-\int_{S^2\setminus S} \sin\theta d\theta \wedge d\varphi\right) \tag{3.49}$$

where the direction of the normal of S' and the direction in which $\mathbf{C}$ is traversed are again given by the right-hand rule. This means that the normal of S' points into the sphere S^2, ($\mathbf{C} = \partial S'$) and the normal of S points out of the sphere S^2, ($\mathbf{C} = \partial S$). In the equation (3.49) we denote by $S^2\setminus S$ the surface with the same area as S' but whose normal points out of the sphere S^2. We rewrite (3.4a) in the form:

$$\gamma_k(\mathbf{C}) = k\left(\int_{S^2} \sin\theta d\theta \wedge d\varphi - \int_S \sin\theta d\theta \wedge d\varphi\right)$$

As the integral of $d\Omega$ over the whole unit sphere is 4π and the integral over S is given by (3.47) we obtain:

$$\gamma_k(\mathbf{C}) = +k\,(4\pi - \Omega(\mathbf{C})) \quad \text{modulo} \quad 2\pi \tag{3.50}$$

From this we conclude that k is integer or half integer, because from comparing (3.50) and (3.48) it follows

$$-k\Omega(\mathbf{C}) = 4\pi k - k\Omega(\mathbf{C}) \quad \bmod \quad 2\pi \tag{3.51}$$

which can only be fulfilled if

$$k = 0, \pm\frac{1}{2} \pm 1, \pm\frac{3}{2}, \pm 2 \cdots. \tag{3.52}$$

If the basis vectors (3.11) and (3.19) are also eigenvectors of $\mathbf{J}^2$ then the possible values of k are already obtained from (3.6). But if these vectors are not $\mathbf{J}^2$-eigenvectors — as is the case for the molecule — then (3.52) is a new result.[5])

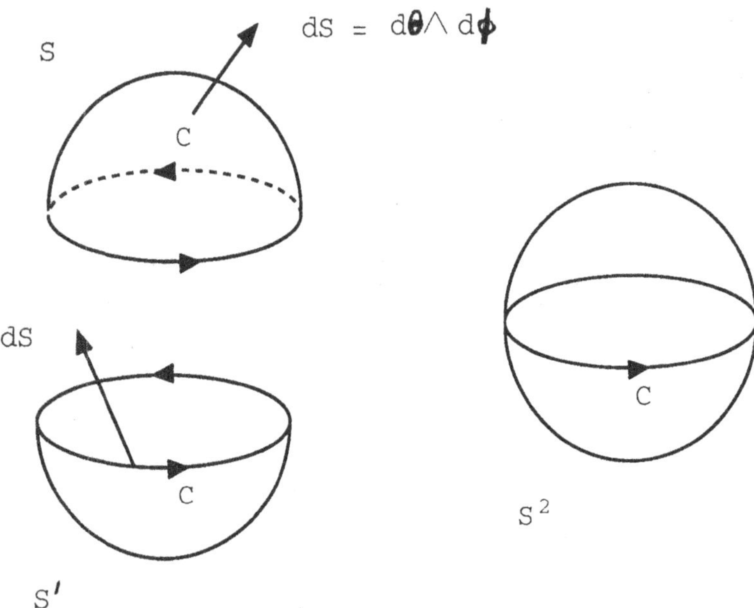

Figure 3.2. The difference of the line integrals of A and A' can be transformed, using Stokes' theorem, into an integral over the closed 2-surface $S \cup S'$.

So far $\mathbf{C}$ could have been any closed path on the unit sphere. For the special path given by (3.4) and shown in Fig. 3.1.

$$\mathbf{C}_1 : R(\theta(t), \varphi(t)) = R(\theta = const., \ \varphi = \omega t) \tag{3.53}$$

the Berry phase (3.48) is

$$\gamma_k(\mathbf{C}_1) = -k \int_0^{2\pi} \int_0^{\theta} \sin\theta d\theta d\varphi = -k2\pi(1 - \cos\vartheta). \tag{3.54}$$

where θ is the constant angle shown in Fig. 3.1. We shall use this expression below when we consider non-adiabatic cyclic evolution.

If $F_{\theta\varphi} = -F_{\varphi\theta}$ is the only non-zero spherical covariant component of F then the only non-zero spherical orthonormal component is $\hat{F}_{\theta\varphi} = -\hat{F}_{\varphi\theta}$. The 2-form F is given by

$$F = F_{\theta\varphi} d\theta \wedge d\varphi = -k \sin\theta d\theta \wedge d\varphi = \hat{F}_{\theta\varphi}(r d\theta)(r \sin\theta d\varphi) = \vec{F} \cdot d\vec{S} \, ,$$

and we immediately read off the value of $\hat{F}_{\theta\varphi}$:

$$\hat{F}_{\theta\varphi} = -\frac{k}{r^2} \qquad \text{or} \qquad \vec{F} = -\frac{k}{r^2}\,\hat{\mathbf{e}}_r = -\frac{k}{r^2}\,\hat{\mathbf{R}}(\theta,\varphi) \tag{3.55}$$

where $\hat{\mathbf{e}}_\mathbf{r}$ is the unit vector in the radial direction which we had denoted by $\hat{\mathbf{R}}(\theta,\varphi)$.

The gauge field (3.55) has a very familiar form, it is the form of the magnetic field of a monopole. An imaginary electromagnetic system, with the vector potential given by (3.35) or (3.40) and the magnetic field given by (3.55), had been envisioned by Dirac more than 50 years ago[6] and has intrigued several generations of theoreticians and experimentalists.[7] It consists of a charge e and a magnetic monopole with magnetic monopole strength g with the vector $\hat{\mathbf{R}}$ pointing from g to e. The magnetic field $\mathbf{B}^{\text{mon.}}$ of this charge-monopole system is given by $e\mathbf{B}^{\text{mon.}} = e\frac{g}{4\pi}\frac{\hat{\mathbf{R}}}{r^2}$ and the electromagnetic vector potential $e\mathbf{A}^{\text{mon.}}$ is given by (3.35), (3.40)) with $k \to -\frac{eg}{4\pi}$. For this magnetic monopole the field and the vector potential are supposed to be electromagnetic.

If one such magnetic monopole exists with magnetic charge g, then all electric charges must be multiples of $2\pi/g$ because k is integer or half integer according to (3.52). This would be a remarkable explanation of one of the basic facts in nature. Unfortunately, in spite of all efforts, such electromagnetic type of monopoles have never been found.

The field $\mathbf{F}$ of (3.55) and the vector potential $\mathbf{A}$ of (3.35) and (3.40) are not electromagnetic in nature. They have their origin in a quantum system spinning with fixed angular momentum component k around the axis $\hat{\mathbf{R}}(t)$. The analogy between classical mechanical systems of this type and the charge-monopole system has been known for some time.[8)] The Hamiltonian $h(R(t))$ of (3.1) also appears for quantum systems in a quantal environement, e.g. for diatomic molecules, because there is a magnetic field in the direction of the internuclear axis $\hat{\mathbf{R}}(t)$ which changes while the molecule as a whole rotates. As a consequence, the fast (electronic) motion about the body axis "induces" the gauge potential (3.35), where k is the "fast" angular momentum along the internuclear axis. So the dynamics of the collective motion contains the monopole vector potential (3.35). (The canonical momenta: P_i of the collective motion go into the gauge-covariant momenta $P_i \to P_i - A_i$). As a consequence, the effective Hamiltonian of the slow collective motion (the quantal environment) is governed – in addition to the radial potential – by the monopole vector potential.[9)] But this is not an electromangetic vector potential because the electromagnetic constant $\frac{eg}{4\pi}$ is replaced by the motion constant k. Thus there "is'" a "motional monopole" in the molecule. But it cannot come out of it because the monopole is the consequence of the fast (electronic) motion which is the other part of the molecule, without which it cannot "be".

The Mead-Berry potential of the molecule can be derived directly from the Schrödinger equation of the molecule if one does not make the drastic assumption of the Born-Oppenheimer approximation.[9)] The same arguments should hold for any other quantum physical system that can be visualized as a spinning bead which can slide along a rod that rotates slowly in space.

3.3 The Exact Solution of the Schrödinger Equation

So far we have not discussed the time evolution of a state vector $\psi(t)$ for our example (3.1), i.e. the solutions of the Schrödinger equation

$$i\frac{\partial\psi}{\partial t} = h(t)\psi(t) \quad ; \quad h(t) = b\hat{\mathbf{R}}(\theta, \omega t)\cdot\mathbf{J} \tag{3.56}$$

In particular we are interested in solutions which describe cyclic evolutions (2.34):

$$|\psi(T)\rangle\langle\psi(T)| = |\psi(0)\rangle\langle\psi(0)|. \tag{3.57}$$

The period T in here could be any time period. In the adiabatic approximation T would be given by $T = \frac{2\pi}{\omega}$.

It is not clear that solutions of (3.56) exist which fulfill the condition (3.57), however for a periodic Hamiltonian, like (3.1), we would expect that solutions with the same period exist. In addition we shall see that other cyclic evolutions exist and that even a time-independent Hamiltonian can have cyclic states.

We start with the adiabatic approximation and obtain the expression for the state vector by inserting the results, in particular (3.54) and (3.5) into (2.43)

$$\psi(t) \overset{\substack{\text{adiabatic}\\ \downarrow}}{=} e^{-ibtk}\, e^{i\gamma_k(t)}\ |k;\, \theta,\, \omega t\rangle \ \ , \quad \omega = \frac{2\pi}{T}, \quad \gamma_k(t) = \gamma_k(\mathbf{C})\frac{t}{T}. \tag{3.58}$$

This formula is an approximate equality indicated by the suffic "adiabatic" and, in general, not an equality (except for $\omega = 0$ in which case it is the stationary state of a time independent Hamiltonian).

For the state vector after a period T we obtain from (2.51):

$$\psi(T) \overset{\substack{\text{adiabatic}\\ \downarrow}}{=} e^{-i2\pi\frac{b}{\omega}k}\, e^{i\gamma_k(\mathbf{C})}\psi(0) \ \ ; \ \ \psi(0) = |k; \theta,\, 0\rangle\,. \tag{3.59}$$

These expressions for the state vector are, as we showed in general in section 2, incompatible with the Schrödinger equation (3.56) and can only be approximation.

But as all solutions of (3.56) are known[1)] we can see whether there are cyclic solutions and select those. The exact solution of (3.56) with the initial state $\psi(0)$ is given by[1)2)]

$$\psi(t) = U^{\dagger}(t)\psi(0) = e^{-it\omega J_3}e^{-it\Omega\mathbf{e}\cdot\mathbf{J}}\psi(0) \tag{3.60}$$

The time evolution thus consists of a rotation of the sate $\psi(0)$ about $\mathbf{e}$ by Ωt followed by a rotation about $\mathbf{e}_3$ by ωt:

The unit vector $\mathbf{e}$ is shown in Fig. 3.1 and given by

$$\mathbf{e} = \hat{\mathbf{R}}(\theta,0) - \frac{\omega}{\Omega}\mathbf{e}_3 = \frac{b}{\Omega}\Big(\cos\theta - \frac{\omega}{b}\Big)\mathbf{e}_3 + \frac{b}{\Omega}\sin\theta\mathbf{e}_1 \equiv \cos\tilde{\theta}\mathbf{e}_3 + \sin\tilde{\theta}\mathbf{e}_1 = \hat{\mathbf{R}}(\tilde{\theta},0) \tag{3.61}$$

The new angle $\tilde{\theta}$ is shown in Fig. 3.1, and is defined by equation (3.61):

$$\cos\tilde{\theta} = \frac{b}{\Omega}(\cos\theta - \frac{\omega}{b}) \qquad \sin\tilde{\theta} = \frac{b}{\Omega}\sin\theta. \tag{3.62}$$

The angular frequency Ω is obtained from the normalization condition $\mathbf{e}^2 = 1$:

$$\Omega = b\Big(1 + \frac{\omega}{b}\Big(-2\cos\theta + \frac{\omega}{b}\Big)\Big)^{\frac{1}{2}} \tag{3.63}$$

These relations between (θ,φ) and $(\tilde{\theta},\varphi)$ define a map $F : M = S^2 \to S^2$, which in cartesian coordinates corresponding to the global orthonormal basis system $(\mathbf{e}_1, \mathbf{e}_2, \mathbf{e}_3)$ is given by

$$F_\nu(x^1,x^2,x^3) = \frac{(x^1, x^2, (x^3 - \frac{\omega}{b}))}{\sqrt{(x^1)^2 + (x^2)^2 + (x^3 - \frac{\omega}{b})^2}} \tag{3.64}$$

It depends upon the physical parameter

$$\nu \equiv \frac{\omega}{b} \tag{3.65}$$

and is given in terms of the local spherical coordinates by

$$F_\nu(R(\theta,\varphi)) = F_\nu(\cos\theta,\varphi) = \Big(\frac{\cos\theta - \nu}{\sqrt{\nu^2 - 2\cos\theta\nu + 1}},\varphi\Big) = (\cos\tilde{\theta},\varphi) \equiv R(\tilde{\theta},\varphi) \tag{3.66}$$

The map F_ν is for $\frac{\omega}{b} < 1$ a homeomorphism of $\mathcal{S}^2$ and consist of a translation in the $\mathbf{e}_3$ direction followed by a projection along $\mathbf{R}$ onto $\mathcal{S}^2$.

$F_\nu(\cos\theta,\varphi)$ is a continuous function over $(-1 \le \cos\theta \le 1, 0 \le \varphi < 2\pi)$ as long as $\nu \equiv \frac{\omega}{b} < 1$. If $\nu \ge 1$ the map F_ν becomes discontinuous. We shall therefore restrict ourselves here to $\nu < 1$;[10)] the adiabatic approximation is fulfilled for $\nu \ll 1$ which means that the external angular velocity ω for the precession of $\hat{\mathbf{R}}(\theta,\omega t)$ is small compared to the Larmor frequence b.

The Larmor frequency b is the angular velocity with which the state $\psi(t)$ rotates about the direction of the magnetic field $\hat{\mathbf{R}}(\theta,0)$ when $\omega \equiv \frac{2\pi}{T} = 0$ or $\nu = 0$:

$$\psi(t) = e^{-itb\hat{\mathbf{R}}(\theta,0)\cdot\mathbf{J}}\psi(0) \qquad \text{for} \qquad \omega = 0 \qquad \text{or} \qquad \nu = 0 \tag{3.67}$$

This can be seen immediately if one rewrites (3.60) using (3.61) and (3.65):

$$\psi(t) = e^{-itb\nu J_3} \quad e^{-itb(\mathbf{R}(\theta,0)\cdot\mathbf{J} - \nu J_3)}\psi(0) \tag{3.60'}$$

Equation (3.67) gives the time evolution of a state vector $\psi(0)$ for the time independent Hamiltonian:

$$h = b\hat{\mathbf{R}}(\theta, 0) \cdot \mathbf{J} \qquad ; \qquad \theta \doteq \text{constant} \tag{3.68}$$

In order to select from all the solutions (3.60) those, which describe cyclic evolution, (3.57), we have to choose in (3.60) for the initial state vector $\psi(0)$ an eigenvector of the operator

$$U^\dagger(T) = e^{-iT\omega J_3} e^{-iT\Omega\hat{\mathbf{R}}(\tilde{\theta},0)\cdot\mathbf{J}} \tag{3.69}$$

Its eigenvalue $e^{-i\alpha_\psi}$ will then be the total phase factor in the cyclic evolution (3.57):

$$U^\dagger(T)\psi(0) = e^{-i\alpha_\psi}\psi(0) = \psi(T) \tag{3.70}$$

If $\psi(0)$ is eigenvector of $U^\dagger(T))$ with eigenvalue $e^{-i\alpha}$ it must also be eigenvector of $U(T) = e^{iT\Omega\vec{R}\cdot\vec{J}} e^{iT\omega J_3}$ with eigenvalue $e^{i\alpha}$. This requires that $e^{-iT\omega J_3}$ and $e^{-iT\Omega\mathbf{R}(\tilde{\theta},0)\cdot\mathbf{J}}$ commute:

$$\left[e^{-iT\omega J_3} \quad , \quad e^{-iT\Omega\hat{\mathbf{R}}(\tilde{\theta},0)\cdot\mathbf{J}}\right] = 0 \tag{3.71}$$

This can be fulfilled under the following two conditions:

$$\text{CLASS A}: \qquad \omega T = 2\pi n; \ \text{evolution period}: \qquad T = \frac{2\pi}{\omega} \tag{3.72}$$

$$\text{CLASS B}: \qquad \Omega T = 2\pi n; \ \text{evolution period}: \qquad T = \frac{2\pi}{\Omega} \tag{3.73}$$

For both classes there will be a discrete set of cyclic solutions. A special case is $\omega = 0$ in which case we have a time-independent Hamiltonian (3.68) with evolution (3.67). The class A cyclic solutions have the same period as the Hamiltonian and the adiabatic evolution. The Class B cyclic solutions have a period which is given by the intrinsic properties of the quantum system (the Larmor frequency b modified by the precession frequency ω).

We shall discuss here only the Class A cyclic evolutions because they are directly connected with the adiabatic cyclic changes.

The Class A cyclic solutions are obtained by finding the eigenvectors of the operator $U_A^\dagger(T) = e^{-i2\pi J_3} e^{-iT\Omega\hat{\mathbf{R}}(\tilde{\theta},0)\cdot\mathbf{J}}$. These eigenvectors of $U^\dagger(T)$ must be eigenvectors of both $e^{-iT\Omega\hat{\mathbf{R}}(\tilde{\theta},0)\mathbf{J}}$ and of $e^{-i2\pi J_3}$. From this it follows immediately that they must be eigenvectors of the operator:

$$\frac{\Omega}{b} h(\tilde{\theta}, 0) \equiv \Omega\hat{\mathbf{R}}(\tilde{\theta}, 0) \cdot \mathbf{J} \qquad \text{for} \qquad \tilde{\theta} \neq \theta \quad \text{i.e. for } \omega \neq 0: \tag{3.74}$$

These eigenvectors we call ϕ_k:

$$\frac{\Omega}{b}h(\tilde{\theta},0)\phi_k = \Omega k\phi_k \quad ; \quad \mathbf{e}\cdot\mathbf{J}\phi_k = k\phi_k. \tag{3.75}$$

From (3.11) and (3.12) with θ replaced by $\tilde{\theta}$ and φ by 0 it follows that ϕ_κ is (up to a phase factor if ϕ_κ is assumed to be normalized) given by

$$\phi_k = |k;\tilde{\theta},0\rangle = U(\tilde{\theta},0)|k;\mathbf{e}_3\rangle = e^{-i\tilde{\theta}J_2}|k;\mathbf{e}_3\rangle. \tag{3.76}$$

To obtain the eigenvalue of $e^{-i2\pi J_3}$ we calculate using (3.15) for $\varphi = 2\pi$:

$$\begin{aligned} e^{-i2\pi J_3}\phi_k &= e^{-i2\pi J_3}|k;\tilde{\theta},0\rangle = e^{-i2\pi J_3}e^{-i\tilde{\theta}J_2}\,|k;0,0\rangle \\ &= e^{-i\tilde{\theta}J_2}e^{-i2\pi J_3}\,|k;0,0\rangle \\ &= e^{-i2\pi k}|k;\tilde{\theta},0\rangle\,. \end{aligned} \tag{3.77}$$

The initial state vectors which lead to a cyclic evolution with period $T = \frac{2\pi}{\omega}$ are thus labelled by the integer or half-integer number k and given by (3.76):

$$\psi(0) = \phi_k = |k;\tilde{\theta},0\rangle \tag{3.78}$$

(after an arbitrary phase factor has been fixed). The total phase of these cyclic evolutions, i.e. the eigenvalues $e^{-i\alpha_\psi}$ of (3.70) are also labelled by k and given by

$$\psi(T) = U^\dagger(T)\psi(0) = e^{-i\alpha_k}\psi(0) = e^{-i2\pi k}e^{-i2\pi\frac{\Omega}{\omega}k}\psi(0) \tag{3.79}$$

Thus the possible initial states of a cyclic evolution are very similar to the initial states of an adiabatic evolution which are given by the eigenvectors $|k;\theta,0\rangle$ of $h(0) = b\hat{\mathbf{R}}(\theta,0)\cdot\mathbf{J}$. The initial states of the real cyclic evolution are instead given by the eigenvectors (3.76) of the operator

$$\tilde{h}(\theta,0) \equiv \frac{\Omega}{b}h(\tilde{\theta},0) \equiv \Omega\hat{\mathbf{R}}(\tilde{\theta},0)\cdot\mathbf{J} \tag{3.80}$$

They are thus state with the component k of angular momentum along the direction $\mathbf{e} = \hat{\mathbf{R}}(\tilde{\theta},0)$ rather than along the direction $\hat{\mathbf{R}}(\theta,0)$ of the initial magnetic field, see Fig. 3.1. For large values of $\omega, \omega \approx b$ these two directions can be very different, but for $\nu = \frac{\omega}{b} \to 0$ they become the same. We now have to determine the analogue of the single-valued eigenvectors $|k;\mathbf{R}(t)\rangle = |k;\theta(t),\varphi(t)\rangle$ which are used in the calculation of the Berry connection and Berry phase. These vectors $\phi_k(t)$ lie on a curve above the closed curve

$$\mathcal{C}: t \to |\psi(t)\rangle\langle\psi(t)| \quad ; \quad |\psi(T)\rangle\langle\psi(T)| = |\psi(0)\rangle\langle\psi(0)| \tag{3.81}$$

They must fulfill:

$$\phi_k(t) = \text{phase factor} \cdot \psi(t); \qquad \phi_k(T) = \phi_k(0) \tag{3.82}$$

There are many such vectors all differing by a phase transformation

$$\phi_k(t) \to \phi'_k(t) = e^{i\zeta_k(t)}\phi_k(t) \quad , \quad \zeta_k(0) = \zeta_k(T) \qquad \text{mod } 2\pi \tag{3.83}$$

One, and the by far most obvious choice for $\phi_k(t)$ is

$$\phi_k(t) \equiv U(\tilde{\theta}, \omega t)|k; \mathbf{e}_3\rangle = |k; \tilde{\theta}, \omega t\rangle = e^{-i\omega t\, J_3}\, e^{-i\tilde{\theta}\, J_2}\, e^{i\omega t k}|k, \mathbf{e}_3\rangle\,. \tag{3.84}$$

These single-valued basis vectors $|k; \tilde{\theta}, \varphi = \omega t\rangle$ are eigenvectors of the operator

$$\tilde{h}(\theta, \varphi) \equiv \frac{\Omega}{b} h(\tilde{\theta}, \tilde{\varphi}) = \frac{\Omega}{b} h(F_\nu(\theta, \varphi)) = \Omega \hat{\mathbf{R}}(\tilde{\theta}, \varphi)) \cdot \mathbf{J} \tag{3.80a}$$

where F_ν is the function (3.66). The operator $\tilde{h}(R) = \frac{\Omega}{b} h(F_\nu(R))$ is not the Hamiltonian but another parameter dependent operator which can as well serve to define a basis system for the space $\mathcal{H}$. In terms of these known vectors $|k; \tilde{\theta}, \varphi\rangle$ the cyclic solution of the Schrödinger equation (3.56) is given by

$$\psi(t) = e^{-i\omega t k} e^{i\Omega t k} |k; \tilde{\theta}, \omega t\rangle \tag{3.85}$$

Therewith we have expressed the state vectors for cyclic time evolutions in terms of the known quantities ω, Ω (of (3.63)) and $|k, \tilde{\theta}, \omega t\rangle$ (of (3.84). The equation (3.85) is the generalization of the adiabatic "equality" (3.58) and (3.79) is the generalization of (3.59). However the phase factor in (3.85) and (3.79) is not in the form of a product of the dynamical and the geometrial part. In order to obtain that we have to calculate the dynamical or the geometrical phase factor separately.

3.4 The Dynamical and the Geometrical Phase Factor for Non-Adiabatic Evolution

The splitting of the phase factor into a "dynamical and a "geometrical" part can be approach from two different points of view. Either one gives an argument why a certain part is geometrical and obtains the dynamical phase as the difference between the total and geometrical phase, or one defines the dynamical phase and obtains the geometrical phase as the difference between the total and dynamical phase. We will define here the dynamical phase α_k^{dyn} and obtain the geometrical phase as a derived quantity.

$$\alpha_k^{geom.} = \alpha_k - \alpha_k^{dyn.} \tag{3.86}$$

The dynamical phase for general cyclic evolution was defined by Aharonov and Anandan[3] by

$$\alpha_k^{dyn.} \equiv \int_0^T \langle\psi(t)|h(t)|\psi(t)\rangle dt \tag{3.87}$$

This is reasonable because one can show that the phase for the evolution of a stationary state (2.25) is given by (3.87) and because (3.87) goes into the dynamical phase for adiabatic cyclic evoluton given by (2.31).

It is straightforward to calculate the dynamical phase (3.87) for the Hamiltonian (3.1):

$$\begin{aligned}\alpha_k^{dyn} &\equiv \int_0^T \langle\psi(t)|h(t)|\psi(t)\rangle dt \\ &= k2\pi\Big(\frac{\Omega}{\omega} + \cos\tilde{\theta}\Big).\end{aligned} \tag{3.88}$$

Using (3.86) we then calculate from (3.79)

$$\alpha_k^{geom.} = k2\pi(1-\cos\tilde{\theta}) = k2\pi\Big(1 + \frac{\omega}{\Omega} - \frac{b}{\Omega}\cos\theta\Big). \tag{3.89}$$

This has the same form as (3.54) for the adiabatic case except that θ is replaced by $\tilde{\theta}$ of (3.62).

In the adiabatic approximation, $\nu = \frac{\omega}{b} \ll 1$, we can expand the expressions in (3.62) and (3.63) with respect to $\frac{\omega}{b}$:

$$\begin{aligned}&\frac{\Omega}{b} \approx 1 - \frac{\omega}{b}\cos\theta + \Big(\frac{\omega}{b}\Big)^2 \\ &\sin\tilde{\theta} \approx \sin\theta + \frac{1}{2}\frac{\omega}{b}\sin 2\theta \quad ; \quad \cos\tilde{\theta} \approx \cos\theta + \frac{1}{2}\frac{\omega}{b}(\cos 2\theta - 1)\end{aligned} \tag{3.90}$$

From this we obtain that the adiabatic approximation of the geometrical phase angle for general cyclic evolution (3.89) is identical with the Berry phase angle:

$$-\alpha_k^{\rm geom} \approx -2\pi k(1-\cos\theta) = \gamma_k^{\rm Berry}. \tag{3.91}$$

These results indicate that (3.86) with (3.87) is the appropriate choice of the geometrical phase also for the general cyclic evolution.

We now want to show that this phase angle can be obtained in the same way as Berry's phase for the adiabatic approximation from a gauge potential (connection).

In analogy to (2.40), (2.42) we define the connection one-form by

$$A^{\phi_k} = i\langle\phi_k(t)|d|\phi_k(t)\rangle = \langle k;\tilde{\theta},\omega t|d|k;\tilde{\theta},\omega t\rangle = \langle k;\mathrm{e}_3|U^\dagger(\tilde{\theta},\omega t)\frac{d}{dt}U(\tilde{\theta},\omega t)|k;\mathrm{e}_3\rangle dt. \tag{3.92}$$

Here we have used in place of the single-valued eigenvectors of $h(t)$ in (2.42) the basis vectors (3.84) which are eigenvectors of another operator $\tilde{h}(R)$ of (3.80a). From the results (3.30) (with $\varphi = \omega t$) we obtain immediately:

$$A^{\phi_k} = \langle k;\ \mathbf{e}_3| - (J_1 \cos\varphi + J_2 \sin\varphi)\sin\tilde{\theta} + J_3(\cos\tilde{\theta} - 1)|k;\ \mathbf{e}_3\rangle d(\omega t) \quad (3.94)$$

or

$$A^{\phi_k} = -k(1 - \cos\tilde{\theta})d\omega t = -k(1 - \frac{b}{\Omega}\cos\theta + \frac{\omega}{\Omega})d\varphi(t) \quad (3.95)$$

The curvature (field strength) that follows from this connection (gauge potential) is

$$F = d\,A = -k\frac{\left(1 - \frac{\omega}{b}\cos\theta\right)}{\left(\frac{\Omega}{b}\right)^3}\sin\theta\, d\theta \wedge d\varphi$$

or

$$F_{\theta\varphi} = -k\frac{1 - \frac{\omega}{b}\cos\theta}{\left[1 - \frac{\omega}{b}\cos + \left(\frac{\omega}{b}\right)^2\right]^{3/2}}\sin\theta; \qquad \mathbf{F} = -\frac{k}{r^2}\frac{\left(1 - \frac{\omega}{b}\cos\theta\right)}{\left(\frac{\Omega}{b}\right)^3}\hat{\mathbf{R}}(\theta,\varphi) \quad (3.96)$$

This is again the "monopole" type field as in (3.55) but with modified "monopole strength".

The connection one-form (3.95) has the same form as (3.34) for the adiabatic connection, except that here we have the angle $\tilde{\theta}$ which is not the angle θ for the direction of the magnetic field (parameter of the Hamiltonian).

We can now use the analog of the formula (2.54′) and (3.46) for the adiabatic approximation and define a "geometric phase" angle also for general cyclic evolution by

$$\gamma_k \equiv \oint_{\mathbf{C}} A^{\phi_k} = \oint i\langle\phi_k(t)|d|\phi_k(t)\rangle \quad (3.97)$$

Then we obtain with (3.95)

$$\gamma_k = -k\int_0^T (1 - \cos\tilde{\theta})d\omega t = -k2\pi(1 - \cos\tilde{\theta}) = -2\pi k\left(1 - \frac{b}{\Omega}\cos\theta + \frac{\omega}{\Omega}\right). \quad (3.98)$$

This agrees with the result (3.89) obtained from (3.86), (3.87).

With these results we can rewrite the cyclic solution of the Schrödinger (3.85) in a form which completely resembles the adiabatic approximation (3.58):

$$\psi(t) = e^{-i\omega tk(1-\cos\tilde{\theta})}\, e^{-i\omega\, tk\left(\frac{\Omega}{\omega}+\cos\tilde{\theta}\right)}|k;\tilde{\theta},\varphi\rangle = e^{i\gamma_k}\, e^{-i\alpha_k^{\text{dyn}}}|k;\tilde{\theta},\varphi(t)\rangle\,. \quad (3.99)$$

The concepts introduced in this section for the cyclic evolution (3.81) are the analogs of the concepts for the adiabatic approximation. The single-valued basis vectors ϕ_k are the analogs of the Hamiltonian eigenvectors $|k;\theta,\varphi\rangle$. The connections A^{ϕ_k} of (3.94) are generalizations of the adiabatic connections (2.42). The geometric phase (3.97) is a generalization of the Berry phase (2.48).

The distinction between (3.99) and (3.58) is that in (3.99) we have an equality sign. This means that the the state vector $\psi(t)$ indeed "rotates" along the curve $t \rightarrow \phi_k(t)$ described by the curve $t \rightarrow (\tilde{\theta}, \omega t)$ in parameter space, whereas the curve $t \rightarrow (\theta, \omega t)$, describes the path of the external magnetic field, not of a state. For slow enough ω (however before ω becomes zero which is the stationary state development with $\gamma_k = 0$) the curve (3.58) is "close enough" to (3.99), in order to provide an acceptable approximation for the geometric phase. But an exact adiabatic cyclic evolution does not exist.

As we mentioned above in addition to the cyclic evolutions of a pure state with period $T = \frac{2\pi}{\omega}$ (ω=precession frequency of the external magnetic field) there exist other cyclic evolutions with period $T = \frac{2\pi}{\Omega}$ where Ω is given by (3.63) and therewith mainly determined by the Larmor frequency b of the physical state (for $\omega = 0$, which is the cyclic evolution in a constant magnetic field with time-independent Hamiltonian one has exactly $\Omega = b$). We shall not discuss these evolutions here.

The geometrical phase (3.98) is not purely geometrical in the way that (3.91) is, which only depends upon θ and is given solely by the path in the parameter space. The geometrical phase for general cyclic evolution also contains the parameters ω, b of the Hamiltonian. To fully justify the name geometric phase also for the phase (3.98) of non-adiabatic cyclic evolution we have to go from the geometry of the parameter space to the geometry of the space of physical states $\mathcal{P}(\mathcal{H})$.

References

1. I.I. Rabi, N.F. Ramsey, J. Schwinger, Rev. Mod. Phys. **26**, 167 (1954).
2. D.J. Fernandez, L.M. Nieto, L.M. del Olmo, M. Santander, Journal of Physics A (1992); S.J. Wang, Phys. Rev. A**42**, 5107 (1990).
3. See reference 9, section 1.
4. L.C. Biedenham and J.D. Louck, *Angular Momentum in Quantum Physics*, Addison-Wesley, 1981, Chapter 3.5 eq. (3.40)
5. The number $C_1 = -\frac{1}{2\pi}\int_{\mathcal{S}^2} F$ is called the first Chern number of the monopole bundle ($U(1) = \mathcal{S}^1$ bundle over the base space $M = \mathcal{S}^2$). The fiber bundles behind the geometric phase will be discussed in section 4. The Chern number characterizes the bundle and takes the integer $2k$ as shown above. If k is the component of angular momentum in an irreducible representations of the J_i then k must be one of the integers or half integers $-j, -j+1, \ldots +j$. But there is also another proof which shows that k must be integer or half integer, even if $|k; e_3\rangle$ is not an eigenvector of J^2. R_i are the eigenvalues of the operators X_i which together with J_i are the generators of the group E_3. Its unitary

irreducible representations (see Appendix V.3 of reference 1 section 2) are characterized by the numbers (k, ε) with $\varepsilon^2 = R_i R_i (= 1)$ and k = eigenvalue $\frac{1}{\varepsilon} X_i J_i$, where ε is any real positive number and k is positive or negative integer or halfinteger including 0. There are no unitary irreducible representations of E_3 for $2k \neq$ integer.

6. P.A.M. Dirac, Proc. Roy. Soc. **133**, 60(1931).
7. P. Goddard, D.I. Olive Rep. Prog. Phys. **41**: 1357 (1978); L.C. Biedenham, J.D. Louck, (reference 4) Chapter 5.2; S. Coleman in *The Unity of the Fundamental Interactions* ; H. Zichichi (editor) Plenum Press (1983) (Erice lecture (1981); A.B. Balachandran, G. Marmo, B.S. Skagerstam, A. Stern, *Classical Topology and Quantum States* Part I; World Scientific Publishers, (1991).
8. J.M. Leinaas, Physica Scripta **17**, 483 (1978).
9. See reference 3 of section 1 and M.V. Berry and R. Lim, J. Phys. A**23**, L655 (1990); B. Zygelman, Phys. Rev. Lett. **64**, 256 (1990), T. Pacher, C.A. Mead, L.S. Cederbaum, H. Köppel, J. Chem. Phys. **91**, 7057 (1989); A. Bohm, B. Kendrick, Mark E. Loewe and L.J. Boya, *J. Math. Phys.* **33**, 977 (1992); H.K. Lee and M. Rho, Hanyang University preprint HYUPT-92-07 (1992).

4. THE ANANDAN-AHARANOV PHASE FOR GENERAL CYCLIC EVOLUTION AND ITS RELATION TO BERRY'S PHASE — $U(1)$ Bundles Over Parameter Space and Over the Space of Physical States

In this section we shall uncover the general pattern that underlies the results which we obtained for the specific example of the time-dependent Hamiltonian (3.1).

For a time-dependent cyclic Hamiltonian $h(t) = h(x(t)$ with a period T, whose time dependence may be given by a closed path $\mathbf{C}$ in the parameter space M:

$$\mathbf{C} : x(0) \to x(t) \to x(T) = x(0)\ , \tag{4.1}$$

There exists a number of cyclic paths

$$\mathcal{C} : W(0) \to W(t) = |\psi(t)\rangle\langle\psi(t)| \to W(T) = W(0) \tag{4.2}$$

in the space of pure physical states $\mathcal{P}(\mathcal{H})$ with the same period T. (There may be in addition other cyclic paths with different period.) In our example the number of cyclic path is countable and labelled by the half integer k,

$$\mathcal{C}_k : W(t) = W_k(t) = |\phi_k(t)\rangle\langle\phi_k(t)| \tag{4.3}$$

with $\phi_k(t)$ given by (3.84).

Associated with the curve $\mathcal{C}$ (or curves $\mathcal{C}_k$) in $\mathcal{P}(\mathcal{H})$ are three different kind of curves in the Hilbert space $\mathcal{H}$.

1. The curve

$$C : \psi(0) \to \psi(t) \to \psi(T) = e^{-i\alpha_\psi}\psi(0) \tag{4.4}$$

is the curve of solutions of the Schrödinger equation with initial condition that leads to a cyclic solution.

2. The curve

$$C^{closed} : \phi(x(0)) \to \phi(x(t)) \to \phi(x(T)) = \phi(x(0)) \tag{4.5}$$

is a curve of smooth single-valued functions (continuously differentiable) of vectors in $\mathcal{H}$ — that have the property $W(t) = |\phi(t)\rangle\langle\phi(t)|$. These single-valued vector functions are only determined up to a gauge transformation

$$\phi(x(t)) \to \phi'(x(t)) = e^{i\zeta(x(t))}\phi(x(t));\ e^{i\zeta(\tau)} = e^{i\zeta(0)}. \tag{4.6}$$

In our example this closed curve of vectors in the Hilbert space $\mathcal{H}$ associated with the curve $\mathcal{C}_k$ in $\mathcal{P}(\mathcal{H})$ is

$$C_k^{\text{closed}} :\ |k;\tilde{\theta},0\rangle \to \phi_k(t) = |k;\tilde{\theta},\omega t\rangle \to |k;\tilde{\theta},\omega T\rangle = |k;\tilde{\theta},0\rangle\,. \tag{4.7}$$

where the vectors $\phi_k(t)$ are those of (3.84) and the gauge transformation (4.6) is the one in Eq. (3.83). The curve (4.4) in our example is given by the vectors (3.99).

In addition to the curves C^{closed} and C in $\mathcal{H}$ associated with a closed curve of pure physical states (solutions of the von Neumann-Schrödinger Eq. (2.3b)) $\mathcal{C}$ in $\mathcal{P}(\mathcal{H})$ we define[1)]

3. The curve

$$\tilde{C} : \tilde{\psi}(0) \to \tilde{\psi}(t) = e^{i\int_0^t \langle\psi(t)|h(t')|\psi(t')\rangle dt'}\psi(t) \to \tilde{\psi}(T) \tag{4.8}$$

where $\psi(t)$ is solution in (4.4).

All three curves in $\mathcal{H}$ have the property

$$|\psi(t)\rangle\langle\psi(t)| = |\phi(t)\rangle\langle\phi(t)| = |\tilde{\psi}(t)\rangle\langle\tilde{\psi}(t)| = W(t) \text{ in } \mathcal{P}(\mathcal{H}). \tag{4.9}$$

For this reason we call these three curves in $\mathcal{H}$ lifts of the curve $\mathcal{C}$ in $\mathcal{P}(\mathcal{H})$, imagining that $t \to W(t)$ is a curve in some base space and $t \to \psi(t), t \to \phi(t), t \to \tilde{\psi}(t)$ are curves lying above these base space points. This is depicted in Fig. 4.1. The curve C will be called the dynamical lift. The curve C^{closed} defined by (4.8) will be called the closed lift. The curve $\tilde{C}$ defined by (4.8) will be called the A-A lift. As we mentioned already, the closed lift is not uniquely determined but only up to a gauge transformation (4.6). The dynamical lift of a given curve in the space of physical states $\mathcal{P}(\mathcal{H})$ is also not uniquely determined as we shall discuss instantly. However, the A-A lift of a given curve $\mathcal{C}$ is uniquely defined, it is identical with the horizontal lift of differential geometry, as we shall show below.

In our example (3.1) the A-A lift is given by

$$\tilde{\psi}(t) = e^{i\int_0^t \langle\psi(t')|h(t')|\psi(t')\rangle dt'}\psi(t) = e^{i\alpha_k^{\text{dyn}}}\psi(t) = e^{i\Omega k(1+\frac{\omega}{\Omega}\cos\tilde{\theta})t}\psi(t)\,. \tag{4.10}$$

It is not a closed curve but has the property

$$\tilde{C}_k \;:\; \tilde{\psi}(0) \to \tilde{\psi}(t) \to \tilde{\psi}(T) = e^{i\gamma_k}\tilde{\psi}(0)\,, \tag{4.11}$$

where $\gamma_k(T) = -\alpha_k^{geom}$ is the geometrical phase (3.89), (3.98).

The relation between the A-A lift, the dynamical lift and the closed lift in our example is given by:

$$\tilde{\psi}_k(t) = e^{i\alpha_k^{dyn}}\psi(t) = e^{i\gamma_k(t)}|k;\tilde{\theta},\varphi(t)\rangle \tag{4.12}$$

It is a straightforward calculation (Problem 4.1) * to show that the A-A lift

$$\tilde{\psi}(t) = e^{i\int_0^t \langle\psi(t')|h(t')|\psi(t')\rangle dt'}\psi(t). \tag{4.13}$$

---

* *Problem 4.1.* Show that the vector $\tilde{\psi}(t)$ defined by (4.13) fulfills the equations (4.14) if $\psi(t)$ fulfills the Schrödinger equation $i\frac{d\psi(t)}{dt} = h(t)\psi(t)$.

in general fulfills the following equation and initial condition if $\psi(t)$ fulfills the Schrödinger Eq. (2.3):

$$i\frac{d\tilde{\psi}(t)}{dt} = (h(t) - \langle\psi(t)|h(t)|\psi(t)\rangle\mathbf{1})\tilde{\psi}(t) \qquad (4.14a)$$

$$\tilde{\psi}(0) = \psi(0) \qquad (4.14b)$$

Taking the scalar product of (4.14) with $\tilde{\psi}(t)$ or $\psi(t)$ we obtain

$$\langle\tilde{\psi}(t)|\frac{d}{dt}|\tilde{\psi}(t)\rangle = \langle\psi(t)|\frac{d}{dt}|\tilde{\psi}(t)\rangle = 0\,. \qquad (4.15)$$

This means that the tangent vector $\frac{d}{dt}\tilde{\psi}(t)$ of $\tilde{\psi}(t)$ is orthogonal (in the Hilbert space sense) to $\psi(t)$ and to $\tilde{\psi}(t)$.

The dynamical lift C is uniquely determined by the Hamiltonian $h(t)$, but not uniquely determined by the physical problem. A simple substitution

$$h(t) \to h'(t) = h(t) - \kappa(t)\mathbf{1} \qquad \text{with } \kappa(t)\epsilon\mathbb{R} \qquad (4.16)$$

leads to a new Hamiltonian $h'(t)$ which describes the same physics. It has the same (closed) curves of physical states $t \to W(t) = |\psi(t)\rangle\langle\psi(t)|$ in the projective Hilbert space $\mathcal{P}(\mathcal{H})$ as $h(t)$ (Problem 4.2).* But $h(t)$ and $h'(t)$ have different dynamical lifts $t \to \psi(t)$ and $t \to \psi'(t)$ (Problem 4.3).* Also, two Hamiltonians $h(t)$ and $h'(t)$ which have the same curves of physical states $t \to |\psi(t)\rangle\langle\psi(t)|$ and $t \to |\psi'(t)\rangle\langle\psi'(t)|$ differ by a multiple of the unit operator $\kappa(t)\mathbf{1}$. We are looking for a lift which is unique for a given physical problem, i.e. a lift that is uniquely associated to a closed curve

* *Problem 4.2.* Show that the closed path $\mathcal{C} : t \to |\psi(t)\rangle\langle\psi(t)|$ in the space of physical states $\mathcal{P}(\mathcal{H})$ which is generated by a Hamiltonian $h(t)$ is not affected by a change of the Hamiltonian $h(t) \to h'(t) = h(t) - \kappa(t)1$ where $\kappa(t)\epsilon\mathbb{R}$.

Problem 4.3. Show that the dynamical lift of the closed path $\mathcal{C}$ in $\mathcal{P}(\mathcal{H})$ changes under the substitution $h(t) \to h'(t) = \kappa(t)1$ and calculate the phase difference between the dynamical lift $\psi(t)$ belonging to $h(t)$ and the dynamical lift $\psi'(t)$ belonging to $h'(t)$.

Problem 4.4. Show that the horizontal lift $\tilde{C}$ of a closed path $\mathcal{C}$ in $\mathcal{P}(\mathcal{H})$, defined in terms of the dynamical lift $\psi(t)$ by

$$\tilde{\psi}(t) = \exp\Big(i\int_0^t \langle\psi(t')|h(t')|\psi(t')\rangle dt'\Big)\psi(t),$$

is not affected by the substitution

$$h(t) \to h'(t) = h(t) - \kappa(t)1.$$

$t \to |\psi(t)\rangle\langle\psi(t)|$ in $\mathcal{P}(\mathcal{H})$. Such a lift happens to be the A-A lift $\tilde{C} : t \to \tilde{\psi}(t)$. This can be seen in the following way: In addition to $\tilde{\psi}(t)$ defined by (4.13) using $h(t)$ and $\psi(t)$, (where $\psi(t)$ is the solution of the Schrödinger equation (2.3a) with Hamiltonian $h(t)$), one defines $\tilde{\psi}'(t)$ by

$$\tilde{\psi}'(t) = e^{i\int_0^t \langle\psi'(t')|h'(t')|\psi'(t')\rangle dt'}\psi'(t) \tag{4.13'}$$

where $\psi'(t)$ is a solution of the Schrödinger equation with $h'(t)$:

$$i\frac{d\psi'(t)}{dt} = h'(t)\psi'(t). \tag{2.3a'}$$

Then one can show that $\tilde{\psi}'(t) = \tilde{\psi}(t)$ (Problem 4.4).*

We will now discuss the generalizations of (4.7), the closed lifts of the curve $\mathcal{C}$. As seen from (3.97) the $\phi_k(t)$ are the quantities in terms of which the geometric phase can be calculated if one does not have a solution $\psi(t)$ of the Schrödinger equation. There are various ways to obtain these single valued vectors $\phi(t)\epsilon\mathcal{H}$.[2] In general they are curves of local "sections". Sections are defined as continuously differentiable maps of an open area $O \subset \mathcal{PH}$ into $\mathcal{H} : \phi : O \to \mathcal{H}$. We will assume that our curve $\mathcal{C}$ lies in such an open area O. Then for any closed curve $\mathcal{C} : t \to W(t)$ one has the single valued lift

$$t \to \phi(t) = \phi(W(t)), \quad W(t) = \mid \phi(t)\rangle\langle\phi(t) \mid ; \qquad \phi(T) = \phi(0) \tag{4.17}$$

(determined only up to a gauge transformation (4.6)). The problem is, of course, to determine a "section" or one of these single-valued lifts for a given time dependent Hamiltonian. But if one knows ϕ one can use it to calculate the geometric phase.

Since $\psi(t)$, $\tilde{\psi}(t)$ and any $\phi(t)$ are lifts of the same closed curve $W(t)$, there must exist a phase factor $\omega(t)$ such that

$$\tilde{\psi}(t) = \omega(t)\phi(t)\,. \tag{4.18}$$

We now calculate $\omega(t)$. Taking the derivative of (4.18), we get,

$$\frac{d\tilde{\psi}(t)}{dt} = \frac{d\omega(t)}{dt}\phi(t) + \omega(t)\frac{d\phi(t)}{dt}\,.$$

Taking the scalar product of this with $\hat{\psi}(t)$, we obtain

$$\omega(t)\frac{d\omega(t)}{dt} + \omega^2(t)\Big(\phi(t), \frac{d\phi(t)}{dt}\Big) = \Big(\tilde{\psi}(t), \frac{d\tilde{\psi}(t)}{dt}\Big) = 0$$

where the last equality holds because of (4.15). Thus

$$\frac{1}{\omega(t)}\frac{d\omega(t)}{dt} = -(\phi(t), \frac{d\phi(t)}{dt}).$$

Integrating this we obtain

$$\frac{\omega(t)}{\omega(0)} = e^{-\int_0^t (\phi(t'),\frac{d\phi(t')}{dt'})dt'} . \tag{4.19}$$

This phase factor we call $e^{i\gamma(t)}$ and write (4.18) as

$$\tilde{\psi}(t) = \omega(0)e^{i\int_0^t i(\phi(t'),\frac{d}{dt'}\phi(t'))dt'}\phi(t) \equiv \omega(0)\, e^{i\gamma(t)}\phi(t) . \tag{4.20}$$

If we choose the arbitrary constant phase $\omega(0)$ again such that $\tilde{\psi}(0) = \psi(0) = \phi(0)$, i.e. choose $\omega(0) = 1$, then we have

$$\tilde{\psi}(t) = e^{i\gamma(t)}\phi(t) , \tag{4.21}$$

which is the generalization of (4.12). With the general definition (4.13) of the A-A lift we then have

$$\psi(t) = e^{-i\int_0^t (\psi(t'),h(t')\psi(t'))dt'}\, e^{i\gamma(t)}\, \phi(t). \tag{4.22}$$

This is the general relation between the solution of the Schrödinger equation $\psi(t)$ and the closed lift $\phi(t)$. The equation (2.43) is the adiabatic approximation of this, and (4.12) is the special case of this for the spinning quantum system in the precessing external magnetic field. The adiabatic approximation uses in place of the single-valued lifts $\phi(t)$ the single-valued eigenvectors of $h(R(t)$ which are more easily accessible than the $\phi(t)$ (by solving (2.4)). For the closed path $\mathcal{C}$ in $\mathcal{P}(\mathcal{H})$ the phase angle is (according (4.20))

$$\gamma(T) = \gamma(\mathcal{C}) = \oint_0^T i\langle\phi(t)|\frac{d}{dt}|\phi(t)\rangle dt = \oint i\langle\phi|d|\phi\rangle \bmod 2\pi. \tag{4.23}$$

In here ϕ is any of the closed lifts of $\mathcal{C}$.

The phase angle $\gamma(T)$ is independent of the choice of the parameterization (the speed with which $\phi(t)$ traverses its closed path). It is gauge invariant. It is independent of the choice of the Hamiltonian as long as these Hamiltonians described the same closed path $\mathcal{C}$ in $\mathcal{P}(\mathcal{H})$. It depends upon the closed curve $\mathcal{C}$. It also depends upon the class of Hamiltonians that are connected to each other by the transformation (4.16), but only to the extent that the possible curves $\mathcal{C}$ are determined by $h(t)$. It can thus be considered a "geometric" property of the curve $\mathcal{C}$ in the space of physical states $\mathcal{P}(\mathcal{H})$. Thus the name geometric phase.

The one-form, defined in analogy to the adiabatic connection (2.42) and in analogy to (3.92)

$$A^{\phi} = i\langle\phi|d|\phi\rangle \tag{4.24}$$

transforms under a gauge transformation (4.6) as (Problem 4.5):*

$$A^{\phi} \to A^{\phi'} = A^{\phi} - d\zeta\,. \tag{4.25}$$

This one-form is called the (non-adiabatic) connection form. The formula (4.24) for the connection was obtained from the requirement that $\tilde{\psi}(t)$ was the A-A lift i.e. the lift fulfilling (4.15) (which in turn followed from its definition (4.13)). This was the only possible definition of a lift which depends only upon the physics of the problem (and not upon the arbitrary choice of the Hamiltonian within the class connected by the transformation (4.16)) and which is uniquely determined by the closed curve $\mathcal{C}$ in the space of physical states. The A-A lift $\tilde{\psi}$ is therefore uniquely determined by the requirement that it be a property of the physical quantities $W(t)$ only.

From (4.21) and (4.17) it follows immediately that

$$\tilde{\psi}(T) = e^{i\gamma(\mathcal{C})}\tilde{\psi}(0) \;=\; e^{i\gamma(\mathcal{C})}\psi(0). \tag{4.25}$$

Using (4.13) one obtains then for the cyclic evolution of a state vector:

$$\psi(T) = e^{-i/\hbar\int_0^T \langle\psi(t)|h(t)|\psi(t)\rangle dt'}\, e^{i\gamma(\mathcal{C})}\,\psi(0)\,. \tag{4.26}$$

This is the general relation of which (2.51) is the adiabatic approximation.

As the physics is not in ψ (which depends upon some arbitrary choice of the Hamiltonian) but in $\tilde{\psi}$ (which is uniquely determined by the physical state $|\psi(t)\rangle\langle\psi(t)|$), the so-called geometrical phase $\gamma(\mathcal{C})$ is really the phase that is of physical importance.

A graphical representation of the above described situation for a general cyclic evolution is shown in Fig. 4.1. It shows a *"base space"* with a closed curve in it. The closed curve $\mathcal{C}$ represents the cyclic evolution of a pure state $W(t)$, and the base space represents the space of physical states $\mathcal{P}(\mathcal{H})$. Above the closed curve are shown the three lifts C^{closed}, $\tilde{C}$ and C in the Hilbert space $\mathcal{H}$ (depicted in the figure by the three-dimensional space). Also shown is one "fibre" above the base point $W(0)$ (depicted by the positive z-axis with the z coordinate representing the phase angle modulo 2π or the element $e^{i\zeta(0)}$ of the gauge group $U(1)$). The *"fibre"* in this case (though it has been drawn as a straight line) is given by the circle of unit radius S^1 or by the group $U(1)$. We can attach a copy of this S^1 not only to the point $W(0)$ but to every one-dimensional projection operator $\Lambda \in \mathcal{P}(\mathcal{H})$. In this way we get a *bundle* of $U(1)$ fibres attached (which means loosely associated) to each point of $\mathcal{P}(\mathcal{H})$.

---

* *Problem 4.5.* Find the transformation of the connection one-form $A^{\phi} \equiv i\langle\phi|d|\phi\rangle$ when the section ϕ undergoes the phase transformation $\phi(t) \to \phi'(t) = e^{i\zeta(t)}\phi(t)$.

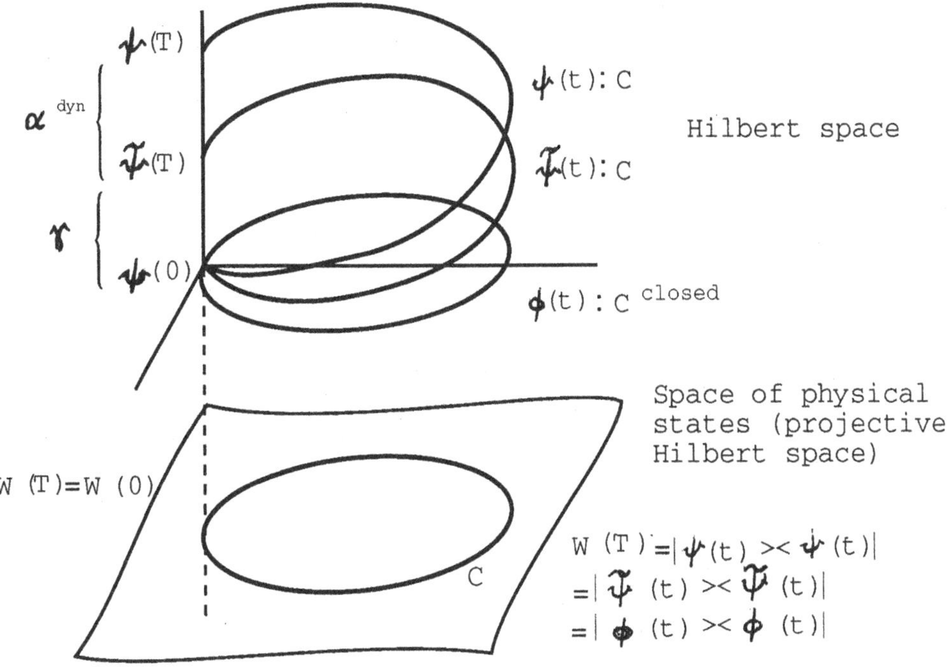

Figure 4.1. Closed path in the space of physical states and its lifts.

The mathematical structure that we have encountered here is an example of a *principal fibre bundle.* The formal definition of a fibre bundle and a principal fibre bundle will be given in the Appendix. Here we will adapt the mathematical notions defined and developed in the Appendix to the $U(1)$ bundles underlying the geometric phase.[3]

In general a fibre bundle consists of a total space, E, (topological space or differentiable manifold), a base space M, a fiber space F, a structure group G acting on the fiber and a projection map π

$$E : F \to E \xrightarrow{\pi} M$$

Often, but not always, one denotes the fibre bundle and the total space by the same letter E. If the structure group G is a Lie group of diffeomorphisms of F with G being the same manifold as F itself (G acting on itself by left or right translations) then the bundle is called a principal fiber bundle.

The set of all one-dimensional projection operators of $\mathcal{H}$ forms the projective space $\mathcal{P}(\mathcal{H}) = \mathbb{C}P^{\infty}$ (If $\mathcal{H}$ is finite dimensional, i.e. the N dimensional space $\mathbb{C}^N$, then $\mathcal{P}(\mathcal{H})$ is the projective space $\mathbb{C}P^{N-1}$). The set of all state vectors is the Hilbert space[4)] without the zero vectors $\mathcal{H}-\{0\}$. If we restrict ourselves to normalized state vectors, which we can do for unitary time development, then we have to consider only the unit sphere $\mathcal{S}(\mathcal{H}) = \mathcal{S}^{\infty} = \{\psi \in \mathcal{H}; \|\psi\| = 1\}$ in $\mathcal{H}$ (If $\mathcal{H}$ is N dimensional then the unit vectors form $\mathcal{S}^{2N-1}$). To each state $|\psi\rangle\langle\psi| \in \mathcal{P}(\mathcal{H})$ there corresponds a set of vectors $\{e^{i\zeta}\psi; \psi \in \mathcal{S}^{\infty};\ e^{i\zeta} \in U(1)\}$ which have all the same projection operator: $|e^{i\zeta}\psi\rangle\langle e^{i\zeta}\psi| = |\psi\rangle\langle\psi|$. Therefore we have a map, the projection map

$$\pi : \mathcal{S}(\mathcal{H}) \to \mathcal{P}(\mathcal{H}) \qquad (\text{or} \qquad \pi : \mathcal{S}^{2N-1} \to \mathbb{C}P^{N-1}), \tag{4.27}$$

with the property

$$\pi(e^{i\zeta}\psi) = \pi(|\psi\rangle) = |\psi\rangle\langle\psi| \tag{4.28}$$

The inverse image $\pi^{-1}(|\psi\rangle\langle\psi|)$ is called the fiber over $|\psi\rangle\langle\psi|$. For any $|\chi\rangle\langle\chi| \in \mathcal{P}(\mathcal{H})$ the fibres $\pi^{-1}(|\chi\rangle\langle\chi|), \chi \in (\mathcal{H})$ are essentially the same. They are given by $\{|\chi\rangle e^{i\zeta}, 0 \leq \zeta < 2\pi\} = F_{|\chi\rangle}$, the fiber over $|\chi\rangle$. The fibres are all homeomorphic (related by a continuous map whose inverse is also continuous) or diffeomorphic to $U(1) = \{e^{i\zeta}\}$, called the typical fibre. Thus we have

$\mathcal{P}(\mathcal{H}) = \mathbb{C}P^{\infty}$	(or $\mathbb{C}P^{N-1}$)	as the base space M,
$\mathcal{S}(\mathcal{H}) = \mathcal{S}^{\infty}$	(or $\mathcal{S}^{2N-1}$)	as the total space E,
π of (4.28)		as the projection map π, and
$U(1)$		as the fibre F
$U(1)$		also as the structure group G.

The aggregate

$$\eta : U(1) \to \mathcal{S}(\mathcal{H}) \xrightarrow{\pi} \mathcal{P}(\mathcal{H}) \qquad (\text{or } U(1) \to \mathcal{S}^{2N-1} \to \mathbb{C}P^{N-1}) \tag{4.29}$$

is our principal fibre bundle (principal bundle because $F = G = U(1)$.

The projection map π leads from a curve $t \to p(t) \in E$ in the bundle to a curve $t \to m(t) = \pi(p(t))$ in the base space. The curve $p(t)$ is called a lift of the curve $m(t)$ in the base space M. Thus the curves (4.4), (4.5) and (4.8) are examples of different lifts of the curve (4.2) in $\mathbb{C}P^\infty$. We mentioned that of these three kind of lifts only the lift (4.8), which we called the A-A lift, was uniquely defined. We shall now show that this lift is related to the geometrical notion of parallelism which at first sight has nothing to do with the definition (4.13) (by Anandan and Aharonov)[1]. Indeed it will turn out that $\tilde{\psi}(t) \in \mathcal{H}$ is the point obtained from $\tilde{\psi}(0) = \psi(0)$ by "parallel transport" along the curve $\mathcal{C}$ in $\mathcal{P}(\mathcal{H})$. This is also expressed by saying that the curve $\tilde{\mathcal{C}}$ is the "horizontal lift" of the curve $\mathcal{C}$.

In order to discuss parallelism between a smooth curve in the base space $\mathcal{C} : t \to |\chi(t)\rangle\langle\chi(t)|$ and one of its lifts $C : t \to \chi(t)\rangle$ we have to define an additional geometric structure for the fibre bundle, called a connection on η. For a given bundle one can define different connections and get different meanings of parallel transport. For the $U(1)$-bundle over the projective space $\mathcal{P}(\mathcal{H})$, η, the mathematically natural connection is the Stiefel connection which has been defined in mathematics long before[5] the lift (4.8) was introduced in quantum physics.[1] That the lift which is horizontal with respect to the Stiefel connection is identical to the A-A lift (4.13), (which provides the quantal phase factor, (4.25), that can be measured in an interference experiment) is one of the more remarkable example of the "unreasonable effectiveness of mathematics in natural science" (Wigner). This was probably the main reason for the popularity which Berry's phase acquired among mathematical physicists.

In order to define the connection and therewith the horizontal lift for the fibre bundle E, we consider the tangent space T_pE at the point $p \in E$. This tangent space can be decomposed into two subspaces: The vertrical subspace is defined as the subspace of those vectors which are tangent to the fibre through p (tangent vectors to all possible curves along the fibre *only*). This subspace is denoted by V_pE; it is determined by the fiber. For the principal bundle it is isomorphic to the Lie algebra of the structure group G (in our case of $G = U(1)$ it is generated by i) and has the dimension of the fibre (= structure group G).

The horizontal subspace, denoted by H_pE, is the space chosen to be the supplementary linear subspace in T_pE to V_pE:

$$T_pE = V_pE \dotplus H_pE \tag{4.30}$$

where $\dotplus$ is the direct sum of linear spaces. In other words, every $\tau \in T_pE$ is given by

$$\tau = ver(\tau) + hor(\tau) \qquad ; \qquad ver\ \tau \in V_pE,\ hor\ \tau \in H_pE \tag{4.31}$$

The horizontal subspace is isomorphic to the tangent space $T_{\pi(p)}M$ on the base space M at the point to which p projects, $\pi(p)$.

The assignment to each point $p \in E$ of a horizontal subspace H_pE of T_pE such that (4.30) and (4.31) is fulfilled, is called a *connection*.

After the choice of a connection the *horizontal lift* can be defined: Let $t \to m(t)$ be a curve in M then its horizontal lift $p(t)$ is the curve in E for which $\pi(p(t)) = m(t)$ and for which all vectors tangent to $p(t)$ are in the horizontal space $H_{p(t)}E$.

We will now choose the connection for our particular bundle η of (4.29). This means we have to assign the horizontal subspace $H_{|\psi\rangle}S(\mathcal{H})$ to every $|\psi\rangle \in \mathcal{S}(\mathcal{H})$. We use the scalar product of $\mathcal{H}$ to make this assignment.

In our case $\psi(t)$ is in the fiber over $|\psi(t)\rangle\langle\psi(t)|$ for every fixed t. The vector $\frac{d}{dt}\psi(t)$ is the tangent vector: $\frac{d}{dt}\psi(t) \in T_{\psi(t)}\mathcal{S}(\mathcal{H})$. We now define the *horizontal lift* of (4.2):

$$\hat{\psi}(t) \in \pi^{-1}(|\psi(t)\rangle\langle\psi(t)|) = \pi^{-1}(W(t)) \text{ is a horizontal lift}$$

i.e.,

$$\frac{d}{dt}\hat{\psi}(t) \in H_{\hat{\psi}(t)}S(\mathcal{H}) \tag{4.32a}$$

iff

$$\langle\hat{\psi}(t)|\frac{d}{dt}|\hat{\psi}(t)\rangle = \langle\psi(t)|\frac{d}{dt}|\hat{\psi}(t)\rangle = 0 \tag{4.32b}$$

In other words: the curve $t \to \hat{\psi}(t)$ is a horizontal lift of the curve $t \to W(t)$ if the vector $\frac{d}{dt}\hat{\psi}(t) \in \mathcal{H}$ is orthogonal in the Hilbert space sense to the elements of the fiber.

Note that $\hat{\psi}(t)$ are elements of $\mathcal{S}(\mathcal{H}) \subset \mathcal{H}$, and $\frac{d\hat{\psi}(t)}{dt}$ are elements of $T\mathcal{S}(\mathcal{H})$. For a general fiber bundle the elements of TE are not elements of E. But in the particular case of (4.29) the $\frac{d}{dt}\hat{\psi}(t)$ (as the derivative of a parameter dependent Hilbert space vector) is again an element of $\mathcal{H}: T\mathcal{S}(\mathcal{H}) \subset \mathcal{H}$. Thus (4.32), relating elements of $T\mathcal{S}(\mathcal{H})$ with elements of $\mathcal{S}(\mathcal{H})$ makes sense. The assignment of a horizontal subspace by (4.32) is called the Stiefel connection.[5)]

For a given curve $m(t)$ in the base space and any initial point $p_0 = p(0)$ in E (with $\pi(p_0) = m(0)$) there exists exactly one horizontal lift p(t). For a closed curve $t \to m(t)$; $m(0) = m(T) = m_0$, the horizontal lift starting from $p(0)$ will end at a point $p(T) = \pi^{-1}(m_0)$ in the fiber above m_0. As every point of the fiber can be reached from $p(0)$ by a transformation with exactly one element g of the structure group G we must have that $p(T) = p(0)g$. The set of all the g for a given starting point $p(0)$ forms a subgroup of G which is called the holonomy group of $p(0)$. In some cases, which we shall restrict ourselves to, the holonomy group of some point $p_0 \in E$ will be identical to the structure group G. Then (if M is connected) the holonomy group is the same for all $p \in E$. In this case any two points of E can be joined by a horizontal curve in E. A connection, for which the holonomy groups of any $p \in E$ is the structure group G, is called irreducible connection.

After this review of general differential geometrical results we will turn to our specific fiber bundle η and the closed curve (4.2) in the space of physical states

$\mathcal{P}(\mathcal{H})$. We want to find the horizontal lift $t \to \hat{\psi}(t)$ with respect to the Stiefel connection. According to the general results above $\hat{\psi}(t)$ is uniquely determined for a given starting point $\hat{\psi}(0)$ for which we will usually choose the initial state vector $\hat{\psi}(0) = \psi(0)$. The phase factor $e^{i\hat{\gamma}}$:

$$\hat{\psi}(T) = e^{i\hat{\gamma}}\hat{\psi}(0) \tag{4.33}$$

is then an element of the holonomy group (which here is identical with the structure group $U(1)$). The phase factor $e^{i\hat{\gamma}}$ is uniquely determined for a given $\hat{\psi}(0)$. It is called the holonomy (anholonomy*) of the closed path $\mathcal{C} : \hat{\gamma} = \hat{\gamma}(\psi(0)) = \hat{\gamma}(\mathcal{C})$.

Another description of the connection than the assignment (4.32a) of a horizontal subspace $H_{\hat{\psi}(t)}S(\mathcal{H})$ is the description of the connection in terms of a connection-one-form or the connection coefficients (gauge potential, see appendix). A connection form is a one-form $\mathcal{A}$ with values in the Lie algebra of the structure group G (which is isomorphic to the vertical subspace V_pE) such that the holonomy can be written as

$$\hat{\gamma} = \oint_{\mathcal{C}} \mathcal{A} \qquad (\hat{\gamma} = T\oint \mathcal{A} \text{ if G is non} - \text{Abelian}). \tag{4.34}$$

where $\mathcal{C}$ is the closed path (4.2) in the base space. In order to obtain the connection one-form $\mathcal{A}$ that leads by (4.34) to the holonomy in (4.33) we start with a section. A section of a bundle (Appendix) is a continuous (C^∞) map $\phi : M \to E$ which assigns a preferred point $\phi(m)$ on the fibre $\pi^{-1}(m)$ to every $m \in M$. Thus the section undoes in a continuous way what the projection has done: $\pi\phi(m) = m$ for all $m \in M$. Global sections (for all $m \in M$) of principal bundles only exist if the bundle is trivial (i.e. homeomorphic to $M \times G$). For the nontrivial bundle η there exist only local sections. A local section over a patch $O \subset M$ is a map $\phi : O \to E$ such that $\pi\phi(m) = m$ for $m \in O$. For η (like for every principal bundle) there is always a covering $\{O_\alpha\}$ of M such that every O_α admits a local section $\phi^{(\alpha)}$. Given a local section $\phi^{(\alpha)}$, one can define another section $\phi^{(\alpha)'}$ by

$$\phi^{(\alpha)'}(m) = \phi^{(\alpha)}(m)g(m) \tag{4.35}$$

where $g(m) \in G =$ structure group. This is the general gauge transformation which has occurred in various specific forms before, cf (4.6), (3.83), (3.19), (2.12). A local section ϕ of O maps a closed path $t \to W(t)$ in $\mathbb{C}P^\infty$ ($W(0) = W(T)$) into a closed path in $\mathcal{S}(\mathcal{H})$:

$$t \to \phi(t)\ ;\ \phi(0) = \phi(T). \tag{4.36}$$

A local section thus provides a closed lift (4.5) of the curve $\mathcal{C}$ in $\mathcal{P}(\mathcal{H})$.

We now obtain (a local representation of) the connection one-form $\mathcal{A}$ in terms of the local section.

We choose a local section $\phi = \phi(W)$ for the patch that contains the closed path $\mathcal{C}$ of (4.2). The problem in each particular case is, of course, to find such a local section for a given time-dependent Hamiltonian $h(t)$. (In our example of section 3 we obtained the section $|k;\tilde{\theta},\varphi\rangle$ of (3.84) (local in the northern hemisphere) essentially by solving the problem explicitly.) With this section we lift the closed path $\mathcal{C}$ into a closed path of this section

$$C^{closed}: \phi(t) = \phi(W(t)) \quad ; \quad \phi(T) = \phi(0) \tag{4.37}$$

As $\hat{\psi}(t)$ is the horizontal lift of the same path (4.2) we can express the open path $t \to \hat{\psi}(t)$ in $\mathcal{S}(\mathcal{H})$ in terms of the closed path $t \to \phi(t)$ in $\mathcal{S}(\mathcal{H})$ by

$$\hat{\psi}(t) = e^{if(t)}\phi(t) \tag{4.38}$$

where $f(t)$ is a smooth (single-valued) functon. As $\phi(T) = \phi(0)$ we obtain

$$\hat{\psi}(T) = e^{i(f(T)-f(0))}\hat{\psi}(0) \tag{4.39}$$

Comparing this with the definition of the holonomy (4.33) we see that

$$\hat{\gamma} = f(T) - f(0) \tag{4.40}$$

This we can calculate now by substituting (4.38) into the defining relation (4.32b) of the horizontal lift and integrating. This yields:

$$\hat{\gamma} = \int_0^T i\langle\phi(t)|\frac{d}{dt}|\phi(t)\rangle dt = \int_0^T i\langle\phi|d|\phi\rangle \tag{4.41}$$

Thus the connection one-form given by the assignment of the horizontal subspace (4.32) is:

$$\mathcal{A} = i\langle\phi|d|\phi\rangle\ . \tag{4.42}$$

This connection one-form is also called the Stiefel connection (of the $U(1)$ bundle η over $\mathbb{C}P^{N-1}$ or $\mathbb{C}P^{\infty}$). The bundle space $\mathcal{S}(C^N) = S^{2N-1}$ (for dimension of $\mathcal{H} = N$) is called the Stiefel manifold, denoted $V_{1,N}(\mathbb{C}) = U(N)/U(N-1) = \mathcal{S}(\mathbb{C}^N)$. It is the manifold of 1-frames in the space $\mathcal{H} = \mathbb{C}^V$. We will call the bundle (4.29) with the connection (4.42) the Stiefel bundle. It is the special case, $\mathcal{N} = 1$, of a whole class of Stiefel bundles:

$$\eta_{\mathcal{N}}: U(\mathcal{N}) \to V_{\mathcal{N},N}(\mathbb{C}) \to Gr_{\mathcal{N},N}\ . \tag{4.43}$$

Here $Gr_{\mathcal{N},N} = U(N)/U(\mathcal{N}) \times U(N-\mathcal{N})$ is the Grassmanian manifold of $\mathcal{N}$-dimensional subspaces (or $\mathcal{N}$-dimensional projection operators) in the Hilbert space of dimension N $\mathcal{H} = \mathbb{C}^N$, and $V_{\mathcal{N},N}(\mathbb{C}) = U(N)/U(N-\mathcal{N})$ is the Stiefel manifold of N-frames in $\mathcal{H} = \mathbb{C}^N$. The structure group of this bundle is $U(\mathcal{N})$. We will not explain the cases $\mathcal{N} > 1$ any further here[6] but continue with the case $\mathcal{N} = 1$.

The Stiefel connection (4.42) or (4.32) is the mathematically natural definition of a connection for the bundle $\eta = \eta_1$. This connection is given without any reference to a Hamiltonian or an equation of motion. For a given closed curve $\mathcal{C}$ ($\mathcal{C}$ is determined by the Hamiltonian and the dynamical equation (2.3b)) the holonomy is a purely geometric property of the Stiefel bundle; the gauge group is the structure group of this bundle.

Comparing (4.41) with (4.23) and (4.33) with (4.25) we see that the Anandan-Aharanov phase angle $\gamma(\mathcal{C})$ is identical with the holonomy $\hat{\gamma}$ and the A-A lift $\tilde{\psi}(t)$ is identical with the horizontal lift of the Stiefel connection:

$$\tilde{\psi}(t) = \hat{\psi}(t) \quad ; \quad \gamma(\mathcal{C}) = \hat{\gamma} \,. \tag{4.44}$$

Also, the non-adiabatic connection form (4.24) is the same as the Stiefel connection, $A = \mathcal{A}$. Of course, this is what we should have expected, because the consequence (4.15) of the definition (4.13) is identical with the definition (4.32b).

Thus, the Stiefel bundle is the mathematical structure that underlies the theory of the quantum geometric phase.

We started our investigation (in section 2) with a geometric phase, (2.50), which is not the holonomy of a closed path $\mathcal{C}$ of physical states. Instead we had chosen a closed path (2.7) = (4.1) in another base space, the parameter space M of the observables. The bundle that underlies the adiabatic geometric phase (Berry's phase) is therefore a $U(1)$ fibre bundle over the parameter space M:

$$\lambda : U(1) \to \lambda \to M \tag{4.45}$$

We shall now discuss the relation between the bundle λ, which Simon[7] recognized as the mathematics behind Berry's phase factor, and the Stiefel bundle of the Anandan-Aharonov approach.

As the Berry-Simon and the Anandan-Aharonov approach use different base spaces we first have to establish a relation between the base space M and the base space $\mathcal{P}(\mathcal{H})$. Then we use the classification theorem for principal fibre bundles to obtain all the bundles λ that relate to the bundle η in (4.29). The problem is thus to find the right map:

$$M \;\to\; \mathcal{P}(\mathcal{H})(= \mathbb{C}P^{N-1} \text{ for } N = \text{finite}) \,. \tag{4.46}$$

There always exists a smooth map f_n which is determined by the Hamiltonian $h(x)$ and defined by

$$M \ni x \to f_n(x) = \Lambda_n(x) = |n; x>< n; x| \in \mathbb{C}P^{N-1} \tag{4.47}$$

(for every given value of the energy quantum number n). This smooth map associates with the closed path $\mathbf{C}$ in M a closed curve of (one-dimensional) eigenprojections of $h(x)$ in $\mathcal{P}(\mathcal{H})$

$$\mathcal{C}^{\Lambda} : \Lambda_n(x(0)) \to \Lambda_n(x(t)) = |n; x(t)\rangle\langle n; x(t)| \to \Lambda_n(x(T)) = \Lambda_n(x(0)) \tag{4.48}$$

As already mentioned in section 2 (following eq. (2.33)), the closed curve (4.48) is a path of observables not of physical states. Any path of states $t \to W'(t)$ with the initial condition $W'(0) = \Lambda_n(x(0))$ where $\Lambda_n(x(0))$ is an eigenprojector of $h(x(0))$, is not closed. Thus the smooth map f_n does not relate closed curves in M to closed curves of physical states $\mathcal{C}$. Only in the adiabatic approximation (2.24) do the closed curves of eigenprojectors (4.48) *approximate* a physical state.

The Berry phase $\gamma_n(\mathbf{C})$ is according to (2.51) and (2.49) the holonomy of the closed curves $\mathcal{C}^{\Lambda}$ of eigenprojectors (4.48). The Berry connection form (2.42) is the Stiefel connection, however it is *not* taken on the closed path $\mathcal{C}$ of physical states (4.2), *but* on the closed path $\mathcal{C}^{\Lambda}$ of eigenprojectors (4.48) in $\mathcal{P}(\mathcal{H})$ or on the closed path $\mathbf{C}$ of parameters (4.1) in M.

The single-valued eigenvectors $|n; x(t)\rangle$ of $h(x(t))$, (2.4), are sections, however they are *not* taken on a closed path above the cyclic path $\mathcal{C}$ of physical states (4.2) but they are taken on a closed path above $\mathcal{C}^{\Lambda}$ or C in M. For the example (3.1) the sections $\phi_k(W)$ are taken above the closed path $t \to |\psi(t)\rangle\langle\psi(t)| = |k, \tilde{\theta}, \varphi(t)\rangle\langle k; \tilde{\theta}, \varphi(t)|$ and given by (4.7): $\phi_k(W(t)) = |k; \tilde{\theta} = const. \varphi = \omega t\rangle$. But the single-valued eigenvectors of $h(R(t))$ are taken above the path $R(\theta = const, \varphi = \omega t)$ of (3.53) and given by (3.11). These two sets of functions are related by the map (3.64):

$$|k; \tilde{\theta}, \varphi\rangle = |k; F_\nu(\theta, \varphi)\rangle \qquad ; \nu = \frac{\omega}{b} \tag{4.49}$$

The Berry (adiabatic) connection on the path (3.53) in parameter space is given by

$$A^k(\theta, \varphi) = i\langle k; \theta, \varphi|d|k; \theta, \varphi\rangle = -k(1 - \cos\theta)d\varphi(t) \tag{4.50}$$

and the Anandan-Aharonov connection on the same path $R(\theta, \omega t)$ in parameter space is

$$\begin{aligned} A^{\phi_k}(\theta, \varphi) &= i\langle \phi_k|d|\phi_k\rangle = i < k; F_\nu(\theta, \varphi)|d|k; F_\nu(\theta, \varphi)\rangle \\ &= -k(1 - \frac{b}{\Omega}\cos\theta + \frac{\omega}{\Omega})d\varphi \end{aligned} \tag{4.51}$$

which leads to the two different expressions (3.54) and (3.98) for the adiabatic and exact geometric phase connected with the same closed path $\mathbf{C}_i$ in the parameter space M.

In addition to the $U(1)$-Stiefel bundle over the space of pure physical states (or of one-dimensional eigenprojectors $\mathcal{P}(\mathcal{H}) = \mathbb{C}P^{N-1}$) we have thus a $U(1)$ bundle over the parameter manifold M as base space.

The Stiefel bundle (4.29) is the universal classifying $U(1)$-bundle.[5)] The classification theorem states that any $U(1)$ principal fibre bundle ξ is isomorphic to the pull-back bundle $f^*(\eta)$ for some continuous function $f : M \to \mathcal{P}(\mathcal{H})$. (Appendix)

The pullback bundle $f^*(\eta)$ is a new bundle which one can define for each smooth map f and original bundle η. It has the same fibre F as the original bundle and is defined in the following way: $f^*(\eta)$ is the subspace of $M \times \eta = M \times \mathcal{S}(\mathcal{H})$ which consists of the points $(x, \psi) \in M \times \eta$ such that $f(x) = \pi(\psi)$. Thus

$$f^*(\eta) = \{(x, \psi) \in M \times \mathcal{S}(\mathcal{H}); \Lambda_n(x) = |\psi >< \psi|\} \tag{4.52a}$$

if we choose for f the map $f_n = \Lambda_n$ of (4.47) and for the projection the map $\pi(|\psi >)$ of (4.28). The fibre $\mathcal{F}_x$ of $f^*(\eta)$ is a copy of the fibre $\mathcal{F}_{f(x)}$ of η: $\mathcal{F}_x = \mathcal{F}_{f(x)}$ which in our case means that

$$\mathcal{F}_x = \mathcal{F}_{\Lambda_n(x)} = \mathcal{F}_{|\psi><\psi|} = \{|\psi >; |\psi >= e^{i\zeta}\psi_0\} \doteq \{e^{i\zeta}\} = U(1)$$

The projection map of the pullback bundle $f^*(\eta) \xrightarrow{\pi_f} M$ is defined by $\pi_f :$ $(x, \psi) \to x$. If we define $f_* : f^*(\eta) \to \eta$ by $f_* : (x, \psi) \to \psi$ then the pair of maps (f_*, f) is a bundle morphism between the two bundles $f^*(\eta)$ and η. This means

$$f \circ \pi_f = \pi \circ f_* \, , \tag{4.53_π}$$

or, in other words the following diagram is commutative:

$$\begin{array}{ccc} f^*(\eta) \cong \xi & \xrightarrow{f_*} & \eta \\ \hat{\phi}\uparrow\downarrow\pi_f & & \pi\downarrow\uparrow\phi \\ M & \xrightarrow{f} & \mathbb{C}\mathrm{P}(\infty) \end{array} \tag{4.52b}$$

The commutativity of the diagram (4.52b) follows because

$$\begin{aligned} f \circ \pi_f(x, \psi) &= f(x) = \Lambda_n(x) = |\psi\rangle\langle\psi| \\ \pi \circ f_*(x, \psi) &= \pi(\psi) = |\psi\rangle\langle\psi| \end{aligned}$$

Sections pull back, too. Thus if $\phi : \mathcal{P}(\mathcal{H}) \to \eta$ is a section of η then the pull-back section $\hat{\phi} : M \to f^*(\eta)$ is defined by

$$\hat{\phi}(x) \equiv (x, \phi(\Lambda_n)) = (x, \phi(f(x)) = |n; x\rangle \tag{4.54}$$

The diagram (4.52) with the sections is also commutative in the sense

$$f_* \circ \hat{\phi} = \phi \circ f \tag{4.53_ϕ}$$

This commutativity follows since

$$f_* \circ \hat{\phi}(x) = f_*(x, \phi(\Lambda_n)) = \phi(\Lambda_n) = \phi(f(x))$$

Two homotopic maps $f_1 : n \to \mathcal{P}(\mathcal{H})$ and $f_2 : M \to \mathcal{P}(\mathcal{H})$ (Appendix) lead to isomorphic bundles $f_1^*(\eta)$ and $f_2^*(\eta)$.

Thus there is a one-to-one correspondence between $U(1)$ bundles over M and the homotopy classes of maps of $f : M \to \mathbb{C}P(\infty)$. For every homotopy class $[f]$ of maps there exist one $U(1)$ bundle $f^*(\eta)$ (up to equivalence) that is determined by this map $[f]$.

In the construction of the pullback bundle in (4.52) we used the map f of (4.47) given by the eigenprojectors Λ_n of the Hamiltonian. From our example (3.1) we know that the closed path of eigenprojectors $f = \Lambda_k(\theta, \varphi) = |k; \theta, \varphi\rangle\langle k; \theta, \varphi| \in \mathcal{P}(\mathcal{H})$ which is the image under f of the closed path $t \to R(\theta, \omega t)$ in M is not a closed path of physical states. Thus the $|\psi\rangle\langle\psi|$ and ψ in (4.52a) are not solutions of the dynamical equations (2.3b) and (2.3a) respectively. From the results (2.32) we know tht this is generally true. The map (4.47) does not map closed paths in M into closed paths of physical states. We thus have to find in place of the map f another map $\tilde{f} : M \to \mathcal{P}(\mathcal{H})$ that maps closed paths in the parameter space M into closed paths of physical states (solutions of (2.3b)) in $\mathcal{P}(\mathcal{H})$. To find such maps we will use our example (3.1) as a guide. The closed path of physical states that corresponds to the closed path $t \to R(\theta, \omega t)$ in parameter space is given by

$$\begin{aligned} t \to |k; \tilde{\theta}, \omega t\rangle\langle k; \tilde{\theta}, \omega t| &= |k; F_\nu(\theta, \omega t)\rangle\langle k; F_\nu(\theta, \omega t)| \\ &\equiv \tilde{\Lambda}_k(\theta, \omega t) \end{aligned} \tag{4.55}$$

where

$$|k; \tilde{\theta}, \varphi\rangle \equiv e^{-i\varphi J_3} e^{-i\tilde{\theta} J_2} e^{i\varphi J_3} |k; \vec{e}_3\rangle = |k; F_\nu(\theta, \varphi)\rangle \tag{4.56}$$

are the sections of (3.84). (F_ν and $\tilde{\theta}$ are given by (3.66) and (3.62) respectively). The projection operator $\tilde{\Lambda}_k(\theta, \omega t)$ is no more an eigenprojector of the time-dependent Hamiltonian $h(R(\theta, \omega t))$, it is the eigenprojector of an other operator defined by

$$\tilde{h}(\theta, \varphi) = h(\tilde{\theta}, \tilde{\varphi}) = h(F_\nu(\theta, \varphi)) = b\vec{R}(\tilde{\theta}, \varphi) \cdot \vec{J} \tag{4.57}$$

However this projection operator $\tilde{\Lambda}_k$ is a cyclically evolving state (solution of the dynamical equation (2.3)):

$$\tilde{\Lambda}_k(\theta, \omega t) = |\psi(t)\rangle\langle\psi(t)| = W(t)$$

where $\psi(t)$ is the solution (3.99) of the equation (3.56). Therefore we choose in place of the map f in the diagram (4.52) the map

$$\tilde{f}_k = f_k \circ F_\nu : \mathcal{S}^2 \ni R \to \tilde{\Lambda}_k(R) \in \mathbb{C}P(\infty), \qquad \nu < 1 \tag{4.58}$$

This map $\tilde{f}_k$ associates with every $\mathbf{R}(\theta,\varphi) \in \mathcal{S}^2$ an eigenstate $\tilde{\Lambda}_k$ of the operator $\mathbf{R}(\tilde{\theta},\varphi) \cdot \vec{J}$ with eigenvalue k (k being the component of angular momentum along $\mathbf{R}(\tilde{\theta},\varphi)$ not along the direction of the magnetic field $\mathbf{R}(\theta,\varphi)$. The restriction $\nu < 1$ is necessary so that F_ν and therewith $\tilde{f}_k$ is continuous.

We take this example as our guiding principle for the general case: There exist classes of time-dependent periodic Hamiltonians for which the dynamical equation (2.3) has cyclic non-degenerate solutions with the same period as the period ω of the Hamiltonian.[8)] (We want to restrict ourselves here to the case of non-degenerate solutions, in the case of an $\mathcal{N}$-fold degeneracy the $U(1)$ principal bundle has to be replaced by the Stiefel bundle with structure group $U(\mathcal{N})$.[6)]) These classes are characterized by the existence of a homeomorphism (diffeomorphism)

$$F : M \rightarrow M$$

which has the following properties:

1.) The operator $\tilde{h}(x)$ defined by

$$\tilde{h}(x) \equiv h(F(x)) \qquad \text{has eigenprojectors} \qquad \tilde{\Lambda}_n(x) = \Lambda_n(F(x)) \tag{4.59}$$

For a closed path (4.1) these eigenprojectors are cyclic solutions of (2.3b): $\Lambda_n(F(x(t)) = W(t)$.

2.) F is continuously related to the identity $F(x,\nu) \rightarrow x$ for $\nu \rightarrow 0$, where ν are parameters of the Hamiltonian $h(x(t)) = h_\nu(x(t))$ with $[h_{\nu=0}(t), h_{\nu=0}(t')] = 0$ for $t \neq t'$.

In our example (3.1), the parameter is given by $\nu = \omega/b$ where ω is the precession frequency and b is the Larmor frequency.

The map

$$\tilde{f} = f \circ F : M \ni x \rightarrow \tilde{\Lambda}_n(x) \in \mathbb{C}P(\infty) = \mathcal{P}(\mathcal{H}) \tag{4.60}$$

where f is given by (4.47), associates to every $x \ni M$ an eigenstate $\tilde{\Lambda}_n(x) \in \mathcal{P}(\mathcal{H})$ of $\tilde{h}(x)$. This map $\tilde{f}$ also associates to a closed curve (4.1) in M a closed curve of *states* $W(t) = \tilde{\Lambda}(x(t))$ in $\mathcal{P}(\mathcal{H})$. With this map $\tilde{f}$ we construct the pullback bundle over the parameter space using the classifying theorem of $U(1)$ principal fibre bundles. The classifying theorem of fibre bundles thus becomes the central mathematical tool in establishing the relation between the Stiefel bundle of the Anandan-Aharonov approach and the parameter space $U(1)$ bundle over M in the Berry-Simon approach.

The $U(1)$ bundle over M, that associates closed paths in M (4.1) with closed paths of physical states (4.2) is given by the pullback $\tilde{f}^*$ of (4.60):

$$
\begin{array}{ccccc}
\tilde{f}^*(\eta) = \tilde{\lambda} & \xleftarrow{\tilde{f}^*} & \eta & & \\
\tilde{\phi}\uparrow\downarrow\pi_f & & \pi\downarrow\uparrow\phi & & \tilde{f}(x) = \tilde{\Lambda}_n(x). \\
\mathrm{M} & \xrightarrow{\tilde{f}} & \mathcal{PH} & &
\end{array}
\tag{4.61}
$$

Thus for every value n we obtain a map $\tilde{f}$ and a $U(1)$ bundle over parameter space $\tilde{\lambda}$

To see how Berry connection and Berry phase are obtained, we start with the universal classifying bundle η of (4.29). The bundle η carries as a natural connection the Stiefel connection. We choose a (local) section $|\phi(\tilde{\Lambda})\rangle$ of $\mathcal{P}(\mathcal{H}) \supset O \rightarrow S(\mathcal{H})$ with the property $|\phi(\tilde{\Lambda})\rangle\langle\phi(\tilde{\Lambda})| = \tilde{\Lambda}_n$. The Stiefel connection is then (locally) represented by the $U(1)$-valued one-form

$$\mathcal{A} \equiv i\langle\phi|d|\phi\rangle \tag{4.62}$$

The local section $\tilde{\phi}$ in the fibre bundle $\tilde{f}^*(\eta)$ is given by

$$\tilde{\phi}(x) = \phi(\tilde{f}(x) = \phi(f(F(x))) = |n; F(x)\rangle \tag{4.63}$$

The $|n;\ \tilde{x}\rangle = |n; F(x)\rangle$ are eigenvectors of $\tilde{h}(x)$ and are determined up to a gauge transformation (4.6).

The canonical connection one-form on the bundle $\tilde{\lambda} = \tilde{f}^*(\eta)$ is obtained from (4.62):

$$A(x) \equiv \tilde{f}^*\mathcal{A} = \langle n; F(x)|\frac{\partial}{\partial x^\mu}|n; F(x)\rangle dx^\mu \tag{4.64}$$

The geometric phase factor (holonomy) is then given by

$$
\begin{aligned}
\exp i\gamma(\mathcal{C}) &= \exp i \oint_{\mathcal{C}=\tilde{f}C} \mathcal{A} = \exp(-\oint_{\mathcal{C}=\tilde{f}C} \langle\phi|d|\phi\rangle) \\
&= \exp i \oint_C \tilde{f}^*\mathcal{A} = \exp(-\oint_C \langle n;\ F(x)|\frac{\partial}{\partial x^\mu}|n;\ F(x)\rangle dx^\mu).
\end{aligned}
\tag{4.65}
$$

The path in the first two integral is over the closed curve (4.2) in $\mathcal{P}(\mathcal{H}), \mathcal{C} = \tilde{f} \circ C$. The path in the second two integral C is the closed curve (4.1) in M. In the third equality the property of the pullback was used. This formula (4.65) says that the geometric phase can be either calculated using the canonical (Stiefel) connection $\mathcal{A}$ in the universal bundle η or using its pullback $\tilde{f}^*\mathcal{A}$ in the induced bundle. This means the geometric phase factor $e^{i\gamma(\mathcal{C})}$ (the holonomy) acquired after parallel transport with respect to the Stiefel connection along the closed curve $\mathcal{C}$ in $\mathcal{P}(\mathcal{H})$ is the same as the geometric phase factor acquired after parallel transport with respect to $A(x) \equiv \tilde{f}^* \mathcal{A}$ along the closed curve C in the parameter space M.

The expression in terms of the integral over $\mathcal{C}$ in $\mathcal{P}(\mathcal{H}) = \mathbf{C}P(\infty)$ is the standard expression for the Aharonov-Anandan phase. Thus the Aharonov-Anandon connections and therewith the geometric phase can also be obtained from the eigenvectors of the Hamiltonian, if F is known. However this phase was introduced originally not through the Stiefel connection but as the difference between the total phase and the dynamical phase defined by $\exp(-i \oint \langle \psi(t)|h(t)|\psi(t)\rangle dt)$.

The Berry connection and phase is obtained from (4.64) and (4.65) in the adiabatic approximation in which F is replaced by the identity

$$A = \langle n;\ x|\frac{\partial}{\partial x^\mu}|n; x\rangle dx^\mu \tag{4.66}$$

$$e^{i\gamma(C)} = \exp\Big(-\oint_C \langle n; x|\frac{\partial}{\partial x^\mu}|n; x\rangle dx^\mu\Big) \tag{4.67}$$

where C is the closed curve (4.1) which also appears in (4.45).

The B-S bundle (4.45) $\lambda = f^*(\eta)$ and the bundle $\tilde{\lambda} = \tilde{f}^*(\eta)$ of (4.61) are related by the bundle morphism F^*:

$$\begin{array}{ccc} \tilde{\lambda} & \xleftarrow{F^*} & \lambda \\ \downarrow & & \downarrow \\ \mathrm{M} & \xrightarrow{F} & M \end{array} \tag{4.68}$$

As F is a diffeomorphism continuously related to the identity, the two maps $f = \tilde{f} \circ F^{-1}$ and $\tilde{f}$ belong to the same homotopy class in $[M, \mathcal{C}P(\infty)]$. Consequently $\lambda, \tilde{\lambda}$ have the same topology.[9)]

In this section 4 we gave a general description of the structures which we discovered in section 3 for the specific example given by (3.1). We introduced the geometric phase for general cyclic (non-adiabatic) evolution and showed that the adiabatic phase introduced in section 2 is a limit of the general geometric phase. But we also showed that exact cyclic adiabatic evolution is always stationary, i.e. the closed path is a point in $\mathcal{P}(\mathcal{H})$.

We uncovered the mathematical structure behind the gauge theory of the geometric phase and showed that this involved two fibre bundles with the structure group $U(1)$: the $U(1)$ bundle over the parameter space M (differentiable manifold) of the parameter dependent Hamiltonian in the adiabatic approximation considered by Berry and Simon and the $U(1)$ Stiefel bundle in the Anandan Aharonov approach where the base space is the space of (pure) physical states $\mathcal{P}(\mathcal{H}) = \mathbb{C}P(\infty)$.

We then related the time evolution (4.2) of a cyclic quantum state $W(t)$ (considered in the Aharanov-Anandan approach) to the change of a parameter dependent Hamiltonian and its eigenprojectors along a closed path (4.1) in parameter space, (considered in the B-S approach). Since the Stiefel bundle is the universal classifying bundle of the $U(1)$ bundles over manifolds M, the classifying theorem provides all $U(1)$ bundles over M (characterized by the first Chern class). Bundles and Chern class are labeled by the quantum number n.

The classifying Stiefel bundle has a natural connection (4.42) given by differential geometry (Stiefel connection) which is identical to the connection obtained from the Aharanov-Anandan approach by separating from the total phase in a natural way a dynamical phase. The connection in the parameter space bundle for exact (non-adiabatic) evolution is the pullback (4.64) of the Stiefel connection (4.62). This pullback connection goes into Berry's connection (4.66) in the adiabatic limit.

The mathematics of the fibre bundles was created and the classification theorem of principal bundles was established[5)] long before the quantum geometric phase was considered to be of any importance in physics. Though we have accepted the fact that physics is described by mathematical structures, one usually faces the situation that the mathematics is not quite ready for some new physical theory and has to be created along with, or adapted specifically for, this new physics. It is very rare that the mathematical structure is already all there and the physical quantities need just to be mapped upon the elements of this mathematical structure. This was the situation for the quantum geometric phase and was therefore the cause of great excitement and exultation. The role which the mathematics of fibre bundles plays for the physics connected with the quantal geometric phase is one of the more spectacular examples of what Wigner calls "the unreasonable effectiveness of mathematics in natural science."

References

1. Y. Aharonov and J. Anandan, Phys. Rev. Lett. **58**, 1593 (1987); J. Anandan and Y. Aharonov, *Phys. Rev.* D**38**, 1863 (1988).
2. The method used in section 3 can be generalized in a straight forward way for Hamiltonians that are linear functions of generators of any Lie algebra. For more details and for other calculational methods see reference 8 below.
3. The generalization to $U(\mathcal{N})$ bundles describing the non-abelian gauge theory for $\mathcal{N}$-fold degenerate states is straightforward and causes few new principal problems; cf. reference 8 of section 1.
4. More precisely, it is a dense subspace of $\mathcal{H}$ for $N \to \infty$, reference 12 of section 1. The projective space $\mathbb{C}P(\infty)$ is the inductive limit of the $\mathbb{C}P^N$ for $N \to \infty$. To avoid all questions connected with $N \to \infty$ we want to assume here that N is arbitrary large but finite. For the infinite dimensional case we shall always assume that we are in one particular $\mathbb{C}P^{N-1}$ with N finite. This is always possible if the Hamiltonian is in the enveloping algebra of a compact group of which (3.1) is the simplest example.
5. What we call Stiefel connection here is usually called the universal connection

because the bundle (4.29) is the universal classifying bundle. M. Narasinhan and S. Ramanan, Amer. J. Math. **83**, 563 (1961). See also Ch. J. Ishan. *Modern Differential Geometry for Physicists*, World Scientific Publ. (1989) section 3.2.

6. The non-abelian case $\mathcal{N} > 1$ is discussed in reference 8 of section 1.
7. Reference 11 of section 1.
8. D.J. Moore, Phys. Rep. **240**, No. 1 (1992) sect. 3,4 and 5.
9. For our example (3.1) the map $F = F_\nu(x)$ of (3.66) becomes discontinuous for $\nu \geq 1$. As a consequence the bundles $\tilde{\lambda}$ and λ are no m ore isomorphic for $\nu > 1$. For details see: A Mostafazadeh, A. Bohm. *Topological Aspects of the Non-Adiabatic Berry Phase.*

Acknowledgement:

I should like to express my gratitude to my friends and colleagues who taught me and worked with me on this subject: Ali Mostafazadeh, Mark Loewe, L.J. Boya, B. Kendrick, J. Lemke and G. Rudolph. I am grateful to M.V. Berry, C.A. Mead and J. Anandan who read parts of earlier stages of the manuscript and made suggestions and gave me advice. This paper is dedicated to W. Drechsler on the occasion of his coming birthday, who was the first to teach me about fibre bundles.

Mathematical Appendix:

"A Brief Review of Fibre Bundles and Their Classification"

by

Ali Mostafazadeh

A.I. Fibre Bundles:

Definition: A fibre bundle [(1),(4)] is a collection (E, M, π, G) of two (smooth) manifolds E and M, an onto continuous (smooth) map $\pi : E \longrightarrow M$ and a Lie group $G \cdot E, \eta, \pi$, and G are called the "total space," the "base manifold," the "projection map" and the "structure group", respectively. For any $x \in M$, the set $\pi^{-1}(x)$ is a (smooth) manifold and $\pi^{-1}(x) =: F_x$ is called the fibre over x. As a point set $E = \bigcup_{x \in M} F_x$. Furthermore, all fibres look alike, *i.e.* there exist a (smooth) manifold F such that F_x are homeomorphic (diffeomorphic) to F. F is called the "typical fibre". The structure group G is a group of diffeomorphisms of F and it has a global right (or left) action on E. The collection (E, M, π, G), which sometimes collectively called E, is said to have a fibre bundle structure if for any $x \in M$ there exist an open neighborhood O_i of x such that

$$\pi^{-1}(O_i) \simeq O_i \times F \tag{A.1}$$

Moreover, the (right) action of G on E is such that $\forall p \in E$, $\forall g \in G$:

$$\pi(p.g) = \pi(p)$$

i.e. the action of G moves the points along the fibres. Furthermore, G acts on fibres freely and transitively.

Definition: If F happens to be identical to G (as manifolds), then the fibre bundle is called a "principal fibre bundle".

Definition: If $E = \pi^{-1}(M) \simeq M \times F$ then the fibre bundle is said to be "trivial" or a "product" bundle. Not every fibre bundle is trivial. The degree in which a fibre bundle differs from being trivial is measured by a set of functions called the "transition functions" $\{g_{ij}\}$. Let $\{O_i\}$ be on open covering of $M = \bigcup_i O_i$ such that

$$\varphi_i : \pi^{-1}(O_i) \to O_i \times F$$

are the diffeomorphisms of (A.1). Then one can view E as a collection of patches $O_i \times F$ over O_i. The global or topological structure of E is characterized by gluing these patches in their intersections. Let O_i and O_j have a nonempty intersection then the gluing map is defined by:

$$\begin{aligned} g_{ij} &:= \varphi_j \circ \varphi_i^{-1} : \varphi_i(O_i \cap O_j) \longrightarrow \varphi_j(O_i \cap O_j) \\ &: \ (O_i \cap O_j) \times F \longrightarrow (O_i \cap O_j) \times F \end{aligned}$$

Then $\forall x \in O_i \cap O_j$

$$\begin{aligned} g_{ij}(x) &: \{x\} \times F \to \{x\} \times F \\ &: F \longrightarrow F \end{aligned}$$

and

$$g_{ij}(x) \in G.$$

Example: A nontrivial and quite relevant$^{(2)}$ example of a principal bundle is the Hopf bundle: $(S^3, S^2, \pi, U(1))$ also denoted by:

$$S^1 = U(1) \longrightarrow S^3 = SU(2) \longrightarrow S^2 = SU(2)/U(1).$$

where the fibres are $U(1)$ orbits in $SU(2)^{(3)}$.

A.II: Sections, Lifts, Connections and Holonomy:

Let (E, M, π, G) be a fibre bundle.

Definition: For any open submanifold $O \subset M$, a continuous (smooth) map $S : O \to E$ is called a "section" over O if

$$\pi \circ S = id\Big|_O$$

If $O = M$ then S is said to be a global section of E.

Remark: A principal bundle is trivial if and only if it has a global section.

Definition: Let $\mathbf{C} : [0, T] \to M$ be a continuous (smooth) curve in M. Then a curve $C : [0, T] \to E$ in E is said to be a lift of $\mathbf{C}$ if

$$\pi \circ C = \mathbf{C}$$

If a principal bundle P is endowed with a connection (a geometric structure), one has a well-defined notion of parallel transport.[4] Then, a point $p_0 \in P$ can be parallelly transported along the curve $\mathbf{C}$, to define a curve

$$C : p_0 \longrightarrow p(t) \rightarrow p(T)$$

in P which is called the "horizontal lift" of $\mathbf{C}$ associated with the connection on P.

Alternatively, a connection on P may be defined in terms of horizontal lifts. More precisely, a connection is realized as a rule which associates a lift C to any pair $(\mathbf{C},\ p_0)$. The tangent vectors $w_t := \frac{d}{ds}C(s)|_{s=t}$ to $C(t)$ which project to the tangent vectors v_t to $\mathbf{C}(t)$ are called "horizontal vectors". Note that $w_t \in T_{C(t)}P$, and $v_t \in T_{\mathbf{C}(t)}M$ where $T_{C(t)}P$ denotes the tangent space of the bundle P at $C(t)$ and $T_{\mathbf{C}(t)}M$ denotes the tangent space of the base space M at the point $\mathbf{C}(t)$. Since one can reconstruct C from the knowledge of the horizontal vectors, one has[4]:

Definition (1): A connection on (P, M, π, G) is a linear mapping $\sigma_p : T_xM \rightarrow T_pP, \forall_p \in P$, such that:

i) $\pi' \circ \sigma_p : T_xM \rightarrow T_xM$ is the identity map.
ii) σ_p depends smoothly on $p \in P$.
iii) $\sigma_{p\cdot g} = R'_g \sigma_p$

where, $\pi' : T_pP \rightarrow T_{\pi(p)}M$ is the push forward map corresponding to $\pi : P \rightarrow M$, and $R'_g : T_pP \rightarrow T_pP$ is the push forward map corresponding to the right action map of $g \in G$, $Rg : P \rightarrow P$, which is denoted by $R_g(p) = p \cdot g, \forall_p \in P$ and $g \in G$. The right action of G on P is defined as follows: Let $g \in G, x \in O_i \subset M, p \in \pi^{-1}(x) \subset P$, and $\varphi_i(x) : \{x\} \times \pi^{-1}(x) \rightarrow G$ be the restriction of $\varphi_i : \pi^{-1}(O_i) \rightarrow O_i \times F$ (of A.I) onto the fibre over x, $\pi^{-1}(x)$. Clearly, $\varphi_i(x)$ is a diffeomorphism identifying the fibre $\pi^{-1}(x)$ with G. The right action map $R_g : P \rightarrow P$ is defined by

$$R_g(p) = p \cdot g := \varphi_i^{-1}[\varphi_i(p) \cdot g]$$

where $\varphi_i(p) \in G$ and the product in the bracket is the group product.

The horizontal vectors span a subspace of $T_pP, \forall_p \in P$ which is called the "horizontal subspace H_pP of T_pP. Hence, a connection leads to a, not necessarily orthogonal, decomposition of T_pP into:

$$T_pP = H_pP \dot{+} V_pP$$

V_pP is called the "vertical subspace". One can view a connection as a rule which determines $H_pP, \forall_p \in P$:

Definition (2): A connection on (P, M, π, G) is a collection of vector spaces $H_pP \subset T_pP$, such that

i) $\pi|_{H_pP} : H_pP \rightarrow T_{\pi(p)}M$ is an isomorphism.

ii) H_p depends smoothly on $p \in P$.
iii) $H_{p\cdot g} = R'_g H_p$.

The vertical subspaces V_pP are actually canonically isomorphic to the Lie algebra $\mathcal{G}$ of G. This is seen as follows: For all $X = \frac{dg_{(s)}}{ds}|_{s=0} \in T_eG = \mathcal{G}$, let $v(p) \in V_pP$ be defined by

$$v(p) := \frac{d(p \cdot g(s))}{ds}\Big|_{s=0} \tag{A.2}$$

Then, (A.2) establishes an isomorphism between $\mathcal{G}$ and V_pP. This isomorphism allows one to have a more practical definition of a connection on a principal bundle.

Definition (3): A connection on (P, M, π, G) is a 1-form ω on P with values in $\mathcal{G}$, such that

i) $\forall v_p \in V_pP, \omega_p(v_p) = X$ where $X \in \mathcal{G}$ is related to V_p as dictated by (A.2).
ii) ω_p depends smoothly on $p \in P$.
iii) $\omega_{p\cdot g}(R'_g v_p) = Ad(g^{-1})\omega_p(v_p)$ where Ad denotes the adjoint representation of G on $\mathcal{G}$. Given ω the horizontal subspaces are obtained by

$$H_p := \{w_p \in T_pP : \omega_p(w_p) = 0\}$$

The three definitions (1), (2), and (3) can be shown to be equivalent.[(4)] In the physics literature one usually encounters a 1-form A on M with values in $\mathcal{G}$. This is obtained by pulling back the connection one-form ω on M, via a local section

$$S_i : O_i \longrightarrow P$$

Then, locally

$$A_{(i)} := S_i^*(\omega)$$

where the subscript i refers to the local patch over $O_i \subset M$ and S_i^* denotes the pullback. If P is a trivial bundle, as is the case in many gauge theories, S_i can be taken to be a global section, *i.e.* $O_i = M$.

It turns out that the 1-form $A_{(i)}$ is well-defined if one requires a particular transformation rule which reads:

$$A_{(i)} = g_{ij}^{-1}(x) \cdot A_{(j)} \cdot g_{ij}(x) + g_{ij}^{-1}(x) \cdot g'_{ij}(x) \tag{A.3}$$

In (A.3), $g_{ij}(x) \in G$ is the transition function connecting the patch O_i to O_j, $g_{ij(x)}^{-1}$ is its (group) inverse, and $g'_{ij}(x)$ is the push forward map: $g'_{ij}(x) : T_xM \to \mathcal{G}$, corresponding to $g_{ij} : M \to G$. Here we have assumed G to be a matrix group for simplicity. The second term in the right hand side of (A.3) is known as the pullback of "Maurer-Cartran" 1-form on G defined by $g_{ij} : M \to G$. Another notation for

$g'_{ij}(x)$ is $dg_{ij}(x)$. To motivate this notation, let $\frac{d}{dS} \in TM_{x(s)}$ and act $g'_{ij}(x(s))$ on $\frac{d}{ds}$. By definition of push forward map[1,4] one has:

$$g'_{ij}(x(s))\frac{d}{ds} = \frac{d}{ds}(g_{ij}(x(s)))$$

In a local coordinate frame (x^μ) this becomes

$$g'_{ij}(x(s))\Big(\frac{dx^\mu(s)}{ds}\frac{\partial}{\partial x^\mu}\Big) = \frac{dx^\mu(s)}{ds}\frac{\partial}{\partial x^\mu}(g_{ij}(x(s))$$

which together with the fact that $g'_{ij}(x)$ is a linear map suffices to see

$$g'_{ij}(x)\Big(\frac{\partial}{\partial x^\mu}\Big) = \frac{\partial}{\partial x^\mu}(g_{ij}(x))$$

Hence, in general one has:

$$g'_{ij}(x) = \frac{\partial}{\partial x^\mu} g_{ij}(x) dx^\mu = dg_{ij}(x)$$

and (A.3) is then written as:

$$A_{(i)} = g_{ij}^{-1}(x)A_{(j)}g_{ij}(x) + g_{ij}^{-1}(x)dg_{ij}(x) \qquad (A.3)'$$

It is remarkable that (A.3′) coincides with the transformation rule for the gauge potential of particle physics.

Consider a closed curve, a loop, $\mathbf{C} : [0,T] \to M$ in M and let $p_\circ \in \pi^{-1}(\mathbf{C}(0))$ and C be the horizontal lift associated to a connection 1-form ω on P. Then, since $\mathbf{C}(0) = \mathbf{C}(T), p(T) := C(T) \in \pi^{-1}(\mathbf{C}(T)) = \pi^{-1}(\mathbf{C}(0))$, *i.e.* $p(T)$ belongs to the same fibre as $p_\circ$ does. On the other hand, G has a transitive action on the fibres, thus, there exists a $g \in G$, such that

$$p(T) = p_\circ \cdot g = p(0) \cdot g$$

since $p(T)$ is the parallel transport of $p(0)$, g is determined by the one-form $A = A_\mu dx^\mu = A^a_\mu dx^\mu \otimes J_a$, as:

$$g = \mathbb{P}\exp\Big[-\oint_{\mathbf{C}} A\Big] = \mathcal{T}\exp\Big[-\int_0^T A_\mu(x(t))\frac{d^\mu x(t)}{dt}dt\Big] \qquad (A.4)$$

where J_a are generators of $\mathcal{G}$, and $\mathbb{P}$ and $\mathcal{T}$ stand for the path ordered and time ordered products, respectively.[1]

Definition: The set of all elements g of G given by (A.4), for fixed p_0, form a subgroup of G which is called the "holonomy" group of ω associated to $p_\circ \cdot g$ is called a holonomy element.

As is seen from (A.3), A does not transform as a tensor. However, one can introduce a tensorial quantity for a given connection. This is called the curvature two-form Ω of the connection ω. Ω is pulled back on M by a local section to define a Lie algebra valued two-form F on M. The latter is in perfect analogy to the field strength associated to a gauge potential. F is related to A by the following local expression:

$$F_{(i)} = dA_{(i)} + \frac{1}{2}\Big[A_{(i)}, A_{(i)}\Big]$$

$F_{(i)}$ has the following tensorial transformation rule:

$$F_{(i)} = g_{ij}^{-1}(x) \cdot F_{(j)} \cdot g_{ij}(x).$$

Finally (A.4) is also written in terms of F:

$$g = \mathbb{P}\exp\Big[-\int_S F\Big]$$

where Stokes theorem is used and S is any two-surface in M bounded by **C**.

A. III.Classification Theorem for Principal Fibre Bundles:

Definition: Let $(\tilde{E}, \tilde{M}, \pi, G)$ be a fibre bundle, and M a (smooth) manifold. Let $f : M \to \tilde{M}$ be a continuous (smooth) map of manifolds. Then one can induce a (G-) bundle structure over M using f.

For any $x \in M$, the fibre F_x is defined to be $\pi^{-1}(f(x))$. The "pullback bundle" is

$$f^*(\tilde{E}) := \cup_{x \in M} \pi^{-1}(f(x)).$$

as a point set. The bundle structure is then fixed by requiring the transition functions $g_{ij}(x)$ of $E = f^*(\tilde{E})$ to be given by

$$g_{ij}(x) = \tilde{g}_{ij}(f(x)) \qquad \forall x \in M$$

where $\{\tilde{g}_{ij}\}$ are transition functions of $\tilde{E}$ corresponding to an open covering $\{\tilde{0}_i\}$ of $\tilde{M}$, and $\{g_{ij}\}$ correspond to the open covering $\{f^{-1}(\tilde{O}_i)\}$ of M.

Definition: Let M and $\tilde{M}$ be two topological spaces (smooth manifolds), and f_1 and f_2 two continuous maps

$$f_i : M \to \tilde{M} \qquad\qquad i = 1, 2$$

f_1 and f_2 are said to be "homotopic" if there exists a continuous map:

$$\mathcal{F} : M \times [0,1] \to \tilde{M}$$

such that $\mathcal{F}\big|_{M\times\{0\}} = f_1$ and $\mathcal{F}\big|_{M\times\{1\}} = f_2$. The existence of $\mathcal{F}$ is equivalent to the statement that f_1 can be continuously deformed into f_2 and vice versa.

Theorem: For any Lie group G, there exist a principal G-bundle

$$\eta : (\eta(G), BG, \tilde{\pi}, G)$$

such that any principal G-bundle (P, M, π, G) is obtained from η as a pullback bundle *i.e.* there exist a continuous (smooth) map $f : M \to BG$ such that

$$P \simeq f^*(\eta(G))$$

Furthermore, homotopic maps pullback (topologically) identical bundles.[5]

The statement of the classification theorem can be illustrated by a commutative diagram:

$$\begin{array}{ccc} P = f^*(\eta(G)) & \xleftarrow{f_*} & \eta(G) \\ \pi \big\downarrow & & \big\downarrow \tilde{\pi} \\ M & \xrightarrow{f} & BG \end{array}$$

BG and $\eta(G)$ are called the "classifying space" and the "classifying bundle", respectively.

For G=U(N), BG can be chosen to be the complex Grassmannian $Gr_{\mathcal{N}} = \bigcup_{n=1}^{\infty} Gr(n, \mathcal{N})$ and η is the so-called "universal (Stiefel) bundle." For $G = U(1), BG = Gr_1 = \mathbb{C}P(\infty)$.

Theorem: There exists a "universal" (Stiefel) connection on η such that any connection on any principal G-bundle (P, M, π, G) can be obtained as a pullback connection one-form from the Stiefel connection.[6]

A. IV. References used in Appendix:

(1) M. Nakahara, "Geometry, Topology and Physics," Adam Hilger, Bristol (1991).

(2) A. Bohm, L.J. Boya, A. Mostafazadeh, and G. Rudolph, "Classification Theorem for Principal Fibre Bundles, Berry's Phase, and Exact Cyclic Evolution," to appear in J. Geometry and Physics.

(3) T. Brocker and T. tom Dieck, "Representations of Compact Lie Groups," Springer-Verlag (1985).

(4) Y. Choquet-Bruhat, C. DeWitt-Morette, and M. Dillard-Bleick, "Analysis, Manifolds and Physics," Vol. 1, North-Holland (1989).

(5) N. Steenrod, "The Topology of Fibre Bundles," Princeton University Press (19XX).

(6) M. Narasimhan and S. Ramanan, Amer. J. Math. **83**, 563 (1961).

QUANTUM MECHANICS ON PHASE SPACE.

M. Gadella

Departamento de Física Teórica
Universidad de Valladolid
Valladolid, Spain

Introduction.

The origin of a formulation of Quantum Mechanics (QM) on phase space lies on a paper by Moyal [1] written in the forties. As a consequence, this is called the Moyal formalism or sometimes the Weyl-Wigner-Moyal formalism of Q.M.. It does not represent a new interpretation ot quantum mechanics, but rather an alternative approach that, providing a new insight, must yield the same results given by the usual formalism of Q.M. on Hilbert space. It has, of course, advantages and disadvantages with respect to the latter. Among the advantages, it counts:

a).- The formalism uses functions instead of operators and, therefore, most of the dificult problems concerning operator theory are circumvented.

b).- In most situations, taking semiclassical limit $\hbar \to 0$ is easier than in the Hilbert space formalism.

c).- One can study the theory of quadratic Hamiltonians as a whole and obtaining general results concerning spectrum and other physical data. The approach seems simpler than the usual in many cases.

The Moyal formalism has traditionally included quantum mechanics of quantum simple particle systems with no spin and in absence of relativity or other constraints. The possibility of describing particles with spin as well as relativistic particles with the formalism of Moyal, was made possible only recently.

In order to construct this formalism, Moyal made use of two main concepts: the Weyl correspondence [2] and the Wigner function [3]. The Weyl correspondence is a mapping that associates an operator on the Hilbert space $L^2(\mathbb{R}^n)$ to any function (or generalized function) on the phase space $\mathbb{R}^{2n}$ belonging to a wide class. If $f(\boldsymbol{u})$, where $\boldsymbol{u} = (\boldsymbol{q}, \boldsymbol{p})$ is one of these functions, its corresponding operator via the Weyl correspondence is given by:

$$W(f) = \left(\frac{1}{2\pi\hbar}\right)^n \int_{\mathbb{R}^{2n}} \mathcal{F}(f)(\boldsymbol{\sigma}, \boldsymbol{\eta})\, e^{i\{\boldsymbol{\sigma}\cdot\boldsymbol{Q}+\boldsymbol{\eta}\cdot\boldsymbol{P}\}}\, d\boldsymbol{\sigma}\, d\boldsymbol{\eta} \qquad (1.1)$$

where n is the dimension of the configuration space. The Weyl mapping W is obviously linear with respect to sums of functions and products by complex numbers.

L. A. Ibort and M. A. Rodríguez (eds.), Integrable Systems, Quantum Groups, and Quantum Field Therapy 417–428.

However, it is not linear with respect to the products ordinarely defined on both algebras. If we want that W be an homomorphism between algebras, we need to define a noncommutative product $\times$ on the space of the functions on $\mathbb{R}^{2n}$ such that

$$W(f \times g) = W(f)\, W(g) \tag{1.2}$$

Since W es invertible, (1.2) defines $\times$ uniquely. From (1.1) and (1.2), one obtains

$$(f \times g)(\boldsymbol{u}) = \left(\frac{1}{2\pi\hbar}\right)^n \int_{\mathbb{R}^{4n}} f(\boldsymbol{v})\, g(\boldsymbol{w})\, e^{i\{{}^t\boldsymbol{u} J \boldsymbol{v} + {}^t\boldsymbol{v} J \boldsymbol{w} + {}^t\boldsymbol{w} J \boldsymbol{u}\}}\, d\boldsymbol{v}\, d\boldsymbol{w} \tag{1.3}$$

where $\boldsymbol{u} = (\boldsymbol{q}, \boldsymbol{p})$; $\boldsymbol{v} = (\boldsymbol{q}', \boldsymbol{p}')$; $\boldsymbol{w} = (\boldsymbol{q}'', \boldsymbol{p}'')$ and

$$J = \begin{pmatrix} 0 & I \\ -I & 0 \end{pmatrix}$$

The product $\times$ is called the twisted product of functions on $\mathbb{R}^{2n}$. It can be defined for any pair of functions in the domain of W [4]. The twisted product of two functions in the domain of W is always in it [4,5].

The Weyl mapping plays the role of "quantizer" in the sense that it relates a classical observable, viewed as a real function of the variables position $\boldsymbol{q}$ and momentum $\boldsymbol{p}$, with its corresponding quantum observable.

The Weyl correspondence is best defined with the aid of the Grossmann-Royer (G.R.) operators [6,7], which are defined as follows:

$$\boldsymbol{\Pi}(\boldsymbol{q}, \boldsymbol{p})\, \Psi(\boldsymbol{\zeta}) = 2^n \exp\left[i\boldsymbol{p} \cdot (\boldsymbol{\zeta} - \boldsymbol{q})\right] \Psi(2\boldsymbol{q} - \boldsymbol{\zeta}) \tag{1.4}$$

where $\Psi(\boldsymbol{\zeta}) \in L^2(\mathbb{R}^n)$. One can show that (1.4) and (1.1) yield

$$W(f) = \left(\frac{1}{2\pi\hbar}\right)^n \int_{\mathbb{R}^{2n}} f(\boldsymbol{u})\, \boldsymbol{\Pi}(\boldsymbol{u})\, d\boldsymbol{u} \tag{1.5}$$

Note that there exists a G.R. operator for each point in phase space. The G.R. operators are bounded and self-adjoint. They have infinite trace. However, if we define a generalized notion of trace as:

$$\operatorname{tr} A = \int_{\mathbb{R}^n} \langle \boldsymbol{q} | A | \boldsymbol{q} \rangle\, d\boldsymbol{q} \tag{1.6}$$

where the $|\boldsymbol{q}\rangle$ form a complete set of eigenkets for the multiplication operator on $L^2(\mathbb{R}^n)$, one has

$$\operatorname{tr}\{\boldsymbol{\Pi}(\boldsymbol{u})\} = 1 \qquad \operatorname{tr}\{\boldsymbol{\Pi}(\boldsymbol{u})\, \boldsymbol{\Pi}(\boldsymbol{v})\} = (4\pi\hbar)^n\, \delta(\boldsymbol{u} - \boldsymbol{v}) \tag{1.7}$$

The second formula in (1.7) is useful to write the inversion formula for (1.5) as

$$f(\boldsymbol{u}) = \text{tr } \{W(f)\,\boldsymbol{\Pi}(\boldsymbol{u})\} \tag{1.8}$$

This new trace keeps up some of the most important properties of the ordinary trace. In particular, tr AB = tr BA, tr $A^+A \geq 0$ and tr $I = \infty$, where I is the identity on $L^2(\mathbb{R}^n)$. It generalices the old trace in the sense that, if B is a trace class operator, its trace evaluated with (1.6) concides with its ordinary "old" trace.

The Wigner function..- In order to make use of the techniques of Classical Statistical Mechanics (CSM) to perform calculations in Q.M., we need a rule associating a distribution function on phase space to every quantum state. A minimal requirement for a distribution function to represent $\rho(\boldsymbol{u})$ to represent on phase space a quantum state ρ is that

$$\text{tr } \boldsymbol{\rho}\, W(f) = \int_{\mathbb{R}^{2n}} f(\boldsymbol{u})\, \rho(\boldsymbol{u})\, d\boldsymbol{u} \tag{1.9}$$

for any given classical observable $f(\boldsymbol{u})$. This means that the mean value of the quantum observable $W(f)$ in the state ρ ought to be equal to the mean value of the classical observable $f(\boldsymbol{u})$ evaluated with the distribution function $\rho(\boldsymbol{u})$. There is not a unique solution for $\rho(\boldsymbol{u})$ fulfilling (1.9). One possible solution was given by Wigner: if $\psi(\boldsymbol{y})$ is the wave function for a certain quantum state, its distribution function on phase space can be given by

$$\rho(\boldsymbol{q},\boldsymbol{p}) = \left(\frac{1}{2\pi\hbar}\right)^n \int_{\mathbb{R}^n} e^{i\boldsymbol{p}\cdot\boldsymbol{y}/\hbar}\, \Psi^*(\boldsymbol{q}+\boldsymbol{y}/2)\, \Psi(\boldsymbol{q}-\boldsymbol{y}/2)\, d\boldsymbol{y} \tag{1.10}$$

We call $\rho(\boldsymbol{q},\boldsymbol{p})$ the Wigner function of the state $\psi(\boldsymbol{y})$. This is not only the simplest function fulfilling (1.9) for any $f(\boldsymbol{u})$ but also the only one that satisfies certain plausible physical requirements [8].

It was Moyal who first realized that the Wigner function of a quantum state is, save for a trivial constant, the result of applying the inverse of the Weyl mapping to the corresponding density operator. In fact, one obtains with the help of (1.4) and (1.8)

$$\begin{aligned} W^{-1}(|\Psi\rangle\langle\Psi|) &= \text{tr } |\Psi\rangle\langle\Psi|\boldsymbol{\Pi}(\boldsymbol{u}) = \langle\Psi|\boldsymbol{\Pi}(\boldsymbol{u})|\Psi\rangle \\ &= 2^n \int_{\mathbb{R}^n} e^{i\boldsymbol{p}\cdot\boldsymbol{y}/\hbar}\, \Psi^*(\boldsymbol{q}+\boldsymbol{y}/2)\, \Psi(\boldsymbol{q}-\boldsymbol{y}/2)\, d\boldsymbol{y} \end{aligned} \tag{1.11}$$

Now, let us summarize the foundations of the Moyal formulation of QM on phase space. Its basic principles are:

a) The explicit form of a quantum observable on phase space coincides with that of the corresponding classical observable.

b) We define a non commutative product , called the twisted product, on a vector space $\mathcal{M}$ of functions or generalized functions on phase space, which is equivalent to the ordinary product of operators on Hilbert space, via the Weyl mapping. As $\mathcal{M}$ is endowed with this product, it becomes a noncommutative algebra and is contained in $\mathcal{S}'(\mathbb{R}^{2n})$. $\mathcal{M}$ is also the domain of the Weyl mapping.

c) If we define the Moyal bracket as

$$\{f, g\}_M = \frac{1}{i\hbar}\{f \times g - g \times f\} \tag{1.12}$$

one has:

$$\{q_i, q_j\}_M = \{p_i, p_j\}_M = 0 \qquad ; \qquad \{q_i, p_j\}_M = i\hbar\delta_{ij}$$

d) A state is a real normalized distribution function on phase space, which is non positive definite in general. It is, save for a trivial constant, the Wigner function of the corresponding density operator.

e) To evaluate the mean value of an observable on a given state, one proceeds as in Classical Statistical Mechanics.

It is important to remark that the Wigner function of a given quantum state is not positive in general. As a consequence, quantum mechanics is not a particular case of C.S.M.

If A is an operator on $L^2(\mathbb{R}^n)$ belonging to the range of W, $W^{-1}(A)$ is a function (or generalized function) on the phase space $\mathbb{R}^{2n}$. Henceforth, we will call Wigner function of A to $W^{-1}(A)$, even if A does not represent a state.

Moyal propagators and spectral projections

The next step in our formalization is to construct the basic tools to describe dynamics and to obtain quantum spectra of states and observables. The former is the equivalent of the evolution operator. This is called the Moyal propagator and is defined as the inverse image by the Weyl mapping of the evolution operator. For a time independent Hamiltonian, this is

$$\begin{aligned}\Xi_H(\boldsymbol{u},t) &= W^{-1}\left(\exp\left\{\frac{1}{i\hbar}W(H)t\right\}\right)\\ &= 1+\frac{1}{i\hbar}W(H)t+\left(\frac{1}{i\hbar}\right)^2\frac{1}{2}W(H)\times W(H)t^2+\dots\end{aligned} \tag{2.1}$$

where H denotes the classical Hamiltonian on phase space.

When the Hamiltonian is time independent, the Schrödinger equation on the usual formalism of Q.M. on Hilbert space can be written as $i\hbar\frac{\partial}{\partial t}e^{-itW(H)/\hbar} = W(H)\,e^{-itW(H)/\hbar}$. By applying W^{-1} to both members of this identity, one obtains the version of the Schröd- inger equation in the Moyal formalism, which is

$$H\times\Xi_H(\boldsymbol{u},t) = i\hbar\frac{\partial}{\partial t}\Xi_H(\boldsymbol{u},t) \tag{2.2}$$

Similar arguments provide the law of evolution of an observable $f(\boldsymbol{u},t)$ in the Heisenberg picture. This is

$$f(\boldsymbol{u},t) = \Xi_H(\boldsymbol{u},-t)\times f(\boldsymbol{u},0)\times\Xi_H(\boldsymbol{u},t) \tag{2.3}$$

The definition of the Moyal propagator gives $\Xi_H(\boldsymbol{u},-t) = \Xi_H^*(\boldsymbol{u},t)$, where the star denotes complex conjugation.

A new important concept is given with the term named spectral projection. By definition, the spectral projection of a Hamiltonian H is given by the Fourier transform with respect to the time variable t of the Moyal propagator of H. Thus,

$$\boldsymbol{\Gamma}_H(\boldsymbol{u},E) = \frac{1}{2\pi\hbar}\int_{\mathbb{R}}\Xi_H(\boldsymbol{u},t)\,e^{itE/\hbar}\,dt \tag{2.4}$$

E must have dimensions of energy if the exponent in (2.4) is to be dimensionless. Furthermore, it is commonly accepted that the support on E of the spectral projection (2.4), i.e., the closure of the set of points E for which $\boldsymbol{\Gamma}_H(\boldsymbol{u},E)$ does not vanish, is the Hilbert space spectrum of the quantum Hamiltonian $W(H)$ [9]. This would provide a prescription so as to obtain the energy levels of any quantum Hamiltonian. However, this statement has rigurously been proven in two situations only: when the spectrum is purely discrete and also when H is quadratic in the variables $\boldsymbol{q}$ and $\boldsymbol{p}$ [10].

The quadratic Hamiltonians are of special interest, since they have the following property:

$$\{H,f\}_M = i\hbar\,\{H,f\}_P\,; \qquad \forall f \in C^\infty(\mathbb{R}^{2n}) \tag{2.5}$$

where $\{H,f\}_P$ denotes the Poisson bracket of H and f. This means, in particular, that the classical and quantum evolution of a certain density on phase space $\rho(\boldsymbol{u})$ produced by the same quadratic Hamiltonian must coincide. Due to this circumstance, an exact calculation of the Moyal propagators for all quadratic Hamiltonians is rather simple [10]. Quadratic Hamiltonians are the only ones fulfilling (2.5).

The spin

As we have pointed out in the first section of the present lesson, the formalism described so far is valid for spinless nonrelativistic systems of particles only. In order to introduce spin variables, let us consider one particle system in which the only relevant variables are those of spin. Consequently, the Hilbert space of states is $\mathbb{C}^2$, the vector space of all ordered pairs of complex numbers. A phase space description for this situation was sucessfully realized by Gracia-Bondía and Varilly [11]. They proposed the two dimensional sphere S^2 as the phase space. The role of the G.R. operators is here performed by

$$\boldsymbol{\Delta}(\boldsymbol{n}) = \sum_{r,s=+,-} \mathbf{Z}_{r,s}(\mathbf{n})\,|s\rangle\,\langle r| \tag{3.1}$$

where

$$\mathbf{Z}_{r,s}(\boldsymbol{n}) = \sum_{l=0,1} \sqrt{\pi(2l+1)}\,C(1/2,l,1/2;s,(r-s),r)\,\mathbf{Y}_{l,r-s}(\boldsymbol{n}) \tag{3.2}$$

Here, $C(1/2,l,1/2;s,(r-s),r)$ are Clebs-Gordan coefficients and $Y_{l,r-s}(\boldsymbol{n})$ spherical harmonics. $\boldsymbol{n}$ is the unit vector representing a point in S^2. If $f(\boldsymbol{n})$ is any function on S^2, its image by the Weyl mapping is an operator on $\mathbb{C}^2$, i.e., a 2×2 complex matrix given by

$$W(f) = A = \frac{1}{2\pi}\int_{S^2} f(\boldsymbol{n})\,\boldsymbol{\Delta}(\boldsymbol{n})\,d\boldsymbol{n} \tag{3.3}$$

where $d\boldsymbol{n} = \mathrm{sen}\theta\, d\theta\, d\varphi$.

A twisted product of functions on the sphere corresponding to the product of operators on $\mathbb{C}^2$ can now be given as:

$$(f\times h)(\boldsymbol{n}) = \int_{S^2}\int_{S^2} f(\boldsymbol{m})\,h(\boldsymbol{k})\,\mathcal{L}(\boldsymbol{n},\boldsymbol{m},\boldsymbol{k})\,d\boldsymbol{m}\,d\boldsymbol{k} \tag{3.4}$$

where

$$\mathcal{L}(\boldsymbol{n},\boldsymbol{m},\boldsymbol{k}) = \text{tr}\ [\boldsymbol{\Delta}(\boldsymbol{n})\,\boldsymbol{\Delta}(\boldsymbol{m})\,\boldsymbol{\Delta}(\boldsymbol{k})] = \left(\frac{1}{4\pi}\right)\left(1+3(\boldsymbol{n}\cdot\boldsymbol{m}+\boldsymbol{m}\cdot\boldsymbol{k}+\boldsymbol{k}\cdot\boldsymbol{n})+3\sqrt{3}i[\boldsymbol{n},\boldsymbol{m},\boldsymbol{k}]\right) \tag{3.5}$$

The inversion formula is

$$W_A(\boldsymbol{n}) = \text{tr}\ [A\,\boldsymbol{\Delta}(\boldsymbol{n})] \tag{3.6}$$

(3.6) gives a function on the sphere for any operator on $\mathbb{C}^2$. For instance, if $\boldsymbol{S} = \frac{\hbar}{2}\boldsymbol{\sigma} = \frac{\hbar}{2}(\sigma_x, \sigma_y, \sigma_z)$, its corresponding function on the sphere is given by [11]

$$\hbar\frac{\sqrt{3}}{2}\boldsymbol{n} = \hbar\frac{\sqrt{3}}{2}(\cos\theta, \sin\theta\cos\varphi, \sin\theta, \sin\theta\ \sin\varphi) \tag{3.7}$$

It is important to remark that, in this case, W is not one-to-one. However, it is an onto mapping. In this situation, let us pose the following question: For which domain is W an one-to-one mapping?. Note that if $f(\boldsymbol{n})$ is a function in this domain, the expression $W(1\times f) = W(1)\,W(f) = W(f)$ implies that $(1\times f)(\boldsymbol{n}) = f(\boldsymbol{n})$. This equation has as solution

$$f(\boldsymbol{n}) = C + \boldsymbol{B}\cdot\boldsymbol{n} \tag{3.8}$$

where

$$C = \frac{1}{4\pi}\int_{S^2} f(\boldsymbol{k})\cdot d\boldsymbol{k}; \qquad \boldsymbol{B} = \frac{3}{4\pi}\int_{S^2} \boldsymbol{k}\,[f(\boldsymbol{k})\cdot d\boldsymbol{k}]$$

Formula (3.6) always gives a function of the form (3.8). Note that the one-to-one correspondence between 2×2 complex matrices and functions on the sphere of the form (3.8) is obvious since C and the three components of $\boldsymbol{B}$ are complex numbers. The situation can be described as follows: if we start with *any* function $f(\boldsymbol{n})$ on S^2, the action of the Weyl mapping (3.3) gives an operator $A = W(f)$ on $\mathbb{C}^2$. Then, if we apply (3.6) to A, we obtain a function on the sphere of the form (3.8), where the coefficients C and $\boldsymbol{B}$ can be calculated using the original $f(\boldsymbol{n})$ in (3.9). Consequently, we can divide the set of functions on S^2 into conjugacy classes. Each of these classes is the image by W^{-1} of an operator A on $\mathbb{C}^2$ and has one and only one representative of the form (3.8). We call the Wigner function of A to this representative and we denote it as $W^{-1}(A)$.

Now, let us consider

$$\rho = \alpha|+\rangle\langle+| + \beta|-\rangle\langle-|; \quad 0\leq\alpha,\ \beta\leq 1; \quad \alpha+\beta=1 \tag{3.9}$$

where $|+\rangle$ and $|-\rangle$ are the eigenstates of the operator $\boldsymbol{S}\cdot\boldsymbol{a}$ where $\boldsymbol{a}$ is a unit vector.

The Wigner function for this state is:

$$W^{-1}(\boldsymbol{\rho}) = \frac{1}{2} + \frac{\sqrt{3}}{2}(\alpha - \beta)\boldsymbol{a} \cdot \boldsymbol{n} \tag{3.10}$$

Thus, a function on the sphere of the form $C + \boldsymbol{B} \cdot \boldsymbol{n}$ is the Wigner function of a spin state if and only if:

$$C = \frac{1}{2}; \qquad |\boldsymbol{B}| \leq \frac{\sqrt{3}}{2} \tag{3.11}$$

and will represent a pure state if $|\boldsymbol{B}| = \frac{\sqrt{3}}{2}$.

$W^{-1}(\boldsymbol{\rho})$ is positive if and only if

$$|\alpha - \beta| \leq \frac{\sqrt{3}}{3} \tag{3.12}$$

In particular, this formula implies that no pure state of spin can have a positive Wigner function.

To finish this section, let us consider a particle with spin in which the orbital variables are now relevant. In this situation, the phase space is $\mathbb{R}^6 \times S^2$ and the role of the G.R. operators is performed by the following tensor product on the Hilbert space of states $L^2(\mathbb{R}^n) \otimes \mathbb{C}^2$

$$\boldsymbol{\Omega}(\boldsymbol{q}, \boldsymbol{p}, \boldsymbol{n}) = \boldsymbol{\Pi}(\boldsymbol{q}, \boldsymbol{p}) \otimes \boldsymbol{\Delta}(\boldsymbol{n}) \tag{3.13}$$

The Weyl mapping is defined as

$$W(f) = \left(\frac{1}{2\pi\hbar}\right)^n \frac{1}{2\pi} \int_{\mathbb{R}^{2n}} \int_{S^2} f(\boldsymbol{q}, \boldsymbol{p}, \boldsymbol{n})\, \boldsymbol{\Pi}(\boldsymbol{q}, \boldsymbol{p}) \otimes \boldsymbol{\Delta}(\boldsymbol{n})\, d\boldsymbol{q}\, d\boldsymbol{p}\, d\boldsymbol{n} \tag{3.14}$$

for any function $f(\boldsymbol{q}, \boldsymbol{p}, \boldsymbol{n})$ on the phase space for wich the integral in (3.14) converges. If A is an operator on $L^2(\mathbb{R}^n) \otimes \mathbb{C}^2$, its Wigner function is defined as

$$W_A(\boldsymbol{q}, \boldsymbol{p}, \boldsymbol{n}) = \mathrm{tr}\,[A\,\boldsymbol{\Omega}(\boldsymbol{q}, \boldsymbol{p}, \boldsymbol{n})] \tag{3.15}$$

where

$$\mathrm{tr}\; A = tr \left[\int \langle \boldsymbol{q}|A|\boldsymbol{q}\rangle\, d\boldsymbol{q}\right] \tag{3.16}$$

The symbol tr means here the trace of the 2×2 matrix that results of performing the integral between brackets.

Identical Particles

In the present section, we want to describe identical particles whithin the context of the Moyal formalism. We assume that one has N quantum nonrelativistic particles with spin. When one uses the usual Hilbert space formalism, a pure state of N particles is described by the N-particle wave function $\psi(\boldsymbol{x}_1, \boldsymbol{x}_2, \ldots, \boldsymbol{x}_N)$ where $\boldsymbol{x}_i$ accounts for the position and spin coordinates of the i-th particle. The Hilbert space $\mathcal{H}$ of these wave functions splits into a direct sum of three Hilbert spaces

$$\mathcal{H} = \mathcal{H}_+ \oplus \mathcal{H}_- \oplus \mathcal{H}_{ws} \tag{4.1}$$

where $\mathcal{H}_+$ and $\mathcal{H}_-$ are the Hilbert spaces of the symmetric and the antisymmetric wave functions respectively and the functions in $\mathcal{H}_{ws}$ have not this kind of symmetry. In other words, $\mathcal{H}_+$ and $\mathcal{H}_-$ are the Hilbert spaces of the states of N bosons and N fermions respectively. The orthogonal projections onto $\mathcal{H}_+$ and $\mathcal{H}_-$ are:

$$P_+ = \frac{1}{N!} \sum_{\sigma \in S_N} P_\sigma \quad ; \quad P_- = \frac{1}{N!} \sum_{\sigma \in S_N} (-1)^\sigma P_\sigma \tag{4.2}$$

where

$$(P_\sigma \Psi)(\boldsymbol{x}_1, \ldots, \boldsymbol{x}_N) = \Psi(\boldsymbol{x}_{\sigma(1)}, \ldots, \boldsymbol{x}_{\sigma(N)}) \tag{4.3}$$

If B is an observable on $\mathcal{H}$, then $P_\pm B P_\pm$ is the corresponding observable in $\mathcal{H}_\pm$. If B remains invariant after a permutation of the particles, it follows that $P_\pm B = P_\pm B P_\pm = B P_\pm$ and hence, B leaves invariant the spaces $\mathcal{H}_\pm$.

The point of departure of a description à la Moyal of a system of N identical particles is the following postulate

Postulate.- *The Wigner function for a state or observable B of a system of N identical particles is:*

$$W^{-1}(P_\pm B P_\pm) = W^{-1}(P_\pm) \times W^{-1}(B) \times W^{-1}(P_\pm) \tag{4.4}$$

We see that the basic tool are the Wigner functions of the projections P_+ and P_-. To obtain them, we consider two cases:

A) Spinless particles

Since W^{-1} is linear, one has

$$\begin{aligned} p_+(\boldsymbol{u}) &= W^{-1}(P_+) = \frac{1}{N!} \sum_{\sigma \in S_N} W^{-1}(P_\sigma) \\ p_-(\boldsymbol{u}) &= W^{-1}(P_-) = \frac{1}{N!} \sum_{\sigma \in S_N} (-1)^\sigma W^{-1}(P_\sigma) \end{aligned} \tag{4.5}$$

$W^{-1}(P_\sigma)$ can be evaluated by means of the following formula, wich can be deduced as (1.11)

$$W^{-1}(P_\sigma) = \tilde{\sigma}(\boldsymbol{u}_1, \ldots, \boldsymbol{u}_N) = \int \langle \boldsymbol{q} - \frac{\boldsymbol{v}}{2} | P_\sigma | \boldsymbol{q} + \frac{\boldsymbol{v}}{2} \rangle \, e^{i\boldsymbol{p}\cdot\boldsymbol{v}/\hbar} \, d\boldsymbol{v} \tag{4.6}$$

A direct calculation of the integral in (4.5) seems rather difficult. In order to circumvent this problem and obtaining $W^{-1}(P_\sigma)$, we make a detour. First of all, let us note that any permutation can be written as a product of cycles with no common variables. Then, $\sigma = \sigma_1 \sigma_2 \ldots \sigma_k$, where each of the σ_j is a cyclic permutation. Thus,

$$P_\sigma = P_{\sigma_1} \ldots P_{\sigma_k} \tag{4.7}$$

Since any two of these cycles affect to different variables, their respective Wigner functions, obtained as in (4.6), also depend on different variables $\boldsymbol{u}_i$. From (1.3), one sees that the twisted product of two functions depending on a disjoint set of variables reduces to the ordinary product of these two functions. Therefore,

$$\begin{aligned} W^{-1}(P_\sigma) &= W^{-1}(P_{\sigma_1}) \times W^{-1}(P_{\sigma_2}) \times \ldots \times W^{-1}(P_{\sigma_k}) \\ &= W^{-1}(P_{\sigma_1}) W^{-1}(P_{\sigma_2}) \ldots W^{-1}(P_{\sigma_k}) \end{aligned} \tag{4.8}$$

The final step in our argument is based in the fact that, for a cyclic permutation, the integral in (4.6) can be evaluated without difficulties. In this case, the final result is:

$$\tilde{\sigma}(\boldsymbol{u}_1, \ldots, \boldsymbol{u}_M) = 2^{M-1} \exp\left\{ -\frac{2i}{\hbar} \sum_{k=1; l>k}^{M} (-1)^{k+l} \, \boldsymbol{u}_k J \boldsymbol{u}_l \right\} \qquad M \text{ odd} \tag{4.9}$$

$$\begin{aligned} \tilde{\sigma}(\boldsymbol{u}_1, \ldots, \boldsymbol{u}_M) = {} & 2^{M-1} \pi\hbar\delta(\boldsymbol{u}_1 - \boldsymbol{u}_2 + \ldots - \boldsymbol{u}_M) \\ & \exp\left\{ -\frac{2i}{\hbar} \sum_{k=1; l>k}^{M} (-1)^{k+l} \, \boldsymbol{u}_k J \boldsymbol{u}_l \right\} \qquad M \text{ even} \end{aligned} \tag{4.10}$$

which allows for a reconstruction of the Wigner functions (4.4) for any number N.

B) Particles with spin $\frac{1}{2}$

Now, let us assume that we are considering particles with spin $\frac{1}{2}$ in wich the orbital variables are also relevant. This produces a decomposition of the Hilbert space of states as in (4.1). The orthogonal projections onto the spaces $\mathcal{H}_\pm$ are now

$$P_+ = \frac{1}{N!} \sum_{\sigma \in S_N} P_\sigma Q_\sigma \quad ; \quad P_- = \frac{1}{N!} \sum_{\sigma \in S_N} (-1)^\sigma P_\sigma Q_\sigma \tag{4.11}$$

where Q_σ is the operator that produces the permutation σ on the spin variables. We keep the notation P_σ for the operator producing the same permutation on the orbital variables. Since orbital and spin variables are different, the Wigner functions for $P_\pm$ on the phase space $\mathbb{R}^{6N} \times S^2$ are

$$p_+(\boldsymbol{u},\boldsymbol{n}) = \frac{1}{N!}\sum_{\sigma\in S_N} W^{-1}(P_\sigma)W^{-1}(Q_\sigma)$$
$$p_-(\boldsymbol{u},\boldsymbol{n}) = \frac{1}{N!}\sum_{\sigma\in S_N} (-1)^\sigma W^{-1}(P_\sigma)W^{-1}(Q_\sigma) \tag{4.12}$$

with $\boldsymbol{u} = (\boldsymbol{q},\boldsymbol{p})$ and $\boldsymbol{q} = (\boldsymbol{q}_1,\dots,\boldsymbol{q}_N)$, and same for $\boldsymbol{p}$. In (4.12), one has:

$$W^{-1}(Q_\sigma) = \sum Z_{b_{\sigma(1)}\,b_1}(\boldsymbol{n}_1)\dots Z_{b_{\sigma(N)}\,b_N}(\boldsymbol{n}_N) \tag{4.13}$$

where the sumation runs out all possible choices of $b_i = +,-$.

We conclude this section with the following remark: if we want to obtain Moyal propagators and spectral projections for a Hamiltonian H of a system of N identical particles, we can do it and their calculation is rather simple provided that H be time independent and invariant under a permutation of the particles. In this case, $p_\pm \times H \times p_\pm = H \times p_\pm$, where $p_\pm$ are the functions in (4.13). Simple algebraic manipulations give

$$\boldsymbol{\Xi}_\pm(\boldsymbol{u},\boldsymbol{n},t) = \boldsymbol{\Xi}(\boldsymbol{u},\boldsymbol{n},t)\times p_\pm \tag{4.14}$$
$$\boldsymbol{\Gamma}_\pm(\boldsymbol{u},\boldsymbol{n},E) = (\boldsymbol{\Gamma}\times p_\pm)(\boldsymbol{u},\boldsymbol{n},E) \tag{4.15}$$

Since we have already obtain the Moyal prpagators for all quadratic Hamiltonians [10], (4.14) and (4.15) can be useful tools in order to obtain the nergy levels of a few particle system with or without spin, provided that the external potential as well as the interaction between the particles be of quadratic type [12]. This formalism can be extended to spins different from $\frac{1}{2}$ by using the appropiate function in (4.12).

REFERENCES

1. J. E. Moyal, Proc. Cam. Phil. Soc. **45** (1949), 99.
2. H. Weyl, *The Theory of Groups and Quantum Mechanics*, Dover, New York, 1931.
3. E.P. Wigner, Phys. Rev. **40** (1932), 749.
4. J. C. Varilly, J. M. Gracia-Bondía, J. Math. Phys. **29** (1988), 880.
5. J. M. Gracia Bondía and J. C. Varilly, J. Math. Phys. **29** (1988), 869.
6. A. Grossmann, Commun. Math. Phys. **48** (1976), 191.

7. A. Royer, Phys. Rev. A **15** (1977), 449.

8. J.G.Krüger, A. Poffin, Physica **87A** (1977), 132.

9. F. Bayen, M. Flato, C. Fronsdal, A. Lichnerowicz, D. Sternheimer, Ann. Phys. (NY) **111** (1978), 61.

10. M. Gadella, J.M. Gracia-Bondía, L.M. Nieto, J.C. Varilly, J. Phys. A: Math Gen. **22** (1989), 2709.

11. J. C. Varilly, J. M. Gracia-Bondía, Ann. Phys. **190** (1989), 107.

12. M. Gadella, L.M. Nieto, *Moyal Quantization of Identical Particles*, U. of Valladolid Preprint (1992).

LIE GROUPS AND SOLUTIONS OF NONLINEAR PARTIAL DIFFERENTIAL EQUATIONS

P.Winternitz

Centre de recherches mathématiques, Université de Montréal, CP 6128-A, Montréal, H3C 3J7, Québec (Canada).

Abstract. The application of local Lie point transformation groups to the solution of partial differential equations is reviewed. The method of symmetry reduction is presented as an algorithm. Included is the construction of group invariant solutions, partially invariant solutions and also the use of conditional symmetries. The emphasis in on recent developments, including the use of computer algebra. Many examples and applications are treated.

TABLE OF CONTENTS

Lectures delivered at School on Recent Problems in Mathematical Physics, Salamanca, Spain, June 15–27, 1992.

L. A. Ibort and M. A. Rodríguez (eds.), Integrable Systems, Quantum Groups, and Quantum Field Therapy 429–495.

1. INTRODUCTION.

The purpose of these lectures is to show how Lie group theory can help us to obtain exact analytic solutions of partial differential equations. The emphasis will be on nonlinear equations and nonlinear phenomena in physics.

The topic of Lie groups and differential equations is a vast one and has a long history, going back to the classical work of S. Lie. Several modern books on the subject exist [1, ... ,9].

The subject is currently going through rapid developments, mainly as far as applications are concerned. This is partly due to increased motivation. Indeed nonlinear phenomena play a crucial role in most areas of physics, biology and other sciences. Lie group theory provides universal tools for obtaining at least particular solutions of the equations describing such phenomena. Another reason for the recent flourishing of group theoretical methods is the fact that algebraic computing, using symbolic languages such as MACSYMA, REDUCE, MAPLE, MATHEMATICA, etc. can be used very efficiently in this field of activity. Computations, that are algorithmic, but tend to be very cumbersome, can now be done on a computer. Further, developments in abstract Lie algebra and Lie group theory can now be put to good use. These include new results on the classification of finite and infinite dimensional Lie algebras. Recent methods and algorithms for classifying subalgebras of Lie algebras make it possible to systematically generate group invariant solutions. Group theoretical methods can be combined with the recent results on integrability of nonlinear differential equations, in particular with Painlevé analysis [10,11].

In these lectures we shall restrict ourselves to Lie groups of local point transformations. By definition, these will have the form

$$\tilde{x} = \Lambda_g(x,u), \ \tilde{u} = \Omega_g(x,u), \quad x \in \mathbb{R}^p, \ u \in \mathbb{R}^q, \tag{1.1}$$

where x and u are the independent and dependent variables, respectively. The functions Λ and Ω thus depend on x and u, but not on derivatives of u. They also depend on the group parameters g and are well defined for g close to the identity transformation, $g = e$, and for x and u close to some origin in the space $X \times U$ of independent and dependent variables.

Contact transformations, for which Λ and Ω also depend on the first derivatives u_x, or generalized transformations, for which they depend on higher order derivatives, will not be treated here.

The main motivation for studying Lie point transformations in the context of differential equations can be summed up as follows:

1. Symmetry transformations, leaving the solution set of a given differential system invariant, can be applied to a known solution, in order to obtain new solutions. Sometimes one can get families of nontrivial solutions from one trivial one.
2. Symmetry transformations can be used to perform "symmetry reduction". For ordinary differential equations (ODEs) this amounts to reducing the

order of the equations. For partial differential equations (PDE) symmetry reduction provides a reduction of the number of independent variables. If successful, the reduction for ODEs provides the general solution. For PDEs it provides particular solutions, satisfying particularly symmetric boundary conditions.

3. Different equations can be transformed amongst each other by Lie point transformations. Equations can this be classified into equivalence classes. Two equivalent equations have isomorphic symmetry groups. The isomorphism between symmetry groups can be used as a necessary condition for two equations to be transformable into each other.

2. HOW TO DETERMINE THE LIE POINT SYMMETRIES OF A DIFFERENTIAL SYSTEM.

2.1. General Comments.

Let us now consider a completely general system of differential equations

$$\Delta^i(x, u, u^{(1)}, u^{(2)}, \ldots, u^{(n)}) = 0 \tag{2.1}$$
$$x \in \mathbb{R}^p, \;\; u \in \mathbb{R}^q, \;\; i = 1, \ldots, m, \;\; p, q, m, n \in \mathbb{Z}^{>0},$$

where $u^{(k)}$ denotes all partial derivatives of order k of all components of u. We now wish to obtain the local Lie group G of local Lie point transformations (1.1) taking solutions of eq. (2.1) into solutions. In other words, the group G acts on a manifold

$$M \subset X \times U, \quad X \sim \mathbb{R}^p, \; U \sim \mathbb{R}^q \tag{2.2}$$

in such a manner that whenever $u = f(x)$ is a solution of eq. (1.1), then so is $\tilde{u}(\tilde{x}) = g \cdot f(x)$.

Following S. Lie, as presented e.g. in Ref. 1, we shall look for infinitesimal group transformations, i.e. construct the Lie algebra L of the Lie group G.

We shall present the algorithm in a purely utilitarian manner. For all proofs we refer to the literature [1, ... ,4] specially P. Olver's book [1]. The Lie algebra L will be realized in terms of vector fields on M, i.e. differential operators of the form

$$\hat{v} = \sum_{i=1}^{p} \xi^i(x, u)\, \partial_{x_i} + \sum_{\alpha=1}^{q} \phi^\alpha(x, u)\, \partial_{u_\alpha}. \tag{2.3}$$

The functions ξ^i and ϕ^α are to be determined. The fact that they depend on the independent variables x and the dependent variables u, but not on derivatives of u with respect to x, corresponds to our restriction to point transformations.

A basic tool to be used are prolongations. Thus, if we are given a function:

$$f : X \to U, \quad u = f(x), \tag{2.4}$$

then its n-th prolongation

$$\mathrm{pr}^{(n)} f(x) = \{f, f_{x_i}, f_{x_i x_j}, \ldots, f_{x_{i_1}, \ldots, x_{i_n}}\} \tag{2.5}$$

consists of the function and all of its derivatives up to order n. If we are given a point transformation G on M

$$\begin{aligned} G : \{x, u\} &\to (\tilde{x}(x,u), \tilde{u}(x,u)) \in M \\ f(x) &\to f(\tilde{x}) = g \cdot f(x), \end{aligned} \tag{2.6}$$

taking functions into functions, then its prolongation also takes derivatives into derivatives:

$$\begin{aligned} \mathrm{pr}^{(n)} G : &\{x, u = f(x), f_{x_i}(x), \ldots, f_{x_{i_1} \ldots x_{i_n}}(x)\} \to \\ &\{\tilde{x}, \tilde{u} = \tilde{f}(\tilde{x}), \tilde{f}_{\tilde{x}_i}(\tilde{x}), \ldots, \tilde{f}_{\tilde{x}_{i_1} \ldots \tilde{x}_{i_n}}(\tilde{x})\}. \end{aligned} \tag{2.7}$$

We note that all the information about the prolongation $\mathrm{pr}^{(n)} G$ of a point transformation is already contained in the transformation G itself (we assume that all functions involved are sufficiently smooth, at least locally).

If we were to take a global approach to Lie point symmetries, we would set

$$\tilde{u} = \Omega(x, u), \qquad \tilde{x} = \Lambda(x, u) \tag{2.8}$$

and calculate derivatives as

$$\tilde{u}_{\tilde{x}} = (\Omega_u u_x + \Omega_x)(x_{\tilde{x}} + x_{\tilde{u}} \tilde{u}_{\tilde{x}})$$

and solve for $\tilde{u}_{\tilde{x}}$:

$$\tilde{u}_{\tilde{x}} = \frac{(\Omega_u u_x + \Omega_x) x_{\tilde{x}}}{1 - (\Omega_u u_x + \Omega_x) x_{\tilde{u}}}. \tag{2.9}$$

Similarly we can calculate $\tilde{u}_{\tilde{x}\tilde{x}}$ and higher derivatives (these equations are to be taken literally for the case of one independent and one dependent variable, symbolically otherwise). In any case, substituting into the differential system (2.1) and requiring that $\tilde{u}(\tilde{x})$ satisfies the same equations as $u(x)$, we obtain a system of equations for $\Omega(x, u)$ and $\Lambda(x, u)$. The equations are nonlinear if eq. (2.1) is nonlinear. Usually they are totally unmanageable.

The point of Lie algebra theory is that the Lie algebra L actually catches all the local continuous phenomena represented by the group G. If we consider infinitesimal transformations, we put

$$\tilde{u}_\alpha(\tilde{x}) = u_\alpha(x) + \epsilon\, \phi_\alpha(x, u), \quad \tilde{x}_i = x_i + \epsilon\, \xi_i(x, u). \tag{2.10}$$

When evaluating the derivatives $\tilde{u}_{\tilde{x}}$ and higher order derivatives, we only keep terms linear in ϵ. The equations for $\phi_\alpha(x,u)$ and $\xi_i(x,u)$ are then linear by construction.

The results are best formulated in terms of the vector fields (2.3). Below we shall present an algorithm for calculating the coefficients $\xi_i(x,u)$ and $\phi_\alpha(x,u)$. Once they are known, we obtain the group transformations by integrating the vector fields. Each vector field (2.3) provides a one parameter Lie group $G(\lambda)$ obtained by integrating the following system

$$\begin{aligned} \frac{d\tilde{x}^i}{d\lambda} &= \xi^i(\tilde{x},\tilde{u}) \qquad & \tilde{x}^i\big|_{\lambda=0} &= x^i \\ \frac{d\tilde{u}^\alpha}{d\lambda} &= \phi^\alpha(\tilde{x},\tilde{u}) \qquad & \tilde{u}^\alpha\big|_{\lambda=0} &= u^\alpha. \end{aligned} \tag{2.11}$$

Vice versa, given a one parameter group of transformations (2.8), we obtain the coefficients of the vector fields by differentiating

$$\begin{aligned} \frac{d\tilde{x}^i}{d\lambda} &= \frac{d\Lambda^i_\lambda(x,u)}{d\lambda}\Big|_{\lambda=0} = \xi^i(x,u) \\ \frac{d\tilde{u}^\alpha}{d\lambda} &= \frac{d\Omega_\lambda(x,u)}{d\lambda}\Big|_{\lambda=0} = \phi^\alpha(x,u). \end{aligned} \tag{2.12}$$

The entire symmetry group is obtained by composing the one-dimensional subgroups corresponding to the distinct one-dimensional subalgebras of L:

$$g = g(\lambda_1)\, g(\lambda_2) \dots g(\lambda_r),$$

where r is the dimension of the Lie algebra L and Lie group G.

2.2. The Algorithm.

The vector field $\hat{v}$ of equation (2.3) acts on functions $F(x,u)$ of the independent and dependent variables. The n-th prolongation $\mathrm{pr}^{(n)}\hat{v}$ acts on functions $F(x,u,u_x,\dots,u_{nx})$ of x,u and all derivatives of u upto order n. Integrating the vector field as in eq. (2.12) we obtain the group transformations, integrating the prolongation of the vector field, we obtain the prolongation of the group action (see eq. (2.7)). The prolongation of the vector field is again a differential operator that we can write as

$$\mathrm{pr}^{(n)}\hat{v} = \hat{v} + \sum_{\alpha=1}^{q} \sum_{k=1}^{n} \sum_{J} \phi_\alpha^J \frac{\partial}{\partial u_J^\alpha}, \tag{2.13}$$

where J is a multiindex:

$$J \equiv J(k) = (j_1,\dots,j_k), \;\; 1 \le j_k \le p, \;\; k = j_1 + j_2 + \dots + j_k. \tag{2.14}$$

The coefficients ϕ_α^J depend on x, u and the derivatives of u upto order k (for $J = J(k)$). They can be expressed in terms of the functions ξ and ϕ figuring in eq. (2.3) and their derivatives. An explicit formula is given by Olver [1], here we shall reproduce a recursive formula, also given by Olver [1], that we find more convenient to use.

For the first prolongation we have

$$\mathrm{pr}^{(1)}\hat{v} = \hat{v} + \sum_{\alpha=1}^{q} \sum_{i=1}^{p} \phi_\alpha^i(x, u, u_{x_i}) \frac{\partial}{\partial u_{x_i}^\alpha} \tag{2.15}$$

$$\phi_\alpha^i = D_{x_i}\phi_\alpha - \sum_{k=1}^{p} (D_{x_i}\xi^k) u_{x_k}^\alpha, \quad \begin{array}{l} i = 1, \ldots, p \\ \alpha = 1, \ldots, q \end{array}, \tag{2.16}$$

where $D_i \equiv D_{x_i}$ is the total derivative operator:

$$D_{x_i} = \frac{\partial}{\partial x_i} + \sum_\alpha \frac{\partial u_\alpha}{\partial x_i} \frac{\partial}{\partial u_\alpha} + \sum_\alpha \sum_j \frac{\partial u_{\alpha, x_j}}{\partial x_i} \frac{\partial}{\partial u_{\alpha, x_j}} + \ldots \tag{2.17}$$

For the higher order terms in eq. (2.13) we have

$$\phi_\alpha^{J,k} = D_{x_k}\phi_\alpha^J - \sum_{a=1}^{p} (D_{x_k}\xi^a) u_{J, x_a}^\alpha. \tag{2.18}$$

Among the properties of the prolongations of vector fields we mention linearity

$$\mathrm{pr}^{(n)}(a\,\hat{v} + b\,\hat{w}) = a\,\mathrm{pr}^{(n)}\hat{v} + b\,\mathrm{pr}^{(n)}\hat{w} \tag{2.19}$$

and the Lie algebra property

$$\mathrm{pr}^{(n)}[\hat{v}, \hat{w}] = [\mathrm{pr}^{(n)}\hat{v}, \mathrm{pr}^{(n)}\hat{w}]. \tag{2.20}$$

Hence, if the vector fields $\{\hat{v}_i\}$ form a Lie algebra, their prolongations realize an isomorphic Lie algebra.

The algorithm for determining the Lie algebra L of the symmetry group G of the system of eq. (1.1) is now simply

$$\mathrm{pr}^{(n)}\hat{v}\ \Delta^i\big|_{\Delta^l=0} = 0, \quad \begin{array}{l} i = 1, \ldots m \\ l = 1, \ldots m \end{array}. \tag{2.21}$$

In other words, for a system of n-th order equations, the n-th order prolongation of the vector field $\hat{v}$ must annihilate the equations on the solution set of the equations. Condition (2.21) provides a system of linear ordinary differential equations of order

n for the coefficients ξ_i and ϕ_α of the vector field $\hat{v}$. These coefficients depend on x and u only. The expression (2.21) contains derivatives of u_α. The "*determining equations*" for the symmetry operators $\hat{v}$ are obtained by setting equal to zero the coefficient of each linearly independent expression in the derivatives $u_{\alpha,x_{j_1}\dots x_{j_k}}$.

In practice the algorithm for finding the symmetry algebra L boils down to the following steps.

1. Calculate the n-th prolongation (2.13) of the vector field (2.3).
2. View the original system (2.1) as a system of algebraic equations for a set of m of the highest derivatives of u_α. Call these derivatives $v_1, \dots, v_m$. They must be so chosen that no v_i is a derivative of the other ones, that no derivative of v_i figures in the equation and each v_i involves at least one derivative. Moreover, it must be possible to solve eq. (2.1) for $v_1, \dots, v_m$ unambiguously (preferably linearly).
3. Apply $\mathrm{pr}^{(n)}\hat{v}$ to the system (2.1), replace the expressions v_i by their expressions found in Step 2 and equate to zero.
4. Identify all linearly independent expressions in the remaining derivatives of u_α and set the coefficient of each such expression equal to zero. This provides the determining equations. They will be linear, even if the original system (2.1) is nonlinear. Their linearity is due to the infinitesimal approach: this corresponds to keeping only terms linear in ϵ of eq. (2.10).
5. Solve the determining equations and obtain the functions $\xi_i(x,u)$ and $\phi_\alpha(x,u)$.

The above steps $1, \dots, 4$ are entirely algorithmic, but can be very cumbersome and time-consuming, not to mention sensitive to errors. They are best performed on a computer. Numerous routines performing this task exist (in REDUCE [12], MACSYMA [13,14], Mathematica [15] and other languages).

Step 5, solving the determining equations, is less algorithmic. The system is usually overdetermined and the following possibilities occur.

1. The only solution is $\xi_i = 0, \phi_\alpha = 0$. In this case there is no nontrivial symmetry group.
2. The general solution depends on a finite number N of significant integration constants. We then obtain an N-dimensional Lie algebra and N-dimensional symmetry group.
3. The general solution depends on arbitrary functions. In this case the Lie algebra is infinite-dimensional, as is the corresponding Lie group.

The REDUCE package [12] goes quite far in solving the determining equations. The MACSYMA packages [13,14] solve the simplest equations and use the result to simplify the remaining ones. Finite algorithms exist that determine the dimension of the symmetry algebra and its commutation relations, without integrating the determining equations [16]. In any case, the tedious and time consuming part of the exercise is the derivation of the determining equations, and that the computer packages do very well.

2.3. Example of the Variable Coefficient Korteweg-de Vries Equation.

The Korteweg-de Vries equation (KdV)

$$u_t + uu_x + u_{xxx} = 0 \tag{2.22}$$

is an equation describing the propagation of weakly nonlinear weakly dispersive unidirectional waves in shallow water of constant depth and density, either in a narrow channel, or on an unbounded surface. It is the prototype of an "integrable equation" [10,11]. It has soliton and multisoliton solutions, a linear Lax pair that makes it possible to obtain large classes of solutions by linear techniques, infinitely many conservation laws and other nice properties. The KdV equation can be derived from the fundamental equations of hydrodynamics, the Euler equations, under rather drastic simplifying assumptions. If these assumptions are somewhat relaxed, allowing e.g. for variable depth and density, one obtains more general equations, for instance the variable coefficient KdV equation (VCKdV):

$$\Delta \equiv u_t + f(x,t)\, uu_x + g(x,t)\, u_{xxx} = 0. \tag{2.23}$$

Let us investigate the Lie point symmetries of this equation, assuming at first only that

$$f(x,t) \neq 0, \;\; g(x,t) \neq 0 \tag{2.24}$$

in some open region of the (x,t)-plane [17].

We write the vector field of eq. (2.3) in the form

$$\hat{v} = \xi(x,t,u)\, \partial_x + \tau(x,t,u)\, \partial_t + \phi(x,t,u)\, \partial_u. \tag{2.25}$$

Its third prolongation will have the form

$$\begin{aligned} \mathrm{pr}^{(3)}\hat{v} = {} & \xi\partial_x + \tau\partial_t + \phi\partial_u + \phi^x\partial_{u_x} + \phi^t\partial_{u_t} + \phi^{xx}\partial_{u_{xx}} + \phi^{xt}\partial_{u_{xt}} \\ & + \phi^{tt}\partial_{u_{tt}} + \phi^{xxx}\partial_{u_{xxx}} + \phi^{xxt}\partial_{u_{xxt}} + \phi^{xtt}\partial_{u_{xtt}} \\ & + \phi^{ttt}\partial_{u_{ttt}}. \end{aligned} \tag{2.26}$$

We apply $\mathrm{pr}^{(3)}\hat{v}$ to the VCKdV equation, as in eq. (2.21) and obtain

$$\begin{aligned} \mathrm{pr}^3\hat{v}\cdot\Delta\big|_{\Delta=0} = {} & \{g\phi^{xxx} + \phi^t + uf\phi^x + fu_x\phi \\ & + (\xi f_x + \tau f_t)\, uu_x + (\xi g_x + \tau g_t)\, u_{xxx}\}\,\big|_{\Delta=0} = 0. \end{aligned} \tag{2.27}$$

Thus, we need to calculate the coefficients ϕ^t, ϕ^x and ϕ^{xxx} of $\mathrm{pr}^{(3)}\hat{v}$, but do not need the other ones (an intelligent computer program takes this into account). The vanishing of $\mathrm{pr}^3\hat{v}\cdot\Delta$, on the solution surface, rather than vanishing identically, is best assured by setting

$$v_1 \equiv u_{xxx} = -\frac{1}{g}(u_t + fuu_x) \tag{2.28}$$

everywhere in eq. (2.27). Replacing u_t from the equation (2.23) is mathematically equivalent, but would lead to many more terms in intermediate expressions. In particular, terms like $u_{tx}, u_{txx}\ldots$ would have to be calculated from eq. (2.23) and replaced in (2.27). Again, an intelligent computer program will make the appropriately intelligent choice (or the choice will be imputed by the user).

Using eq. (2.16), ... , (2.18) we find

$$\begin{aligned}\phi^t &= \phi_t - \xi_t u_x + (\phi_u - \tau_t)u_t - \xi_u u_x u_t - \tau_u u_t^2 \\ \phi^x &= \phi_x + (\phi_u - \xi_x)u_x - \tau_x u_t - \xi_u u_x^2 - \tau_u u_x u_t\end{aligned} \tag{2.29}$$

$$\begin{aligned}\phi^{xxx} &= \phi_{xxx} + (3\phi_{xxu} - \xi_{xxx})u_x - \tau_{xxx}u_t \\ &+ 3(\phi_{xuu} - \xi_{xxu})u_x^2 - 3\tau_{xxu}u_x u_t \\ &+ (\phi_{uuu} - 3\xi_{xuu})u_x^3 - 3\tau_{xuu}u_x^2 u_t - \xi_{uuu}u_x^4 \\ &- \tau_{uuu}u_x^3 u_t + 3(\phi_{xu} - \xi_{xx})u_{xx} - 3\tau_{xx}u_{tx} \\ &+ 3(\phi_{uu} - 3\xi_{xu})u_x u_{xx} - 6\tau_{xu}u_x u_{xt} \\ &- 3\tau_{ux}u_t u_{xx} - 6\xi_{uu}u_x^2 u_{xx} \\ &- 3\tau_{uu}u_x u_t u_{xx} - 3\tau_{uu}u_x^2 u_{xt} - 3\xi_u u_{xx}^2 \\ &- 3\tau_u u_{tx}u_{xx} + (\phi_u - 3\xi_x)u_{xxx} - 3\tau_x u_{txx} \\ &- \tau_u u_t u_{xxx} - 4\xi_u u_x u_{xxx} - 3\tau_u u_x u_{xxt}.\end{aligned} \tag{2.30}$$

The next step is to substitute ϕ^t, ϕ^x and ϕ^{xxx} into eq. (2.27) and to replace u_{xxx} everywhere, using eq. (2.28).

The coefficients of $u_{xxt}u_x, u_{xxt}, (u_{xx})^2, u_x u_{xx}$, and u_{xx} must vanish separately and we obtain the simplest subset of determining equations

$$\tau_u = 0,\ \ \tau_x = 0,\ \ \xi_u = 0,\ \ \phi_{uu} = 0,\ \ \phi_{xu} - \xi_{xx} = 0. \tag{2.31}$$

Thus, τ and ξ do not depend on u. This means that the corresponding transformations are "fiber preserving". The new independent variables $\tilde{x}$ and $\tilde{t}$ will not depend on the original dependent variable u. Furthermore, ϕ is linear in u and hence the symmetry transformations will be linear. The result (2.31) should now be used to simplify the remaining equations. The simplification is very significant: all terms involving derivatives like τ_{xu}, ϕ_{uu} etc. drop out and the dependence on u is now explicit.

The term u^2 occurs only once and its coefficient yields

$$\phi_{ux} = 0. \tag{2.32}$$

The result obtained so far can be summed up as follows. For $f(x,t)\ g(x,t)$ satisfying (2.24), but otherwise arbitrary, the symmetry algebra will be realized by vector fields of the form (2.25) with

$$\xi = b(t)x + c(t),\ \tau = \tau(t),\ \phi = a(t)u + R(x,t), \tag{2.33}$$

where a, b, c, τ and R are as yet undetermined functions of the indicated arguments.

We now set the coefficients of u_x, u, uu_x, u_t and 1 in eq. (2.27) equal to zero and obtain the remaining determining equations:

$$-\dot{b}x - \dot{c} + Rf = 0 \tag{2.34}$$

$$\dot{a} + fR_x = 0 \tag{2.35}$$

$$(bx + c)\, f_x + \tau f_t + (\dot{\tau} - b + a)\, f = 0 \tag{2.36}$$

$$(bx + c)\, g_x + \tau g_t + (\dot{\tau} - 3b)\, g = 0 \tag{2.37}$$

$$gR_{xxx} + R_t = 0. \tag{2.38}$$

A further analysis depends crucially on the form of the functions $f(x,t)$ and $g(x,t)$ in the eq. (2.23). We postpone a study of this question to Section 8 and restrict to the case of the KdV itself, i.e.

$$f = f_0 = \text{const}, \quad g = g_0 = \text{const}. \tag{2.39}$$

We solve eq. (2.34),...,(2.38) and find that eq. (2.33) restricts to

$$\xi = c_1 + c_3 t + c_2 x,\ \tau = c + 3\, c_2 t,\ \phi = -2\, c_2 u + c_3. \tag{2.40}$$

We see that the solution of the determining equations depends on 4 arbitrary constants. The symmetry algebra of the KdV equation is hence four dimensional. As a basis we can choose the elements corresponding to c_0, c_1, c_2 and c_3 separately, i.e.

$$\begin{aligned} &P_0 = \partial_t \quad P_1 = \partial_x,\ D = x\partial_x + 3\, t\partial_t - 2\, u\partial_u, \\ &B = t\partial_x + \partial_u. \end{aligned} \tag{2.41}$$

The connected component of the symmetry group is obtained by integrating the vector fields as in eq. (2.11) and composing the results. We obtain

$$\begin{aligned}&\tilde{t} = e^{3\lambda}(t - t_0), \quad \tilde{x} = e^{\lambda}[x - x_0 + v(t - t_0)],\\ &\tilde{u}(\tilde{x}, \tilde{t}) = u(x,t) + v.\end{aligned} \tag{2.42}$$

The constants t_0, x_0, λ and v correspond to $P_0, P_1; D$ and B respectively. We see that P_0 corresponds to time translations, P_1 to space translations, D to dilations (with different scales for x, t and u) and B to Galilei boosts. The statement about solutions is: if $u(x,t)$ is a solution of the KdV equation, then so is

$$\tilde{u}(\tilde{x}, \tilde{t}) = u(x,t) + v, \quad t = e^{-3\lambda}\tilde{t} + t_0$$

$$x = e^{-\lambda}\tilde{x} + x_0 - ve^{-3\lambda}\tilde{t}, \tag{2.43}$$

where t_0, x_0, v and λ are arbitrary constants.

Had we used e.g. the MACSYMA program of Ref. 14, it would have directly compiled the determining equations (2.31), (2.32) and (2.34), ... , (2.38). Solving them would be left to the user.

This relatively simple example shows rather clearly why Lie group methods for solving differential equations were underused before computer algebra became available. Now one can use formulas like (2.30) without ever actually seeing them!

Finally, let us write out the commutation relation for the symmetry algebra:

$$\begin{aligned}&[D, P_0] = -3P_0, \quad [D, P_1] = -P_1, \quad [D, B] = 2B,\\ &[B, P_0] = -P_1, \quad\;\; [B, P_1] = 0, \qquad\;\; [P_0, P_1] = 0.\end{aligned} \tag{2.44}$$

We see that the Lie algebra is solvable and has a three-dimensional ideal $\{B, P_0, P_1\}$, isomorphic to the Heisenberg algebra.

2.4. Example of an Infinite Dimensional Symmetry Group: The Kadomtsev-Petviashvili Equation.

Let us now consider another example, namely the Kadomtsev-Petviashvili equation (KP), also known as the two-dimensional KdV equation [18]. We shall write it in the form

$$\left(u_t + \frac{3}{2}\, uu_x + \frac{1}{4}\, u_{xxx}\right)_x + \frac{3}{4}\,\sigma u_{yy} = 0, \quad \sigma = \pm 1. \tag{2.45}$$

For functions u independent of y the KP essentially reduces to the KdV equation. It describes water waves in a similar regime as the KdV equation, but allows the wave crests to have a shape, e.g. "horse-shoe" shaped solitons.

To find the symmetry algebra, this time we make full use of the MACSYMA program [13,14]. The vector field $\hat{v}$ is written as

$$\hat{v} = \xi\partial_x + \eta\partial_y + \tau\partial_t + \phi\partial_u, \tag{2.46}$$

where ξ, η, τ and ϕ depend on x, y, t, u. The original set of determining equations represents a set of 89 linear differential equations for ξ, η, τ and ϕ. The computer solves the 6 simplest equations (they are $\xi_u = \tau_u = \eta_u = 0$, $\tau_x = \tau_y = \eta_x = 0$) and implements the result in the remaining system. This reduces the number of equations to 17 and these are printed out. They are quite easy to solve and more elaborate programs do that [12]. In any case, the result is that the vector field $\hat{v}$ of eq. (2.46), representing a general element of the symmetry algebra, depends on three arbitrary functions of time and can be written as [19]

$$\hat{v} = T(f) + Y(g) + X(h). \tag{2.47}$$

Here $f = f(t), g = g(t)$ and $h = h(t)$ are arbitrary $C^\infty(I)$ functions of time t, where I is an open interval in $\mathbb{R}$ and we have

$$\begin{aligned} T(f) &= f\,\partial_t + \left[\frac{1}{3}x\dot{f} - \frac{2}{9}\sigma y^2\ddot{f}\right]\partial_x \\ &\quad + \frac{2}{3}y\dot{f}\,\partial_y - \left[\frac{4\sigma}{27}y^2\dddot{f} - \frac{2}{9}x\ddot{f} + \frac{2}{3}u\dot{f}\right]\partial_u, \\ Y(g) &= g\,\partial_y - \frac{2}{3}\sigma y\dot{g}\,\partial_x - \frac{4\sigma}{9}y\ddot{g}\,\partial_u, \\ X(h) &= h\,\partial_x + \frac{2}{3}\dot{h}\,\partial_u \end{aligned} \tag{2.48}$$

(the dots indicate time derivatives).

A subalgebra of physically obvious symmetries is obtained by restricting $f(t)$, $g(t)$ and $h(t)$ to be first order polynomials in t. We obtain

$$\begin{aligned} &P_0 = T(1) = \partial_t, \quad P_1 = X(1) = \partial_x, \quad P_2 = Y(1) = \partial_y, \\ &D = T(t) = t\,\partial_t + \frac{1}{3}x\,\partial_x + \frac{2}{3}y\,\partial_y - \frac{2}{3}u\,\partial_u, \\ &R = Y(t) = t\,\partial_y - \frac{2}{3}\sigma\,y\,\partial_x, \\ &B = X(t) = t\,\partial_x + \frac{2}{3}\partial_u, \end{aligned} \tag{2.49}$$

i.e. translations in t, x and y, dilations D, Galilei boosts B and the "pseudorotations" corresponding to R. This last symmetry is a "left over" from the rotational invariance and Galilei invariance (in the y-direction) of the Euler equations, from which eq. (2.45) was derived.

The commutation relations for the complete algebra (2.48) are

$$[T(f_1), T(f_2)] = T(f_1\dot{f}_2 - \dot{f}_1 f_2), \tag{2.50a}$$

$$[Y(g_1), Y(g_2)] = \frac{2}{3}\sigma X(\dot{g}_1 g_2 - g_1\dot{g}_2),$$

$$[Y(g), X(h)] = 0, \quad [X(h_1), X(h_2)] = 0, \tag{2.50b}$$

$$[T(f), Y(g)] = Y(f\dot{g} - \frac{2}{3}\dot{f}g), \ [T(f), X(h)] = X(f\dot{h} - \frac{1}{3}\dot{f}h). \tag{2.50c}$$

The symmetry algebra L thus has a decomposition

$$L = S \triangleright N \quad S = \{T(f)\} \quad N = \{Y(g), X(h)\}, \tag{2.51}$$

where N is an infinite-dimensional nilpotent ideal and S is an infinite-dimensional simple Lie algebra. The algebra S can be identified as a centerless Virasoro algebra, isomorphic to the algebra of the group of diffeomorphisms of a real line. The algebra N is a subalgebra of a centerless Kac-Moody algebra (the loop algebra $\hat{sl}(4, \mathbb{R})$).

Thus, the symmetry algebra L of the Kadomtsev-Petviashvili equation has the structure of a Kac-Moody-Virasoro algebra [20-22]. This property is shared by all known integrable equations involving 3 independent variables. In particular we mention the Davey-Stewartson equation [23], the 3-wave resonant interaction equations [24] and quite a few others [25].

3. SYMMETRY REDUCTION FOR PARTIAL DIFFERENTIAL EQUATIONS.

The most important application of the symmetry group G of Lie point transformations, leaving a system of PDEs invariant, is to perform symmetry reduction. In the case of PDEs that means a reduction of the number of independent variables in the equation. In particular, it may be possible and desirable to reduce to an ODE or even to an algebraic equation.

The basic idea of symmetry reduction is to take some subgroup $G_0 \subseteq G$ of the symmetry group G and look for solutions that are invariant under G_0 (rather than transformed into other solutions). Requiring invariance is equivalent to imposing additional first order linear equations on solutions. These can be solved and the result imputted into the original equations. This provides the reduced systems to be solved. The procedure involves an obvious loss of generality. We obtain particular solutions for which boundary conditions can be imposed on surfaces invariant under the chosen group G_0, rather than on arbitrary surfaces.

The procedure for performing symmetry reduction can be outlined as an algorithm, consisting of the following steps.

Step 1. Find the symmetry group G of local point transformations (1.1), leaving the considered system (2.1) invariant, and obtain the corresponding Lie algebra L of vector fields (2.3). The method for doing this was presented in Section 2 above. As mentioned above, this step has been computerized [12, ... ,16].

Step 2. Identify the symmetry algebra L as an abstract Lie algebra. In Step 1 one obtains a finite or infinite set of linearly independent vector fields. From these one can calculate the structure constants C^l_{ik} of the Lie algebra. The structure constants are basis dependent. The aim is to extract basis independent information from them. The idea is to transform to a "canonical basis", in which all the basis independent properties are obvious. In particular, if the Lie algebra is decomposable, it should be decomposed into the direct sum of indecomposable Lie algebras

$$L = L_1 \oplus L_2 \oplus \cdots \oplus L_n. \tag{3.1}$$

For each indecomposable component L_1 one should obtain its Levi decomposition [26,27]

$$L = S \triangleright R, \quad [S, S] = S, \; [S, R] \subseteq R, \; [R, R] \subseteq R,$$

where S is semisimple and R is the radical of L (maximal solvable ideal). For each solvable Lie algebra R it is useful to identify its nilradical (maximal nilpotent ideal) [26,27].

For low dimensional Lie algebras it is possible to go further and to completely identify their isomorphism class.

Algorithms exist for identifying a Lie algebra from its structure constants [28] and they have been at least partly computerized [29,30]. They are valid for finite-dimensional Lie algebras, and need to be generalized to infinite dimensional ones.

Step 3. Classify the subalgebras of L into conjugacy classes under the action of the Lie group G. This is a classification under inner automorphisms of L, if G is restricted to being the connected component $\exp L$ of G. In some cases it is convenient to enlarge the classification group G to include discrete transformations, leaving the studied equations invariant. This problem will be discussed below in Section 4.

The reason this classification is needed is that each subgroup G_0, corresponding to a different conjugacy class of subalgebras G_0 will give a different type of invariant solution.

Step 4. Consider a subalgebra $L_0 \subseteq L$, representing a class of subalgebras found in Step 3, and the corresponding subgroup $G_0 \subseteq G$. The group G_0 acts on the space $M \sim X \times U$ of independent and dependent variables. Find the invariants of this action, i.e. the functionally independent solutions

$$\tilde{I}_j(x,u), \quad j = 1, \ldots N \tag{3.2}$$

of the set of first order linear PDEs

$$X_i F(x,u) = 0, \quad i = 1, \ldots n_0, \tag{3.3}$$

where X_i are vector fields of the form (2.3) that form a basis of the Lie algebra L_0. The number of invariants N is equal to the codimension of the generic orbits of G_0 in M:

$$N = p + q - d, \tag{3.4}$$

where d is the dimension of these orbits.

The following cases can arise.

(i) Among the invariants $\tilde{I}_j(x,u)$ it is possible to choose q functions $I_j(x,u)$ that provide an invertible mapping to the dependent variables. The Jacobian determinant then satisfies

$$J \equiv \left(\frac{\partial(I_1, \ldots, I_q}{\partial(u_1, \ldots, u_q)} \right), \quad \det J \neq 0. \tag{3.5}$$

The remaining $k = N - q$ invariants can be chosen to depend only on the independent variables and we denote them

$$\xi_1(x), \ldots, \xi_k(x), \quad k < p. \tag{3.6}$$

Now let us restrict to the solution set of eq. (2.1). We consider u_j as functions of x and this can be imposed by setting

$$I_i = F_i(\xi_1, \ldots, \xi_k). \tag{3.7}$$

Using condition (3.5) we solve (3.7) for the dependent variables and obtain

$$u_i(x) = U_i(x, F_i(\xi)). \tag{3.8}$$

Upon substitution into eq. (2.1) we obtain a set of equations involving only the functions $F_i, i = 1 \ldots q$, the variables $\xi_a (a = 1, \ldots, k)$ and derivatives of F_i with respect to ξ_a. Since the original equation is G invariant and eq. (3.2) provides a complete set of G_0 invariants, the noninvariant quantities x in (3.8) must drop out.

Since we have $k < p$, we have reduced the number of independent variables. If the reduced equations can be solved for $F_i(\xi)$, substitution into (3.8) provides solutions of the original system.

(ii) Eq. (3.5) is satisfied, but the complementary variables ξ_i of eq. (3.6) also depend upon u. We proceed as above, however substitution of $F_i(\zeta)$ with $\xi = \xi(x, u)$ yields implicit solutions, rather than explicit ones, i.e. eq. (3.8) is a functional equation for u.

(iii) The condition (3.5) on the Jacobian is not satisfied. Let the rank of the Jacobian J be

$$1 \leq r(J) = q' < q. \tag{3.9}$$

We choose q' of the invariant $\tilde{I}_j(x, u)$ such that we can invert q' relations of the type (3.7) for q' dependent variables, say $u_1, \ldots u_{q'}$. We then have

$$u_i(x) = U_i(x, F_i(\xi)), \quad i = 1, \ldots, q'. \tag{3.10}$$

The remaining variables $u_{q'+1}, \ldots u_q$ depend on all the original ones $x_1, \ldots, x_p$. Substituting into system (2.1) we obtain a system of PDEs in which some of the dependent variables $F_i(\xi)$ depend on fewer variables than the remaining ones. This imposes consistency conditions on the reduced equations. Solutions of the reduced equations then provide "partially invariant solutions" of the original equations. This concept was introduced by Ovsiannikov [4]. Partially invariant solutions very often turn out to coincide with invariant ones, i.e. the compatibility conditions force solutions to be invariant. Genuinely partially invariant solutions do exist in special cases [31,32] and we shall return to this question below in Section 7.

Step 5. Solve the reduced equations. This step is of course less algorithmic than the previous ones. The reduced equation may be integrable, even if the original one was not. Thus, it may be transformable into a linear equation, or solvable by inverse spectral transform techniques [10,11]. If necessary, group theory can be applied once more to the reduced equation. If it is a PDE, we can further reduce the number of independent variables. For ODEs one can reduce the order of the equation. An approach that is often very fruitful is Painlevé analysis: An analysis of the singularity structure of the solutions of the reduced PDE, or ODE [33, … ,36].

Step 6. The last step is to do physics with the obtained solutions: Analyze their stability, their asymptotic behavior, calculate observable quantities, etc.

4. CLASSIFICATION OF SUBALGEBRAS OF FINITE DIMENSIONAL LIE ALGEBRAS.

4.1. General Comments.

Let us consider a Lie algebra L of dimension $N = \dim L < \infty$. The algebra L can be thought of as an algebra of real or complex matrices of some finite dimension $N_0 \times N_0$. Consider a group G of automorphisms of L:

$$GLG^{-1} = L. \tag{4.1}$$

For our purposes G can be either the group of inner automorphisms of L, $G = G_0 = \exp L$, or an extension of G_0 by some discrete group G_D that also leaves the studied equations invariant. In most cases G_D consists of various transformations like reflections in coordinate planes, time reversal, complex conjugation of one or more of the dependent variables, parity, etc.

The task is to obtain a representative list of subalgebras L_i of L such that any subalgebra of L is conjugate to precisely one in the list. Two subalgebras $L_i \subset L$ and $L_i' \subset L$ are conjugate under G if we have

$$GL_i'G^{-1} = L_i, \tag{4.2}$$

i.e., for any element $x' \in L_i'$ there exists an element $x \in L_i$ and an element $g \in G$ such that

$$gx'g^{-1} = x. \tag{4.3}$$

In order to be able to perform a subalgebra classification for an arbitrary Lie algebra L, it is sufficient to be able to handle the following situations:

1. The Lie algebra L is simple.
2. The Lie algebra is a direct sum of two subalgebras

$$L = L_1 \oplus L_2. \tag{4.4}$$

3. The Lie algebra L is a semidirect sum of two subalgebras

$$L = F \triangleright N, \quad [F,F] \subseteq F, \quad [N,N] \subseteq N, \quad [F,N] \subseteq N, \tag{4.5}$$

with $\quad F \neq \emptyset, N \neq \emptyset.$

4.2. Subalgebras of a Simple Lie Algebra.

If L is a simple Lie algebra, i.e. it does not contain any nontrivial ideals, then we start the classification by finding all *maximal* subalgebras $L_M \subset L$. A subalgebra $L_M \subset L$ is maximal if the relation s

$$L_M \subseteq \tilde{L} \subseteq L, \quad [\tilde{L}, \tilde{L}] \subseteq \tilde{L}, \tag{4.6}$$

imply

$$\tilde{L} = L_M, \quad \text{or} \quad \tilde{L} = L. \tag{4.7}$$

To find all maximal subalgebras of L we first consider a finite dimensional faithful irreducible representation $E(L)$ of L by real, or complex matrices, acting on a space V with $\dim V = N_0 < \infty$. (The dimension N_0 should be chosen as small as possible). Any subalgebra $L_0 \subset L$, in particular any maximal subalgebra, will be imbedded in the representation $E(L)$ either reducibly, or irreducibly.

If L_M is imbedded reducibly, then it leaves a subspace $V_0 \subset V$ invariant. To find the matrices representing L_M,and hence L_M itself, it suffices in this case to classify the subspaces $V_0 \subset V$ under the action of G, to choose a representative subspace for each class and then to find the set of matrices leaving the subspace invariant. Moreover, the subspaces V_0 are completely characterized by their dimension, if no G invariant metric on V exists (e.g. for $sl(n, \mathbb{R})$, or $sl(n, \mathbb{C})$). If there is an invariant metric, then V_0 is characterized by its signature (e.g. for $o(p,q)$), the number of positive length, negative length and isotropic vectors in any orthogonal basis of V_0).

If L_M is imbedded irreducibly, then it must be simple, semisimple or reductive (a semisimple algebra is a direct sum of simple ones, a reductive algebra is the direct sum of a semisimple and an abelian one). All semisimple subalgebras of the simple Lie algebras over $\mathbb{C}$ are known [37], for results over $\mathbb{R}$ see Cornwall [38]. The reductive subalgebras L_M are obtained from the semisimple ones by finding their centralizers in L:

$$\text{Cent } (L_M, L) = \{x \in L \mid [x, L_M] = 0\}. \tag{4.8}$$

Example: the Lie algebra of the conformal group of Minkowski space-time, [39,40]. It is isomorphic to $o(4,2)$, the Lie algebra of matrices
$X \in \mathbb{R}^{6\times 6}$ satisfying

$$XK + KX^T = 0, \qquad K = K^T \in \mathbb{R}^{6\times 6}, \tag{4.9}$$

signature $K = (4,2)$, (K is $SL(6,\mathbb{R})$ conjugate to $I_{4,2} = \text{diag}(1,1,1,1,-1,-1)$).

Invariant subspaces that lead to maximal subalgebras have dimensions 1,2 and 3 (higher dimensions need not be considered, since if V_0 is invariant, its orthogonal complement in V is also invariant). The relevant signatures are $(+)$ $(-)$ (0), $(++)$

$(--)$ (00), $(+++)$, $(++-)$. The signatures $(+0)$, $(+-)$, (-0), $(++0)$, $(+-0)$ and $(+00)$) do not lead to maximal subalgebras; no other signatures are possible.

The corresponding maximal subalgebras of $o(4,2)$, imbedded reducibly in the considered representation, are:

1.	$(+)$	$o(3,2)$
2.	$(-)$	$o(4,1)$
3.	(0)	$sim(3,1)$
4.	$(++)$	$o(2) \oplus o(2,2)$
5.	$(--)$	$o(4) \oplus o(2)$
6.	(00)	$opt(3,1)$
7.	$(+++)$	$o(3) \oplus o(1,2)$
8.	$(++-)$	$o(2,1) \oplus o(2,1)$

The simple pseudounitary Lie algebra $su(2,1)$ has a six-dimensional real representation in $\mathbb{R}^{6\times 6}$ leaving a metric with signature

$(+,+,+,+,-,-)$ invariant. Hence it is a subalgebra of $o(4,2)$. There exists an element of $o(4,2)$ that commutes with all elements of $su(2,1)$. Adding it, we obtain the irreducibly imbedded maximal subalgebra $u(2,1)$:

$$9.\quad u(2,1) \sim su(2,1) \oplus u(1).$$

The subalgebra u (2,1) is realized by the matrices

$$\begin{pmatrix} 0 & a & c & d & e & f \\ -a & 0 & -d & c & -f & e \\ -c & d & 0 & b & g & h \\ -d & -c & -b & 0 & -h & g \\ e & -f & g & -h & 0 & -a-b \\ f & e & h & g & a+b & 0 \end{pmatrix} + \begin{pmatrix} 0 & j & & & & \\ -j & 0 & & & & \\ & & 0 & j & & \\ & & -j & 0 & & \\ & & & & 0 & j \\ & & & & -j & 0 \end{pmatrix} \tag{4.10}$$

with $a, \ldots, j \in \mathbb{R}$.

The algebra $sim(3,1)$ is the similitude algebra, i.e. the Poincaré algebra extended by dilations [41,42]. The algebra $opt(3,1)$ is that of the "optical group". It is analyzed in Ref. [43]. For further details and the original work see Ref. [39, 40].

Once the maximal subalgebras are found, we continue in the same manner to find subalgebras of the maximal simple ones ($o(3,2)$ and $o(4,1)$ in the example). For those that are direct sums (No 4,5,7,8,9 in the example), we use the method of Subsection 4.3 below. For semidirect sums ($sim(3,1)$ and $opt(3,1)$ of the example) we use the method of Subsection 4.4 below.

4.3. Subalgebras of Direct Sums: The Goursat method.

Goursat proposed a method for classifying subgroups of direct products of discrete groups [44,45]. The method has been adapted to the classification of subalgebras of direct sums of Lie algebras [39].

Very briefly, the corresponding algorithm consists of several steps. Let us consider the Lie algebra L satisfying

$$L = A \oplus B, \quad \dim A = n_a, \dim B = n_b, 1 \leq n_a < \infty, 1 \leq n_b < \infty, \tag{4.11}$$

where A and B are not necessarily indecomposable (the method can be applied iteratively to sums of several indecomposable Lie algebras). We wish to classify all subalgebras of L into conjugacy classes under the action of the direct product group

$$G = G_A \otimes G_B, \qquad G_A = \exp A, \quad G_B = \exp B. \tag{4.12}$$

We distinguish two types of subalgebras of the direct sum (4.11)

a) *Nontwisted subalgebras.* These are conjugate to direct sums of subalgebras $A_0 \subseteq A$, $B_0 \subseteq B$ and can be represented by direct sums:

$$L_o = A_o \oplus B_o, \quad A_o \subseteq A, \quad B_0 \subseteq B. \tag{4.13}$$

b) *Twisted subalgebras.* Subalgebras that are not conjugate to algebras of the form (4.13). In any basis there will be elements with nonzero projections on to both A and B.

To provide a representative list of all G–conjugacy classes of subalgebras of L, we proceed in 4 steps.

Step 1. Find representatives of all G_A and G_B conjugacy classes of subalgebras of A, and B, respectively. Denote them

$$\begin{aligned} &A_{j,a} \subseteq A, \qquad j = 1, \ldots, n_a, \quad a = 1, 2, \ldots, \\ &B_{k,b} \subseteq B, \qquad k = 1, \ldots, n_b, \quad b = 1, 2, \ldots, \end{aligned} \tag{4.14}$$

where the first subscript is equal to the dimension of the subalgebra, the second one distinguishes nonconjugate algebras of the same dimension. We put

$$A_{0,1} = B_{0,1} = \{\emptyset\}, \qquad A_{n_a,1} = A, \quad B_{n_b,1} = B, \tag{4.15}$$

i.e. the trivial subalgebras are included.

For each representative subalgebra find its normalizer group in G_A, or G_B, respectively:

$$\begin{aligned} \text{Nor}(A_{j,a}, G_A) &= \{g \in G_A \mid g A_{j,a} g^{-1} \subseteq A_{j,a}\}, \\ \text{Nor}(B_{k,b}, G_B) &= \{g \in G_B \mid g B_{k,b} g^{-1} \subseteq B_{k,b}\}. \end{aligned} \tag{4.16}$$

Step 2. Form a representative list of all nontwisted subalgebras of L:

$$S_1 = A_{j,a} \oplus B_{k,b}, \quad j = 0, \dots, n_a,\ k = 0, \dots, n_b,\ a = 1, 2, \dots,\ b = 1, 2, \dots. \quad (4.17)$$

Step 3. Form a representative list S_2 of all twisted subalgebras of L. Two representatives subalgebras $A_{j,a} \subseteq A$ and $B_{k,b} \subseteq B$ can be twisted together (the "Goursat twist") if a homomorphism (that may be an isomorphism) exists from one to the other, say

$$\tau(A_{j,a}) = B_{k,b} \qquad j \geqslant k \geqslant 1. \quad (4.18)$$

If a homomorphism τ exists then choose a basis for $A_{j,a} = \{a_1, \dots, a_j\}$ and construct the most general mapping

$$\tau \ : a_i \to \tau(a_i) \in B_{k,b}. \quad (4.19)$$

A twisted subalgebra is obtained by taking

$$\tilde{L}_{j,a} = \{a_i + \tau(a_i)\}, \qquad i = 1, \dots, j. \quad (4.20)$$

The mapping (4.19) will in general contain free parameters that make their appearance in eq. (4.20). In order to classify the subalgebras $\tilde{L}_{j,a}$
we apply the normalizer group

$$\mathrm{Nor}(A_{j,a}, G_A) \otimes \mathrm{Nor}(B_{k,b}, G_B). \quad (4.21)$$

This will transform the subalgebras $\tilde{L}_{j,a}$ for different values of the subscript a amongst each other and change the values of the free
parameters. We annul as many of the parameters as possible and standardize as many as possible of the remaining ones. If all parameters can be annuled, the algebra is not twisted and is conjugate to one in list S_1.

Step 4. Form a final representative list of all subalgebras of L by merging the lists S_1 and S_2. The final list can always be ordered by dimension and isomorphism class. Moreover, the representative subalgebras can be so chosen, that the final list is "normalized". This means that the normalizer algebra $\mathrm{nor}(L_{j,a}, L)$ of each algebra in the list is also in the list, where

$$\mathrm{nor}(L_{j,a}, L) = \{x \in L \mid [x, L_{j,a}] \subseteq L_{j,a}\}. \quad (4.22)$$

<u>Example</u>. Subalgebras of $o(2,2) \sim o(2,1) \oplus o(2,1)$.

We have

$$A \sim o(2,1) \sim \{a_1, a_2, a_3\}, \quad B \sim o(2,1) \sim \{b_1, b_2, b_3\}$$

$$\begin{aligned} &[a_3, a_1] = a_1, && [b_3, b_1] = b_1, \\ &[a_3, a_2] = -a_2, && [b_3, b_2] = -b_2, \\ &[a_1, a_2] = 2a_3, && [b_1, b_2] = 2b_3. \end{aligned} \tag{4.23}$$

Step 1. The representative subalgebras of A and their normalizer groups in $G_A \sim O(2,1)$ are:

$A_{j,a}$	$\mathrm{Nor}\{A_{j,a}, G_A\}$
$A_{0,1} = \{0\}$	$\exp A_{3,1} \sim G_A$
$A_{1,1} = \{a_3\}$	$\exp A_{1,1} \sim O(1,1) \cup e^{\pi(a_1 - a_2)/2}$
$A_{1,2} = \{\frac{a_1 - a_2}{2}\}$	$\exp A_{1,2} \sim O(2)$
$A_{1,3} = \{a_1\}$	$\exp A_{2,1}$
$A_{2,1} = \{a_3, a_1\}$	$\exp A_{2,1}$
$A_{3,1} = \{a_1, a_2, a_3\}$	$\exp A_{3,1} \sim G_A$

(4.24)

and similarly for B.

Step 2. The list S_1 of nontwisted subalgebras consists of all pairs of the form (4.17) with $A_{j,a}$ from the list (4.24) and $B_{k,b}$ from an equivalent list for B.

Step 3. Construct the homomorphisms $A_{j,a} \to B_{k,b}$

$$\begin{aligned} &A_{1,1}: && \tau(a_3) = \lambda_1 b_1 + \lambda_2 b_2 + \lambda_3 b_3, \\ &A_{1,2}: && \tau(\tfrac{a_1 - a_2}{2}) = \lambda_1 b_1 + \lambda_2 b_2 + \lambda_3 b_3, \quad && \lambda_i \in \mathbb{R}, \\ &A_{1,3}: && \tau(a_1) = \lambda_1 b_1 + \lambda_2 b_2 + \lambda_3 b_3, \\ &A_{2,1}: && \tau(a_3) = b_3 + \lambda b_1, && \lambda \in \mathbb{R}, \\ & && \tau(a_1) = \nu b_1, && \nu \in \mathbb{R}, \end{aligned} \tag{4.25}$$

$$A_{3,1}: \begin{cases} \tau(a_1) = \mu b_1 \\ \tau(a_2) = \frac{1}{\mu} b_2, \ \mu \neq 0, \ \mu \in \mathbb{R}, \\ \tau(a_3) = b_3 \end{cases} \quad \text{and} \quad \begin{cases} \tau(a_1) = \nu b_2 \\ \tau(a_2) = -\frac{1}{\nu} b_2, \ \nu \neq 0, \ \nu \in \mathbb{R}. \\ \tau(a_3) = -b_3 \end{cases}$$

Using the normalizer group (4.21) we simplify the algebras formed as in (4.20) and obtain the list S_2 of twisted subalgebras of $o(2,2)$

$$\begin{aligned}
\tilde{L}_{1,1} &=\{a_3+\lambda b_3\}, && \lambda>0,\\
\tilde{L}_{1,2} &=\{a_3+b_1\},\\
\tilde{L}_{1,3} &=\{a_3+\lambda\frac{b_1-b_2}{2}\}, && \lambda>0,\\
\tilde{L}_{1,4} &=\{\frac{a_1-a_2}{2}+\lambda b_3\}, && \lambda>0,\\
\tilde{L}_{1,5} &=\{\frac{a_1-a_2}{2}+\lambda\frac{b_1-b_2}{2}\}, && \lambda\neq 0,\\
\tilde{L}_{1,6} &=\{\frac{a_1-a_2}{2}+\epsilon b_1\}, && \epsilon=\pm 1,\\
\tilde{L}_{1,7} &=\{a_1+b_3\},\\
\tilde{L}_{1,8} &=\{a_1+\epsilon b_1\}, && \epsilon=\pm 1,\\
\tilde{L}_{1,9} &=\{a_1+\lambda\frac{b_1-b_2}{2}\}, && \lambda\neq 0,\\
\tilde{L}_{2,1} &=\{a_3+b_3, a_1+\kappa b_1\}, && \kappa=0,\pm 1,\\
\tilde{L}_{3,1} &=\{a_3+b_3, a_1+b_1, a_2+b_2\},\\
\tilde{L}_{3,2} &=\{a_3+b_3, a_1-b_1, a_2-b_2\}.
\end{aligned} \tag{4.26}$$

Step 4. Merge the two lists.

4.4. Subalgebras of Semidirect Sums.

The method for classifying subalgebras of semidirect sums was elaborated and applied in Ref. [39,41,42,43 and 8]. Let L have the form of eq. (4.5), i.e. it contains an ideal N and a factor algebra F that is itself a Lie algebra. In particular, this may be a Levi decomposition; then F is semisimple and N is solvable.

We again proceed in a series of steps.

Step 1. Form a representative list $S(F)$ of subalgebras of the factor algebra F, classified under the action of the group $G_F = \exp F$:

$$S(F) = \{F_1 = \{0\}, F_2, \ldots, F_N \equiv F\}. \tag{4.27}$$

The list should be normalized, i.e. for each subalgebra F_i its normalizer algebra in F should also be in the list $S(F)$. For each subalgebra F_i construct its normalizer group in G_F.

Step 2. Classify all *"splitting"* subalgebras of L. A subalgebra of a semidirect sum is called splitting if it is itself a semidirect sum of subalgebras of the two components:

$$L_0 = F_0 \triangleright N_0, \qquad F_0 \subseteq F, \qquad N_0 \subseteq N. \tag{4.28}$$

Splitting subalgebras of semidirect sums are thus analogs of the untwisted subalgebras of direct sums.

The procedure for finding all splitting subalgebras is as follows.

1). For each subalgebra F_i in the list $S(F)$ find all invariant subspaces $\tilde{N}_{i,\alpha}$ that are also subalgebras:

$$[F_i, \tilde{N}_{i,\alpha}] \subseteq \tilde{N}_{i,\alpha}, \qquad \tilde{N}_{i,\alpha} \subseteq N, \qquad [\tilde{N}_{i,\alpha}, \tilde{N}_{i,\alpha}] \subseteq \tilde{N}_{i,\alpha}, \tag{4.29}$$

(if N is abelian then any subspace is a subalgebra). The trivial invariant subspaces $\tilde{N}_{i,0} = \{0\}$ and $\tilde{N}_{i,n} = N$ must be included for each F_i.

2). For each F_i classify all invariant subalgebras $\tilde{N}_{i,\alpha}$
into conjugacy classes under the action of the normalizer group $\mathrm{Nor}(F_i, G_F)$. Choose a representative $\tilde{N}_{i,\alpha}$ of each conjugacy class.

3). Form a representative list of all splitting subalgebras of L

$$S_1(L) = \{L_{i,\alpha} \subseteq L \mid L_{i,\alpha} = F_i \triangleright N_{i,\alpha}\}. \tag{4.30}$$

Note that the list $S(F)$ is contained in $\mathrm{S}_1(L)$.

4). For each subalgebra $L_{i,\alpha}$ in the list $\mathrm{S}_1(L)$ find its normalizer group $\mathrm{Nor}(L_{i,\alpha}, G)$ in the group G. The list $\mathrm{S}_1(L)$ should be normalized: for each $L_{i,\alpha}$ in the list, the normalizer algebra $\mathrm{nor}(L_{i,\alpha}, L)$ in the algebra L, should also be in the list.

We have obtained the list $\mathrm{S}_1(L)$ in such a manner that each splitting subalgebra $L_{i,\alpha} \subseteq L$ has a basis satisfying.

$$L_{i,\alpha} = \{B_a, X_j\}, \quad B_a \in F,\, X_j \in N,\, 1 \le a \le f_i = \mathrm{dim}F_i,\, 1 \le j \le r = \mathrm{dim}N_{i,\alpha}. \tag{4.31}$$

Step 3. Classify all *"nonsplitting"* subalgebras of L. By definition, a nonsplitting subalgebra is not conjugate to any splitting one. Nonsplitting subalgebras of a semidirect sum are generalizations of Goursat twisted subalgebras of direct sums.

The procedure for finding all nonsplitting subalgebras is as follows.

1). Run through the list $\mathrm{S}_1(L)$ of all splitting subalgebras of L. For each member $L_{i,\alpha}$ of the list take a basis as in eq. (4.31). Complement the basis $\{X_j\}$ of $N_{i,\alpha}$ to a basis of N

$$N = \{X_1, \ldots, X_r, Y_1, \ldots, Y_s\}, \qquad r + s = \mathrm{dim}N. \tag{4.32}$$

2). For each splitting subalgebra $L_{i,\alpha}$ form the vector space

$$V = \{B_a + \sum_{\mu=1}^{s} c_{a\mu} Y_\mu, X_j\}, \quad a = 1, \ldots, f_i, j = 1, \ldots, r, \tag{4.33}$$

where the constants $c_{a\mu}$ are such that V is a Lie algebra

$$[V, V] \subseteq V. \tag{4.34}$$

In order to obtain equations for the constants $c_{a\mu}$ following from eq. (4.34) we must now specify the commutation relations for L in the chosen basis:

$$[B_a, B_b] = f_{ab}^c B_c, \quad [B_a, X_k] = \alpha_{ak}^l X_l,$$

$$[B_a, Y_\mu] = \rho_{a\mu}^\nu Y_\nu + \sigma_{a\mu}^m X_m, \quad [X_i, X_j] = \omega_{ij}^m X_m, \tag{4.35}$$

$$[Y_\mu, Y_\nu] = \beta_{\mu\nu}^\sigma Y_\sigma + \gamma_{\mu\nu}^m X_m, \quad [X_i, Y_\mu] = \lambda_{i\mu}^\nu Y_\nu + \tau_{i\mu}^m X_m.$$

The condition (4.34) implies that the constants $c_{a\mu}$ in eq. (4.33) must satisfy a set of algebraic equations

$$c_{b\nu}\rho_{a\nu}^\alpha - c_{a\mu}\rho_{b\mu}^\alpha - c_{c\alpha} f_{ab}^c = -c_{a\mu} c_{b\nu} \beta_{\mu\nu}^\alpha, \tag{4.36}$$

$$c_{a\mu}\lambda_{j\mu}^\nu = 0, \tag{4.37}$$

for all a, b, α, j and ν.

In general, equations (4.36) are bilinear, rather than linear and solving them is a nontrivial task in algebraic geometry. It reduces to linear algebra if the ideal N is such that

$$\beta_{\mu\nu}^\alpha = 0, \quad \forall \mu, \nu, \alpha. \tag{4.38}$$

In the simplest case, when the ideal N is abelian, we have

$$\omega_{ij}^m = \beta_{\mu\nu}^\sigma = \gamma_{\mu\nu}^m = \lambda_{i\mu}^\nu = \tau_{i\mu}^\nu = 0, \tag{4.39}$$

and the equations for $c_{a\mu}$ reduce to a system of homogeneous linear equations

$$c_{b\nu}\rho_{a\nu}^\alpha - c_{a\mu}\rho_{b\mu}^\alpha - c_{c\alpha} f_{ab}^c = 0. \tag{4.40}$$

In mathematical terms eq. (4.40) means that the quantities $c_{a\mu}$ form 1-cocycles.

3). Once the constants $c_{a\mu}$ are obtained as general solutions of the system (4.36), (4.37), or of (4.40) if N is abelian, the vector space V of eq. (4.33) becomes a Lie algebra. The Lie algebras V must now be classified into conjugacy classes under the group

$$\tilde{G} = \mathrm{Nor}(L_{i,\alpha}, G) \rhd \mathrm{Nor}(N_{i,\alpha}, G_N), \tag{4.41}$$

where $G_N = \exp N$.

This classification is again simplified if N is abelian. Then we have $\mathrm{Nor}(N_{i,\alpha}, G_N) = G_N$.

In the abelian case we can generate trivial nonzero cocycles by conjugating the splitting subalgebras by elements of G_N:

$$\exp(\lambda_\mu Y_\mu) B_a \exp(-\lambda_\mu Y_\mu) = B_a + \lambda_\mu [Y_\mu, B_a], \quad \exp(\lambda_\mu Y_\mu) X_j \exp(-\lambda_\mu Y_\mu) = X_j. \tag{4.42}$$

We obtain an algebra conjugate to the splitting one, namely:

$$L_{i,\alpha} \sim \{B_a + \lambda_\mu \rho^\nu_{a\mu} Y_\nu, X_j\}. \tag{4.43}$$

The constants λ_μ can be freely chosen and the trivial cocycles

$$\delta_{a\mu} = \sum_{\alpha=1}^{s} \lambda_\alpha \rho^\mu_{a\alpha} \tag{4.44}$$

are called coboundaries. Any cocycle $c_{a\mu}$ can be replaced by $c_{a\mu} + \delta_{a\mu}$. We choose the constants λ_α to annul as many as possible of the cocycles. If all cocycles can be annuled, the subalgebra is splitting. The remaining cocycles $c_{a\mu}$ (mod $\delta_{a\mu}$) are to be further classified under the action of the normalizer

$\mathrm{Nor}(L_{i\alpha}, G)$.

Step 4. Form the final representative list $S(L) = S_1(L) \cup S_2(L)$ of all G–conjugacy classes of splitting and nonsplitting subalgebras. At this stage it is convenient to rename the subalgebras in a unified may, independently of whether they are splitting or not. Thus, say $S_{j,\alpha}$ will denote an algebra of dimension j. Within a given dimension we should order by isomorphism class [46,...,49]. The list $S(L)$ should be a normalized one.

We note that most of the work in the classification procedure concerns nonsplitting algebras. Which algebras are splitting depends on the decomposition of L and this is usually not unique. Whenever possible, N should be chosen to be abelian. The algorithm starts with the assumption that the subalgebras of F are known. If this is not the case, then one of the methods described above should be applied to F: this is a lower dimensional problem, so the procedure is iterative.

Example: The euclidean Lie algebra $e(3)$.

We have

$$e(3) \sim o(3) \triangleright T(3).$$

$$o(3) \sim \{L_1, L_2, L_3\}, \quad T(3) = \{P_1, P_2, P_3\}. \tag{4.45}$$

$$[L_i, L_k] = \varepsilon_{ikl} L_l, \quad [L_i, P_k] = \varepsilon_{ikl} L_l, \quad [P_i, P_k] = 0.$$

We have $F \sim o(3)$, $N \sim T(3)$ and N abelian (translations).

Step 1. The subalgebras of F and their normalizers in $O(3)$ are:

$F_1 = \{0\}$	$O(3)$
$F_2 = \{L_3\}$	$\exp L_3 \cup D_2$
$F_3 = \{L_1, L_2, L_3\}$	$O(3)$

(4.46)

Here D_2 is the dihedral group: generated by rotations through π about each of the 3 axes.

Step 2.

$$\begin{aligned} F_1 \quad &: \quad N_{1,1} = \{0\}, \\ &\quad\;\; N_{1,2} = \{P_3\}, \\ &\quad\;\; N_{1,3} = \{P_1, P_2\}, \\ &\quad\;\; N_{1,4} = \{P_1, P_2, P_3\}, \\ F_2 \quad &: \quad N_{2,1} = \{0\}, \\ &\quad\;\; N_{2,2} = \{P_3\}, \\ &\quad\;\; N_{2,3} = \{P_1, P_2\}, \\ &\quad\;\; N_{2,4} = \{P_1, P_2, P_3\}, \\ F_3 \quad &: \quad N_{3,1} = \{0\}, \\ &\quad\;\; N_{3,2} = \{P_1, P_2, P_3\}. \end{aligned} \tag{4.47}$$

This provides us with a list of all splitting subalgebras.

Step 3.

Nonsplitting subalgebras.

Let us start with $L_{2,1} = \{L_3\}$, and put

$$\tilde{L}_{2,1} = \{L_3 + c_1 P_1 + c_2 P_2 + c_3 P_3\}. \tag{4.48}$$

Find coboundaries:

$$[L_3, P_1] = P_2, \quad [L_3, P_2] = -P_1, \quad [L_3, P_3] = 0.$$

Hence wave can add $\lambda_1 P_1 - \lambda_2 P_2$ to eq. (4.48). We obtain the nonsplitting subalgebra (putting $\lambda_1 = -c_2$, $\lambda_2 = c_1$):

$$L_{2,1} = \{L_3 + cP_3,\, c \neq 0\}. \tag{4.49}$$

The only other nonsplitting subalgebra is

$$L_{2,1} = \{L_3 + cP_3,\, P_1,\, P_2,\; c \neq 0\}. \tag{4.50}$$

Step 4. The final list is

$$\begin{aligned} S_{0,1} &= \{0\}, & S_{3,1} &= \{L_1, L_2, L_3\}, \\ S_{1,1} &= \{L_3\}, & S_{3,2} &= \{L_3, P_1, P_2\}, \\ S_{1,2} &= \{P_3\}, & S_{3,3} &= \{L_3 + aP_3, P_1, P_2\}, \\ S_{1,3}(a) &= \{L_3 + aP_3\}, & S_{3,4} &= \{P_1, P_2, P_3\}, \\ S_{2,1} &= \{L_3, P_3\}, & S_{4,1} &= \{L_3, P_1, P_2, P_3\}, \\ S_{2,2} &= \{P_1, P_2\}, & S_{6,1} &= \{L_1, L_2, , L_3, P_1, P_2, P_3\}, \end{aligned} \tag{4.51}$$

with $a \neq 0$. For more sophisticated examples, namely the Poincaré group, the similitude group, the optical group and the symmetry group of the heat equation, see Ref. [41], [42], [43] and [8], respectively.

4.5. Finite Dimensional Subalgebras of Infinite Dimensional Lie Algebras.

The classification methods for direct sums and semidirect sums described above have been generalized to the case of infinite dimensional Lie algebras. We shall not present the results here, but refer to the original articles [8,19,23,24,25]. An example will however be given in section 6.3 below.

5. PAINLEVÉ ANALYSIS AND ITS APPLICATIONS.

Before going over to specific physical problems let us briefly review a method that provides exact solutions of certain classes of nonlinear ODEs that very often occur in applications. The equations that we have in mind are those having the *Painlevé property*. We shall say that an ODE has the Painlevé property if its general solution has no movable critical points anywhere in the complex plane. In this context a "critical point" means any singularity, other than a pole (e.g. a branch point, or an essential singularity). "Movable" means that the position of the singularity depends on the initial conditions (or, equivalently, on the integration constants).

Linear equations always have the Painlevé property: they have no movable singularities at all. Singularities, including poles can only occur for $t \to \infty$, or at points where the coefficients in the equation are themselves singular.

As an example, consider the ODE

$$\dot{y} + y^2 = 0. \tag{5.1}$$

Its general solution is

$$y = \frac{1}{x - x_0}. \tag{5.2}$$

Thus, eq. (5.1) has the Painlevé property: the general solution has a movable pole at $x = x_0$, where x_0 is an arbitrary complex number.

A further example is

$$\dot{y} + \frac{1}{3}y^4 = 0. \tag{5.3}$$

The general solution is

$$y = (x - x_0)^{-1/3} \tag{5.4}$$

and has a movable branch point at $x = x_0$. Thus, equation (5.3) does not have the Painlevé property. However the transformation

$$y = w^{1/3} \tag{5.5}$$

takes eq. (5.3) into an equation for w which does have the Painlevé property, namely eq. (5.1).

An equation, the general solution of which has a movable logarithmic branch point is

$$\dot{y} + \ddot{y} = 0, \qquad y = y_0 + \ln(x - x_0). \tag{5.6}$$

The equation

$$(y\ddot{y} - \dot{y}^2)^2 + 4y\dot{y}^3 = 0 \tag{5.7}$$

allows a movable essential singularity

$$y = y_0 e^{1/(t-t_0)}. \tag{5.8}$$

More generally, let $y = y(t)$ satisfy an n-th order ODE, say

$$y^{(n)} = F(t, y, y', y'', \dots, y^{(n-1)}). \tag{5.9}$$

If eq (5.9) has the Painlevé property, then the general solution can be expanded into a convergent Laurent series with a finite number of negative powers. The series, in order to be a general solution, must contain n arbitrary constants. This provides us with an algorithmic test [33], verifying whether an equation can have the Painlevé property (passing the test is necessary, not sufficient).

Indeed, assume that the general solution of eq. (5.9) has the form

$$y(t) = \sum_{k=0}^{\infty} a_k(t - t_0)^{k+\alpha}. \tag{5.10}$$

In order for eq. (5.9) to pass the test it is required that:

(i) α is a negative integer.

(ii) A recursion formula of the form

$$P(k)a_k = \phi_k(t_0, a_0, a_1, \dots, a_{k-1}) \tag{5.11}$$

exists for the expansion coefficients, where $P(k)$ is a polynomial that has $n - 1$ nonnegative integer zeros. These integer values $k = k_r$, for which we have $P(k_r) = 0$, are called resonances. The coefficient a_k for $k = k_r$ is not determined by the recursion relation and is hence a free parameter.

(iii) A compatibility condition, called the resonance condition, must be satisfied identically at each resonance k_r, namely

$$\phi_{k_r}(t_0, a_0, a_1, \dots, a_{k_r-1}) = 0. \tag{5.12}$$

Relation (5.12) must be satisfied identically in t_0 and in the values of a_k at all lower resonances.

If an equation passes the Painlevé test then at least a formal series solution exists, depending on n arbitrary constants. The test is not sufficient, for several reasons. First, the series may diverge everywhere, and thus not represent a local solution. Second, the fact that it depends on n constants does not guarantee that it is a general solution, able to satisfy arbitrary initial conditions.

The first two parts of the test are usually easy to perform. To find α it is sufficient to keep the leading term only

$$y = a_0(t - t_0)^{\alpha} + \dots \tag{5.13}$$

If several values of $\alpha \in \mathbb{Z}^{<0}$ occur, then the test must be performed for each branch. At least one branch must contain $n-1$ resonances. If α is a negative rational number

$$\alpha = -\frac{p}{q}; \qquad p, q \in \mathbb{Z}^{>0} \tag{5.14}$$

then eq. (5.9) does not have the Painlevé property. If the rest of the test is passed, then we conclude that the equation for

$$w = y^q \tag{5.15}$$

will pass the Painlevé test.

Obtaining the resonances is also not difficult. It suffices to write the solution as

$$y(t) = a_0(t - t_0)^\alpha + a_r(t - t_0)^{\alpha+r} + \dots, \tag{5.16}$$

substitute into the equation (α and a_0 are already known) and to collect the coefficient of a_r in the leading terms of the equation (those that gave us α in the first place). This gives us $P(k)$ and we must find its nonnegative integer roots.

The difficult, or rather time consuming part of the test is to verify the resonance condition (5.12).

The original test, proposed by Ablowitz, Ramani and Segur [33] has been implemented as a MACSYMA program [34]. For a further discussion of the Painlevé test see e.g. Kruskal [50], Conte [51], Weiss [52] (for ODEs and PDEs).

The importance of the Painlevé test is in the fact that it very often allows us to solve the considered ODE explicitly. First of all, the Painlevé property is invariant under Möbius transformations. Thus, if an equation for $y(t)$ has the Painlevé property, then so does the equation for $w(x)$ with

$$w(x) = \frac{\alpha(x)y(t) + \beta(x)}{\gamma(x)y(t) + \delta(x)}, \qquad \alpha\delta - \beta\gamma = 1,\ x = \phi(t),\ \phi' \neq 0. \tag{5.17}$$

Hence, equations having the Painlevé property can be classified into equivalence classes under the action of the transformations (5.17).

This has been done in the following cases.

1. *First order equations* [53]

$$\dot{y} = F(y, t), \tag{5.18}$$
$$\dot{y}^2 = F(y, t), \tag{5.19}$$

where F is rational in y and analytic in t. If eq. (5.18) has the Painlevé property, it is equivalent to the Riccati equation

$$\dot{u} = a(t) + b(t)u + c(t)u^2 \tag{5.20}$$

(or in particular a linear equation if $c(t) = 0$).

If eq. (5.19) has the Painlevé property it is either equivalent to the Riccati equation, or to the equation

$$\dot{u}^2 = a(u - u_1)(u - u_2)(u - u_3)(u - u_4) \tag{5.21}$$

where a, u_1, ... , u_4 are constants. If all u_i are different, eq. (5.21) is solved in terms of Jacobi, or Weierstrass elliptic functions. If two or more of the roots coincide we get solutions in terms of elementary functions (e.g. kinks and solitons).

2. *Second order equations*

$$\ddot{y} = F(\dot{y}, y, t), \tag{5.22}$$
$$\ddot{y}^2 = F(\dot{y}, y, t), \tag{5.23}$$
$$\ddot{y}^2 = G(\dot{y}, y, t)\ddot{y} + F(\dot{y}, y, t), \tag{5.24}$$

where F and G are rational in $\dot{y}$ and y and analytic in t.

The class of eq. (5.22) was studied by Painlevé [54] and Gambier [55]. They obtained 50 types of equations. They are solved in terms of one of the 6 Painlevé transcendents $P_I, \ldots, P_{VI}$ (depending on 0,1,4,2,4 and 4 parameters, respectively), or in terms of elliptic functions, elementary functions, the Riccati equation, or reduced to linear equations. In any case, the solutions are all known.

Eq. (5.23) were recently treated in an exhaustive manner by Cosgrove and Scoufis [56]. They have shown that whenever eq. (5.23) has the Painlevé property, it can be transformed by a contact transformation into an equation of the type (5.22). Equations (5.23) are thus solved in terms of the same functions as eq. (5.22) (and the derivatives of these functions).

The more general equations (5.24) were studied in a somewhat less complete manner by Bureau [57]. The conclusions were the same: if eq. (5.24) has the Painlevé property, then its solutions are expressed in terms of solutions of eq. (5.22) (and their derivatives).

3. *Third order equations.*

The study of third order equations

$$\dddot{y} = F(\ddot{y}, \dot{y}, y, t) \tag{5.25}$$

is very incomplete, though some work has been done [58].

Whenever higher order equations with the Painlevé property have occured, it was always possible to find first integrals lowering the order and reducing to known solutions. In particular for special cases of eq. (5.25) it is possible to find an integral of the form

$$K = A(y, \dot{y}, t)\ddot{y}^2 + B(y, \dot{y}, t)\ddot{y} + C(y, \dot{y}, t) \tag{5.26}$$

such that $dK/dt = 0$ is equivalent to eq. (5.25) and (5.26) is one of the known Painlevé type equations (5.22), ..., (5.24).

For further information on Painlevé analysis, see Ref. [11], [36] and [59].

6. GROUP INVARIANT SOLUTIONS OF THE STIMULATED RAMAN SCATTERING EQUATIONS.

6.1. General Comments.

We shall now apply the theory presented in the previous sections to a physical problem of considerable theoretical and practical importance, namely stimulated Raman scathering. This is a phenomenon usually studied in molecular gases irradiated by laser beams. The same type of pheomenon is used in optical fibers, where an optical pump is set up to enhance a signal propagating along a fiber. Stimulated Raman scattering has been studied theoretically by Wang and by Carman *et al* [60,61].

The equations they derived can be written as

$$\frac{\partial A_1}{\partial x} = -X\,A_2, \qquad \frac{\partial A_2}{\partial x} = X^*\,A_1, \qquad \frac{\partial X}{\partial t} + gX = A_1\,A_2^* \tag{6.1}$$

where A_1, A_2 and X are complex wave amplitudes and the stars denote complex conjugation. More specifically A_1, A_2 and X are the pump wave, a Stokes wave and the material excitation, respectively. The independent variable x measures distance along the Raman cell, t is usually called "retarded time", i.e. laboratory time with a position dependent origin. The real constant g satisfies

$$g \geq 0 \tag{6.2}$$

and represents dissipation.

The SRS equations (6.1) for $g = 0$ have a Lax pair, are integrable and have soliton and multisoliton solutions [62,63,64]. Experimentally, however, the solitons are not the most important solutions. On the contrary, the dominant behavior has all the characteristics of self–similar solutions. Here we shall present a group theoretical analysis of the SRS equations [65,66,67]. We first rewrite the equations in a more symmetric form, putting

$$u_1 = i\,A_1^*, \qquad u_2 = A_2\,, \qquad u_3 = X\,, \tag{6.3}$$

so that we have

$$u_{1,x} + i\,u_2^*\,u_3^* = 0, \quad u_{2,x} - i\,u_3^*\,u_1^* = 0, \quad u_{3,t} + g\,u_3 - i\,u_1^*\,u_2^* = 0. \tag{6.4}$$

We also introduce the wave moduli and phases

$$u_k = M_k e^{i\alpha_k}, \qquad k = 1,2,3,\ M_k \geq 0,\ 0 \leq \alpha_k < 2\pi. \tag{6.5}$$

Notice that the SRS equations have a first integral, namely

$$I_1 = |u_1|^2 + |u_2|^2 = I_1(t), \qquad \frac{\partial I_1}{\partial x} = 0, \tag{6.6}$$

corresponding physically to the conservation of the number of photons.

6.2. Symmetry Group of the SRS Equations.

We apply the algorithm of Section 2 to calculate the symmetry algebra. The vector field $\hat{v}$ of eq. (2.3) is particularized to

$$\hat{v} = \tau\,\partial_t + \xi\,\partial_x + \sum_{\mu=1}^{3} \left(\phi_\mu\,\partial_{M_\mu} + \psi_\mu\,\partial_{\alpha_\mu}\right), \tag{6.7}$$

where τ, ξ, ϕ_μ and ψ_μ are real functions of x, t, M_k and α_k.

We skip all details. The calculation was done using the MACSYMA package [14]. It provided 44 determining equations, solved 20 of them and printed out the remaining ones, which we solved by inspection. The solution depends crucially on whether dissipation is absent ($g = 0$) or present ($g \neq 0$).

For $g = 0$ (the integrable case) we obtain a Lie algebra involving two arbitrary functions of time t. A suitable basis is

$$P_1 = \partial_x, \quad V = -\partial_{\alpha_2} + \partial_{\alpha_3}, \quad T(f) = f(t)\partial_t - \frac{1}{2}\dot{f}(t)\left(M_1\partial_{M_1} + M_2\partial_{M_2}\right),$$

$$D = x\partial_x - \frac{1}{2}\left(M_1\partial_{M_1} + M_2\partial_{M_2} + 2M_3\partial_{M_3}\right), \quad U(h) = h(t)\left(-\partial_{\alpha_1} + \partial_{\alpha_2}\right). \tag{6.8}$$

The algebra L is a direct sum of 3 subalgebras

$$L = L_1 \oplus L_2 \oplus L_3, \tag{6.9a}$$

$$L_1 = \{D, P_1\}, \quad L_2 = \{V\}, \quad L_3 = \{T(f), U(h)\}. \tag{6.9b}$$

Since $f(t)$ and $h(t)$ are arbitrary functions (C^∞ in some interval), the algebra L_3, is infinite dimensional. The commutation relations are

$$\begin{aligned} [P_1, D] &= P_1, & [T(f_1), T(f_2)] &= T(f_1\dot{f}_2 - \dot{f}_1 f_2), \\ [T(f), U(h)] &= U(f\dot{h}), & [U(h_1), U(h_2)] &= 0. \end{aligned} \tag{6.10}$$

Thus $\{T(f), U(h)\}$ is a centerless Kac-Moody-Virasoro algebra [8, 20, ..., 24]. The elements P,D and V generate x–translations, dilations and changes of phase:

$$\tilde{x} = e^{\lambda}\,(x - x_0)\,, \qquad \tilde{t} = t, \tag{6.11}$$

$$\tilde{u}_1 = e^{\frac{-\lambda}{2}}\,u_1, \quad \tilde{u}_2 = e^{\frac{-\lambda}{2}}\,e^{-i\,\mu}\,u_2, \quad \tilde{u}_3 = e^{-\lambda}\,e^{i\,\mu}\,u_3.$$

Further, $U(h)$ generates time dependent changes of phase (a gauge transformation)

$$\tilde{x} = x, \quad \tilde{t} = t, \quad \tilde{u}_1 = u_1\,e^{-i\,H(t)}, \quad \tilde{u}_2 = u_2\,e^{i\,H(t)}, \quad \tilde{u}_3 = u_3, \tag{6.12}$$

(with $H(t) = h(t)$). Finally, T(f) generates an arbitrary reparametrization of time:

$$\tilde{t} = F(t), \qquad \tilde{x} = x, \qquad \tilde{v}_1 = \left[F'(t)\right]^{\frac{1}{2}} v_1, \tag{6.13a}$$

$$\tilde{v}_2 = \left[F'(t)\right]^{\frac{1}{2}} v_2\,, \qquad \tilde{v}_3 = v_3\,, \qquad F'(t) \neq 0, \tag{6.13b}$$

$$F(t) = \phi^{-1}\left[\lambda + \phi(t)\right], \quad \phi(t) = \int \frac{dt}{f(t)}.$$

For $g \neq 0$ the determing equations imply an additional condition, namely $\dot{f}(t) = 0$, so the symmetry group is much smaller. A basis for the Lie algebra in this case is

$$P_1 = \partial_x\,, \quad P_0 = \partial_t\,, \quad D = x\,\partial_x - \frac{1}{2}\left(M_1\,\partial_{M_1} + M_2\,\partial_{M_2} + 2\,M_3\,\partial_{M_3}\right),$$

$$V = -\partial_{\alpha_2} + \partial_{\alpha_3}\,, \qquad U(h) = h(t)\,\left(-\partial_{\alpha_1} + \partial_{\alpha_2}\right). \tag{6.14}$$

Thus, all that remains of the reparametrization of time (6.13), corresponding to $T(f)$, is a time translation

$$\tilde{t} = t - t_0\,, \qquad \tilde{x} = x\,, \qquad \tilde{u}_k = u_k. \tag{6.15}$$

Invariance under the x–translations, dilations, changes of phase (6.11) and gauge transformations (6.12) persists in the nonintegrable case with $g \neq 0$.

6.3. Subalgebras of the Symmetry Algebra.

The SRS equations (6.4) are a system of PDEs involving just two independent variables x and t. We wish to perform symmetry reduction and to reduce to a system of ODEs. For this it is sufficient to require that the solution be invariant under any one-dimensional subgroup of the symmetry group, acting in a nontrivial manner on space-time. In other words, we need a classification of the one-dimensional subalgebras of the symmetry algebra into conjugacy classes under the action of the symmetry group of the SRS system. Moreover, we only need algebras for which we have $\tau(t) \neq 0$, or $\xi(t) \neq 0$ in eq.(6.7). Let us consider the cases $g = 0$ and $g \neq 0$ separately.

A. $g = 0$.

A general element of the Lie algebra (6.8) has the form

$$X = \alpha\, D + \beta P_1 + \gamma V + \delta\, T(f) + \mu\, U(h) \tag{6.16}$$

We first consider each component in (6.9) separately, then apply the Goursat method (that trivializes for one-dimensional subalgebras).

L_1: One dimensional subalgebras are

$$A_{1\,,1} = \{D\}\,, \qquad A_{1\,,2} = \{P_1\}. \tag{6.17}$$

L_2: The only one-dimensional subalgebra is

$$B_{1\,,1} = \{V\}. \tag{6.18}$$

L_3: Consider a general element

$$X = T(f) + U(h). \tag{6.19}$$

Let us first assume $f(t) \neq 0$ and perform a transformation (6.12). The vector field X is transformed into

$$\begin{aligned}\tilde{X} =& f(\tilde{t})\, \partial_{\tilde{t}} - \frac{1}{2}\, \dot{f}(\tilde{t}) \left(\tilde{M}_1\, \partial_{\tilde{M}_1} + \tilde{M}_2\, \partial_{\tilde{M}_2}\right) + f(\tilde{t})\, \dot{H}(\tilde{t})\, (-\,\partial_{\alpha_1} + \partial_{\alpha_2}) \\ &+ h\,(\tilde{t})\, (-\,\partial_{\alpha_1} + \partial_{\alpha_2})\,.\end{aligned} \tag{6.20}$$

Choosing

$$\dot{H} = -\frac{h}{f}\,, \tag{6.21}$$

we find that X of eq.(6.19) for $f \neq 0$ is conjugate to $T(f)$, i.e. we can set $h = 0$ with no loss of generality.

Now let us assume $f \neq 0$, $h = 0$ in (6.19) and perform a transformation (6.13). The vector field T(f) is transformed into

$$\tilde{V}\,(f(\tilde{t})) = f\,\dot{F}\,\frac{\partial}{\partial \tilde{t}} + \frac{f\,\ddot{F} - \dot{f}\,\dot{F}}{2\,\dot{F}}\,\left(\tilde{M}_1\, \partial_{\tilde{M}_1} + \tilde{M}_2\, \partial_{\tilde{M}_2}\right). \tag{6.22}$$

Choosing $F(t)$ such that

$$f\dot{F} = 1, \tag{6.23}$$

we obtain a time translation

$$V(f(t)) \longrightarrow V(1) = \partial_t = P_0. \tag{6.24}$$

Equivalently, we can choose

$$f\dot{F} = \tilde{t}, \tag{6.25}$$

and obtain a dilation

$$V(f(t)) \longrightarrow V(t) = t\,\partial_t - \frac{1}{2}\left(M_1\,\partial_{M_1} + M_2\,\partial_{M_2}\right). \tag{6.26}$$

Now consider X of eq.(6.19) with $f = 0$, i.e. $X = U(h)$. Perform a transformation (6.13). For $\dot{h} = 0$ we have $U(1)$ and this is invariant under time reparametrizations. For $h \neq 0$ we obtain

$$U(h(t)) \longrightarrow U\left(h(\phi(\tilde{t}))\right), \qquad \phi(\tilde{t}) = F^{-1}(\tilde{t}). \tag{6.27}$$

Choosing $\phi(\tilde{t})$, i.e. $F(t)$ appropriately, we can transform $h(t)$ into any other function $\tilde{h}(\tilde{t})$ (for $h_t \neq 0$ and $\tilde{h}_{\tilde{t}} \neq 0$). In particular we can put $h(t) = t$. Thus L_3 has the following one-dimensional subalgebras

$$C_{1,1} = \{P_0\}, \qquad C_{1,2} = U(1), \qquad C_{1,3} = U(t). \tag{6.28}$$

Equivalently, we can replace $C_{1,1}$ by a dilation as in (6.26).

Combining the results (6.17), (6.18) and (6.28) together (i.e. using the Goursat method) we obtain the following list of one-dimensional subalgebras of the SRS symmetry algebra for $g = 0$:

$$\begin{aligned}
L_1(\epsilon,a) &= \{D + \epsilon\,T(t) + a\,V\}, & L_2(a,b) &= \{D + a\,U(1) + b\,V\},\\
L_3(\epsilon,a) &= \{D + \epsilon\,U(t) + a\,V\}, & L_4(\epsilon,a) &= \{P_1 + \epsilon\,T(1) + a\,V\},\\
L_5(\kappa,a) &= \{P_1 + \kappa U(1) + a\,V\}, & L_6(\epsilon,\kappa) &= \{P_1 + \epsilon\,U(t) + \kappa\,V\},\\
L_7(\kappa) &= \{P_0 + \kappa\,V\}, & L_8(\kappa) &= \{U(1) + a\,V\},\\
L_9(\kappa) &= \{U(t) + \kappa\,V\}, & L_{10} &= \{V\},
\end{aligned} \tag{6.29}$$

where we put $\epsilon = \pm 1$, $\kappa = 0, \pm 1$, $a, b \in \mathbb{R}$. Algebras L_8 , L_9 and L_{10} are pure gauge transformations, not acting on space-time. They are included for completeness, but will not produce reductions.

B. $g \neq 0$

In this case we have a straightforward application of the Goursat method. The function $h(t)$ in the gauge transformation $U(h)$ cannot be changed. The subalgebras in this case are

$$\begin{aligned} &L_1^g(a\,,b) = \{D + a\,P_0 + b\,V\}, \quad L_2^g(a\,,h) = \{D + a\,V + U(h)\}, \\ &L_3^g(\kappa\,,a) = \{P_1 + \kappa\,P_0 + a\,V\}, \quad L_4^g(\kappa\,,h) = \{P_1 + \kappa\,V + b\,U(h)\}, \end{aligned} \tag{6.30}$$

$$L_5^g(a) = \{P_0 + a\,V\}, \quad L_6^g(h) = \{V + U(h)\}, \quad L_7^g(h) = \{U(h)\}.$$

In (6.30) $h = h(t)$ is an arbitrary function, $\kappa = 0, \pm 1$, $a, b \in \mathbb{R}$. In $L_3^g(\kappa\,,a)$ for $\kappa = 0$ we have $a = 0, \pm 1$.

6.4 Invariant Solutions of the SRS Equations for $g = 0$.

Let us now use the subgroups corresponding to the subalgebras (6.29) to reduce the SRS equations to ODEs and obtain the corresponding solutions. We shall only be interested in genuinely three wave solutions, satisfying

$$u_1\,u_2\,u_3 \neq 0. \tag{6.31}$$

In all cases we follow the same procedure. The basis element of the considered algebra will have the form

$$X = (\alpha\,t + \beta)\,\partial_t + (\gamma\,x + \delta) + \sum_{k=1}^{3} \{\mu_k\,M_k\,\partial_{M_k} + (\nu_k\,t + \sigma_k)\,\partial_{\alpha_k}\}, \tag{6.32}$$

where $\alpha\,,\beta\,,\gamma\,,\delta\,,\mu_k\,,\nu_k$ and σ_k are constants to be specified in each case. The invariants of the corresponding subgroup will satisfy

$$X\,I\,(x\,,t\,,M_1\,,M_2\,,M_3\,,\alpha_1\,,\alpha_2\,,\alpha_3) = 0$$

and will be obtained by solving the characteristic system

$$\begin{aligned} \frac{dt}{\alpha\,t + \beta} = \frac{dx}{\gamma\,x + \delta} &= \frac{dM_1}{\mu_1\,M_1} = \frac{dM_2}{\mu_2\,M_2} = \frac{dM_3}{\mu_3\,M_3} \\ &= \frac{d\alpha_1}{\nu_1\,t + \sigma_1} = \frac{d\alpha_2}{\nu_2\,t + \sigma_2} = \frac{d\alpha_3}{\nu_3\,t + \sigma_3}. \end{aligned} \tag{6.33}$$

For the algebras $L_1, \ldots, L_7$ elementary invariants will have the form

$$I_0 = \xi(x,t), \qquad (6.34)$$
$$I_j = I_j(x,t,M_1,M_2,M_3,\alpha_1,\alpha_2,\alpha_3), \quad j=1,\ldots,6.$$

In each case we shall then put

$$I_j(x,t,M_i,\alpha_i) = \tilde{I}_j(\xi), \quad j=1,\ldots,6. \qquad (6.35)$$

The Jacobian determinant condition (3.5) is always satisfied, so we can solve eq.(6.35) for M_i and α_i. In all cases we got an expression for the dependent variables in the form

$$u_k(x,t) = M_k e^{i\alpha_k} = \omega_k(x,t) f_k(\xi), \quad k=1,2,3. \qquad (6.36)$$

The complex functions $\omega_k(x,t)$ and the real function $\xi(x,t)$ are explicitly known. We shall call eq. (6.36) the "reduction formulas".

The reduction formula is substituted into the SRS equations (6.4). The factors $\omega(x,t)$ drop out and we obtain ODEs for $f_k(\xi)$. We put

$$f_k(\xi) = \varrho_k(\xi) e^{i\phi_k(\xi)}, \qquad 0 \le \varrho_k < \infty, \; 0 \le \phi_k < 2\pi, \qquad (6.37)$$

separate the real and imaginary parts of the equations, decouple and solve. Let us run through the individual cases.

(i) *Subalgebra* $L_1(\epsilon, a)$ *and self-similar solutions.* The reduction formulas (6.36) particularize to

$$\begin{aligned} u_1 &= t^{-\frac{(1+\epsilon)}{2}} \varrho_1 e^{i\phi_1}, \\ u_2 &= t^{-\frac{(1+\epsilon)}{2}} e^{-i\epsilon a \ln t} \varrho_2 e^{i\phi_2}, \\ u_3 &= \frac{1}{x} e^{i a \ln x} \varrho_3 e^{i\phi_3}, \qquad \xi = x t^{-\epsilon}, \end{aligned} \qquad (6.38)$$

with $\varrho_i = \varrho_i(\xi)$, $\phi_i = \phi_i(\xi)$.

The reduced equations are

$$\begin{aligned} \dot{\varrho}_1 &= -\frac{1}{\xi} \varrho_2 \varrho_3 \sin\phi, & \varrho_1 \dot{\phi}_1 &= -\frac{1}{\xi} \varrho_2 \varrho_3 \cos\phi, \\ \dot{\varrho}_2 &= \frac{1}{\xi} \varrho_3 \varrho_1 \sin\phi, & \varrho_2 \dot{\phi}_2 &= \frac{1}{\xi} \varrho_3 \varrho_1 \cos\phi, \\ \dot{\varrho}_3 &= -\epsilon \varrho_1 \varrho_2 \sin\phi, & \varrho_3 \dot{\phi}_3 &= -\epsilon \varrho_1 \varrho_2 \cos\phi. \end{aligned} \qquad (6.39)$$

$$\phi = \phi_1 + \phi_2 + \phi_3 + a \ln \xi.$$

Eq. (6.39) have two first integrals

$$I_1 = \varrho_1^2 + \varrho_2^2, \qquad I_2 = \varrho_1\,\varrho_2\,\varrho_3 \cos\phi - \frac{a}{2}\,\varrho_1^2, \tag{6.40}$$

(where $\dot{I}_1 = \dot{I}_2 = 0$ follows directly from eq. (6.39)).

We can now express $\varrho_2\,,\varrho_3$ and ϕ in terms of ϱ_1:

$$\begin{aligned} \varrho_2 &= \left(I_1 - \varrho_1^2\right)^{\frac{1}{2}}, \\ \varrho_3 \sin\phi &= -\,\xi\,\dot{\varrho}_1\,\left(I_1 - \varrho_1^2\right)^{-\frac{1}{2}}, \\ \varrho_3 \cos\phi &= \frac{1}{\varrho_1}\,[I_2 + \frac{a}{2}\,\varrho_1^2]\,\left(I_1 - \varrho_1^2\right)^{-\frac{1}{2}}. \end{aligned} \tag{6.41}$$

Once ϱ_1 is known, we can obtain the phases $\phi_1\,,\phi_2$ and ϕ_3 by quadratures.

Multiplying the first of equations (6.39) by ξ, differentiating once and substituting from the other equations and from (6.41), we obtain an ODE for ϱ_1:

$$\xi\,\ddot{\varrho}_1 + \dot{\varrho}_1 = -\,\xi\,\frac{\varrho_1\,\dot{\varrho}_1^2}{I_1 - \varrho_1^2} - \frac{\left(I_2 + (a/2)\,\varrho_1^2\right)^2}{\xi\,\varrho_1\,(I_1 - \varrho_1^2)} + \left(I_1\,\epsilon - \frac{a^2}{4\,\xi}\right)\,\varrho_1 - \epsilon\,\varrho_1^3 + \frac{I_2^2}{\xi\,\varrho_1^3}. \tag{6.42}$$

Eq. (6.42) is in the class studied by Painlevé and Gambier [53,54,55] and we can apply the techniques of Section 5. It does not pass the Painlevé test, since we obtain $\alpha = -\frac{1}{2}$ in eq. (5.10). The rest of the test is passed, i.e. we obtain one resonance and the resonance condition (5.12) is satisfied.

This means that the equation for $H(\xi)$ where

$$\varrho_1 = \sqrt{H(\xi)} \tag{6.43}$$

will pass the Painlevé test. We then look for a Möbius transformation (5.17) taking the equation for $H(\xi)$ into one of the 50 standard equations [53]. The final result is that we put

$$\varrho_1(\xi) = (I_1)^{\frac{1}{2}}\left(\frac{W(\xi)}{W(\xi) - 1}\right)^{\frac{1}{2}}, \tag{6.44}$$

where $W(\xi)$ satisfies the P_V equation:

$$\ddot{W} = [\frac{1}{2\,W} + \frac{1}{W-1}]\,\left(\dot{W}\right)^2 - \frac{1}{\xi}\,\dot{W} + \frac{(W-1)^2}{\xi^2}\,[\alpha\,W + \frac{\beta}{W}] + \frac{\gamma}{\xi}\,W + \frac{\delta\,W\,(W+1)}{W-1}, \tag{6.45}$$

where the constants $\alpha\,,\beta\,,\gamma$ and δ are

$$\alpha = -\,\frac{1}{2}\left[\frac{2\,I_2}{I_1} + a\right]^2, \qquad \beta = \frac{2\,I_2^2}{I_1^2}, \qquad \gamma = 2\,\epsilon\,I_1\,, \qquad \delta = 0. \tag{6.46}$$

Notice that the variable ξ is real and so are all the constants in the equation. Thus we have solved the SRS equations in terms of the Painlevé transcendent $P_V(\xi;\alpha,\beta,\gamma,0)$. For $I_2=0\,,a=0$ the equation simplifies and the P_V transcendent can be transformed into a special case of P_{III}, also with real parameters.

A large body of knowledge exists on the Painlevé transcendents [11,36]. Particularly important in the present context are connection formulas, relating the behavior of the functions at various fixed singular points, e.g. $\xi=0$ and $\xi\to\infty$ (in various directions). For these we refer to the literature [11,36,68,69]. For a discussion of the asymptotics of $\varrho_1\,,\varrho_2$ and ϱ_3 for $\xi\to\infty$ and plots of the self–similar solutions, see the original articles [65,66].

The obtained self-similar solution in terms of P_V or P_{III} is particularly robust: large classes of initial data provide solutions that aymptotically approach the P_V or P_{III} ones [66,70].

(ii) *Subalgebra $L_4(\epsilon\,,a)$ and traveling wave solutions.*

The reduction formulas (6.36) reduce to

$$\begin{aligned} u_1 &= \varrho_1\,e^{i\,\phi_1}\,, & u_2 &= e^{-i\,\epsilon\,a\,t}\,\varrho_2\,e^{i\,\phi_2}\,, & u_3 &= e^{i\,a\,x}\,\varrho_3\,e^{i\,\phi_3}\,,\\ \xi &= x-\epsilon\,t\,, & \varrho_i &= \varrho_i(\xi)\,, & \phi_i &= \phi_i(\xi). \end{aligned} \tag{6.47}$$

The reduced equations are

$$\begin{aligned} \varrho_1' &= -\varrho_2\,\varrho_3\,\sin\phi\,, & \varrho_1\,\phi_1' &= -\varrho_2\,\varrho_3\,\cos\phi,\\ \varrho_2' &= \varrho_3\,\varrho_1\,\sin\phi\,, & \varrho_2\,\phi_2' &= \varrho_3\,\varrho_1\,\cos\phi,\\ \varrho_3' &= -\epsilon\varrho_1\,\varrho_2\,\sin\phi\,, & \varrho_3\,\phi_3' &= -\epsilon\,\varrho_1\,\varrho_2\,\cos\phi, \end{aligned} \tag{6.48}$$

$$\phi=\phi_1+\phi_2+\phi_3+a\,\xi.$$

This time 3 first integrals exist:

$$I_1=\varrho_1^2+\varrho_2^2\,,\qquad I_2=\varrho_1^2-\epsilon\,\varrho_3^2\,,\qquad I_3=\varrho_1\,\varrho_2\,\varrho_3\,\cos\phi-\frac{a}{2}\,\varrho_1^2. \tag{6.49}$$

Using eq.(6.49) we express $\varrho_2\,,\varrho_3$ and ϕ in terms of ϱ_1.

From (6.48) we get a first order equation for ϱ_1. The Painlevé test again suggests the transformation

$$\varrho_1(\xi)=\sqrt{Z(\xi)},$$

and for $Z(\xi)$ we obtain an elliptic function equation

$$\left(\dot{Z}\right)^2=-4\,\epsilon(Z-Z_1)(Z-Z_2)(Z-Z_3), \tag{6.50}$$

with

$$\begin{aligned} Z_1 + Z_2 + Z_3 &= I_1 + I_2 - \frac{\epsilon a^2}{4}, \\ Z_1 Z_2 + Z_2 Z_3 + Z_3 Z_1 &= I_1 I_2 + \epsilon a I_3, \\ Z_1 Z_2 Z_3 &= -\epsilon I_3^2. \end{aligned} \tag{6.51}$$

Real finite solutions for ϱ_1 are obtained in the following cases:

a) $\epsilon = 1, \quad Z_1 < 0 < Z_2 \leq Z \leq Z_3$.

$$Z = Z_3 - (Z_3 - Z_2)\,\mathrm{sn}^2(p\xi, k), \quad p = (Z_3 - Z_1)^{\frac{1}{2}}, \quad k^2 = \frac{Z_3 - Z_2}{Z_3 - Z_1}, \tag{6.52}$$

where $\mathrm{sn}\,(p\xi, k)$ is a Jacobi elliptic function [71]. These solutions are periodic.

b) $\epsilon = 1, \quad 0 = Z_1 = Z_2 \leq Z \leq Z_3 = I_1 - \frac{a^2}{4}, \quad I_2 = I_3 = 0,$

$$Z = \frac{Z_3}{\cosh^2 \sqrt{Z_3}\,(\xi - \xi_0.)} \tag{6.53}$$

This is a typical soliton solution, first obtained by Chu and Scott [62]. We have $Z \to 0$ for $\xi \to \pm\infty$, $Z = Z_3$ for $\xi = \xi_0$.

c) $\epsilon = -1, \quad 0 < Z_1 \leq Z \leq Z_2 < Z_3$.

$$Z = Z_1 + (Z_2 - Z_1)\,\mathrm{sn}^2(p\xi, k), \quad p = (Z_3 - Z_1)^{\frac{1}{2}}, \quad k^2 = \frac{Z_2 - Z_1}{Z_3 - Z_1}. \tag{6.54}$$

d) $\epsilon = -1, \quad 0 < Z_1 \leq Z \leq Z_2 = Z_3$.

$$Z = Z_2 - \frac{Z_2 - Z_1}{\cosh^2 \sqrt{Z_2 - Z_1}\,(\xi - \xi_0)}. \tag{6.55}$$

In optics, this is sometimes called a "black soliton",: a "well", or a "hole" on a constant background: $Z \to Z_2$ for $\epsilon \to \pm\infty$, $Z = Z_1 < Z_2$ for $\xi = \xi_0$.

(iii) *Subalgebra* $L_5(\epsilon, a)$ *and phase-wave solutions.*

For $\kappa = 0$ in $L_5(\kappa, a)$ we get trivial solutions. For $\kappa = \epsilon = \pm 1$ the reduction formulas are

$$u_1 = e^{-i\epsilon x}\,\varrho_1\, e^{i\phi_1}, \qquad u_2 = e^{i(\epsilon - a)x}\,\varrho_2\, e^{i\phi_2}, \qquad u_3 = e^{iax}\,\varrho_3\, e^{i\phi_3}. \tag{6.56}$$

The reduced equations are elementary and their solution is

$$
\begin{aligned}
u_1 &= -\,(1-\epsilon\, a)^{\frac{1}{2}}\,(H')^{\frac{1}{2}}\, e^{-i(\epsilon\, x+\epsilon\, H+\phi)},\\
u_2 &= (H')^{\frac{1}{2}}\, e^{i((\epsilon-a)x+\phi)},\\
u_3 &= -\,(1-\epsilon\, a)^{\frac{1}{2}}\, e^{i\,\epsilon\, H},
\end{aligned}
\tag{6.57}
$$

where $H(t)$ and $\phi(t)$ are arbitrary functions, a is a constant, $\epsilon = \pm 1$.

(iv) *Algebra* $L_7(\epsilon)$ *and plane wave solutions.*

The reduction formulas are

$$
u_1 = f_1(x)\,, \qquad u_2 = e^{-i\epsilon t} f_2(x)\,, \qquad u_3 = e^{i\epsilon t} f_3(x), \tag{6.58}
$$

and the reduced equations yield the solution

$$
\begin{aligned}
u_1 &= \alpha\, e^{i\left(-\epsilon\,\beta^2 x+\gamma\right)},\\
u_2 &= \beta\, e^{i\left(\epsilon\, \alpha^2 x-\epsilon\, t+\delta\right)},\\
u_3 &= -\,\epsilon\,\alpha\,\beta\, e^{i\left[\epsilon\left(\beta^2-\alpha^2\right)x+\epsilon\, t-\gamma-\delta\right]}.
\end{aligned}
\tag{6.59}
$$

The plane wave solutions are similar to solutions of linear equations, except that the velocities and the amplitudes of the waves are related.

The example of the SRS equations for $g = 0$ has brought out the importance of the subgroup classification. Each of the 4 considered subgroups has given very different solutions.

The fact that we could always solve in terms of known functions is related to the integrability of the original system (6.1). Indeed, it confirms the Painlevé conjecture [10,11,33] that all reductions of equations solvable by the inverse scattering transform should have the Painlevé property, possibly after some transformation.

6.5 Invariant Solutions of the SRS Equations for $g \neq 0$.

For $g \neq 0$ we restrict ourselves to one example, namely the subalgebra $L_1^g(a\,,b)$, corresponding to self-similar solutions. We shall, moreover choose $b = 0\;, a \neq 0$.

The reduction formulas are

$$
u_1 = e^{-\frac{t}{2a}}\,\varrho_1\, e^{i\,\phi_1}\,, \qquad u_2 = e^{-\frac{t}{2a}}\,\varrho_2\, e^{i\,\phi_2}\,, \qquad u_3 = \frac{1}{x}\,\xi^{a\,g}\,\varrho_3\, e^{i\,\phi_3}, \tag{6.60}
$$

$$
\varrho_i = \varrho_i(\xi)\,, \quad \phi_i = \phi_i(\xi)\,, \quad \xi = x e^{-\frac{t}{a}}\,, \qquad a \neq 0.
$$

The reduced equations are

$$\begin{aligned}
\dot{\varrho}_1 &= -\,\varrho_2\,\varrho_3 \xi^{a\,g-1} \sin\phi\,, & \varrho_1\,\dot{\phi}_1 &= -\,\varrho_2\,\varrho_3\,\xi^{a\,g-1} \cos\phi,\\
\dot{\varrho}_2 &= \varrho_3\,\varrho_1 \xi^{a\,g-1} \sin\phi\,, & \varrho_2\,\dot{\phi}_2 &= \varrho_3\,\varrho_1\,\xi^{a\,g-1} \cos\phi\,, \qquad (6.61)\\
\dot{\varrho}_3 &= -\,a\,\varrho_1\,\varrho_2 \xi^{-a\,g} \sin\phi\,, & \varrho_3\,\dot{\phi}_3 &= -\,a\,\varrho_1\,\varrho_2\,\xi^{-a\,g} \cos\phi\,,
\end{aligned}$$

$$\phi = \phi_1 + \phi_2 + \phi_3.$$

We again have two conserved quantities

$$I_1 = \varrho_1^2 + \varrho_2^2\,, \qquad I_2 = \varrho_1\,\varrho_2\,\varrho_3 \cos\phi\,, \tag{6.62}$$

so that we can put

$$\begin{aligned}
\varrho_2^2 &= I_1 - \varrho_1^2,\\
\varrho_3 \cos\phi &= \frac{I_2}{\varrho_1\sqrt{I_1 - \varrho_1^2}},\\
\varrho_3 \sin\phi &= -\,\frac{\dot{\varrho}_1\,\xi^{-a\,g+1}}{\sqrt{I_1 - \varrho_1^2}}.
\end{aligned} \tag{6.63}$$

For $\varrho_1(\xi)$ we obtain the ODE

$$\ddot{\varrho}_1 = -\,\frac{\varrho_1\,\dot{\varrho}_1^2}{I_1 - \varrho_1^2} + \frac{a\,g-1}{\xi}\,\dot{\varrho}_1 - \frac{I_2^2\,\xi^{2\,(a\,g-1)}}{\varrho_1\,(I_1 - \varrho_1^2)} + \frac{a}{\xi}\,\varrho_1\,(I_1 - \varrho_1^2) + \frac{I_2^2\,\xi^{2\,(a\,g-1)}}{\varrho_1^3}. \tag{6.64}$$

Inspired by the Painlevé test for the case $g = 0$ we put

$$\varrho_1 = \left(I_1\,\frac{W}{W-1}\right)^{\frac{1}{2}}, \tag{6.65}$$

and obtain

$$\begin{aligned}
\ddot{W} &= \left(\frac{1}{2\,W} + \frac{1}{W-1}\right)\dot{W}^2 + \frac{a\,g-1}{\xi}\,\dot{W} + \frac{2\,a}{\xi}\,I_1\,W\\
&\quad + \frac{2\,I_2^2}{I_1^2}\,\xi^{2\,(a\,g-1)}\,(W-1)^2\left(-\,W + \frac{1}{W}\right).
\end{aligned} \tag{6.66}$$

For $g = 0$ eq. (6.66) reduces to the Painlevé equation P_V, as we already know. For $g \neq 0$ the equation does not pass the Painlevé test: it will have complicated movable logarithmic branch points. The "guilty" ξ–dependence cannot be transformed away and we cannot get a solution in terms of any known special functions. We can, however, calculate the asymptotic behavior and we can obtain perturbative solutions, perturbing about the $g = 0$ solution [67].

7. PARTIALLY INVARIANT SOLUTIONS.

7.1 Formulation of the Problem for a Complex Nonlinear Klein–Gordon Equation.

In Section 6 we have seen how the method of symmetry reduction can lead to group invariant solutions. The method was outlined in Section 3 in a step by step manner. In step 4 we mentioned that condition (3.5) on a certain Jacobian determinant is necessary in order to obtain invariant solutions.

This section is devoted to the case when condition (3.5) is not satisfied. We shall look for examples of partially invariant solutions [4,31,32,72,73]. As a vehicle we shall use a complex nonlinear Klein-Gordon equation [31]

$$u_{tt} - u_{xx} = f(|u|)u, \tag{7.1a}$$

where $f(|u|)$ is a so far unspecified real function of the modulus $|u|$. Introducing the modulus $\rho = |u|$ and phase ω of u, we rewrite eq. (7.1) as a pair of real equations

$$\begin{aligned} &\rho_{tt} - \rho_{xx} - \rho(\omega_t^2 - \omega_x^2) = f(\rho)\rho \\ &\rho_t(\omega_{tt} - \omega_{xx}) + 2(\rho_t\omega_t - \rho_x\omega_x) = 0. \end{aligned} \tag{7.1b}$$

For any function $f(\rho)$ eq. (7.1) is invariant under a four dimensional group. Its Lie algebra is given by

$$P_0 = \partial_t, \quad P_1 = \partial_x, \quad K = t\partial_x + x\partial_t, \quad W = \partial_\omega. \tag{7.2}$$

The considered equation is also invariant under the discrete group G_0 generated by time reversal T, parity Π and complex conjugation C:

$$\begin{aligned} &T : x \to x, \quad t \to -t, \quad u \to u \\ &\Pi : x \to -x, \quad t \to t, \quad u \to u \\ &C : x \to x, \quad t \to t, \quad u \to u^* \end{aligned} \tag{7.3}$$

For some specific functions $f(\rho)$ the symmetry group may be larger; e.g. for $f = \rho^n$ it includes dilations.

Let us now classify the subalgebras of the symmetry algebra L into conjugacy classes under the symmetry group $G = G_0 \triangleright \exp L$.

We have

$$L = L_1 \oplus L_2, \quad L_1 = \{K, P_0, P_1\}, \quad L_2 = \{W\}$$
$$[P_0, K] = P_1, \quad [P_1, K] = P_0, \quad [P_0, P_1] = 0. \tag{7.4}$$

Applying the methods of Section 4, we obtained the following result. A representative list of all subalgebras L_0 of L consist of the following Lie algebras:

$$\dim L_0 = 1 \quad \{K\}, \{P_0\}, \{P_1\}, \{P_0 - P_1\}, \{W\}, \{K + aW\},$$
$$\{P_0 + aW\}, \{P_1 + aW\}, \{P_0 - P_1 + W\}, \tag{7.5}$$
$$\dim L_0 = 2 \quad \{P_0, P_1\}, \{K, P_0 - P_1\}, \{K + aW, P_0 - P_1\}, \tag{7.6a}$$
$$\{P_0, W\}, \{P_1, W\}, \{P_0 - P_1, W\}, \{K, W\}, \tag{7.6b}$$
$$\dim L_0 = 3 \quad \{K, P_0, P_1\}, \tag{7.7a}$$
$$\{K + aW, P_0, P_1\}, \{K, P_0 - P_1, W\}, \{P_0, P_1, W\}, \tag{7.7b}$$

with $a \in \mathbb{R}$. Thus, we have an equation (7.1); we know its symmetry group (the direct product of the Poincaré group in two dimensions with a one-dimensional group of global gauge transformations). We also know all the subalgebras of the symmetry algebra.

The next step is to find the invariants of the subgroups in the space $\{x, t, \rho, \omega\}$. For $\dim L_0 = 1$ all subgroups except the one corresponding to W will provide reductions to ODEs and ultimately group invariant solutions. The subalgebras (7.6a) and (7.7a) provide reductions to algebraic equations and invariant, but trivial, solutions.

Those to be investigated in the context of partially invariant solutions are subalgebras (7.6b) and (7.7b).

7.2. Reduction of the NLKGE by the Subalgebra $\{P_0, W\}$.

The invariants of $\{P_0, W\}$ are $I_1 = \xi = x$, $I_2 = \rho$. Thus, the phase ω does not figure among the invariants. The relevant Jacobian (3.5) is

$$J = \begin{pmatrix} \frac{\partial I_1}{\partial \rho} & \frac{\partial I_1}{\partial \omega} \\ \frac{\partial I_2}{\partial \rho} & \frac{\partial I_2}{\partial \omega} \end{pmatrix} = \begin{pmatrix} 0 & 0 \\ 1 & 0 \end{pmatrix}, \qquad \det J = 0. \tag{7.8}$$

We proceed as in eq. (3.9), (3.10); i.e. we express ρ in terms of the other invariant and obtain no information about ω. Hence, instead of the usual reduction formulas, we have

$$\rho = \rho(x), \qquad \omega = \omega(x, t). \tag{7.9}$$

Substituting into eq. (7.1b) we obtain

$$\omega_x^2 - \omega_t^2 = \frac{\rho_{xx}}{\rho} + f(\rho), \tag{7.10a}$$

$$\frac{\omega_{tt} - \omega_{xx}}{\omega_x} = 2\frac{\rho_x}{\rho}. \tag{7.10b}$$

The left hand sides of eq. (7.10) depend on x and t, the right hand sides only on x.

The subalgebra $P_0 + aW$ leads to invariant solutions of the form $\rho = \rho(x)$, $\omega = at + \phi(x)$. In order to have genuinely noninvariant solutions we impose the condition

$$\omega_{tt} \neq 0, \tag{7.11}$$

this also implies $\omega_x \neq 0$. We solve eq. (7.10a) for ω_t, substitute into (7.10b) and obtain

$$\begin{aligned} \omega_x^2 &= \left(\frac{\rho_{xx}}{\rho} + f(\rho)\right)\left[1 - S(t)\rho^4\right]^{-1}, \quad S(t) \geq 0, \\ \omega_t^2 &= S(t)\rho^4\left(\frac{\rho_{xx}}{\rho} + f(\rho)\right)\left[1 - S(t)\rho^4\right]^{-1}, \end{aligned} \tag{7.12}$$

where $S(t)$ is some nonnegative-definite function of t. The compatibility condition for eq. (7.12), $\omega_{tx} = \omega_{xt}$, implies

$$\frac{\dot{S}}{\sqrt{S}} = \frac{\epsilon}{\rho}\left[\frac{4\rho_x}{\rho} + (1 - S\rho^4)\left(\frac{\rho_{xx}}{\rho} + f\right)_x\left(\frac{\rho_{xx}}{\rho} + f\right)^{-1}\right], \quad \epsilon = \pm 1. \tag{7.13}$$

We differentiate (7.13) with respect to t and separate variables

$$\frac{1}{\dot{S}}\left(\frac{\dot{S}}{\sqrt{S}}\right)^{\cdot} = -\epsilon_0\rho^2\left(\frac{\rho_{xx}}{\rho} + f\right)_x\left(\frac{\rho_{xx}}{\rho} + f\right)^{-1} = 4\lambda, \tag{7.14}$$

where λ is a constant and $\dot{S} \neq 0$. From (7.14) and (7.13) we obtain equations for $S(t)$, $\rho(x)$ and $f(\rho)$, namely

$$\dot{S} = 4\sqrt{S}(\lambda S + \mu), \qquad (\lambda, \mu) \neq (0, 0), \tag{7.15}$$

$$\rho_x = \epsilon_0 \frac{\lambda}{\rho(1 + \mu\rho^2)}, \tag{7.16}$$

$$f_\rho + \frac{4\lambda}{\rho(\lambda + \mu\rho^4)} f + \frac{12\mu(\lambda + \mu\rho^4)}{\rho} = 0. \tag{7.17}$$

It follows from eq. (7.17) that partially invariant solutions of eq. (7.1) corresponding to the subalgebra $\{P_0, W\}$, are very rare: they only exist if the function $f(\rho)$ satisfies eq. (7.17), i.e., has the form

$$f(\rho) = c\lambda\rho^{-4} + \mu c - 3\mu\lambda - 3\mu^2\rho^4. \tag{7.18}$$

For $f(\rho)$ as in equation (7.18) we can solve the equations for $\omega(x,t)$ explicitly and obtain the partially invariant solutions. They are

$$
\begin{aligned}
&1)\quad \lambda = 0, \quad \mu \neq 0, \\
&\quad \rho^2 = -\frac{1}{2\epsilon_0 \mu x}, \qquad \omega = \left[-4\mu c(t^2 - x^2)\right]^{1/2}. && (7.19)\\
&2)\quad \lambda \neq 0, \quad \mu = 0, \\
&\quad \rho^2 = 2\epsilon_0 \lambda x, \qquad \omega = \left(\frac{c-\lambda}{16\lambda}\right)^{1/2} \ln \frac{t + \sqrt{t^2 - x^2}}{t - \sqrt{t^2 - x^2}}. && (7.20)\\
&3)\quad \lambda\mu > 0, \\
&\quad \rho^2 = \left(\frac{\lambda}{\mu}\right)^{1/2} \varepsilon_0 \tan(2\sqrt{\lambda\mu z}), \\
&\quad \omega = \left(\frac{c-\lambda}{16\lambda}\right)^{1/2} \ln \left[\frac{C(t) + [C(x-t)C(x+t]^{1/2}}{C(t) - [C(x-t)C(x+t]^{1/2}}\right], && (7.21)\\
&\quad C(z) = \cos(2\sqrt{\lambda\mu}\, z).
\end{aligned}
$$

The constants in (7.19), (7.20) and (7.21) must satisfy some obvious inequalities in order ot have $\rho^4 \geq 0$, $\omega \in \mathbb{R}$.

We see that the t dependence in ω is quite nontrivial and quite different than for invariant solutions.

Eq. (7.1) has no additional symmetries for the "potential" (7.18), unless we have $\lambda = c = 0$, or $\mu = 0$. Even in these particular cases, the additional symmetries do not provide the solutions (7.19),…,(7.21) as invariant solutions.

7.3. Reduction by the Subalgebra $\{P_1, W\}$.

The results are the same as for the subalgebra $\{P_0, W\}$ after the replacement

$$x \leftrightarrow t, \qquad f \leftrightarrow -f. \tag{7.22}$$

7.4. Reduction by the Subalgebra $\{P_0 - P_1, W\}$.

In this case we have

$$\rho = \rho(\xi), \quad \omega = \omega(\xi, \eta), \quad \xi = x + t, \quad \eta = x - t. \tag{7.23}$$

Eq. (7.1) reduces to

$$\omega_\xi \omega_\eta = \frac{1}{4} f(\rho), \qquad \frac{\omega_{\xi\eta}}{\omega_\eta} = -\frac{\rho_\xi}{\rho}. \tag{7.24}$$

We integrate the second equation with respect to ξ, require compatibility and obtain

$$\omega=\sqrt{\frac{2(x-t)}{-\lambda}}\,\frac{1}{\rho(\xi)},\qquad \int\frac{d\rho}{\rho^3 f(\rho)}=\frac{\lambda}{4}\xi,\qquad \lambda=\text{constant}. \tag{7.25}$$

Thus, partially invariant solutions are obtained for any function $f(\rho)$. Once $f(\rho)$ is chosen, $\rho(\xi)$ is obtained by a quadrature.

7.5. Reduction by the Subalgebra $\{K, W\}$.

Using the same reasoning as in Section 7.2 we find that invariant solutions exist only for

$$f=\frac{f_0}{\rho^4}e^{-\frac{4\rho^2}{\lambda}}, \tag{7.26}$$

and that they have the form

$$\rho=\left(2\epsilon\lambda\ln\frac{\xi}{\xi_0}\right)^{1/2},$$

$$\omega=\begin{cases}\left(\frac{f_0\xi_0^2+\lambda^2}{4\lambda^2}\right)^{1/2}\arcsin\frac{\eta-\eta_0}{\ln(\xi/\xi_0)}, & \lambda^2+f_0\xi_0^2>0,\\ \left(-\frac{f_0\xi_0^2+\lambda^2}{4\lambda^2}\right)^{1/2}\operatorname{arccosh}\frac{\eta-\eta_0}{\ln(\xi/\xi_0)}, & \lambda^2+f_0\xi_0^2<0,\end{cases} \tag{7.27}$$

$$\xi=\sqrt{t^2-x^2},\qquad \eta=\operatorname{arctanh}\frac{x}{t}, \tag{7.28}$$

(ξ is the Lorentz invariant variable). The three-dimensional subalgebras (7.7b) only lead to solutions equivalent to invariant ones.

7.6 General Comments.

Let us draw some conclusions from this analysis:

1. Partially invariant solutions that are not invariant under some subgroup of the symmetry group do exist, but they are rare.
2. When such solutions exist, they are just as good as invariant ones. They satisfy different boundary conditions than the invariant ones.

3. A careful subgroup classification is essential in a study of partially invariant solutions, as it is in the study of invariant ones. Without such a classification we have no way of knowing, whether a partially invariant solution is not actually invariant. Moreover, different subgroups lead to different equations, as we see by comparing the potentials $f(\rho)$ in eq. (7.18) and (7.26).

8. GROUP CLASSIFICATION OF EQUATIONS.

8.1. General Comments.

A further application of Lie group theory is to classify differential equations according to their symmetries, or to construct equations having a given symmetry.

For instance it is possible to a priori specify the number of dependent and independent variables in a differential equation as well as the order of the equation, and then to construct the most general equation invariant under a given Lie group G. This approach was recently taken to construct the most general second order scalar equations invariant under Poincaré, similitude and conformal groups in 2 dimensions [74]. Similarly Galilei, Galilei-similitude, and Schrödinger group invariant equations have been constructed [75].

Here we shall review a different approach, in which the general form of the equation is specified, but the coefficients involve arbitrary functions of some variables. The Lie point symmetries of the equation will then depend on the choice of the arbitrary functions [17,76].

It is convenient to proceed in two steps.

(i) Determine the "allowed transformations" leaving the form of the equation invariant, but changing the arbitrary functions. To do this we take a global approach [17] though an infinitesimal one is also possible [77,78].

(ii) Among the allowed transformations find the symmetry transformations that leave the equation invariant. The allowed transformations can be used to simplify the calculations by standarizing certain vector fields and thus simplifying the determining equations.

To illustrate the method, we shall use the example of the variable coefficient Korteweg–de Vries equation (2.22), the analysis of which was started in Section 2. We shall follow Ref. 17.

8.2. Allowed Transformations for the Variable Coefficient Korteweg-de Vries Equation.

Let us return to eq. (2.23) where f and g are nonzero, but otherwise nonspecified. We shall classify these equations into equivalence classes under local fiber preserving point transformation

$$u = U(\tilde{x}, \tilde{t}, \tilde{u}), \quad x = X(\tilde{x}, \tilde{t}), \quad t = T(\tilde{x}, \tilde{t}), \quad \frac{\partial U}{\partial \tilde{u}} \neq 0, \quad \det \frac{\partial(X, T)}{\partial(\tilde{x}, \tilde{t})} \neq 0. \quad (8.1)$$

"Fiber preserving" means that the new independent variables depend only on the old independent variables, not on the original dependent ones.

Allowed transformations are subject to the requirement that $\tilde{u}(\tilde{x}, \tilde{t})$ should satisfy

$$\tilde{u}_{\tilde{t}} + \tilde{f}(\tilde{x}, \tilde{t})\tilde{u}\tilde{u}_{\tilde{x}} + \tilde{g}(\tilde{x}, \tilde{t})\tilde{u}_{\tilde{x}\tilde{x}\tilde{x}} = 0. \quad (8.2)$$

Symmetry transformations in addition satisfy the condition

$$\tilde{f}(\tilde{x},\tilde{t}) = f(\tilde{x},\tilde{t}), \qquad \tilde{g}(\tilde{x},\tilde{t}) = g(\tilde{x},\tilde{t}). \tag{8.3}$$

Using (8.1), calculating the appropriate derivatives, substituting into the VCKdV equation (2.23) and imposing that eq. (8.2) be satisfied, we find that allowed transformations for any f and g must have the form:

$$\begin{aligned} u(x,t) &= A(t)\tilde{u}(\tilde{x},\tilde{t}) + B(x,t), \\ \tilde{x} &= \alpha(t)x + \beta(t), \\ \tilde{t} &= \theta(t), \end{aligned} \tag{8.4}$$

where the functions A, B, α, β and θ satisfy

$$\dot{\alpha}x + \dot{\beta} + fB\alpha = 0, \tag{8.5a}$$

$$\dot{A} + fB_xA = 0, \tag{8.5b}$$

$$B_t + fBB_x + gB_{xxx} = 0, \tag{8.5c}$$

$$A(t) \neq 0, \quad \alpha(t) \neq 0, \quad \dot{\theta}(t) \neq 0. \tag{8.5d}$$

The coefficients in the transformed equation (8.2) satisfy

$$\tilde{f}(\tilde{x},\tilde{t}) = \frac{A(t)\alpha(t)}{\dot{\theta}(t)} f(x,t), \qquad \tilde{g}(\tilde{x},\tilde{t}) = \frac{\alpha^3(t)}{\dot{\theta}(t)} g(x,t), \tag{8.6a}$$

with

$$t = t(\tilde{t}) = \theta^{-1}(\tilde{t}), \qquad x = \frac{\tilde{x} - \beta(t(\tilde{t}))}{\alpha(t(\tilde{t}))}. \tag{8.6b}$$

The result of an analysis of eq. (8.5) can be summed up as follows: Six classes of VCKdV equations exist. They are distinguished by the types of allowed transformations they permit.

1). $f(x,t)$ $g(x,t)$ generic, i.e. not satisfying any of the conditions occuring below. The allowed transformations (8.4) are only dilations of u, dilations of x, translations of x and reparametrizations of t:

$$A = A_0, \quad \alpha = \alpha_0, \quad \beta = \beta_0, \quad B = 0, \quad \theta = \theta(t). \tag{8.7}$$

2). $f_x = 0$, $g(x,t)$ arbitrary. We reparametrize time to set

$$f = 1. \tag{8.8}$$

The generic transformations preserving condition (8.8) are dilations and translations:

$$\tilde{x} = e^{k+d}(x - x_0), \quad \tilde{t} = e^{2k}(t - t_0), \quad u = e^{k-d}\tilde{u}. \tag{8.9}$$

Further, we have two nongeneric transformations, namely Galilei transformations

$$\tilde{x} = x - bt, \quad \tilde{t} = t, \quad u = \tilde{u} - b, \tag{8.10}$$

and projective transformations

$$\tilde{x} = \frac{x}{1 - \lambda t}, \qquad \tilde{t} = \frac{t}{1 - \lambda t}, \qquad u = \frac{\tilde{u} - \lambda x}{1 - \lambda t}. \tag{8.11}$$

3). $f(x,t) = x + q(t)$, $g(x,t)$ *arbitrary.* Generic transformations, taking $x + q(t)$ into $\tilde{x} + \tilde{q}(t)$ are

$$\tilde{x} = \alpha_0 x + x_0, \quad \tilde{t} = A_0 t + t_0, \quad u = A_0 \tilde{u}. \tag{8.12}$$

A nongeneric transformation, also preserving the form of f is

$$\tilde{x} = xe^{\lambda t} + \lambda \int q e^{\lambda t} dt. \tag{8.13}$$

4). $f(x,t) = x^{-1}$, $g(x,t)$ *arbitrary.*
Transformations taking x^{-1} into $\tilde{x}^{-1}$ can be generic

$$\tilde{x} = xe^{(\frac{1}{4}d_1 + d_2)}, \quad \tilde{t} = (t - t_0)e^{d_1}, \quad u = \tilde{u}e^{(\frac{1}{2}d_1 - 2d_2)}, \tag{8.14}$$

or nongeneric

$$\tilde{x} = \frac{x}{\sqrt{1 - \lambda t}}, \quad \tilde{t} = \frac{t}{\sqrt{1 - \lambda t}}, \quad u = \frac{\tilde{u} - \frac{1}{2}\lambda x^2}{1 - \lambda t}. \tag{8.15}$$

5). $f(x,t) = [x + q(t)]^{r(t)}$, $r(t) \neq 0, \pm 1$;
$g(x,t) = \frac{(x+q)^2}{(r^2-1)r}[h(t)(x + q) + (1 - r)\dot{q} - \dot{r}(x + q)\ln(x + q)]$.
Generic transformations in this case satisfy

$$\alpha = \alpha_0, \quad A = A_0, \quad \beta = \beta_0, \quad B = 0, \quad \theta = A_0 \int \alpha_0^{1-r} dt + \theta_0. \tag{8.16}$$

Nongeneric ones satisfy

$$\frac{\dot{A}}{A} = (1 - r)\frac{\dot{\alpha}}{\alpha}, \quad \dot{\beta} = q\dot{\alpha}, \quad B = -\frac{\dot{\alpha}}{A}(x + q)^{1-r}, \quad \dot{\theta} = A\alpha^{1-r}, \tag{8.17}$$

where α is a solution of the equation

$$\left(\frac{\dot{\alpha}}{\alpha}\right) + (r-1)\left(\frac{\dot{\alpha}}{\alpha}\right)^2 - h\frac{\dot{\alpha}}{\alpha} = 0, \quad \dot{\alpha} \neq 0. \tag{8.18}$$

6). $f(x,t) = p(t)e^{xq(t)}$, $g(x,t) = -\frac{\dot{q}}{q^3}x + h(t)$, $q \neq 0, p \neq 0$.
Allowed transformations, preserving the form of f and g satisfy

$$\begin{aligned} &\alpha = \alpha_0, \qquad A = A_0 \exp\left[-\frac{1}{\alpha_0}\int \dot{\beta} q dt\right], \qquad \theta = \theta(t), \\ &B = -\frac{\dot{\beta}}{\alpha_0 p}e^{-qx}, \qquad \dot{\beta} + \frac{q}{\alpha_0}\dot{\beta}^2 - \left(\frac{\dot{p}}{p} + hq^3\right)\dot{\beta} = 0. \end{aligned} \tag{8.19}$$

They are nongeneric if we have $\dot{\beta} \neq 0$.

8.3. Symmetries of the VCKdV Equation.

We now return to the determining equations (2.34),...,(2.38) derived in Section 2.3. They will have different types of solutions, depending on whether $f(x,t)$ and $g(x,t)$ are generic, or satisfy one of the conditions 2,...,6 of Section 8.2. For the analysis we refer to the original articles [17]; here we just present the results in the case when the VCKdV has a symmetry group of dimension $d \geq 3$.

The results are:

(i) The symmetry group of the VCKdV equation is of dimension d with $0 \leq d \leq 4$. The value $d = 4$ is achieved only if the VCKdV is equivalent under allowed transformations to the usual KdV of eq. (2.22). This happens only if we have

$$\begin{aligned} &f_x = g_x = 0, \\ &g(t) = f(t)\left[c_1\int_0^t f(s)ds + c_2\right], \qquad (c_1, c_2) \neq (0,0). \end{aligned} \tag{8.20}$$

This is also the only case whe the VCKdV equation passes the Painlevé test [79,80] and allows infinitely many conserved quantities [81].

(ii) The VCKdV equation has a three-dimensional symmetry algebra precisely if it is equivalent, under allowed transformations, to one of the following cases.

A. *The Lie algebra* $sl(2,\mathbb{R})$.

1. $f = 1$, $g = x$.

$$\begin{aligned} &X_1 = \partial_t, \qquad X_2 = t\partial_t + \tfrac{1}{2}x\partial_x - \tfrac{1}{2}u\partial_u, \\ &X_3 = t^2\partial_t + tx\partial_x + (x - tu)\partial_u. \end{aligned}$$

2. $f = 1/x,\ g = 1/x$.

$$X_1 = \partial_t, \qquad X_2 = t\partial_t + \tfrac{1}{4}x\partial_x - \tfrac{1}{4}u\partial_u,$$
$$X_3 = t^2\partial_t + \tfrac{1}{2}t\partial_x + (\tfrac{1}{2}x^2 - tu)\partial_u.$$

3. $f = \sqrt{x+t}, g = -\frac{4}{3}(x+t)^2$.

$$X_1 = \partial_t - \partial_x, \qquad X_2 = t\partial_t + x\partial_x - \tfrac{1}{2}u\partial_u,$$
$$X_3 = t^2\partial_t + (2tx + t^2)\partial_x + (2\sqrt{x+t} - tu)\partial_u.$$

B. *The decomposable solvable Lie algebra* $\{X_1, X_2\} \oplus \{X_3\}$ *with* $[X_1, X_2] = X_1$.
4. $f = x,\ g = x^3/t$.

$$X_1 = tx\partial_x + \partial_u, \qquad X_2 = -t\partial_t + u\partial_u, \qquad X_3 = x\partial_x.$$

5. $f = 1,\ g = t^2$.

$$X_1 = t\partial_x + \partial_u, \qquad X_2 = -t\partial_t + u\partial_u, \qquad X_3 = \partial_x.$$

C. *The nilpotent Lie algebra satisfying* $[X_1, X_2] = X_3$, $[X_1, X_3] = [X_2, X_3] = 0$.
6. $f = x,\ g = x^3$.

$$X_1 = \partial_t, \qquad X_2 = tx\partial_x + \partial_u, \qquad X_3 = x\partial_x.$$

D. *Indecomposable solvable nonnilpotent Lie algebras satisfying*

$$\begin{pmatrix} [X_1, X_3] \\ [X_2, X_3] \end{pmatrix} = A \begin{pmatrix} X_1 \\ X_2 \end{pmatrix}, \qquad [X_1, X_2] = 0, \qquad A = \begin{pmatrix} a & b \\ c & d \end{pmatrix}, \qquad a, b, c, d \in \mathbb{R}.$$

7. $f = 1,\ g = t^\alpha,\ \frac{1}{2} \le \alpha < \infty,\ \alpha \ne 2$.

$$A = \tfrac{1}{3}\begin{pmatrix} 2-\alpha & 0 \\ 0 & -1-\alpha \end{pmatrix}, \qquad X_1 = \partial_x, \qquad X_2 = t\partial_x + \partial_u,$$
$$X_3 = t\partial_t + \tfrac{1}{3}(2-\alpha)x\partial_x - \tfrac{1}{3}(1+\alpha)u\partial_u.$$

8. $f = 1,\ g = \sqrt{1+t^2}e^{3\alpha \arctan(t)}$.

$$A = \begin{pmatrix} \alpha & 1 \\ -1 & \alpha \end{pmatrix}, \qquad X_1 = \partial_x, \qquad X_2 = t\partial_x + \partial_u,$$
$$X_3 = (1+t^2)\partial_t + tx\partial_x + (x - tu)\partial_u + \alpha(x\partial_x + u\partial_u).$$

9. $f = 1$, $g = e^{3t}$.

$$A = \begin{pmatrix} 1 & 0 \\ -1 & 1 \end{pmatrix}, \qquad X_1 = \partial_x, \qquad X_2 = t\partial_x + \partial_u, \qquad X_3 = x\partial_x + u\partial_u + \partial_t.$$

9. CONDITIONAL SYMMETRIES

9.1. General Comments.

We have shown in Sections 3 and 6 how the method of symmetry reduction can be used to decrease the number of independent variables in a PDE.

Recently Clarkson and Kruskal [82] posed a complementary question: how can one obtain all reductions of PDEs to PDEs with fewer independent variables, or to ODEs? In other words, does Lie's classical method provide all reductions? They studied the Boussinesq equation

$$u_{tt} + uu_{xx} + (u_x)^2 + u_{xxxx} = 0 \tag{9.1}$$

and applied what is now called the "Clarkson-Kruskal direct method". Namely they set

$$u(x,t) = U\left(x, t, w(z)\right), \quad z = z(x,t)$$

and requested that the functions U and z be such that $w(z)$ should satisfy an ODE, once $u(x,t)$ is substituted into equation (9.1). Without using any group theory, the authors solved the problem completely. In addition to known reductions, due to dilational and translational invariance, they obtained quite a few new reductions with no apparent group theoretical origin.

A group theoretical explanation of these new reductions was soon provided [83]. It involved "conditional symmetries" related to the "non-classical method" of Bluman and Cole [2,84] and to the "side-conditions" of Olver and Rosenau [85,86].

"Conditional symmetries" of a differential system are transformations that leave only a subset of the solution set of the system invariant. Other solutions, not in the considered subset, are transformed out of the solution set.

To formulate this concept precisely, consider an n-th order scalar equation in two independent variables

$$\Delta^n(x, t, u, u_x, u_t, u_{xx}, u_{xt}, u_{tt}, \ldots) = 0. \tag{9.2}$$

The symmetry algebra of equation (9.2) will be realized by vector fields of the form

$$\widehat{v} = \xi(x,t,u)\partial_x + \tau(x,t,u)\partial_t + \phi(x,t,u)\partial_u, \tag{9.3}$$

satisfying (see Section 2)

$$pr^n \widehat{v} \Delta^n \big|_{\Delta^n=0} = 0. \tag{9.4}$$

Conditional symmetries can be obtained by adding a further equation (the condition) to be satisfied simultaneously with equation (9.2)

$$C^m(x, t, u, u_x, u_t, u_{xx}, u_{xt}, u_{tt}, \dots) = 0 \tag{9.5}$$

and requiring that the overdetermined system (9.2), (9.5) be invariant

$$\begin{aligned} pr^n \widehat{v} \Delta^n \big|_{\Delta^n=0, C^m=0} &= 0, \\ pr^m \widehat{v} C^m \big|_{\Delta^n=0, C^m=0} &= 0. \end{aligned} \tag{9.6}$$

The Lie algebra determined by equation (9.6) can be quite different than that determined by equation (9.4). Indeed, the corresponding transformations have to leave two equations invariant, but they can only be applied to a smaller set of solutions.

The first question is how to choose the condition (9.5) in order to constrain the vector field $\widehat{v}$ as little as possible. The answer is suggested by the Bluman and Cole nonclassical method [2,84]. Namely, consider a first order equation ($m = 1$ in equation (9.5)) that is adapted to the vector field that we are looking for

$$\Delta^1(x, t, u, u_x, u_t) = \xi(x, t, u)u_x + \tau(x, t, u)u_t - \phi(x, t, u) = 0, \tag{9.7}$$

where ξ, τ and ϕ are the same functions as in equation (9.3). We then have

$$pr^{(1)} \widehat{v} \Delta^1 = -(\xi_u u_x + \tau_u u_t - \phi_u)\Delta^1, \tag{9.8}$$

and hence

$$pr^{(1)} \widehat{v} \Delta^1 \big|_{\Delta^1=0} = 0 \tag{9.9}$$

is satisfied identically for any functions τ, ξ and ϕ. Thus, the condition (9.9) poses no restrictions on the vector field $\widehat{v}$ of eq. (9.3).

The determining equations for the conditional symmetries can now be obtained in exactly the same manner as those for ordinary symmetries. Namely, we put

$$pr^{(n)} \widehat{v} \Delta^n \big|_{\Delta^n=0, \Delta^1=0} = 0. \tag{9.10}$$

We also impose equation (9.9), but that is satisfied automatically.

If we have $\tau(x, t, u) \neq 0$, we can, in the search for conditional symmetries, set $\tau = 1$, i.e.:

$$\widehat{v} = \partial_t + \xi \partial_x + \phi \partial_u, \tag{9.11}$$

$$\Delta^1 = u_t + \xi u_x - \phi = 0. \tag{9.12}$$

Equation (9.12) and its differential consequences allow us to eliminate u_t and all higher derivatives like u_{tt}, u_{tx}, u_{txx}, etc.. The highest purely x derivative is eliminated using equation (9.2). The determining equations are then read off from equation (9.10) and solved.

In general there will be fewer determining equations than for ordinary symmetries, since there will be fewer linearly independent expressions in the derivatives.

The case $\tau = 0$, $\xi \neq 0$ must be considered separately. We can then put $\xi = 1$ and have

$$\widehat{v} = \partial_x + \phi \partial_u, \tag{9.13}$$

$$\Delta^1 = u_x - \phi = 0. \tag{9.14}$$

Thus, we can eliminate u_x using equation (9.14) and the highest purely t derivative using equation (9.2).

Before going over to the example of the Boussinesq equation, let us make several general comments on conditional symmetries.

1. The determining equations for conditional symmetries, as opposed to those for ordinary symmetries, will be nonlinear, since the condition (9.7) involves the functions ξ, τ and ϕ, to be obtained from the determining equations.
2. Each conditional symmetry generator $\widehat{v}$ comes with its own condition (9.7). Hence conditional symmetries do not form a Lie algebra, nor even a vector space. For instance, the sum of two conditional symmetry operators is, in general, not a symmetry operator at all. The same goes for the commutator of two conditional symmetry operators. Neither is the sum of an ordinary symmetry operator and a conditional one, a symmetry operator of any kind.
3. The vector fields representing conditional symmetries can be integrated to give group transformations. These transformations are not useful for generating solutions from solutions: they either take solutions out of the solution set, or leave them invariant individually.
4. The conditional symmetries are just as good as ordinary symmetries from the point of view of symmetry reduction for PDEs. Moreover, the symmetry reduction is performed in exactly the same manner.

9.2. Example of the Boussinesq Equation.

First of all, the ordinary symmetries of the Boussinesq equation are well known [87]. The symmetry algebra is three-dimensional and involves only dilations and translations

$$D = x\partial_x + 2t\partial_t - 2u\partial_u, \quad P_1 = \partial_x, \quad P_0 = \partial_t. \tag{9.15}$$

The Boussinesq equation (9.1) is also invariant under parity and time reversal.

Let us now derive the conditional symmetries [83]. We restrict ourselves here to the case $\tau \neq 0$ in equations (9.3) and (9.7). The case $\tau = 0$ was studied by Lou [88].

We apply the MACSYMA program [14] to the pair of equations (9.1) and (9.12), setting $v_1 = u_{xxxx}$ and $v_2 = u_t$. The program will eliminate u_{xxxx}, u_t, u_{tx}, u_{tt}, u_{txx}, etc. from expression (9.10). The determining equations are obtained from the coefficients of terms of the type

$$(u_x)^{n_1} (u_{xx})^{n_2} (u_{xxx})^{n_3}.$$

For ordinary symmetries (rather than conditional ones) we would, in principle, have more determining equations, namely the coefficients of

$$(u_x)^{n_1} (u_{xx})^{n_2} (u_{xxx})^{n_3} (u_t)^{n_4} (u_{tt})^{n_5} (u_{tx})^{n_6} (u_{txx})^{n_7} (u_{ttx})^{n_8} \\ \times (u_{ttt})^{n_9} (u_{txxx})^{n_{10}} (u_{ttxx})^{n_{11}} (u_{tttx})^{n_{12}} (u_{tttt})^{n_{13}}.$$

(n_i are nonnegative integers). The program finds the determining equations, solves some of them and finally prints out a reduced system. Altogether, we obtain 14 equations for ξ and ϕ (having set $\tau = 1$). From these equations we find that the conditional symmetry operator is

$$\begin{aligned} \widehat{v} &= \partial_t + [\alpha(t)x + \beta(t)]\partial_x - [2\alpha(t)u + 2\alpha(\dot{\alpha} + 2\alpha^2)x^2 + 2(\alpha\dot{\beta} + \dot{\alpha}\beta + 4\alpha^2\beta)x \\ &\quad + 2\beta(\dot{\beta} + 2\alpha\beta)]\partial_u, \end{aligned} \tag{9.16}$$

where $\alpha(t)$ and $\beta(t)$ are solutions of the ODEs

$$\ddot{\alpha} + 2\alpha\dot{\alpha} - 4\alpha^3 = 0, \tag{9.17a}$$

$$\ddot{\beta} + 2\alpha\dot{\beta} - 4\alpha^2\beta = 0. \tag{9.17b}$$

Equation (9.17a) will be solved below in terms of elliptic functions, or elementary ones. Equation (9.17b) is then a linear equation for $\beta(t)$.

The conditional symmetry operator $\widehat{v}$ can be used directly to reduce the Boussinesq equation (9.1) to an ODE. Indeed, the corresponding invariants satisfy

$$\widehat{v}\phi(x,t,u) = 0,$$

for which the characteristic equation is

$$dt = \frac{dx}{\alpha x + \beta} = -\frac{du}{2[\alpha u + \alpha(\dot{\alpha} + 2\alpha^2)x^2 + (\alpha\dot{\beta} + \dot{\alpha}\beta + 4\alpha^2\beta)x + \beta(\dot{\beta} + 2\alpha\beta)]}. \tag{9.18}$$

Solving (9.18) we obtain two invariants z and w, and the reduction formula

$$u(x,t) = w(z)K^2(t) - (\alpha x + \beta)^2, \tag{9.19a}$$

$$z(x,t) = xK(t) - \int_0^t \beta(s)K(s)ds, \tag{9.19b}$$

where

$$K(t) = \exp\left(-\int_0^t \alpha(s)ds\right). \tag{9.19c}$$

We substitute (9.19) into the Boussinesq equation and obtain the *ODE*

$$w'''' + ww'' + (w')^2 + (Az+B)w' + 2Aw = 2(Az+B)^2, \tag{9.20}$$

where the primes are derivatives with respect to z and we have

$$A = \frac{\alpha^2 - \dot{\alpha}}{K^4}, \quad B = \frac{\alpha\beta - \dot{\beta}}{K^3} + \frac{\alpha^2 - \dot{\alpha}}{K^4}\int_0^t \beta(s)K(s)ds. \tag{9.21}$$

Using equations (9.17) (it is not necessary at this stage to solve them), we verify

$$\frac{dA}{dt} = 0, \qquad \frac{dB}{dt} = 0, \tag{9.22}$$

i.e., A and B are constants.

Equation (9.20) was also obtained by Clarkson and Kruskal [82], so we see that the results of the two methods (the direct method and conditional symmetries) coincide.

To proceed further we must solve equations (9.17). The first equation has the Painlevé property and is actually P_X of Ince [53]. To integrate it we put

$$\alpha = \frac{\dot{H}}{2H}, \tag{9.23}$$

with $H(t)$ satisfying

$$\dot{H}^2 = h_0 H^3 + h_1, \qquad h_0,\ h_1 \quad \text{constants}. \tag{9.24}$$

For $\dot{H} \neq 0$ we obtain β from (9.17b) as

$$\beta = \beta_1 \frac{\dot{H}}{H} + \beta_2 \frac{\dot{H}}{H}\int_0^t \frac{H(s)}{\dot{H}^2(s)}ds. \tag{9.25}$$

Let us now solve equation (9.24)

(i) $h_0 = h_1 = 0$

$$\alpha = 0, \quad \beta = \beta_0 + \beta_1 t, \quad K = 1, \quad A = 0, \quad B = -\beta_1, \tag{9.26}$$

(β_0 and β_1 are constants).
For $\beta_1 = 0$ we have a classical symmetry

$$\widehat{v} = \partial_t + \beta_0 \partial_x \tag{9.27}$$

(translations).

For $\beta_1 \neq 0$ we can simplify using ordinary symmetries (translations and reflections) to obtain

$$\widehat{v} = \partial_t + t\partial_x - 2t\partial_u, \tag{9.28}$$

$$z = x - \frac{1}{2}t^2, \qquad u = w(z) - t^2, \tag{9.29}$$

$$w''' + ww' - w = 2z + c_1. \tag{9.30}$$

Equation (9.30) is solved in terms of the Painlevé transcendent P_{II}. Notice that the Boussinesq equation is not Galilei invariant, but the conditional symmetry operator (9.28) corresponds to a Galilei transformation.

(ii) $h_0 \neq 0, \quad h_1 = 0$

$$\alpha = -\frac{1}{t}, \qquad \beta = \beta_1 t^4 + \frac{\beta_2}{t}, \qquad K = t, \qquad A = 0, \qquad B = -5\beta_1. \tag{9.31}$$

We substitute into the vector field $\widehat{v}$ of (9.16), translate in x to set $\beta_2 \to 0$, dilate in order to obtain $\beta_1 = 1$, or $\beta_1 = 0$. Finally

$$\widehat{v} = \partial_t + \left(-\frac{x}{t} + \beta_1 t^4\right)\partial_x + \left(\frac{2}{t}u + \frac{6}{t^3}x^2 - 2\beta_1 t^2 x - 4\beta_1{}^2 t^2\right)\partial_u,$$

$$\beta_1 = 0 \quad \text{or} \quad \beta_1 = 1, \quad z = xt - \frac{1}{6}\beta_1 t^6, \quad u(x,t) = w(z)t^2 - \left(\frac{x}{t} - \beta_1 t^4\right)^2. \tag{9.32}$$

For $\beta_1 = 0$ the reduced equations can be integrated to

$$w'' + \frac{1}{2}w^2 = c_1 z + c_0. \tag{9.33}$$

For $c_1 = 0$ equation (9.33) is solved in terms of elliptic functions. For $c_1 \neq 0$, (9.33) is the equation for the first Painlevé transcendent P_I.
For $\beta_1 = 0$, equation (9.20) is integrated to

$$w''' + ww' - 5w = 50z + c_0 \tag{9.34}$$

integrable in terms of the transcendent P_{II}.

(iii) $h_0 = 0, \quad h_1 \neq 0$

$$\alpha = \frac{1}{2t}, \quad \beta = \beta_1 t + \frac{\beta_2}{t}, \quad K = \frac{1}{\sqrt{t}}, \quad A = \frac{3}{4}, \quad B = 0. \tag{9.35}$$

For $\beta_1 = 0$ we obtain an ordinary symmetry, namely dilations. We can set $\beta_2 = 0$ by an x translation and dilate β_1 into $\beta_1 = 1$ (assuming $\beta_1 \neq 0$). We obtain

$$\hat{v} = \partial_t + \left(\frac{x}{2t} + t\right)\partial_x - \frac{1}{t}(u + 2x + 4t^2)\partial_u, \tag{9.36}$$

$$z = \frac{x}{\sqrt{t}} - \frac{2}{3}t^{\frac{3}{2}}, \qquad u = \frac{1}{t}w(z) - \left(\frac{x}{2t} + t\right)^2, \tag{9.37}$$

$$w'''' + ww'' + (w')^2 + \frac{3}{4}zw' + \frac{3}{2}w = \frac{9}{8}z^2. \tag{9.38}$$

Equation (9.38) can be solved in terms of P_{IV}.

(iv) $h_0 \neq 0, \quad h_1 \neq 0$.

We obtain α in terms of a Weierstrass elliptic function $\mathcal{P}$ as

$$\alpha = \frac{1}{2}\frac{\dot{\mathcal{P}}}{\mathcal{P}}, \quad \beta = \beta_1 \frac{\dot{\mathcal{P}}}{2\mathcal{P}} + \beta_2 \frac{\dot{\mathcal{P}}}{2\mathcal{P}} \int_0^t \frac{\mathcal{P}(s)ds}{\left(\dot{\mathcal{P}}(s)\right)^2}, \tag{9.39}$$
$$\dot{\mathcal{P}}^2 = 4\mathcal{P}^3 - g_3, \qquad \mathcal{P} = \mathcal{P}(t, 0, g_3).$$

We can x-translate β_1 to $\beta_1 = 0$. We then have

$$K = (\mathcal{P}(t))^{-\frac{1}{2}}, \quad A = -\frac{3g_3}{4}, \quad B = 0,$$

$$\hat{v} = \partial_t + \frac{1}{2}\left(\frac{\dot{\mathcal{P}}}{\mathcal{P}}x + \beta_2 \frac{\dot{\mathcal{P}}}{\mathcal{P}} W\right)\partial_x$$
$$- \left[\frac{\dot{\mathcal{P}}}{\mathcal{P}}u + 3\dot{\mathcal{P}}x^2 + \frac{\beta_2}{2}\left(\frac{1}{\mathcal{P}} + 12\dot{\mathcal{P}}W\right)x + \frac{1}{2}\beta_2{}^2 W\left(\frac{1}{\mathcal{P}} + 6\dot{\mathcal{P}}W\right)\right]\partial_u,$$

$$z = x(\mathcal{P}(t))^{-\frac{1}{2}} + \frac{1}{3}\beta_2 g_3^{-1}(\mathcal{P}(t))^{-\frac{1}{2}}\int_0^t \mathcal{P}(s)ds,$$

$$u(x,t) = w(z)\mathcal{P}^{-1} - \left(\frac{1}{2}\frac{\dot{\mathcal{P}}}{\mathcal{P}}x + \beta_2 \frac{\dot{\mathcal{P}}}{2\mathcal{P}} W\right)^2, \tag{9.40}$$

where we have put

$$W(t) = \int_0^t \frac{\mathcal{P}(s)ds}{\left(\dot{\mathcal{P}}(s)\right)^2}.$$

The reduced equation is

$$w'''' + ww'' + (w')^2 - \frac{3}{4}g_3 w' - \frac{3}{2}g_3 w = \frac{9}{8}g_3^2 z^2. \tag{9.41}$$

Equation (9.41) is again solved in terms of the Painlevé transcendent P_{IV} [82].

We see that in the case of the Boussinesq equation the direct method and conditional symmetries give the same result. We also see that ordinary symmetries give less than half of the reductions presented here.

9.3. Further Comments on Conditional Symmetries.

Since [82] and [83] were published, the direct method and the method of conditional symmetries have been applied to numerous physically interesting equations [89,...,94]. In some cases reductions were obtained that cannot be obtained by ordinary symmetries. In other cases, conditional symmetries give nothing new with respect to ordinary ones.

Recently it was shown, in a rather general setting [94,95], that the method of conditional symmetries will always yield all the results of the direct method, as it is currently formulated. Moreover, if the ratio ξ/τ of the coefficients of the vector field (9.3), realizing the conditional symmetry, depends on u, then conditional symmetries give more general results than the direct method [92,94]. The additional solutions are usually implicit ones, since the "symmetry variable" z will, in this case, depend on both the independent and dependent variables: $z = z(x, t, u)$.

10. CONCLUSIONS.

This review concentrated on one aspect of the application of Lie group theory to differential equations: the use of Lie point symmetries to solve partial differential equations, specially nonlinear ones. We have reviewed several group theoretical tools available for obtaining solutions. These include: symmetry reduction leading to group invariant solutions, the calculation of partially invariant solutions and the use of conditional symmetries. The example of stimulated Raman scattering clearly brought out two very general aspects of the group analysis of differential equations. First of all, a classification of subgroups of the symmetry group is an essential part of the algorithm for calculating solutions: different subgroups provide very different solutions. Secondly, the exact analytic solutions obtained by group theory are important not simply because they exist and are explicit. They very often are physically particularly relevant. Indeed, they are stable and asymptotically, they attract solutions, that originally do not satisfy the initial conditions imposed on invariant solutions.

For lack of space, several important aspects of group analysis of differential equations have been left out in this presentation. This includes a study of the application of Lie groups to ODEs. Higher symmetries, in particular contact symmetries, were not treated at all. Regretably, recent applications of Lie group theory to the analysis and solution of differential-difference equations [96,...,99] could also not be included.

Though the theme of group theory and differential equations is an old one, much remains to be done in this field. That concerns basic theory, the development and computerization of algorithms, generalizations to other types of equations and further applications.

ACKNOWLEDGEMENTS.

The final version of these lectures was written during the authors visit to the Departamento de Física Teórica, Universidad de Valladolid, where the manuscript was also prepared. The author is much indebted to members of the Departamento, specially to M.A. del Olmo, for hospitality and for great help in preparing the manuscript. The authors research is supported by research grants from NSERC of Canada and FCAR du Qubec.

REFERENCES

1. Olver P.J., *Applications of Lie Groups to Differential Equations*, (Springer, New York, 1986).
2. Bluman G.W. and Cole J.D., *Similarity Methods for Differential Equations*, (Springer, New York. 1974).
3. Bluman G.W. and Kumei S.,*Symmetries and Differential Equations*, (Springer, Berlin 1989).
4. Ovsiannikov, L.V., *Group Analysis of Differential Equation*, (Academic, New York, 1982).
5. Ibragimov N.H., *Transformation Groups Applied to Mathematical Physics*, (Reidel, Boston, 1985).
6. Anderson R.L. and Ibragimov N.H., *Lie-Backlund Transformations in Applications*, (SIAM, Philadelphia, 1979).
7. Ames W.F., *Nonlinear Partial Differential Equations in Engineering*, (Academic, New York, 1972).
8. Winternitz P., "Group Theory and Exact Solutions of Partially Integrable Differential Systems", in R. Conte and N. Boccara (Editors) *Partially Integrable Evolution Equations in Physics*, (Kluwer Academic Publishers, Netherlands, 1990).

9. Levi, D. and Winternitz, P. (Editors), *Symmetries and Nonlinear Phenomena*, (World Scientific, Singapore, 1988).
10. Ablowitz M.J., and Segur H., *Solitons and the Inverse Scattering Transform*, (SIAM, Philadelphia, 1981).
11. Ablowitz M.J., and Clarkson P.A., *Solitons, Nonlinear Evolution Equations and Inverse Scattering*, (Cambridge University Press, Cambridge, 1991).
12. Schwarz F., Computing **34**, 91 (1985).
13. Champagne B. and Winternitz P., Preprint CRM-1278, Montral, 1985.
14. Champagne B., Hereman W. and Winternitz P., Comp. Phys. Commun. **66**, 319 (1991).
15. Brub D. and de Montigny M., Preprint CRM-1822, Montral, 1992.
16. Reid G.J., Euro. J. of Appl. Math., **2**, 293 (1991); **2**, 319 (1991).
17. Gazeau J.P. and Winternitz P., Phys. Lett. **A167**, 246 (1992); J. Math. Phys. **33**, 4087 (1992).
18. Kadomtsev V.V. and Petviashvili V.I., Sov. Phys. Dokl. **15**, 539 (1970).
19. David D., Kamran N., Levi D. and Winternitz P., Phys. Rev. Lett. **55**, 2111 (1985); J. Math. Phys. **27**, 1225 (1986).
20. Kac V., *Infinite Dimensional Lie Algebras*, (Birkhauser, Boston, 1983).
21. Kass S., Moody R.V., Patera J., and Slansky R., *Affine Lie Algebras, Weight Multiplicities and Branching Rules*, (University of California Press, 1990).
22. Goddard P. and Olive D., Int. J.Mod. Phys. **A1**, 303 (1986).
23. Champagne B. and Winternitz P., J. Math. Phys, **29**, 1 (1988).
24. Martina L. and Winternitz P., Ann. Phys. (N.Y.) **196**, 231 (1989).
25. Winternitz P. in Ref. 9.
26. Jacobson N., *Lie Algebras*, (Dover, New York, 1979).
27. Zassenhaus H., *Lie Groups, Lie Algebras and Representation Theory*, (Presses de l'Universit de Montral, 1981).
28. Rand D., Winternitz P. and Zassenhaus H., Lin. Alg. Appl. **109**, 197 (1988).
29. Rand D., Comp. Phys. Commun. **41**, 105 (1987); **46**, 311 (1987).
30. Rand D., Winternitz P. and Zassenhaus H., Comp. Phys. Commun. **46**, 297 (1987).
31. Martina L. and Winternitz P., J. Math. Phys. **33**, 2718 (1992).
32. Martina L., Soliani G. and Winternitz P., J. Phys. A: Math. Gen. **A 25**, 4425 (1992).
33. Ablowitz M.J., Ramani A. and Segur H., J. Math. Phys. **21**, 715 (1980); **21**, 1006 (1980).
34. Rand D. and Winternitz P., Comp. Phys. Commun. **42**, 359 (1986).
35. Weiss J., Tabor M. and Carnavale G., J. Math. Phys. **24**, 522 (1983).

36. Levi D. and Winternitz P. (Editors)., *Painlevé Transcendents, their Asymptotics and Physical Applications*, (Plenum, New York, 1992).
37. Dynkin E.B., Amer. Math. Soc. Transl. Ser. 2 ,**6**, 245 (1957).
38. Cornwell J.F., *Group Theory in Physics*, Vol. II, (Academic Press, 1984).
39. Patera J., Sharp R.T., Winternitz P. and Zassenhaus H., J. Math. Phys. **18**, 2259 (1977).
40. Beckers J., Harnad J., Perroud M. and Winternitz P., J. Math. Phys. **19**, 2126 (1978).
41. Patera J., Winternitz P. and Zassenhaus H., J. Math. Phys. **16**, 1597 (1975).
42. Patera J., Winternitz P. and Zassenhaus H., J. Math. Phys. **16**, 1615 (1975).
43. Burdet G., Patera J., Perrin M. and Winternitz P., J. Math. Phys. **19**, 1758 (1978).
44. Goursat E., Ann. Sci. Ec. Norm. Sup. (3). , **6**, 9 (1889).
45. DuVal P., *Homographies, Quanternions and Rotations*, (Clarendon Press, Oxford, 1964).
46. Patera J., Sharp R.T., Winternitz P. and Zassenhaus H., J. Math. Phys. **17**, 986 (1976).
47. Rubin J., Winternitz P., J. Phys. Math. Gen. **A26**, 1123 (1993).
48. Ndogmo J.C. and Winternitz P., Preprint CRM-1864, Montral,(1993).
49. Turkowski P., J. Math. Phys. **31**, 1344 (1990).
50. Kruskal M.D., in Ref. 36.
51. Conte R., in Ref. 36.
52. Weiss J., in Ref. 36.
53. Ince E.L., *Ordinary Differential Equations*, (Dover, New York, 1956).
54. Painlevé P., Acta Math. **25**, 1 (1902).
55. Gambier B., Acta Math **33**, 1 (1910).
56. Cosgrove C. and Scoufis G., Studies Appl. Math **88**, 25 (1993).
57. Bureau F., Ann. Mat. Pura Appl. (IV). **91**, 163 (1972); Bureau F., Garcet A. and Goffar J., Ann. Mat. Pura Appl. (IV). **92**, 177 (1972), and article in Ref. 36.
58. Bureau F., Acad. Roy. Belg. Bull, Cl. Sc. (5). **73**, 335 (1987).
59. Kruskal M.D. and Clarkson P.A., Stud. Appl. Math. **86**, 87 (1992).
60. Wang C.S., Phys. Rev. **182**, 482 (1969).
61. Carman R.L., Shimizu F., Wang C.S. and Bloembergen N., Phys. Rev. **A2**, 60 (1970).
62. Chu F.Y.F. and Scott A.C., Phys. Rev. **A12**, 2060 (1975).
63. Kaup D.J., Physica **6 D**, 143 (1983).
64. Kaup D.J. and Menyuk C.R., Phys. Rev.**A42**, 1712 (1990).
65. Levi D., Menyuk C.R. and Winternitz P., Phys. Rev. **A 44**, 6057 (1991).

66. Menyuk C.R., Levi D. and Winternitz P., Phys. Rev. Lett. **69**, 3048 (1992).
67. Levi D., Menyuk C.R. and Winternitz P., to be published.
68. Its A.R. and Novokshenov V.Yu., *The Isomonodronic Deformation Method in the Theory of Painlevé Equations.* (Springer, Berlin, 1986).
69. Kitaev A.V., Math. USSR, Sbornik **62**, 421 (1989).
70. Menyuk C.R., Phys. Rev. **A47**, 2235 (1993).
71. Byrd P.F. and Friedman M.D., *Handbook of Elliptic Integrals for Engineers and Scientists.* (Springer, Berlin, 1971).
72. Becker D.A. and Richter E.W., Z. Naturforsch. **A45**, 1219 (1990).
73. Meshkov A.G., Sov. Phys. J. **33**, 571, (1991).
74. Rideau G. and Winternitz P., J. Math. Phys. **31**, 1095 (1990).
75. Rideau G. and Winternitz P., J. Math. Phys. **34**, 558 (1993).
76. Gagnon L. and Winternitz P., Preprint CRM-1851, Montral, (1993).
77. Ibragimov N.H. and Torrisi M., J. Math. Phys. **33**, 3931 (1992).
78. Akhatov I.Sh., Gazizov R.K. and Ibragimov N.H., J. Sov. Math.**55**, 1401 (1991).
79. Joshi N., Phys. Lett. **A125**, 456 (1987).
80. Hlavaty V., Phys. Lett. **A128**, 335 (1987).
81. Abellanas L. and Galindo A., Phys. Lett. **A125**, 456 (1985).
82. Clarkson P.A. and Kruskal M.D., J. Math. Phys. **30**, 2201 (1989).
83. Levi D. and Winternitz P., J. Phys. A. Math. Gen. **22**, 2915 (1989).
84. Bluman G.W. and Cole J.D., J. Math. Mech. **18**, 1025 (1969).
85. Olver P. and Rosenau Ph., Phys. Lett. A114 107, (1986).
86. Olver P. and Rosenau Ph., SIAM J. Appl. Math. **47**, 263 (1987).
87. Nishitani T. and Tajiri M., Phys. Lett. **A89**, 179 (1982).
88. Lou S.Y., Phys. Lett. **A151**, 133 (1990).
89. Clarkson P.A., J. Phys. A. Math. Gen. **22**, 2355 (1989); **22**, 3821 (989); Eur. J. Appl. Math. **1**, 279 (1990); Nonlinearity **5**, 453, (1992).
90. Clarkson P.A. and Winternitz P., Physica **49D**, 257 (1991).
91. Clarkson P.A. and Hood S., J. Phys. A.: Math. Gen. **26**, 133 (1993).
92. Nucci M.C. and Clarkson P.A., Phys. Lett. **164**, 49 (1992).
93. Lou S.Y. and Ni G.J., Commun. Math. Phys. **15**, 465 (1991).
94. Pucci E., J. Phys. A.: Math. Gen. **25**, 2631 (1992).
95. Pucci E. and Saccomandi G., J. Math. Anal. Appl. **163**, 588 (1992).
96. Levi D. and Winternitz P., Phys. Lett. **A152**, 335 (1991).
97. Levi D. and Rodriguez M.A., J. Phys. A.: Math. Gen. **25**, L975 (1992).
98. Quispel G.R.W., Capel H.W. and Sahadevan R., Phys. Lett. **A170**, 379 (1992).
99. Levi D. and Winternitz P., J. Math. Phys. **34**, (1993), to appear.

INDEX

The manufacturer's authorised representative in the EU is Springer Nature Customer Service Centre GmbH, Europaplatz 3, 69115 Heidelberg, Germany. If you have any concerns regarding our products, please contact ProductSafety@springernature.com

Printed and bound by CPI Group (UK) Ltd, Croydon, CR0 4YY
23/04/2026
02095629-0007